18

Springer Series on Fluorescence

Methods and Applications

Series Editor: Martin Hof

Springer Series on Fluorescence

Series Editor: Martin Hof

Recently Published and Forthcoming Volumes

More information about this series at http://www.springer.com/series/4243

Fluorescence in Industry

Volume Editor:
Bruno Pedras

With contributions by

J. Avó · E. L. Bastos · M. Bedoya · M. N. Berberan-Santos ·
D. J. S. Birch · H. D. Burrows · L. Della Ciana ·
M. da Silva Baptista · F. B. Dias · J. J. Díaz-Mochón ·
E. Garcia-Fernandez · J. A. González-Vera · G. Hawa ·
G. Hungerford · G. Ielasi · C. A. T. Laia · K. Lobato ·
J. J. Lowe · V. B. Malkov · L. F. Marques · L. M. Martelo ·
D. McLoskey · G. Orellana · A. Orte · T. Palmeira ·
B. Pedras · S. Pernagallo · M. J. Ruedas-Rama · A. Ruivo ·
K. Sagoo · J. M. Serra · J. A. Silva · I. Urriza-Arsuaga ·
A. M. Vallêra · P. Yip

 Springer

Volume Editor
Bruno Pedras
CQFM-IN and iBB-Institute for
Bioengineering and Biosciences
Instituto Superior Técnico
Universidade de Lisboa
Lisboa, Portugal

ISSN 1617-1306   ISSN 1865-1313   (electronic)
Springer Series on Fluorescence
ISBN 978-3-030-20032-9   ISBN 978-3-030-20033-6   (eBook)
https://doi.org/10.1007/978-3-030-20033-6

This Springer imprint is published by the registered company Springer Nature Switzerland AG.
The registered company address is: Gewerbestrasse 11, 6330 Cham, Switzerland

Series Editor

Martin Hof

Academy of Sciences of the Czech Republic
J. Heyrovsky Institute of Physical Chemistry
Department of Biophysical Chemistry
Dolejskova 3
16223 Prague 8
Czech Republic
martin.hof@jh-inst.cas.cz

Aims and Scope

Fluorescence spectroscopy, fluorescence imaging and fluorescent probes are indispensible tools in numerous fields of modern medicine and science, including molecular biology, biophysics, biochemistry, clinical diagnosis and analytical and environmental chemistry. Applications stretch from spectroscopy and sensor technology to microscopy and imaging, to single molecule detection, to the development of novel fluorescent probes, and to proteomics and genomics. The *Springer Series on Fluorescence* aims at publishing state-of-the-art articles that can serve as invaluable tools for both practitioners and researchers being active in this highly interdisciplinary field. The carefully edited collection of papers in each volume will give continuous inspiration for new research and will point to exciting new trends.

Preface

Every now and then, scientists feel that their research topic may lead to an important practical application. This vision, not always clear, is frequently not pursued, owing to lack of adequate partners or insufficient funding, but is mostly impaired by unawareness of the real potential of the project.

In the field of fluorescence (or broadly speaking, luminescence, since phosphorescence can also be involved), which has experienced a swift growth in the last decades, this mindset is particularly relevant given the numerous areas of industrial and commercial application of its methods and techniques.

This volume is aimed at filling a gap in the literature devoted to fluorescence, by providing several examples of its industrial applications, described by both industry-based researchers and academics who hold strong collaborations with industrial partners. After Professor Martin Hof kindly invited me for this challenging task, my first guideline for choosing topics and authors was to reach a broad audience, having in mind what the reader can expect to learn from such a specific book. Hence, the purpose of this volume is to serve a wide variety of readers coming from both university laboratories (no matter how small or economically constrained) and companies with substantial funding for R&D activities. The former might have incomplete acquaintance with industrial applications and processes, in spite of their large experience in fundamental research; likewise, the latter may want to expand or update their knowledge on fluorescence/luminescence science. In both cases, this volume is intended to serve as a reference work and to stimulate an interplay of expertise.

The edition of the present volume implied a major effort from all the parts involved and has passed by many stages until it reached the present form. I am obliged to all the authors for their outstanding contributions, which reflect a fragment of the state-of-the-art industrial work on fluorescence. Likewise, I would like to express my deep gratitude to the Series Editor, Prof. Dr. Martin Hof, for his kind invitation and for our fruitful discussions. Finally, my sincere appreciation to Dr. Steffen Pauly, Dr. Andrea Schlitzberger, and Ms. Ramya Venkitachalam from Springer, for their collaboration and constant support.

Lisboa, Portugal   Bruno Pedras
February 2019

Contents

Luminescence-Based Sensors for Bioprocess Applications

Idoia Urriza-Arsuaga, Guido Ielasi, Maximino Bedoya,
and Guillermo Orellana

Contents

Abstract Reliable robust sensor systems are essential to monitor in situ and allow controlling of the evolution of bioprocesses in order to maximize the product yields and guarantee their quality. The more chemical and biological information is collected, the better strategies can be applied to generate highly productive cell lines and successfully determine the target production profiles. While the most common sensors employed in cell cultivation processes are those that measure (in-line) physical parameters (temperature, liquid level, conductivity, redox potential, etc.), ruggedized chemical monitors such as pH and gas phase/dissolved O_2 and pCO_2 are essential to determine the status of the cultured microorganisms. Luminescent chemical sensors for these three parameters have demonstrated in the last

I. Urriza-Arsuaga, G. Ielasi, M. Bedoya, and G. Orellana (✉)
Chemical Optosensors & Applied Photochemistry Group (GSOLFA), Department of Organic
Chemistry, Faculty of Chemistry, Complutense University of Madrid, Madrid, Spain
e-mail: orellana@quim.ucm.es

© Springer Nature Switzerland AG 2019   1
B. Pedras (ed.), *Fluorescence in Industry*, Springer Ser Fluoresc (2019) 18: 1–38,
https://doi.org/10.1007/4243_2019_10, Published online: 21 March 2019

few years their superiority over the traditional electrochemical sensors for in situ continuous real-time bioprocess monitoring, particularly those based on disposable analyte-sensitive patches attached to single-use bioreactors combined with emission lifetime measurements. This chapter reviews the progress achieved in luminescent sensing of O_2, CO_2, and pH for bioprocess applications that have led them to be the devices of choice for many manufacturers and customers in most situations.

Keywords Bioprocess monitoring · Bioreactors · Carbon dioxide · CO_2 · Fluorescence · Luminescence · O_2 · Oxygen · pH · Probes · Sensors

1 Introduction

Reliable analytical systems are essential to monitor and allow controlling of the evolution of a bioprocess' conditions in order to maximize yields of the biologic products and guarantee their quality [1]. The more chemical/biological information is collected, the better optimization strategies can be applied to generate highly productive cell lines and successfully determine the target production profiles [2].

Cell cultivation ("upstream process") in bioreactors is a complex three-phase system comprising a gas phase (aerated head space), a liquid phase (the culture medium), and a solid phase (suspended biomass) [3]. Therefore, sensors must monitor chemical, physical, and biological parameters in all the three phases during the bioprocess. Biological variables such as suspended cell density, microorganism viability, cell morphology, and protein or DNA contents are difficult to measure so that they are predominantly used in research environments because many of them require sampling and laboratory analyses. The most common sensors employed in cell cultivation processes are those that measure (in-line) physical parameters such as temperature, liquid level, conductivity, pressure, redox potential, foaming, etc. [4]; ruggedized versions of these sensors have been available for many years [1]. Essential chemical parameters such as pH and gas phase/dissolved O_2 and pCO_2 are measured in-line with dedicated electrochemical or optical sensors; dissolved relevant species such as glucose, lactate, or metabolites are measured in-line with biosensors (electrochemical or optical) or at-line with liquid or gas chromatographs with absorption, fluorescence, or mass spectrometric detectors [5].

In addition to being able to measure under the harsh conditions of the culture media, particularly related to sensor biofouling, external contamination by the monitoring devices must be avoided by all means. Therefore, disposable (single-use) chemical and biological sensors are currently preferred fitted to disposable bioreactors [6, 7], and this is one of the leading advantages of optical sensors: window patches or spots adhered to pre-sterilized transparent (also disposable) plastic or glass bioreactors can be interrogated from outside, minimizing the risk of contamination in automatic (robotic) optimization systems (50–250 mL) and production bioreactors (>25 L) alike. Single-use bioreactor technology offers higher flexibility with lower investment and energy costs and, particularly, a streamlined production under the Good Manufacturing Practice (GMP) requirements because cumbersome labor-intensive cleaning procedures are avoided [6].

Not surprisingly, molecular oxygen, pH, and carbon dioxide sensors have been the most developed optical sensors to realize the advantages of the latter outside the laboratory applications. If, at the onset of the applications of optical sensing to bioprocess monitoring, pH was the most tested sensor due to the long-term availability of a large number of colorimetric indicator dyes, luminescent sensors quickly took over. This was due to the superior sensitivity and selectivity (plus stability, if emission lifetime-based) of the latter [8, 9], together with the significant technological advances in low-cost long-life light sources (light-emitting diodes or LEDs and diode lasers), photon detectors (e.g., avalanche photodiodes or APDs), fiber optics (multimode polymer-coated silica or PCS, low-attenuation plastic optical fibers or POF), digital electronics, and fast computing. This chapter focuses exclusively on the luminescence-based O_2, CO_2, and pH sensors for bioprocess applications, because these three parameters are arguably the most important in that field and because the corresponding luminescent sensors have attained enough maturity to reach the market [1, 3–7].

The term *luminescence* used throughout this chapter encompasses both the short-lived *fluorescence* emission (typically up to 50 ns) after a short excitation pulse and the longer-lived *phosphorescence* emission (typically above 200 ns) that display some molecules after photoexcitation with light of a shorter wavelength. Luminescent sensors can operate in the *intensity* mode, where the emission intensity at a selected wavelength is monitored, or in the *lifetime* mode, by which the average time that the photoexcited indicator molecule spends in its lowest-lying (electronic) excited state is measured. The luminescence lifetime is the reciprocal of the first-order (or pseudo-first-order, in the case of quenching) kinetic rate constant of the excited-state deactivation process. The latter can be measured by the *pump-probe* method (the luminescence decay curve is rapidly sampled while returning to the ground state) or, more commonly, by the *phase-resolved* method using a sinusoidally modulated excitation light and measuring the phase shift of the (sinusoidally) modulated emission with respect to the excitation phase [10]. The intensity mode is always subject to drift due to the excitation light source or detector aging, luminescent indicator dye bleaching and leaching, or fiber-optic bending. Some of these factors can be corrected with intensity-*ratiometric* measurements at two-excitation/single-emission or single-excitation/two-emission wavelengths; however, all of them are eliminated if the luminescence lifetime mode is used instead. The latter is nowadays readily implemented thanks to the performance of the LED sources (that can be modulated up to 20 MHz) and to the power and small size of computing microprocessors.

2   Luminescent Sensors for Molecular Oxygen

2.1 O_2 Sensing

Molecular oxygen (O_2) plays an essential role in all aerobic bioprocesses if we consider that it is the key substrate for most microorganism and cell growth, maintenance, and metabolite production. A large variety of aerobic processes are driven in viscous aqueous media in the presence of salts and organic substances, in which oxygen is sparingly soluble. When a temporary depletion of dissolved O_2 takes place, an irreversible cell damage and a reduction in production yield may occur [11]. Therefore, a properly designed bioreactor must ensure an adequate O_2 supply, regulated by continuous O_2 monitoring, independently of the bioreactor dimensions (laboratory, pilot, or industry scale) and the operation mode (batch, semicontinuous, or continuous), in such a way that optimal growth conditions and product formation are guaranteed.

As a result, O_2 monitoring is an issue of utmost importance in different industrial bioprocesses. For instance, in the pharmaceutical sector where the production of high-quality products, such as biotechnology drugs, vaccines, or antibiotics, is extraordinarily common, O_2 levels play an important role in the final product yield and quality. The O_2 content is also essential for beer-brewing companies. In the wort fermentation process, where the sugars are transformed into alcohol and carbon dioxide, the O_2 concentration plays a vital role in the yeast effectiveness. Both O_2 excess and deficiency can result in undesirable clarity, color, and taste properties of the final beer [12]. In the environmental field where biological systems have been used to transform objectionable materials into more environmentally friendly substances or high-value-added products, dissolved oxygen (DO) concentration is also relevant. For instance, water is cleaned in various steps in municipal wastewater treatment plants (WWTPs). Organic matter is digested by bacteria and other microorganisms through the biological process for water purification, where DO is essential to improve plant efficiency (energy consumption) and to prevent undesirable odors (https://www.emerson.com/documents/automation/application-data-dissolved-oxy gen-measurement-in-wastewater-treatment-en-68468.pdf. Accessed 10 Sept 2018). When it comes to the generation of high-value-added products, energy (biogas) can be produced by the anaerobic fermentation of residual organic matter. This process takes place in bioreactors in the presence of anaerobic organisms, whose activity and growth levels must be sustained by an accurate control of many variables including O_2 (https://www.barbenanalytical.com/-/media/ametekbarbenanalytical/downloads/ application_notes/biogas_an_reva.pdf. Accessed 10 Dec 2018). The O_2 concentration in the generated biomethane final product must also be monitored continuously, in real time and online, regardless if the sustainable biofuel is intended for injection into the natural gas distribution grid or for automotive purposes [13].

So far, in O_2 determination, the traditional electrochemical devices (galvanic for measurements in the gas phase and amperometric for DO determinations) have been the most widely employed, in spite of suffering from drawbacks such as the short life of the electrolyte or the membrane that limits the long-term stability and the regular need for recalibration and technical inspection that increase both the cost and the maintenance time. In contrast, optical sensors have become an advantageous alternative, especially when applied to small-scale reactors, since they are sterilizable, easily miniaturized, inexpensive, rugged, resistant to biofouling, reproducible, and easy to handle with almost no maintenance (only change of the sensitive film). Moreover, they can be interrogated contactless through the glass reactor wall or through a window. Over the last decade, many optical O_2 sensors have been developed and commercialized, and their performance has been tested in different applications and conditions [14]. Most of them are based on the dynamic quenching of the luminescence of a photoexcited indicator dye by the O_2 molecules. Collisions between them lead to an energy transfer that decreases the indicator luminescence quantum yield (approximated by the intensity at a particular wavelength) and lifetime [8]. This collisional quenching is described by the Stern-Volmer equation [10], where L_0 (τ_0) and L (τ) are the luminescence intensities (lifetimes) in the absence and the presence of oxygen, K_{SV} is the so-called Stern-Volmer constant, k_q is the bimolecular quenching constant, and [O_2] is the oxygen concentration (Eq. 1).

$$L_0/L = \tau_0/\tau = 1 + k_q\tau_0[O_2] = 1 + K_{SV}[O_2] \tag{1}$$

In this type of sensors, the luminophore, which is commonly a Pd(II) or Pt (II) porphyrin, or a Ru(II)-polypyridyl complex, with long-lived (μs) luminescent properties [15, 16], is usually immobilized in a polymer matrix and integrated into bioreactors in different formats depending on the final application. Several reviews of the state-of-the-art O_2 sensors applied to bioprocess monitoring can be found in the literature, from more traditional culture vessels (shake flask and bioreactors) [17] to microreactors and microfluidic chips [18–20]. Hereunder, some examples of luminescent O_2 sensors applied to various biological processes are described, covering not only those sensors used in research laboratories but also those commercially available. Although they will be classified according to their format (sensor probes, sensor spots and layers, and sensor particles), examples of their application to O_2 monitoring in different bioprocess vessels (bioreactors, shake flasks, well plates, or microfluidic chips) are also illustrated.

2.2 Fiber-Optic Probe-Type O_2 Sensors

Early attempts to introduce optical O_2 sensors in bioprocess applications started by immersing a fiber-optic probe in the cultivation media, the tip of which was coated with the sensitive fluorophore immobilized in an oxygen-permeable matrix [21]. Several examples can be found in the literature [22–24], where a luminophore (polycyclic aromatic hydrocarbon or Ru(II) polypyridyl complex) was embedded in a silicone or polystyrene membrane. Optical O_2 sensing was applied to *E. coli* or yeast cultures, comparing their behavior to that of a traditional electrode but displaying particular advantages (no analyte consumption, fast response, absence of electrical interferences, resistance to biofouling). Further on, luminescent O_2 sensors tended to be miniaturized, and significant efforts were made to improve their long-term stability, for example, by placing a biocompatible polymer overcoat, additionally protecting the sensor from biofouling and chemical degradation [25]. Likewise, those optical probes ("optodes") started to be commercially available from different manufacturers for different applications (see Table 1).

Generally, these devices consist of three parts: a sensor head connected to an optoelectronic unit or containing itself all the integrated optoelectronics; a dissolved O_2 sensor cap, which contains the sensitive luminophore; and a sensor shaft that connects the oxygen meter and the sensing cap (Fig. 1). The diameter (1.6–40 mm) and immersion length (60–450 mm) of these commercial sensor probes vary depending on the brand. They are able to operate in a wide range of temperatures (typically 0–50°C, but up to 140°C in some models) and pressures (typically 0–10 bar, but up to 200 bar in specific models for environmental applications) and require very little maintenance (typically, just exchange of the sensor cap). Their sensitivity and accuracy are optimal for industrial applications, as well as their response time, which can vary anywhere from 2 to 60 s. The longer response time is often due to the aforementioned overcoat, used as an additional protection layer from biofouling or from mechanical abrasion.

Although it was demonstrated that optical wand-type sensors improved some limitations of traditional electrodes (long-term stability, calibration drifts, CO_2 interferences, and reliability at low O_2 concentrations), the increasingly widespread use of small bioreactors required not only more miniaturized sensors but also contactless measurement systems.

2.3   Oxygen Sensor Layers and Spots

In this approach, the luminescent sensor film is affixed to the bottom or the side wall of a glass or plastic vessel, placing the optical fiber and the instrumentation outside the culture media, in a contactless measurement mode. In this way, by embedding the sensitive dyes in an O_2-permeable polymer matrix, the sensor remains stationary during the bioprocess causing the least possible disruption and leaving valuable space for other required devices (e.g., stirrer, reagents, or gas feedthroughs). This sensor format has made miniaturization of the sensor possible, to their use not only in bioreactors and shake flasks but also in small-scale plastic vessels (typically 50–250 mL), and from microwell plates to microfluidic devices (Fig. 2). Many examples are described in the literature for different applications, and, as of today, these oxygen sensor formats are commercially available from various manufacturers.

For instance, Tolosa et al. [26] have described a DO luminescent thin sensing layer, based on a platinum complex as indicator dye, immobilized in a polymer and

Table 1 Some commercial luminescent O_2 sensors

Sensor	Operating range	Accuracy (%)	t_{90} (s)	$\varnothing$ (mm)	Application	Ref.
Mettler Toledo InPro 6870i	8 ppb–60%	±[1% + 8 ppb]	<20	12	Cell cultures	https://www.mt.com/dam/non-indexed/po/pro/pdf/td/TD_InPro6000_Series_en_52206266_Dec15.pdf. Accessed 10 Sept 2018
PreSens OIM PSt3	0.03–100%	±0.05 at 0.2% O_2 / ±0.4 at 20.9% O_2	<6	–	Cell cultures	https://www.presens.de/products/detail/oxygen-probe-for-in-line-measurement-oim-pst3.html. Accessed 10 Sept 2018
Ocean Optics FOXY probes	0.01–100%	±0.01 at 5% O_2 / ±0.02 at 20% O_2	<2 / <60	1.587 3.175 6.35	Cell cultures	https://oceanoptics.com/product-category/oxygen-probes/. Accessed 10 Sept 2018
Hamilton Visiferm DO	0–62.85% 4 ppb–25 ppm	±0.05 at 1% O_2 / ±0.2 at 21% O_2	<30	12	Cell cultures Brewery	https://cercell.com/media/8917/manual_visiferm_do_en_lowres.pdf. Accessed 10 Sept 2018
Mettler Toledo InPro 6960i	8 ppb–25 ppm	±[1% + 8 ppb]	<20	12	Brewery	https://www.mt.com/dam/non-indexed/po/pro/pdf/td/TD_InPro6000_Series_en_52206266_Dec15.pdf. Accessed 10 Sept 2018
Hach M1100-S10H	15 ppb–40 ppm	±0.02 ppm or 3%[a] whichever is greater	<50	12 28	Brewery	https://www.hach.com/orbisphere-m1100-high-range-12-mm-luminescent-dissolved-oxygen-sensor-compatible-with-pg-13-5-mm-stationary-housing/product-details?id=19301038888&callback=pf. Accessed 10 Sept 2018
Pentair OGM WLO	0.1–45 mg/L	<5%[a]	<30	25	Brewery	https://foodandbeverage.pentair.com/en/products/haffmans-in-line-optical-o2-meter-ogm. Accessed 10 Sept 2018
PreSens OXYBase series	0.02–22.5 mg/L		<30	–	WWTP	https://www.presens.de/products/detail/oxybaser-wr-rs485-wr-rs485m.html. Accessed 10 Sept 2018

(continued)

Table 1 (continued)

Sensor	Operating range	Accuracy (%)	t_{90} (s)	$\varnothing$ (mm)	Application	Ref.
Endress +Hauser Oxymax COS61D	0–20 mg/L	±0.01 mg/L or 1%[a] (<12 mg/L) ±2%[a] (from 12 to 20 mg/L)	<50	40	WWTP	https://www.de.endress.com/en/field-instruments-overview/liquid-analysis-product-overview/oxygen-optical-sensor-cos61d. Accessed 10 Sept 2018
YSI 6150 ROX	0–50 mg/L	±0.1 mg/L or 1%[a] (<20 mg/L) ±15%[a] (from 20 to 50 mg/L)	<30	12	WWTP	https://www.ysi.com/rox. Accessed 10 Sept 2018
Barben Analytical 4401OXY	0.002–4.2%	±0.002% or 3%,[a] whichever is greater	<6	12	Biogas	https://www.barbenanalytical.com/-/media/ametekbarbenanalytical/downloads/datasheets/4401oxy_datasheet_reve.pdf. Accessed 10 Sept 2018

[a]Percentage of the measured value

Fig. 1 Fiber-optic O_2 sensor with the corresponding user-replaceable sensor cap for bioprocess monitoring applications. The picture was kindly provided by PreSens Precision Sensing GmbH

attached to the bottom of a 250-mL shake flask. After being autoclaved, the sensor was applied to yeast and *E. coli* cultures for O_2 measurements, in the 0–100% concentration range, by phase-shift luminescence detection. The sensor layers showed suitable sensitivity and stability (<3% drift in 6 days) during the experiments, compared to Clark-type electrodes, which suffered from mechanical failure due to the continuous shaking. John et al. [27] manufactured two ratiometric optical sensor spots and applied them to microbial cultivation in 96-well plates. The first sensor was based on the $[Ru(dpp)_3]^{2+}$ O_2-sensitive indicator dye and safranin as the reference dye, which were co-immobilized on silica gel and embedded into a silicon matrix, whereas the second sensing spot used polystyrene particles with attached sulforhodamine and incorporated Pt-porphyrin as reference and sensing dyes, respectively, in a poly(2-hydroxyethyl methacrylate) matrix. The sensor spots were deposited onto the bottom of the microtiter plate wells, and dissolved O_2 was measured by luminescence intensity determinations. Jin et al. [28] developed an O_2 sensor based on the hydrophobic luminescent copolymer poly(Pt-TPP-TFEMA), which consists of a Pt(II) tetraphenylporphyrin (TPP) complex into the highly permeable matrix trifluoroethyl methacrylate (TFEMA) for its use in microreactors. The combination of the hydrophobicity and permeability of the sensor layer, which enhance the quenching process and the sensitivity, was added to the NIR-luminescent indicator dye which reduces scattering and sample interferences.

Different sensor spots of various sizes are also commercially available for bioprocess monitoring (Table 2). Most typically the luminophore is immobilized in a polymer matrix, coated onto a foil layer and glued to the bottom of a culture vessel. In the outside, an optical fiber connects the sensor spot with the interrogation optoelectronics working in the luminescence phase-shift detection mode. For instance, an application note from Ocean Optics describes the performance of their RedEye® oxygen patches to monitor an *E. coli* fermentation broth in a small-scale bioflask for 21 h [29].

Furthermore, sensor spots have been used for simultaneous measurements of DO in shake flasks and well plates, as a tool for bioprocess optimization procedures

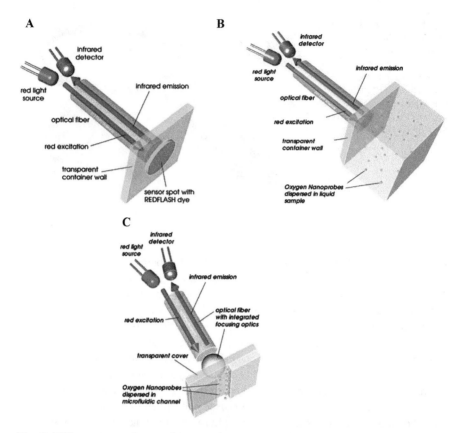

Fig. 2 Different sensor formats for contactless oxygen measurements, applied to cultivation systems in bioprocesses: sensor spots (**a**), sensor nanoparticles (**b**), sensor nanoparticles in microfluidic channel (**c**). These figures were kindly provided by PyroScience GmbH (Germany)

where multiple experiments must be carried in varying conditions to establish the best culture condition, or for automated parallel cell cultivation processes. As an example, Ge et al. [30] reported the evaluation of a multianalyte detection unit that consists of disposable DO and pH luminescent sensors glued to the bottom of 12 mini stirred-tank bioreactors (Fig. 3a), working simultaneously during a mammalian cell culture. The high-throughput bioreactor has a turntable containing all mini reactors, which were placed sequentially over the luminescence phase-shift detector (Fig. 3b). Each bioreactor can be rotated by a motor and be positioned above the detector for DO/pH measuring. Repeatability studies in the 0–100% concentration range carried out for six different sensor patches gave standard deviation less than 8.2% and an average relative error of less than 9.0% from the actual value.

As a further example, PreSens GmbH (Regensburg, Germany) offers a Shake Flask Reader (SFR) (https://www.presens.de/products/detail/sfr-shake-flask-reader.html. Accessed 10 Sept 2018), which is able to simultaneously monitor DO and pH in up to nine Erlenmeyer flasks, cultivation tubes, or T-flask of different sizes (Fig. 4a).

Table 2 Some commercial optical O_2 sensor spots with application in cell culture

Sensor	Operating range (%)	Accuracy (%)	Drift (at 0% O_2)	t_{90} (s)	$\varnothing$ (mm)	Ref.
PreSens SP-PSt3-YAU	0.03–100	±0.05 at 0.2% O_2 ±0.7 at 20.9% O_2	<0.03%/ 30 day	<6	–	https://www.presens.de/products/detail/oxygen-sensor-spot-sp-pst3-yau.html. Accessed 10 Sept 2018
Ocean Optics RedEye® FOXY	0.003–100	±0.05 at 5% O_2 ±0.2 at 20% O_2	0.01%/h	–	4 8	https://oceanoptics.com/product/redeye-oxygen-sensing-patches/. Accessed 10 Sept 2018
PyroScience OXSP5	0.02–50	±0.02 at 1% O_2 ±0.2 at 20% O_2	–	<15	5	http://www.pyro-science.com/contactless-fiber-optic-oxygen-sensor-spots.html. Accessed 10 Sept 2018

All the disposable plastic vessels are equipped with pre-calibrated sensor spots, whereas reusable glass flasks contain autoclavable oxygen sensors. The SFR was applied by Schneider et al. [31] to parallel online DO measurements in nine 250-mL pre-sterilized disposable plastic shake flasks, with *E. coli*, *Corynebacterium glutamicum*, and *Saccharomyces cerevisiae* cultures. They observed good reproducibility between shake flasks, being the standard deviations of DO measurements below 3%. Similarly, PreSens also commercializes the OxoDish® OD24 (https://www.presens.de/products/detail/oxodishr-od24.html. Accessed 10 Sept 2018) which is a disposable polystyrene 24-well plate, for online O_2 monitoring (Fig. 4b). Each well is equipped with a calibrated oxygen sensor film, which is read out contactless by the SensorDish® Reader (SDR). Naciri et al. [32] applied such sensor to mammalian cell cultures in different serum concentrations, taking DO readings every 10 min for a total culture duration of 168 h, allowing in this way an accurate and rapid control of the bioprocess.

Even though such sensor spots and films seem to be a good alternative for DO measuring in bioprocesses, some problems stemming from the device assembly, which can provoke changes in the sensor performance attributed to the use of adhesive tape or thermal bonding, can occur. Additionally, it appeared that the sensor spot size is not small enough for its use in microfluidic devices, potentially causing dead zones, with low oxygen concentrations, air bubbles, and difficult to removed or undesired flow patterns into the channels. Such problems seem to have been solved with O_2 sensing particles, which are described below.

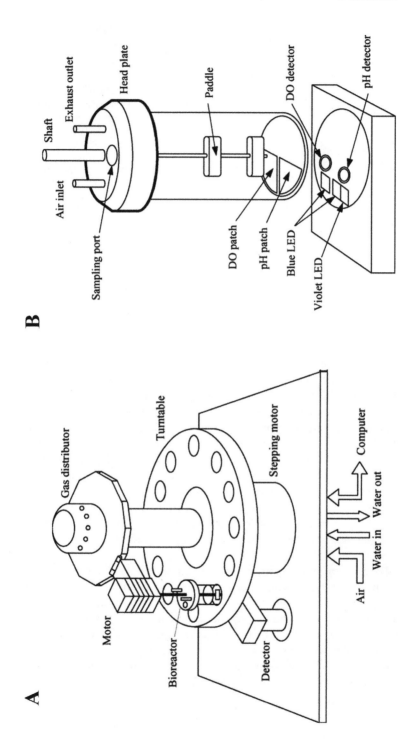

Fig. 3 High-throughput bioreactor system with up to 12 stirred-tank bioreactors (**a**) and a bioreactor system equipped with the DO and pH monitoring system (**b**) (Reprinted from ref. [30], with permission of Elsevier. ©Elsevier, 2006)

A   B

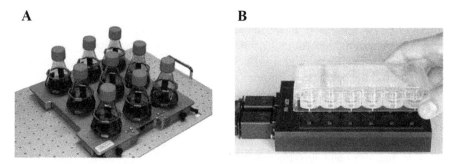

Fig. 4 Optical sensor systems for online simultaneous monitoring in shake flasks (**a**) and well plates (**b**). The figure was kindly provided by PreSens Precision Sensing GmbH

2.4 Oxygen Sensing Particles

O_2 sensing nano- and microparticles are based on polymers stained with indicator dyes. These particles are incorporated into a cell culture forming stable dispersions and providing key advantages over the aforementioned sensor formats. Firstly, the sensor particles are small enough to be used in any cell culture vessel, including microfluidic devices, without causing any blocking of the system or interferences in the flow characteristics. Moreover, in these micrometer systems, particles can be introduced into the system after it has been assembled and sterilized, avoiding the problems arising from those procedures. Furthermore, particles are characterized for their low response time, their higher surface area, and their higher signal, and, as they are not still, they offer the possibility of measuring the O_2 distribution within the cell culture.

Several examples have been published in the literature. Cywinski et al. [33] compared ratiometric polystyrene-based O_2-sensitive nanoparticles and layers, based on Pt(II)-*meso*-tetra(pentafluorophenyl)porphine (PtTFPP) complex as indicator dye and *N,N'*-bis(1-hexylheptyl)perylene-3,4:9,10-bis-(dicarboximide) as the reference dye. Layers and nanoparticle sensor systems were compared in terms of response and sensitivity in aqueous solution and in yeast culture (5 mL medium). Although nanoparticles were more sensitive in aqueous media, in yeast culture the contrary was found; moreover, the nanoparticle sensitivity was affected by the color and turbidity of the culture medium. Both the excitation and emission radiation can be biased due to stray light and absorption phenomena in such complex media. Unlike nanoparticles, which are dispersed in the culture media, sensor layers are placed adjacent to the vessel wall, minimizing the latter effects. Attempting to minimize background fluorescence and interferences with cells, NIR-emitting oxygen sensor probes were investigated as well. However, it should be pointed out that commonly used photodetectors exhibit low sensitivity in NIR spectral region. As an example, Cao et al. [34] used the O_2-sensitive dye Pt(II) 5,10,15,20-*meso*-

tetraphenyltetrabenzoporphyrin (emission maximum at 778 nm) for the preparation of polystyrene O_2 indicator nanoparticles. They used the nanoparticles in nL-sized droplets for bacterial growth monitoring by luminescence intensity measurements. In a similar way, Horka et al. [35] described the preparation and application of O_2 sensing nanoparticles, based on Pt(II) *meso*-tetra(4-fluorophenyl)tetra-benzoporphyrin (emission maximum at 700 nm) embedded in poly(styrene-*block*-vinylpyrrolidone), and their application to monitoring microcultures of *E. coli* and *M. smegmatis*.

Considering that sensor particles suffer from interference with the sample, sensitivity to turbidity, and low light intensity, the use of magnetic nanoparticles as O_2-sensitive materials became an attractive alternative. This approach combines the advantages of both sensor layers and particles. On the one hand, in situ formation of high brightness sensor spots by magnetic separation during a short period of time reduces the sample interference. On the other hand, the procedure benefits from the particle features as their small size, as well as their easy introduction into the system. In the magnetic approach, selection of the particle size is a key issue. The higher the particle diameter, the better the separation and accumulation properties, but the more probable the blocking of microfluidic components and the disruption of microfluidic flow characteristics [36]. Thus, Chojnacki et al. [37] reported preparation of aspherical magnetic O_2-sensitive particles, based on $[Ru(dpp)_3]^{2+}$, as sensitive dye, using tetramethylsilane with dispersed Fe_3O_4 and TiO_2 nanobeads. The magnetic sensor particles were applied to a bacterial growth monitoring in a baffled shake flask. When DO measurements were about to be taken, particles were magnetically separated from the suspension forming a spot on the flask side. Such measurements were successfully compared to the results obtained from industrial oxygen sensor spots. Sensor nanoparticles are also commercially available (e.g., OXNANO, PyroScience GmbH, Aachen, Germany) (http://www.pyro-science.com/oxygen-nanoprobes.html. Accessed 10 Sept 2018), which were applied to different cell cultures in shake flasks [38], microtiter plates [39], and droplet microfluidic-based cell [40].

Finally, a more recent strategy has carried out oxygen sensing based on magnetic nanoparticles with upconversion luminescence properties [41, 42]. The composite particles combine the advantages of NIR excitation and emission, reducing autofluorescence problems and enhancing penetration into biological matter, with the magnetic properties outlined above at the expense of a slightly more complex setup.

3 Luminescent CO_2 Sensors

3.1 Carbon Dioxide Sensing

Carbon dioxide (CO_2) sensing is of paramount importance in many fields, including environmental monitoring [43], clinical analysis [44, 45], food industry [46, 47], and public and personal safety [48, 49]. In particular, the continuous and precise monitoring of CO_2 is essential in bioprocesses because of its role in respiration and fermentation [50] and its effects on cell growth, morphology, and metabolism [51–53]. Moreover, CO_2 is used to maintain the pH of cell culture media and can pass through cell membranes and influence the pH inside cells [54].

Two systems are commonly employed for CO_2 detection: the infrared (IR) detector, usually for sensing this analyte in the gas phase, and the Severinghaus electrode, mainly for sensing water-dissolved CO_2 [55]. The latter is made of a pH-sensitive glass electrode immersed in a bicarbonate solution covered by a thin polymer membrane [56]. This coating is permeable to carbon dioxide while not permeable to water and electrolyte solutes. The CO_2 present in the liquid sample equilibrates through the membrane, and the pH meter measures the resulting changes of the acidity of the internal bicarbonate solution (reversibly) generated upon hydrolysis of the incoming CO_2 gas from the sample.

The latter kind of sensor is widely used for bioprocess monitoring, but it has some weakness due to its susceptibility or interference to the presence of electromagnetic fields and basic or acidic gases, significant fouling of the membrane or the liquid solution, and contamination of the reference electrode. For these and other applications, a superior alternative are the optical CO_2 sensors based on the same principle but where the pH-sensitive electrode is replaced by a pH-sensitive colorimetric or luminescent indicator dye.

3.2 Luminescence-Based CO_2 Sensors

Optodes, allowing remote sensing of disposable CO_2-sensitive patches through transparent windows or via small-sized sterilizable ("clean-in-place," CIP, and "steam-in-place," SIP) optical fibers, are very suitable for bioprocess monitoring. The first luminescence-based CO_2 sensor was probably that reported by Lubbers and Opitz [57] using methylumbelliferone in a $NaHCO_3$ solution separated from the sampled medium by a gas-permeable membrane (Fig. 5a). The possibility of turning it into a fiber-optic chemical sensor (FOCS) was presumably shown for the first time by Zhujun and Seitz [58]. In their work, they used trisodium 8-hydroxypyrene-1,3,6-trisulfonate (HPTS) as fluorescent pH indicator dye, which has been the most used one in bioprocesses since then (Table 3) and for two commercialized devices (see Sect. 3.3 below).

A key component of the above-described CO_2 sensors is the hydrogencarbonate solution reservoir within the indicator layer; as a result, these systems are often bulky. Moreover, water (or moisture) evaporation affects the optosensors' prolonged use and shelf life, and the need of a protective external coating increases their response and recovery time and adds manufacturing complexity. An improvement of the bicarbonate solution-based sensors has been the so-called "dry" sensors (Fig. 5b): the aqueous hydrogel/overcoat membrane combination is replaced by a polymer containing a lipophilic hydrated quaternary ammonium hydroxide, which acts as phase-transfer agent [69]. The latter improves the sensitive phase because it

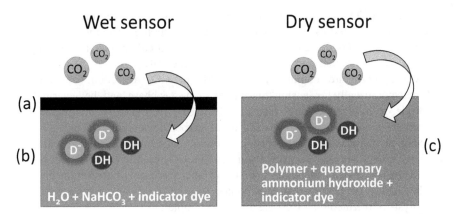

Fig. 5 Schematic illustration of the main difference between "wet" and "dry" luminescence-based CO_2-sensitive layers: (**a**) gas-permeable, ion-impermeable membrane; (**b**) internal solution phase; (**c**) solid phase containing the phase-transfer agent and moisture. DH and D^- are the protonated and deprotonated indicator dye molecules, respectively, which display distinct luminescence features

allows incorporation of some water into the polymer (required for the acid-base reaction) and helps dissolving the indicator molecules into the polymer. Many sensors based on this principle have been reported; a selected number of representative examples can be found in refs [70–76]. A demonstrated drawback of "dry" sensors is their special sensitivity to other acid vapors as well as to the sample humidity. As an improvement, some authors have shown the advantages of *ratiometric* fluorescent detection [77–80], based on the collection of fluorescence from the two maxima corresponding to the protonated and deprotonated forms. Unfortunately, not all dyes show emission from both forms (e.g., only the deprotonated excited species of the much used HPTS indicator dye displays fluorescence in the useful pH interval around its ground-state pK_a of 7.2, Fig. 6). Fluorescence sensing of CO_2 has been demonstrated with HPTS by monitoring the ratio between the emission intensity of the immobilized dye upon excitation at 405 nm and that measured upon excitation at 460 nm, wavelengths of the absorption maximum of ground-state HPTS and its deprotonated form, respectively (Fig. 6b) [65]. Modern ratiometric versions using red-luminescent Zn-azaphthalocyanine dyes have been reported [81].

Another strategy has been exploitation of the *inner filter effect* due to the spectral overlap of the CO_2-induced absorbance change of a pH indicator and the emission of a pH-insensitive luminescent dye. Examples of it are α-naphtholphthalein coupled with tetraphenylporphyrin (TPP) [82] or platinum octaethylporphyrin (PtOEP) [83] and di-OH-aza-BODIPY with silanized Egyptian blue plus di-butoxy-aza-BODIPY-complex [84]. By the way, this sensing chemistry shares the same weakness of the CO_2 sensors based on monitoring of the absolute luminescence intensity: they may be prone to signal drift due to indicator dye photobleaching/leaching or changes in the excitation intensity and detector sensitivity.

Table 3 Characteristics of some CO_2 optosensors used for bioprocess monitoring

Dye (support)	Measurement	Format and use	Analytical features			Strengths	Weaknesses	Ref.
			Dynamic range	Sensitivity	Response time			
HPTS in saturated bicarbonate buffer (Gore-Tex™ coated)	Ratiometric (excitation at 470 nm/405 nm)	Fiber-optic for fermentation medium (wort)	0–0.15 atm	$\pm 5 \times 10^{-4}$ atm (0–0.1 atm); $\pm 5 \times 10^{-3}$ atm (>0.1 atm)	t_{90} 1 min (0–0.5 atm) t_{-90} 30 min (0.5–0 atm)	Autoclavable	Affected by ionic strength	[59]
Sulforhodamine 101 (fluorophore), thymol blue or m-cresol purple (acceptors), and TOAH (EC coated with PDMS)	Fluorescence lifetime (FRET from the pH-insensitive to the pH-sensitive dye)	Patch designed for bioprocess monitoring (but actually not used)	n.a.	0.2° for 0.002% (low conc.) 0.2° for 0.06% (high conc.)	$t_{10-90\%}$ 7.1 s $t_{90-10\%}$ 19.2 s	10 h operation Shelf life in humidor: 2 weeks Fluorescence lifetime-based, phase-sensitive detection	Actually based on absorption Shelf life in air: 2 days	[60]
Acryloylfluorescein in saturated bicarbonate buffer (polyHEMA hydrogel coated with a polysiloxane membrane)	Ratiometric emission (490 nm/430 nm)	Fiber-optic sensor for fermentation medium (wort)	0.0–10.0%	$\pm 0.2\%$ CO_2	n.a.	Three parameters (pH/CO_2/O_2) in an optical fiber	Requires a microscope Not tested for sterilization	[61]
Sulforhodamine 101 (fluorophore), m-cresol purple, and CTMAOH (2-component highly hydrophobic Pt-cure silicone, covered with white PTFE gas-permeable membrane)	Fluorescence lifetime (FRET from the pH-insensitive to the pH-sensitive dye)	Culture pumped from the E. coli bioreactor to the cuvette containing the sensing patch	0–20%	Phase shift 43° (from N_2-saturated H_2O to 20% CO_2-saturated H_2O)	t_{90} 1 min (0–6.7% CO_2-saturated H_2O) t_{90} 1.5 min (6.7–3.3% CO_2-saturated H_2O)	3 weeks operation in liquid Fluorescence lifetime-based, phase-sensitive detection Autoclavable Not affected by pH (6.0–8.0) or osmotic pressure	Actually based on absorption	[62]
YSI 8500 CO_2 monitor (based on ref. [59])	Ratiometric (upon exc. at 459/405)	Fiber-optic sensor for mammalian cell culture in spinner flask (controlling dCO_2 using N_2 sparging)	1–25%	0.5% CO_2 (3.6 mmHg)	t_{90} (5–10% CO_2) 6 min	Stable for 2 months Autoclavable and resistant to CIP and SIP. Not sensitive to pH (2.0–8.0 range) and many metabolites	Affected by temperature (automatically corrected) Long response time (though adequate for most bioprocesses)	[63]

(continued)

Table 3 (continued)

Dye (support)	Measurement	Format and use	Analytical features			Strengths	Weaknesses	Ref.
			Dynamic range	Sensitivity	Response time			
HPTS and CTMAOH (2-component highly hydrophobic Pt-cure silicone)		Patch for monitoring *E. coli* growth in shake flask	0–20%	LOD 0.03% CO_2	t_{90} (0–1.1%) 0.9 min; t_{90} (2.3–0%) 2.2 min	Autoclavable or sterilizable with 70% EtOH. Not affected by ionic strength (0–0.2 M) or pH (5.6–8.0)	Has to be kept in water or reconditioned in water before use	[64, 65]
Same as ref. [65], but different catalyst (PC074)						Improved operation life: 10 days of continuous operation		[66]
Tris(4,7-diphenyl-1,10-phenanthroline) ruthenium(II)dichloride in PDMS + pyrogenic silica (10% w/w)	Lifetime-based with phase-sensitive detection of O_2	Fiber-optic sensor for tracking the microalgae CO_2 fixation activity in a photoreactor	20–200 mM	7.6 mM	t_{90} (0–60 mM HCO_3^-) 11 min; t_{90} (60–0 mM HCO_3^-) 18 min	Indirect, avoids the problems related to classical pH-based CO_2 sensors	Long t_{90} (though adequate for the process) Biofouling (requires automated cleaning)	[67]
$[Ru(pzth)_3]^{2+}$ in PTFE (PDMS overcoated)	Intensity- and lifetime-based luminescence with phase-sensitive detection	Fiber-optic sensor online CO_2 monitoring in the gas effluent of a composter		$I_{N2}/I_{30\% CO2}$ 1.75 LOD 1×10^{-5} MPa	t_{90} (0–30%) 2.0 min; t_{90} (30–0%) 11.7 min	≥ 4 months shelf life	To be stored in buffer	[68]

A **B**

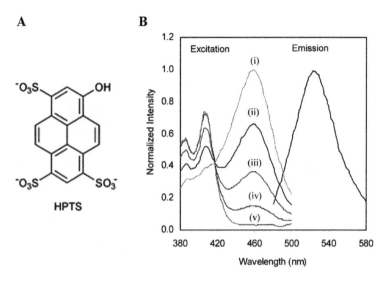

Fig. 6 (**a**) Chemical structure of the widely used HPTS fluorescent indicator dye (protonated form sold as water-soluble trisodium salt). (**b**) Normalized excitation and emission spectra of a HPTS/silicone rubber film when exposed to water purged with gas mixtures containing different CO_2 concentrations in N_2: (i) 0, (ii) 0.13, (iii) 0.37, (iv) 0.58, and (v) 18.15% (reprinted with permission from ref. [65]. ©Elsevier, 2002)

Luminescence lifetime-based CO_2 sensing has been made possible by taking advantage of *Förster resonance energy transfer* (FRET) between a pH-insensitive luminescent dye and a pH-sensitive indicator. In this way, sensors with luminescence lifetimes either in the ns or in the ms range have been manufactured, for instance:

- By using sulforhodamine 101 in tandem with thymol blue or *m*-cresol purple pH indicator dyes [85]
- By combining tris(4,4-diphenyl-2,2-bipyridine)ruthenium(II) dication (abbreviated $[Ru(dpb)_3]^{2+}$) as donor and thymol blue as pH-dependent acceptor [78] or tris(4,7-diphenyl-1,10-phenanthroline)ruthenium(II) dication (abbreviated $[Ru(dpp)_3]^{2+}$) as donor and Sudan III as pH-sensitive acceptor [86]

Even if based on luminescence lifetime measurements, quenching by FRET depends on the concentration of the acceptor dye; therefore, it does not solve the typical problem of the luminescence intensity-based sensors, namely, signal drift due to the indicator leaching or photobleaching. Any loss of the pH-sensitive acceptor dye due to the latter phenomena will affect the sensor response. Nevertheless, FRET photochemistry is the basis of one of the last commercialized optical CO_2 sensors (see Sect. 3.3 below).

An intrinsically drift-proof system can be achieved with luminescence lifetime-based sensing and non-FRET photochemical quenching (e.g., proton transfer to/from the photoexcited indicator dye). Szmacinski and Lakowicz [87] and

Fig. 7 Energy-minimized chemical structure of the acidity-sensitive phosphorescent [Ru (pzth)$_3$]$^{2+}$ complex. The pink-colored blister on the nitrogen atom of the pyrazine (pz) moiety shown in the top of the drawing represents the unshared electron pair that undergoes competitive protonation in the excited state but not in the ground state (photoexcitation leading to the triplet emissive metal-to-ligand charge-transfer state increases ca. 10^7-fold the basicity of that *N*-atom)

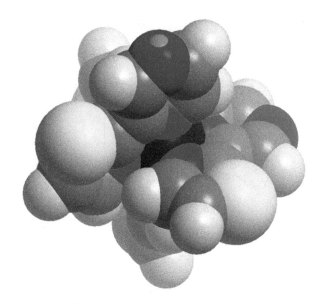

Thompson et al. [88] demonstrated the feasibility of exploiting the pH dependency of the seminaphthofluoresceins fluorescence decay time for developing CO_2 optodes. The main drawback is that these indicator dyes display a rather short emission lifetime (0.5–5 ns) that requires demanding electronics. A longer-lived acidity-sensitive luminophore, the tris[2-(2-pyrazinyl)thiazole]ruthenium (II) dication (abbreviated [Ru(pzth)$_3$]$^{2+}$, Fig. 7), has been shown to provide useful CO_2 optosensors upon immobilization in water-saturated Sephadex films with a gas-permeable thin overcoat [89]. This dye undergoes proton-transfer quenching from the Brönsted acid present in the indicator film [90], the concentration of which varies reversibly upon hydrolysis of the incoming CO_2 [68]. A limitation of this sensor was shown to be the rather long response time due to the thickness of the indicator layer and the kinetics of CO_2 hydrolysis. Incorporation of carbonic anhydrase proved to be instrumental in the decrease of the response time by accelerating the latter process [91]. The benefit of the [Ru(pzth)$_3$]$^{2+}$-based luminescent CO_2 sensor lies in that it can use the very same instrumentation marketed for O_2 luminescent optosensing based on Ru(II) polypyridyl dyes (see the corresponding section above). The potential interference of O_2 depends on its level in the sample; nevertheless, if it occurs, it may be readily corrected with a luminescent sensor for this gas.

Another CO_2 sensing method that can exploit the instrumentation commercialized for luminescent phase shift-based O_2 sensor technology is based on the sometimes called "dual luminophore referencing" (DLR) [92, 93]. This method converts the intensity signal of an analyte-*sensitive* (ns) fluorophore into the phase domain of a long-lived analyte-*insensitive* phosphor ("reference") by co-immobilizing them in the sensing film. For the method to work, the two dyes must display similar spectral features, i.e., both can be excited with the same light source and their emission must be collected together. This is the basis of one of the most recently commercialized devices (see Sect. 3.3 below).

Finally, some strategies to monitor CO_2 with alternative techniques (i.e., not based on pH changes) have been reported. Related to bioprocesses, Haigh-Flórez et al. have sensed CO_2 continuously in real time into a microalgae photoreactor exploiting the physical competition between CO_2 and O_2 for the headspace of a measuring chamber containing an O_2-sensitive luminescence phase-shift sensor [67]. Indirect methods have the advantage of bypassing some of the difficulties illustrated above, but at the expense of being less specific; therefore, their fields of application may be more limited.

Of the above-described luminescent CO_2 sensors, those that have been used for bioprocess monitoring in research applications are gathered in Table 3.

3.3 Marketed Sensors

Despite the large number of scientific works and patents [94–101] in the 40+ years that have passed since the first reported luminescence-based CO_2 sensor, only few of them have reached the commercialization stage.

Probably the most notorious commercial unit – although nowadays out of the market – is the YSI 8500 CO_2 monitor (Yellow Springs Instrument Co., Youngstown, OH). Based on the work of Uttamlal and Walt [59], it uses the fluorescent HPTS dye interrogated with ratiometric detection (see above). Its strengths were the relatively small size (12 mm diameter with a 70–320 mm insertion depth), a dynamic range apt for many bioprocesses (1–25% CO_2), fair accuracy ($\pm 5\%$), and, particularly, its disposable sensor capsules. The downsides it showed were the need of frequent recalibration due to about 2% signal drift per week and the relatively long response time ($t_{90} \approx 7$ min) and even longer recovery time [63].

A CO_2 sensor from Ocean Optics (FCO2-R fiber-optic sensor), based on a dry-type sensor with HPTS encapsulated in a proprietary sol-gel film (made of aluminum silicate copolymer) and covered by a black hydrophobic silicone coating, has been marketed in the past [102] and used for gas transport in fruit studies [103]. The sensor is not currently sold by the manufacturer. The most relevant features of that system were its dynamic range (0–25% CO_2), a limit of detection of 0.003% CO_2, resolution of at least 0.03% at high concentrations, the small probe diameter (1.6 mm), and its ability to be autoclaved. The main difference with the YSI was the detection mode, based on monitoring the whole fluorescence spectrum with a miniature spectrophotometer. However, it did not solve the main weaknesses of the HPTS-based CO_2 sensors, namely, important signal drift (5% of the fluorescence intensity over 72 h), long response time (>10 min), and demanding storage conditions (in water for at least 2 days before use and subsequently in water at all times).

The only luminescence-based fiber-optic CO_2 monitor commercially available at the time of this review is manufactured by PreSens (Regensburg, Germany) [https://www.presens.de/products/co2/sensors.html. Accessed 20 Sept 2018; 104]. Its current sensing chemistry has not been disclosed. The analytical features reported on their website and brochures (https://www.presens.de/products/detail/co2-sensor-spot-sp-cd1.html. Accessed 19 Dec 2018) show a dynamic range (1–25% CO_2 at atmospheric pressure) and accuracy (±5% of the reading after multipoint calibration) similar to the two marketed systems described above, while its response time (t_{90} < 3 min for a 2–5% CO_2 step change) and shelf life (12 months into its original package) are improved compared to the older devices.

Unfortunately, unlike the O_2 optosensors and despite the progress pledged by luminescence-based CO_2 sensors, the field of carbon dioxide measurement is still largely dominated by the Severinghaus electrode (for dissolved CO_2 measurements, e.g., the Mettler Toledo InPro 5000i sensor for fermentations in the pharmaceutical and biotech industries) and infrared spectroscopy (for gas-phase determinations, e.g., the Sartorius Stedim Biotech BioPAT® Xgas O_2/CO_2 sensor for bioprocess online off-gas analysis). The main reason for this fact is probably the real hurdles of the optical technology itself. Significant research and development are still needed to overcome the aforementioned weaknesses.

4 Luminescent pH Sensors

The chemical term "pH" was introduced in 1909 by the Danish chemist Søren P. L. Sørensen [105], and it is defined as the minus logarithm of the hydronium ion (H_3O^+) activity in water (Eq. 2) [106]:

$$pH = -\log_{10}a_{H_3O^+} \tag{2}$$

The pH value is one of the most important chemical parameters to be monitored and precisely controlled in situ, continuously, and in real time. It plays an important role in a large variety of areas such as water bodies quality assessment, water treatment plants, (bio)process control, food processing, fish farming, bedside clinical monitoring of blood, gastroesophageal reflux disease diagnostics, and swimming pool maintenance, to name a few [107, 108].

As pH is a critical parameter, a precise control of it is the key to success in countless industrial and biotechnological processes where uncontrolled variations of this parameter lead to the production of chemicals without adequate physical or chemical properties and loss of quality and yield or directly result in loss of production, especially in bioprocesses. Variations of the medium pH point out changes in metabolic activity in the process. The pH value is controlled in bioprocess operations because enzymatic activities, and therefore the associated metabolism, are sensitive to pH changes [4].

Setting up automation and control systems in bioprocesses such as cell cultures, tissue engineering, bioprocess development, biologics production, regenerative medicine, and so on is essential for keeping those processes under control.

Successful bioprocess development with the highest yield requires a fine control of the pH of the medium, so that all research-, pilot-, and production-scale bioreactors are fitted with sensing devices to monitor this parameter at all times [109, 110].

Sensors to monitor pH can be coupled to a bioreactor in different configurations: the sensors can be placed in the medium (in situ), in the bioreactor outlet or inlet (online sensors), delivering a continuous stream of information if display short response times, or ex situ, when sampling is needed to measure the pH [111].

Currently, the most widely used pH sensor is the "combination" electrode, made of Ag/AgCl and glass membrane electrodes in the same body [112]. This pH electrode is reliable and easy to operate and has a short response time and a relatively low cost. However, it shows some disadvantages like its large size, mechanical fragility, strong susceptibility to electrical interference, signal drift, and impossibility of sterilization, and it can readily undergo biofouling and protein contamination that calls for a frequent maintenance [113, 114]. Some of these drawbacks can be overcome with the so-called pH-sensitive ion field transistors (ISFETs) [4]. For those reasons, bioreactors represent a particular challenge for pH monitoring with electrodes or ISFETs, but optical sensors may be better suited for this task [115].

The two most common optical methods to measure pH are those based on polymer swelling or in optical indicator dyes. The first one uses materials (typically hydrogels) that swell as a function of the solution pH [116]. Their response is immune to photobleaching, decomposition, or leaching, because they are devoid of indicator dye. Nevertheless, they display a significant ionic strength dependence and can undergo delamination from the fiber or optical window during long-term measurements. On the other hand, optical pH sensors made of a colorimetric or fluorometric indicator dye immobilized into a hydrophilic polymer thin film have nowadays come true [117, http://celltainer.com/single-use-bioreactor-bag/; https://www.gelifesciences.co.jp/catalog/pdf/28952058ab.pdf; https://www.pbsbiotech.com/pbs-500.html; https://www.sartorius.es/sartoriusES/es/EUR/bioreactors-fermentors. Accessed 20 Nov 2018]. They are readily miniaturized and manufactured as either peel-and-stick sensing spots compatible with disposable plastic bioreactors [111] or non-disposable as fiber-optic microsensors [118, 119].

The indicator-based optical pH response is a function of the concentration of the acidic and basic forms of the dye molecule [120], determined by its acidity constant (K_a) according to well-known Henderson-Hasselbalch sigmoidal function. The latter normally spans ca. 1–1.5 pH units at each side of the indicator (ground state) pK_a, excited-state pK_a (in the case of equilibrium therein), or apparent pK_a (from the inflection point of the fluorescence titration curve). Therefore, the optosensor dynamic range can be tuned to the sample features by selecting the appropriate dye [121–137].

Nevertheless, the optical pH sensors are not limited to monitoring a narrow pH variation process. The sensitive layer can be manufactured by mixing several indicator dyes, with overlapping pH response ranges, at particular ratios to widen the sensor range. The latter is not only the oldest known approach since the development of the so-called "universal" pH papers in 1927 [138], but it has also been implemented in a variety of ways by combining multiple dyes with suitably different pK_a values to achieve a wide-responding pH optode [113, 139–147]. Alternatively, multi-protogenic molecules with progressive acidity constants can be used as pH-sensitive indicator dyes [148–150].

Among the pH optosensors, fluorescence-based ones show advantages due to their sensitivity and the possibility to operate them in the lifetime-based mode (see above). A pH-sensitive luminophore immobilized in a polymer matrix may undergo either a spectrum, intensity, or luminescence lifetime change upon (reversible) protonation or deprotonation. In particular, optical pH sensors based on emission lifetime measurements [121, 122, 148, 151], dual-lifetime referencing [123, 128, 129, 133, 152, 153], or wavelength ratiometric methods [130, 131, 143, 148, 154] display determining advantages over absolute luminescence intensity-based devices. In addition to the high sensitivity of the luminescence techniques, the effects of lamp and detector fluctuations or drift and the indicator leaching or bleaching are suppressed [8, 9].

The protonation or deprotonation of the pH indicator dye modifies the radiative and non-radiative kinetics of its excited-state deactivation, originating variations in the emission lifetime ("turn-on" or quenching) or in the spectrum, leading to brightness and/or color changes. Figure 8 depicts the relevant acid dissociation chemistry and the different excited-state deactivation pathways as a function of the acidity of the medium. It must be underlined that, at the pH of the sample medium, the equilibrium in the photoexcited dye molecule might not be established if its proton-transfer rate constants (from the H_3O^+ to the OH^- or from/to any buffer species present in the solution) are not competitive with the deactivation rate constants of the basic or acidic forms of the excited dye [126]. Observation of an inflection point in the luminescence intensity or lifetime pH titration curve does not indicate that an equilibrium in the excited state is attained. For the same reason, it is not unusual that the inflection point of such sigmoidal curves depends on the buffer used to set the different pH values along the titration or even on its concentration.

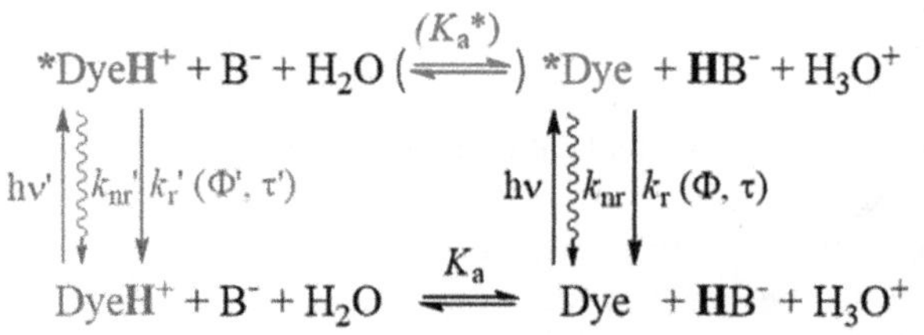

$$*DyeH^+ + B^- + H_2O \overset{(K_a^*)}{\rightleftharpoons} *Dye + HB^- + H_3O^+$$

$$hv' \quad k_{nr}' \quad k_r' (\Phi', \tau') \qquad hv \quad k_{nr} \quad k_r (\Phi, \tau)$$

$$DyeH^+ + B^- + H_2O \overset{K_a}{\rightleftharpoons} Dye + HB^- + H_3O^+$$

Fig. 8 Acid-base reactions relevant to a luminescent pH-sensitive dye and its excited-state deactivation pathways. B represents any buffer or basic species present in the solution at a relevant concentration (i.e., that may also deactivate the excited state of the dye by proton transfer). K_a represents the acid dissociation constant of the dye; the equivalent parameter in the excited state (K_a^*) might not exist if equilibration is not competitive with the corresponding decay rates of the acid or base forms of the excited dye (determined by their radiative and non-radiative deactivation rate constants, k_r and k_{nr})

Such dependency typically reflects a fast protonation or deprotonation of the excited indicator dye that prevents any establishment of equilibrium between the excited forms of the dye (Fig. 7), even for rather long-lived luminophores [90, 155]. Obviously, such dependency invalidates the potential luminophore as pH indicator dye. This is the unfortunate situation with far too many luminescent "pH indicator" dyes reported in the literature, because the proposers have either tested the candidate in a single buffer or they have used a "universal" buffer mixture, at just one concentration in any case.

As described above for the CO_2 fluorescent sensors (Fig. 6), the 8-hydroxy-1,3,6-pyrene trisulfonic acid (HPTS) dye has been widely used as a fluorescent pH indicator dye in the neutral region [131, 133, 156]. This is due to its high quantum yield and the possibility of ratiometric measurements of the green fluorescence of the photoexcited deprotonated form of the dye (the only one fluorescent in the pH 5.5–8.5 range [157]) upon excitation in the violet (~455 nm, absorption maximum of the deprotonated dye) and in the blue (~405 nm, absorption maximum of the protonated dye). The pH sensitivity of HPTS is due to the presence of a hydroxyl moiety on the pyrene ring of its structure. Its pK_a is significantly dependent on the ionic strength of the medium [157]; while this parameter can be controlled in the CO_2 sensors due to the separation of the indicator phase from the sample by a gas-permeable membrane, it may represent a source of variability of the pH sensor response in bioprocesses, unless the solution osmolarity is kept constant by the addition of an excess of an external salt.

In addition to the direct deactivation of the photoexcited dye by proton transfer depicted in Fig. 8, alternative ways to monitor pH by luminescent methods have been reported. For instance, the luminescence of a dye may be intramolecularly quenched by photoinduced electron transfer (PET) from a non-conjugated electron donor (typically, an amine moiety placed at a suitable distance from the luminophore). Protonation of the amine suppresses the intramolecular PET, turning "on" the indicator emission with decreasing pH [135].

A popular way to fabricate luminescent pH sensors that has made it into the market, particularly with phase-sensitive detectors, is to combine a long-lived luminophore with a colorimetric indicator dye, provided the emission of the former (partially) overlaps with the absorption band of only one of the acid/base species of the latter (Fig. 9). Such spectral overlap allows for an efficient Förster resonance energy transfer (FRET) process that quenches the emission intensity and lifetime of the luminophore if at least 50% of the acceptor molecules are within ≤8 nm of the donor ones ("Förster radius") [158]. Since the extent of the overlap depends on the solution acidity, the dye couple may be used as luminescent pH sensor [159, 160]. If the FRET distance requirement cannot be realized, a similar pH sensing scheme may still be possible by the so-called "external filter" effect of the indicator dye absorption on the luminophore emission; however, in this case, the emission lifetime of the latter will not be affected by the pH changes [161].

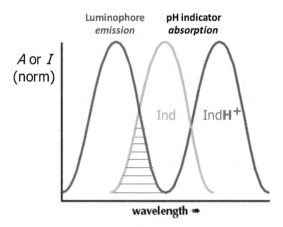

Fig. 9 Luminescent pH sensors based on Förster resonance energy transfer (FRET) from a pH-insensitive luminescent dye (red spectrum) to a nearby pH-sensitive colorimetric indicator molecule (green and blue spectra). Depending on the structure of the pH-sensitive dye, its protonated form may absorb at longer or at shorter wavelengths than the corresponding deprotonated species. For the sake of simplicity, only one situation is depicted

The pH of the culture medium is a key parameter to monitor for proper control of cell growth in bioprocesses. Despite the enormous amount of research having done on fluorescence-based pH sensor development, only a small fraction of novel sensors is suitable for using in bioprocesses. Culture media are a particularly harsh environment for any sensor; the latter are often poorly tested with the many substances that may interfere with sensor readings, and biofouling can invalidate their performance. Furthermore, sensors used in situ must be sterilizable (a challenging feature for pH sensors), and their components must not leach out into the culture. Another feature is that bioprocesses may run for several weeks, during which the sensor must be able to deliver without any recalibration [17, 111, 162]. Some of the characteristics, advantages, and disadvantages of the luminescent pH sensors described in the literature for bioprocess monitoring are summarized in Table 4.

Various luminescent pH optical sensors have been developed and tested in different cultures and fermentation processes. Most of them show the same problems when employed to monitor mammalian and bacterial cell cultures. Contribution from background fluorescence from the cells, ambient light interference, and expensive instrumentation are common problems. However, the use of disposable patches or low-cost sensitive terminals for online pH monitoring over the physiological range (typically 5.5–8.5) has led to the development of several commercially available sensors for bioprocesses. Some examples are described below.

Finesse Solutions (Santa Clara, CA; recently acquired by Thermo Fisher) have designed single-use sensors to measure the four fundamental bioprocess parameters:

Table 4 Luminescent sensors for pH monitoring used in bioprocesses

Material	Principle	Advantages	Disadvantages	Application(s)	Ref.
6-Carboxynaphthofluorescein covalently attached to tris (1,10-phenanthroline)ruthenium(II)	Long luminescence lifetime FRET-based pH probe	µs rapid lifetime determination (RLD) Response time <5 s (5 < pH < 10)	Expensive instrumentation (PLIM). Long-term drift	pH sensor layers in microcavities for pH sensing in *E. coli* cultures	[160]
Proprietary (prototype PreSens fluorometric sensors)	Ratiometric DLR	Stability and precision (72 h runs)	Fluorescence from medium interferes Complex media reduce resolution and dynamic range	Anaerobic batch fermentations of *Clostridium acetobutylicum*	[163]
Methacrylate-based microbeads stained with HPTS	Modified frequency-domain DLR	Internally self-referencing Dual pH/pO$_2$ determination with one sensitive terminal	Calibration is not as easy as in intensity- or lifetime-based procedures. pH measurement depends on the precision of the O$_2$ sensor	Dual pH/pO$_2$ Monitors the growth of *P. putida* cultures in 24-well microplates employing PreSens SensorDish technology	[164]
Proprietary (PreSens)	Not reported	"Rapid, reliable, and accurate" identification of the state of bioreactions	pH and O$_2$ sensor plates are not interchangeable Interference of ambient light	Cultures of Chinese hamster ovary (CHO) and mouse-myeloma derived NS0 6A1 cells (24-well plate PreSens SensorDish technology)	[32]
Proprietary MCR 8pH 8O$_2$ v1 (PreSens)	Not reported	More than 13 g L^{-1} dry cell mass can be achieved in pH-controlled batch cultivations	Parallel pH measurements are not as accurate as with a standard stirred-tank 1-L bioreactor	Bioprocess optimization and scaling up from 10 mL to 20 L of *E. coli* and *S. cerevisiae*	[165]
6,8-Dihydroxypyrene-1,3-disulfonic acid disodium salt (DHPDS)	Not reported (Fluorometrix, MA)	Unaffected by cell density No drift "Good" reproducibility between different pH sensors	Interference of ambient light	Clone screening with rocking T-flasks with sensor patch. SP2/0 myeloma/mouse hybridoma cell cultures	[30, 166]
N-Allyl-7-hydroxyquinolinium bromide (on polyethylene glycol dimethacrylate)	Dual-excitation ratiometric	Independent of ionic strength and temperature Sterilizable and no probe leaching	Deviation between online and offline pH measurements due to the background fluorescence from the cells	Online measurements of the pH of an *S. cerevisiae* (yeast) fermentation	[167]

(continued)

Table 4 (continued)

Material	Principle	Advantages	Disadvantages	Application(s)	Ref.
HPTS (on Dowex strong basic anion-exchanger resin/ PEG hydrogel)	Dual-excitation ratiometric	Reversibility, linearity, and sensitivity over the physiological pH range (5–8)	Optical isolation backing is needed to avoid interference from the nutrient broth and cells fluorescence	Online measurements of the pH of E. coli cultures	[168]
6,8-Dihydroxypyrene-1,3-disulfonic acid disodium salt (DHPDS) (covalently immobilized to hydrogel/white backing composite layer)	Dual-excitation ratiometric	Triple disposable optical sensors in a shake flask	Optical isolation backing is strictly required	Disposable sensors in shake flasks to monitor E. coli cultures	[64]

pH, dissolved oxygen, temperature, and headspace pressure. Thermo Fisher TruFluor® pH sensor consists of a disposable sheath, an optical reader, and a transmitter, but only the disposable sheath is inserted via a standard port into the stirred-tank bioreactor for single-use systems. This sensor is able to measure in the typical physiological pH interval (5.5–8.5), shows a good accuracy (±0.05 pH units after initial 2-point calibration), and a drift lower than ±0.05 pH in 21 days. It is sterilized by standard gamma irradiation and requires daily one-point standardization by the user.

GE Healthcare Bio-Sciences (Uppsala, Sweden) commercializes the Cellbag™ bioreactors, a pre-sterilized single-use bag for noninvasive mixing of culture medium and cells with electrical rocking systems (https://www.gelifesciences.co.jp/catalog/pdf/WAVECellbag_28951136AG.pdf. Accessed 22 Nov 2018). Their optical sensor, a luminescent dye "attached" to a polycarbonate backing, is embedded in the underside of the Cellbag bioreactor, enabling pH measurements (4.5–8.5, with ±0.05 to ±0.1 units) and control (6.0–8.0, with an accuracy of ±0.05 units vs the set point) using CO_2 or acid/base addition.

PreSens Precision Sensing (Regensburg, Germany) has developed and marketed luminescent pH sensors for a very wide variety of monitoring systems, from invasive (micro)sensors to disposable spot sensors optimized for physiological solutions and culture media (https://www.presens.de/products/ph.html. Accessed 22 Nov 2018). The target media pH can be measured in a noninvasive way from the outside using peel-and-stick spots, mounted in transparent vessels made of plastic or glass that are ready-to-use because they are beta-irradiated and pre-calibrated. The technical specifications reported by PreSens include a 5.5–8.5 pH measuring range, with an accuracy of ±0.1 pH for a calibrated sensor batch. A high concentration of small molecules displaying fluorescence in the visible range and large ionic strength variations are potential interferents.

Finally, Sartorius Stedim Biotech (Aubagne, France) commercializes several types of optically sensorized single-use bioreactors, from 125-mL Erlenmeyer flask up to 600-L bioreactor bags, suitable for cultivation of mammalian, insect, and plant cells (https://www.nature.com/articles/nmeth.f.273. Accessed 22 Nov 2018). Nevertheless, the same type of luminescent pH sensor is installed in all bioreactors. The sensor patches are previously calibrated at the factory, and the calibration data must be transferred either manually to the software or automatically by a barcode scanner. These single-use sensors patches have a dynamic response range in the 6.0–8.5 pH range, with an accuracy of ±0.1 pH units with one-point calibration and a 3-min (t_{90}) response time. Furthermore, they display a slight effect of the culture medium ionic strength but significant cross-sensitivity to the presence of dissolved (small) fluorescent molecules.

The luminescence-based spots/patches are nowadays the pH sensors of choice for bioprocess industrial applications. Their proprietary (photo)chemistries do not allow for a deeper account of their features. However, they all display the advantages of being pre-sterilized unbreakable disposable sensors, with a significant imperviousness to biofouling compared to the traditional pH electrodes. Further developments will surely address their pH response range (3.5–8.5 would be desirable for some cultures), long-term measuring stability (several weeks with low drift), and the possibility of sterilization in multiple ways.

5 Conclusions

The luminescent O_2, CO_2, and pH sensors have already displayed their superiority for in situ continuous real-time bioprocess monitoring, particularly those based on disposable patches attached to single-use bioreactors and emission lifetime measurements. While there is probably little room for improving dissolved or gas-phase luminescent oxygen sensors, luminescence-based CO_2 and pH devices still require (applied) research efforts of the scientific and technical communities to meet the current challenges and variety of bioprocesses. In particular, CO_2 sensing layers demand an enhancement of their stability (even "dry-type" ones!) to provide drift-free measurements. Luminescent pH-sensitive layers would require an extended dynamic range (ca. 3.5–8.5) to be able to monitor, for instance, acidophile and microalgae cultures. Common optoelectronics for interrogating all three sensors (single-excitation LED source, same emission region, and identical technique to extract the luminescence lifetimes of the sensor films) is always a bonus to increase the value for money of the optical devices and supply the customer the most compact devices. The latter is a must in automatic small-volume bioreactors for process optimization.

Acknowledgments The authors are indebted to the many public institutions, both national and European, and private companies that have funded our research on luminescent O_2, CO_2, and pH sensor technology and tailored indicator dyes over the last 20 years: the European Commission Framework Programs; EU Funds for Regional Development; the Spanish Ministries of Science, Technology, Industry, and Competitiveness; the Madrid Autonomous Region Government; Complutense University of Madrid; Santander Bank; Interlab Group; TGI; Agilent Technologies; CESA; TAP Biosystems (currently part of Sartorius Stedim); and Gas Natural Fenosa (now Naturgy). Our research in the 2016–2018 period was supported by the MINECO CTQ2015-69278-C2-2-R research grant and Naturgy SmartGreenGas project.

References

1. Classen J, Aupert F, Reardon KF, Solle D, Scheper T (2017) Spectroscopic sensors for in-line bioprocess monitoring in research and pharmaceutical industrial application. Anal Bioanal Chem 409:651–666
2. Roch P, Mandenius C-F (2016) On-line monitoring of downstream bioprocesses. Curr Opinion Chem Eng 14:112–120
3. Biechele P, Busse C, Solle D, Scheper T, Reardon K (2015) Sensor systems for bioprocess monitoring. Eng Life Sci 15:469–488

4. Rodríguez-Duran LV, Torres-Mancera MT, Trujillo-Roldán MA, Valdez-Cruz NA, Favela-Torres E, Saucedo-Castañeda G (2017) Standard instruments for bioprocess analysis and control. In: Larroche C, Sanroman M, Du G, Pandey A (eds) Current developments in biotechnology and bioengineering: bioprocesses, bioreactors and controls. Elsevier, Amsterdam, pp 593–626

5. Holzberg TR, Watson V, Brown S, Andar A, Ge X, Kostov Y, Tolosa L, Rao G (2018) Sensors for biomanufacturing process development: facilitating the shift from batch to continuous manufacturing. Curr Opinion Chem Eng 22:115–127

6. Busse C, Biechele P, de Vries I, Reardon KF, Solle D, Scheper T (2017) Sensors for disposable bioreactors. Eng Life Sci 17:940–952

7. O'Mara P, Farrell A, Bones J, Twomey K (2018) Staying alive! Sensors used for monitoring cell health in bioreactors. Talanta 176:130–139

8. Orellana G (2006) Fluorescence-based sensors. In: Baldini F, Chester AN, Homola J, Martellucci S (eds) Optical chemical sensors, NATO Sci Ser II, vol 224. Springer-Kluwer, Amsterdam, pp 99–116

9. Demchenko AP (2015) Introduction to fluorescence sensing, 2nd edn. Springer, Cham

10. Valeur B, Berberan-Santos MN (2012) Molecular fluorescence: principles and applications, 2nd edn. Wiley-VCH, Weinheim

11. Garcia-Ochoa F, Gomez E, Santos VE, Merchuk JC (2010) Oxygen uptake rate in microbial processes: an overview. Biochem Eng J 49:289–307

12. Anastasova S, Milanova M, Todorovsky D (2008) Photoluminescence response of Ru (II) complex immobilized in SiO_2-based matrix to dissolved oxygen in beer. J Biochem Biophys Methods 70:1292–1296

13. Urriza-Arsuaga I, Bedoya M, Orellana G (2019) Luminescent sensor for O_2 detection in biomethane streams. Sens Actuators B Chem 279:458–465

14. Wang X, Wolfbeis OS (2014) Optical methods for sensing and imaging oxygen: materials, spectroscopies and applications. Chem Soc Rev 43:3666–3761

15. Quaranta M, Borisov SM, Klimant I (2012) Indicators for optical oxygen sensors. Bioanal Rev 4:115–157

16. Orellana G, García-Fresnadillo D (2004) Environmental and industrial optosensing with tailored luminescent Ru(II) Polypyridyl complexes. In: Narayanaswamy R, Wolfbeis OS (eds) Optical sensors. Springer, Berlin, pp 309–357

17. Demuth C, Varonier J, Jossen V, Eibl R, Eibl D (2016) Novel probes for pH and dissolved oxygen measurements in cultivations from milliliter to benchtop scale. Appl Microbiol Biotechnol 100:3853–3863

18. Schäpper D, Alam MN, Szita N, Lantz AE, Gernaey KV (2009) Application of microbioreactors in fermentation process development: a review. Anal Bioanal Chem 395:679–695

19. Grist SM, Chrostowski L, Cheung KC (2010) Optical oxygen sensors for applications in microfluidic cell culture. Sensors 10:9286–9316

20. Gruber P, Marques MPC, Szita N, Mayr T (2017) Integration and application of optical chemical sensors in microbioreactors. Lab Chip 17:2693–2712

21. Orellana G, López-Gejo J, Pedras B (2014) Silicone films for fiber-optic chemical sensing. In: Tiwari A, Soucek MD (eds) Concise encyclopedia of high-performance silicones. Scrivener-Wiley, Beverly, pp 339–354

22. Kroneis HW, Marsoner HJ (1983) A fluorescence-based sterilizable oxygen probe for use in bioreactors. Sens Actuators B Chem 4:587–592

23. Bambot SB, Holavanahali R, Lakowicz JR, Carter GM, Rao G (1993) Phase fluorometric sterilizable optical oxygen sensor. Biotechnol Bioeng 43:1139–1145

24. Kohls O, Scheper T (2000) Setup of a fiber optical oxygen multisensor-system and its applications in biotechnology. Sens Actuators B Chem 70:121–130

25. Navarro-Villoslada F, Orellana G, Moreno-Bondi MC, Vick T, Driver M, Hildebrand G, Liefeith K (2001) Fiberoptic luminescent sensors with composite oxygen-sensitive layers and anti-biofouling coatings. Anal Chem 73:5150–5156
26. Tolosa L, Kostov Y, Harms P, Rao G (2002) Noninvasive measurement of dissolved oxygen in shake flasks. Biotechnol Bioeng 80:594–597
27. John GT, Klimant I, Wittmann C, Heinzle E (2003) Integrated optical sensing of dissolved oxygen in microtiter plates: a novel tool for microbial cultivation. Biotechnol Bioeng 81:829–836
28. Jin P, Chu J, Miao Y, Tan J, Xhang S, Zhu W (2013) A NIR luminescent copolymer based on platinum porphyrin as high permeable dissolved oxygen sensor for microbioreactors. AICHE J 59:2743–2752
29. Guenther D, Mattley Y (2015) Monitoring oxygen and pH during E. coli fermentation. Ocean optics application notes. https://oceanoptics.com/monitoring-oxygen-and-ph-during-e-coli-fermentation/. Accessed 10 Sept 2018
30. Ge X, Hanson M, Shen H, Kostov Y, Brorson KA, Frey DD, Moreira AR, Rao G (2006) Validation of an optical sensor-based high-throughput bioreactor system for mammalian cell culture. J Biotechnol 122:293–306
31. Schneider K, Schütz V, John GT, Heinzle E (2010) Optical device for parallel online measurement of dissolved oxygen and pH in shake flask cultures. Bioprocess Biosyst Eng 33:541–547
32. Naciri M, Kuystermans D, Al-Rubeai M (2008) Monitoring pH and dissolved oxygen in mammalian cell culture using optical sensors. Cytotechnology 57:245–250
33. Cywinski PJ, Moro AJ, Stanca SE, Biskup C, Mohr GJ (2009) Ratiometric porphyrin-based layers and nanoparticles for measuring oxygen in biosamples. Sens Actuators B Chem 135:472–477
34. Cao J, Nagl S, Kothe E, Köhler JM (2014) Oxygen sensor nanoparticles for monitoring bacterial growth and characterization of dose-response functions in microfluidic screening. Microchim Acta 182:385–394
35. Horka M, Sun S, Ruszczak A, Garstecki P, Mayr T (2016) Lifetime of phosphorescence from nanoparticles yields accurate measurement of concentration of oxygen in microdoplets, allowing one to monitor the metabolism of bacteria. Anal Chem 88:12006–12012
36. Ungerböck B, Fellinger S, Sulzer P, Abel T, Mayr T (2014) Magnetic optical sensor particles: a flexible analytical tool for microfluidic devices. Analyst 139:2551–2559
37. Chojnacki P, Mistlberger G, Klimant I (2007) Separable magnetic sensors for the optical determination of oxygen. Angew Chem Int Ed 46:8850–8853
38. Flitsch D, Ladner T, Lukacs M, Büchs J (2016) Easy to use and reliable technique for online dissolved oxygen tension measurement in shake flasks using infrared fluorescent oxygen-sensitive nanoparticles. Microb Cell Factories 15:45
39. Ladner T, Flitsch D, Schlepütz T, Büchs J (2015) Online monitoring of dissolved oxygen tension in microtiter plates based on infrared fluorescent oxygen-sensitive nanoparticles. Microb Cell Factories 14:161
40. Mahler L, Tovar M, Weber T, Brandes S, Rudolph MM, Ehgartner J, Mayr T, Figge MT, Roth M, Zang E (2015) Enhanced and homogeneous oxygen availability during incubation of microfluidic droplets. RSC Adv 5:101871–101878
41. Scheucher E, Wilhelm S, Wolfbeis OS, Hirsch T, Mayr T (2015) Composite particles with magnetic properties, near-infrared excitation, and far-red emission for luminescence-based oxygen sensing. Microsys Nanoeng 1:15026
42. Gorris HH, Resch-Genger U (2017) Perspectives and challenges of photon-upconversion nanoparticles – part II bioanalytical applications. Anal Bioanal Chem 409:5875–5890
43. Mishra D, Tiwari R, Dwivedi AK (2016) Environmental impact assessment of thermal power plants- CO_2 emission and control. Pollut Res 35:127–130
44. Ward KR, Yealy DM (1998) End-tidal carbon dioxide monitoring in emergency medicine, part 1: basic principles. Acad Emerg Med 5:628–636

45. Ward KR, Yealy DM (1998) End-tidal carbon dioxide monitoring in emergency medicine, part 2: clinical applications. Acad Emerg Med 5:637–646
46. Puligundla P, Jung J, Ko S (2012) Carbon dioxide sensors for intelligent food packaging applications. Food Control 25:328–333
47. Meng X, Kim S, Puligundla P, Ko S (2014) Carbon dioxide and oxygen gas sensors-possible application for monitoring quality, freshness, and safety of agricultural and food products with emphasis on importance of analytical signals and their transformation. J Korean Soc Appl Biol Chem 57:723–733
48. Emmerich S, Persily A (2001) State-of-the-art review of CO2 demand controlled ventilation technology and application. Technical report, National Institute of Standards and Technology, California Energy Commission, NISTIR 6729
49. Permentier K, Vercammen S, Soetaert S, Schellemans C (2017) Carbon dioxide poisoning: a literature review of an often forgotten cause of intoxication in the emergency department. Int J Emerg Med 10:14–17
50. Purves WK, Sadava D, Orians GH (2004) Life, the science of biology, 7th edn. Sinauer Associates and W. H. Freeman, Sunderland, pp 139–140. ISBN 978-0-7167-9856-9
51. Dezengotita VM, Kimura R, Miller WM (1998) Effects of CO_2 and osmolality on hybridoma cells: growth, metabolism and monoclonal antibody production. Cytotechnology 28:213–227
52. Godoy-Silva R, Berdugo C, Chalmers JJ, Flickinger MC (2010) Aeration, mixing and hydrodynamics. Animal cell bioreactors. In: Flickinger MC (ed) Encyclopedia of industrial biotechnology: bioprocess, bioseparation, and cell technology. Wiley, New York
53. Nienow AW, Flickinger MC (2009) Impeller selection for animal cell culture. Encyclopedia of industrial biotechnology. Wiley, New York
54. Dixon NM, Kell DB (1989) The control and measurement of CO_2 during fermentations. J Microbiol Methods 10:155–176
55. Zosel J, Oelssner W, Decker M, Gerlach G, Guth U (2011) The measurement of dissolved and gaseous carbon dioxide concentration. Meas Sci Technol 22:072001
56. Severinghaus JW, Bradley AF (1958) Electrodes for blood pO_2 and pCO_2 determination. J Appl Physiol 13:515–520
57. Lubbers DW, Opitz N (1975) The pCO_2-/pO_2-optode: a new probe for measurement of pCO_2 or pO_2 in fluids and gases. Z Naturforsch C 30:532–533
58. Zhujun Z, Seitz WR (1984) A carbon dioxide sensor based on fluorescence. Anal Chim Acta 160:305–309
59. Uttamlal M, Walt DR (1995) A fiber-optic carbon dioxide sensor for fermentation monitoring. Bio/Technology 13:597–601
60. Sipior J, Randers-Eichhorn L, Lakowicz JR, Carter GM, Rao G (1996) Phase fluorometric optical carbon dioxide gas sensor for fermentation off-gas monitoring. Biotechnol Prog 12:266–271
61. Ferguson JA, Healey BG, Bronk KS, Barnard SM, Walt DR (1997) Simultaneous monitoring of pH, CO_2 and O_2 using an optical imaging fiber. Anal Chim Acta 340:123–131
62. Chang Q, Randers-Eichhorn L, Lakowicz JR, Rao G (1998) Steam-sterilizable, fluorescence lifetime-based sensing film for dissolved carbon dioxide. Biotechnol Prog 14:326–331
63. Pattison RN, Swamy J, Mendenhall B, Hwang C, Frohlich BT (2000) Measurement and control of dissolved carbon dioxide in mammalian cell culture processes using an in situ fiber optic chemical sensor. Biotechnol Prog 16:769–774
64. Ge X, Rao G (2012) Real-time monitoring of shake flask fermentation and off gas using triple disposable noninvasive optical sensors. Biotechnol Prog 28:872–877
65. Ge X, Kostov Y, Rao G (2003) High-stability non-invasive autoclavable naked optical CO_2 sensor. Biosens Bioelectron 18:857–865
66. Ge X, Kostov Y, Rao G (2005) Low-cost noninvasive optical CO_2 sensing system for fermentation and cell culture. Biotechnol Bioeng 8:329–334

67. Haigh-Flórez D, Cano-Raya C, Bedoya M, Orellana G (2015) Rugged fibre-optic luminescent sensor for CO_2 determination in microalgae photoreactors for biofuel production. Sens Actuators B Chem 221:978–984

68. Xavier MP, Orellana G, Moreno-Bondi MC, Diaz-Puente J (2000) Carbon dioxide monitoring in compost processes using fibre optic sensors based on a luminescent ruthenium(II) indicator. Quim Anal 19:118–126

69. Mills A, Chang Q, Mcmurray N (1992) Equilibrium studies on colorimetric plastic film sensors for carbon dioxide. Anal Chem 64:1383–1389

70. Mills A, Chang Q (1993) Fluorescence plastic thin-film sensor for carbon dioxide. Analyst 118:839–843

71. Mills A, Yusufu D (2016) Highly CO_2 sensitive extruded fluorescent plastic indicator film based on HPTS. Analyst 141:999–1008

72. Ratterman M, Shen L, Klotzkin D, Papautsky I (2014) Carbon dioxide luminescent sensor based on a CMOS image array. Sens Actuators B Chem 198:1–6

73. Borisov SM, Seifner R, Klimant I (2011) A novel planar optical sensor for simultaneous monitoring of oxygen, carbon dioxide, pH and temperature. Anal Bioanal Chem 400:2463–2474

74. Ertekin K, Klimant I, Neurauter G, Wolfbeis OS (2003) Characterization of a reservoir-type capillary optical microsensor for pCO_2 measurements. Talanta 59:261–267

75. Wolfbeis OS, Kovács B, Goswami K, Klainer SM (1998) Fiber-optic fluorescence carbon dioxide sensor for environmental monitoring. Microchim Acta 129:181–188

76. Burke CS, Markey A, Nooney RI, Byrne P, McDonagh C (2006) Development of an optical sensor probe for the detection of dissolved carbon dioxide. Sens Actuators B Chem 119:288–294

77. Parker JW, Laksin O, Yu C, Lau M, Klima S, Fisher R, Scott I, Atwater BW (1993) Fiber-optic sensors for pH and carbon dioxide using a self-referencing dye. Anal Chem 65:2329–2334

78. Neurauter G, Klimant I, Wolfbeis OS (1999) Microsecond lifetime-based optical-carbon dioxide sensor using luminescence resonance energy transfer. Anal Chim Acta 382:67–75

79. He X, Rechnitz GA (1995) Linear response function for fluorescence-based fiber-optic CO_2 sensors. Anal Chem 67:2264–2268

80. Wencel D, Moore JP, Stevenson N, McDonagh C (2010) Ratiometric fluorescence-based dissolved carbon dioxide sensor for use in environmental monitoring applications. Anal Bioanal Chem 398:1899–1907

81. Lochman L, Zimcik P, Klimant I, Novakova V, Borisov SM (2017) Red-emitting CO_2 sensors with tunable dynamic range based on pH-sensitive azaphthalocyanine indicators. Sens Actuators B Chem 246:1100–1107

82. Amao Y, Nakamura N (2004) Optical CO_2 sensor with the combination of colorimetric change of α-naphtholphthalein and internal reference fluorescent porphyrin dye. Sens Actuators B Chem 100:347–351

83. Pérez de Vargas-Sansalvador IM, Carvajal MA, Roldán-Muñoz OM, Banqueri J, Fernández-Ramos MD, Capitán-Vallvey LF (2009) Phosphorescent sensing of carbon dioxide based on secondary inner-filter quenching. Anal Chim Acta 655:66–74

84. Fritzsche E, Gruber P, Schutting S, Fischer JP, Strobl M, Muller JD, Borisov SM, Klimant I (2017) Highly sensitive poisoning-resistant optical carbon dioxide sensors for environmental monitoring. Anal Methods 9:55–65

85. Sipior J, Bambot S, Romauld M, Carter GM, Lakowicz JR, Rao G (1995) A lifetime-based optical CO_2 gas sensor with blue or red excitation and stokes or anti-stokes detection. Anal Biochem 227:309–318

86. von Bültzingslöwen C, McEvoy AK, McDonagh C, MacCraith BD (2003) Lifetime-based optical sensor for high-level pCO_2 detection employing fluorescence resonance energy transfer. Anal Chim Acta 480:275–283

87. Szmacinski H, Lakowicz JR (1993) Optical measurements of pH using fluorescence lifetimes and phase-modulation fluorometry. Anal Chem 65:1668–1674

88. Thompson RB, Frisoli JK, Lakowicz JR (1992) Phase fluorometry using a continuously modulated laser diode. Anal Chem 64:2075–2078
89. Orellana G, Moreno-Bondi MC, Segovia E, Marazuela MD (1992) Fiber-optic sensing of carbon dioxide based on excited-state proton transfer to a luminescent ruthenium(II) complex. Anal Chem 64:2210–2215
90. Marazuela MD, Moreno Bondi MC, Orellana G (1998) Luminescence lifetime quenching of a ruthenium(II) polypyridyl dye for optical sensing of carbon dioxide. Appl Spectrosc 52:1314–1320
91. Marazuela MD, Moreno Bondi MC, Orellana G (1995) Enhanced performance of a fibre-optic luminescence CO_2 sensor using carbonic anhydrase. Sens Actuators B Chem 29:126–131
92. Lakowicz JR, Castellano FN, Dattelbaum JD, Tolosa L, Rao G, Gryczynski I (1998) Low-frequency modulation sensors using nanosecond fluorophores. Anal Chem 70:5115–5121
93. von Bültzingslöwen C, McEvoy AK, McDonagh C, MacCraith BD, Klimant I, Krausec C, Wolfbeis OS (2002) Sol-gel based optical carbon dioxide sensor employing dual luminophore referencing for application in food packaging technology. Analyst 127:1478–1483
94. Yafuso M, Suzuki JK (1989) Gas sensors. US patent 4,824,789
95. Alderete JE, Olstein AD, Furlong SC (1998) Optical carbon dioxide sensor and associated methods of manufacture. US Patent 5,714,121
96. Adrian W, Mark BS (2002) Optical carbon dioxide sensors. US Patent 6,338,822
97. Furlong SC (1997) Simultaneous dual excitation/single emission fluorescent sensing method for pH and pCO_2. US Patent 5,672,515
98. Bretscher KR, Baker JA, Wood KB, Nguyen MT, Hamer MA, Rueb CJ (1997) Novel emulsion for robust sensing of glass. WO Patent 9,719,348
99. Bentsen JG, Wood KB (1995) Sensor with improved drift stability. US Patent 5,403,746
100. Klainer SM, Goswami K, Herron NR, Simon SJ, Eccles LA (1990) Reservoir fiber optic chemical sensors. US Patent 4,892,383
101. Orellana G, Moreno-Bondi, MC (1992) Sensor óptico. ES Patent 2,023,593
102. Mills A, Hodgen S (2005) Fluorescent carbon dioxide indicators. In: Geddes CD, Lakowicz JR (eds) Topics in fluorescence spectroscopy. Advanced concepts in fluorescence sensing part a: small molecule sensing, vol 9. Springer, New York, pp 119–161
103. Ho QT, Verboven P, Verlinden BE, Herremans E, Wevers M, Carmeliet J, Nicolaï BM (2011) A three-dimensional multiscale model for gas exchange in fruit. Plant Physiol 155:1158–1168, and references therein
104. Long C, Anderson W, Finch C, Hickman J (2012) CO_2 measurement in microfluidic devices. Facilitating biological applications using a flow-through cell. In: Genetic engineering & biotechnology news. https://www.genengnews.com/gen-articles/cosub2sub-measurement-in-microfluidic-devices/4565. Accessed 15 Dec 2018
105. Sørensen SPL (1909) Etudes enzymatiques; II. Sur la mesure et l'importance de la concentration des ions hydrogene dans les reactions enzymatiques. Compt Rend Lab Carlsberg 8:1
106. Buck RP, Rondinini S, Covington AK, Baucke FGK, Brett CMA, Camoes MF, Milton MJT, Mussini T, Naumann R, Pratt KW, Spitzer P, Wilson GS (2002) Measurement of pH. Definition, standards, and procedures. Pure Appl Chem 74:2169–2200
107. McMillan GK, Cameron RA (2005) Advanced pH measurement and control, 3rd edn. ISA, Research Triangle Park
108. Orellana G, Cano-Raya C, López-Gejo J, Santos AR (2011) Online monitoring sensors. In: Wilderer P (ed) Treatise on water science, vol 3. Academic Press, Oxford, p 221
109. Harms P, Kostov Y, Rao G (2002). Curr Opin Biotechnol 13:124
110. Lam H, Kostov Y (2009) Optical instrumentation for bioprocess monitoring. In: Rao G (ed) Optical sensor systems in biotechnology, vol 116. Springer, Berlin, pp 1–4
111. Glindkamp A, Riechers D, Rehbock C, Hitzmann B, Scheper T, Reardon KF (2009) Sensors in disposable bioreactors status and trends. Adv Biochem Eng Biotechnol 115:145–169
112. Scholz F (2011) From the Leiden jar to the discovery of the glass electrode by Max Cremer. J Solid State Electrochem 15:5–14

113. Wencel D, Abel T, McDonagh C (2014) Optical chemical pH sensors. Anal Chem 86:15–29
114. Demuth C (2014) Chemische Sensoren in der Bioprozessanalytik. Chem Unserer Zeit 48:60–67
115. Riley M (2005) Instrumentation and process control. In: Ozturk S, Hu WS (eds) Cell culture technology for pharmaceutical and cell-based therapies. CRC Press, Boca Raton, pp 249–298
116. Richter A, Paschew G, Klatt S, Lienig J, Arndt KF, Adler HJ (2008) Review on hydrogel-based pH sensors and microsensors. Sensors 8:561–581
117. Wang XD, Wolfbeis OS (2016) Fiber-optic chemical sensors and biosensors (2013–2015). Anal Chem 88:203–227
118. Mohamad F, Tanner MG, Choudhury D, Choudhary TR, Wood HAC, Harringtone K, Bradley M (2017) Controlled core-to-core photo-polymerisation – fabrication of an optical fibre-based pH sensor. Analyst 142:3569–3572
119. Kocincova AS, Borisov SM, Krause C, Wolfbeis OS (2007) Fiber-optic microsensors for simultaneous sensing of oxygen and pH, and of oxygen and temperature. Anal Chem 79:8486–8493
120. Kahlert H, Scholz F (2013) Acid-base diagrams. Springer, Berlin
121. Murtaza A, Chang Q, Rao G, Lin H, Lakowicz JR (1997) Long-lifetime metal-ligand pH probe. Anal Biochem 247:216–222
122. Clarke Y, Xu W, Demas JN, DeGraff BA (2000) Lifetime-based pH sensor system based on a polymer-supported ruthenium(II) complex. Anal Chem 72:3468–3475
123. Vasylevska GS, Borisov SM, Krause C, Wolfbeis OS (2016) Indicator-loaded permeation-selective microbeads for use in fiber optic simultaneous sensing of pH and dissolved oxygen. Chem Mater 18:4609–4616
124. Borisov SM, Mayr T, Klimant I (2008) Poly(styrene-block-vinylpyrrolidone) beads as a versatile material for simple fabrication of optical nanosensors. Anal Chem 80:573–582
125. Bowyer WJ, Xu W, Demas JN (2009) Determining proton diffusion in polymer films by lifetimes of luminescent complexes measured in the frequency domain. Anal Chem 81:378–384
126. Tormo L, Bustamante N, Colmenarejo G, Orellana G (2010) Can luminescent Ru (II) polypyridyl dyes measure pH directly? Anal Chem 82:5195–5204
127. Mistlberger G, Koren K, Borisov SM, Klimant I (2010) Magnetically remote-controlled optical sensor spheres for monitoring oxygen or pH. Anal Chem 82:2124–2128
128. Schröder CR, Weidgans BM, Klimant I (2005) pH fluorosensors for use in marine systems. Analyst 130:907–916
129. Borisov SM, Gatterer K, Klimant I (2010) Red light-excitable dual lifetime referenced optical pH sensors with intrinsic temperature compensation. Analyst 135:1711–1717
130. Xia T, Zhu F, Jiang K, Cui Y, Yang Y, Qian G (2017) A luminescent ratiometric pH sensor based on a nanoscale and biocompatible Eu/Tb-mixed MOF. Dalton Trans 46:7549–7555
131. Acquah I, Roh J, Ahn DJ (2017) Dual-fluorophore silica microspheres for ratiometric acidic pH sensing. Macromol Res 25:950955
132. Shen L, Lu X, Tian H, Zhu W (2011) A long wavelength fluorescent hydrophilic copolymer based on naphthalenediimide as pH sensor with broad linear response range. Macromolecules 44:5612–5618
133. Wencel D, Higgins C, Klukowska A, MacCraith BD, McDonagh C (2007) Novel sol-gel derived films for luminescence-based oxygen and pH sensing. Mater Sci Poland 25:767–779
134. Vuppu S, Kostov Y, Rao G (2009) Economical wireless optical ratiometric pH sensor. Meas Sci Technol 20:045202
135. Aigner D, Borisov SM, Orriach-Fernández FJ, Fernández-Sánchez JF, Saf R, Klimant I (2012) New fluorescent pH sensors based on covalently linkable PET rhodamines. Talanta 99:194–201
136. Turel M, Cajlakovic M, Austin E, Dakin JP, Uray G, Lobnik A (2008) Direct UV-LED lifetime pH sensor based on a semi-permeable sol-gel membrane immobilized luminescent Eu^{3+} chelate complex. Sens Actuators B Chem 131:247–253

137. Kateklum R, Gauthier-Manuel B, Pieralli C, Mankhetkorn S, Wacogne B (2017) Improving the sensitivity of amino-silanized sensors using self-structured silane layers: application to fluorescence pH measurement. Sens Actuators B Chem 248:605–612
138. Kolthoff JM (1927) Ueber die Anwendung von Mischindicatoren in der Acidimetrie und Alkalimetrie. Biochem Z 189:26–32
139. Chauhan VM, Burnett GR, Aylott JW (2011) Dual-fluorophore ratiometric pH nanosensor with tuneable pK$_a$ and extended dynamic range. Analyst 136:1799–1801
140. Hashemi P, Zarjani RA (2008) A wide range pH optical sensor with mixture of Neutral Red and Thionin immobilized on an agarose film coated glass slide. Sens Actuators B Chem 135:112–115
141. Martinez-Olmos A, Capel-Cuevas S, López-Ruiz N, Palma AJ, de Orbe I, Capitán-Vallvey LF (2011) Sensor array-based optical portable instrument for determination of pH. Sens Actuators B Chem 156:840–848
142. Devadhasan JP, Kim S (2015) An ultrasensitive method of real time pH monitoring with complementary metal oxide semiconductor image sensor. Anal Chim Acta 858:55–59
143. Gotor T, Ashokkumar P, Hecht M, Keil K, Rurack K (2017) Optical pH sensor covering the range from pH 0–14 compatible with mobile-device readout and based on a set of rationally designed indicator dyes. Anal Chem 89:8437–8444
144. Shamsipur M, Abbasitabar F, Zare-Shahabadi V, Shahabadi, Akhond M (2008) Broad-range optical pH sensor based on binary mixed-indicator doped sol-gel film and application of artificial neural network. Anal Lett 41:3113–3123
145. Capel-Cuevas S, Cuéllar MP, de Orbe-Payá I, Pegalajar MC, Capitán-Vallvey LF (2011) Full-range optical pH sensor array based on neural networks. Microchem J 97:225–233
146. Safavi A, Bagheri M (2003) Novel optical pH sensor for high and low pH values. Sens Actuators B Chem 90:143–150
147. Ma X, Cheng J, Liu J, Zhou X, Xiang H (2015) Ratiometric fluorescent pH probes based on aggregation-induced emission-active salicylaldehyde azines. New J Chem 39:492–500
148. Malins C, Glever HG, Keyes TE, Vos JG, Dressick WJ, MacCraith BD (2000) Sol-gel immobilised ruthenium(II) polypyridyl complexes as chemical transducers for optical pH sensing. Sens Actuators B Chem 67:89–95
149. Sánchez-Barragán I, Costa-Fernández JM, Sanz-Medel A (2005) Tailoring the pH response range of fluorescent-based pH sensing phases by sol-gel surfactants co-immobilization. Sens Actuators B Chem 107:69–77
150. Cui H, Chen Y, Li L, Wu Y, Tang Z, Fu H, Tian Z (2014) Hybrid fluorescent nanoparticles fabricated from pyridine-functionalized polyfluorene-based conjugated polymer as reversible pH probes over a broad range of acidity-alkalinity. Microchim Acta 181:1529–1539
151. Gonçalves HMR, Maule CD, Jorge PAS, Esteves da Silva JCG (2008) Fiber optic lifetime pH sensing based on ruthenium(II) complexes with dicarboxybipyridine. Anal Chim Acta 626:62–70
152. Kasik I, Mrazek J, Martan T, Pospisilova M, Podrazky O, Matejec V, Hoyerova K, Kaminek M (2010) Fiber-optic pH detection in small volumes of biosamples. Anal Bioanal Chem 398:1883–1889
153. Liebsch G, Klimant I, Krause C, Wolfbeis OS (2001) Fluorescent imaging of pH with optical sensors using time domain dual lifetime referencing. Anal Chem 73:4354–4363
154. Hiruta Y, Yoshizawa N, Citterio D, Suzuki K (2012) Highly durable double sol-gel layer ratiometric fluorescent pH optode based on the combination of two types of quantum dots and absorbing pH indicators. Anal Chem 84:10650–10656
155. Higgins B, DeGraff BA, Demas JN (2005) Luminescent transition metal complexes as sensors: structural effects on pH response. Inorg Chem 44:6662–6669
156. Kumar R, Yadav R, Kolhe MA, Bhosale RS, Narayan R (2018) 8-Hydroxypyrene-1,3,6-trisulfonic acid trisodium salt (HPTS) based high fluorescent, pH stimuli waterborne polyurethane coatings. Polymer 136:157–165

157. Barnadas-Rodríguez R, Estelrich J (2008) Effect of salts on the excited state of pyranine as determined by steady-state fluorescence. J Photochem Photobiol A Chem 198:262–267
158. Medintz I, Hildebrandt N (2014) FRET – förster resonance energy transfer: from theory to applications. Wiley-VCH, Weinheim
159. Bambot SB, Sipior J, Lakowicz JR, Rao G (1994) Lifetime-based optical sensing of pH using resonance energy transfer in sol-gel films. Sens Actuators B Chem 22:181–188
160. Poehler E, Pfeiffer SA, Herm M, Gaebler M, Busse B, Nagl S (2016) Microchamber arrays with an integrated long luminescence lifetime pH sensor. Anal Bioanal Chem 408:2927–2935
161. Borisov SM, Klimant I (2013) A versatile approach for ratiometric time-resolved read-out of colorimetric chemosensors using broadband phosphors as secondary emitters. Anal Chim Acta 787:219–225
162. Marose S, Lindemann C, Ulber R, Scheper T (1999) Optical sensor systems for bioprocess monitoring. Trends Biotechnol 17:30–34
163. Janzen NH, Schmidt M, Krause C, Weuster-Botz D (2015) Evaluation of fluorimetric pH sensors for bioprocess monitoring at low pH. Bioprocess Biosyst Eng 38:1685–1692
164. Kocincova AS, Nagl S, Arain S, Krause C, Borisov SM, Arnold M, Wolfbeis OS (2008) Multiplex bacterial growth monitoring in 24-well microplates using a dual optical sensor for dissolved oxygen and pH. Biotechnol Bioeng 100:430–438
165. Kusterer A, Krause C, Kaufmann K, Arnold M, Weuster-Botz D (2008) Fully automated single-use stirred-tank bioreactors for parallel microbial cultivations. Bioprocess Biosyst Eng 31:207–215
166. Vallejos JR, Micheletti M, Brorson KA, Moreira AR, Rao G (2012) Optical sensor enabled rocking T-flasks as novel upstream bioprocessing tools. Biotechnol Bioeng 109:2295–2305
167. Badugu R, Kostov Y, Rao G, Tolosa L (2008) Development and application of an excitation ratiometric optical pH sensor for bioprocess monitoring. Biotechnol Prog 24:1393–1401
168. Kermis HR, Kostov Y, Harms P, Rao G (2002) Dual excitation ratiometric fluorescent pH sensor for noninvasive bioprocess monitoring: development and application. Biotechnol Prog 18:1047–1053

Fluorescence in Pharmaceutics
and Cosmetics

Maurício da Silva Baptista and Erick Leite Bastos

Contents

Abstract Fluorescence is a visually appealing phenomenon that revolutionized many fields of science and technology. The abilities to *see* biological processes as they occur, monitor drug delivery and action, stain specific cells, and highlight microscopic structures are some of the applications of fluorescence that are changing our daily lives. This chapter discusses the use of the fluorescence phenomenon in pharmaceuticals and cosmetics focusing on the toxicology and the use of task-specific fluorescent materials and development of sensitive and selective analytical and imaging methods. Noninvasive bioimaging using endogenous fluorophores as probes, emerging techniques related to fluorescence, and a brief description of selected fundamental concepts in photoscience are provided for completeness.

M. da Silva Baptista (✉) and E. L. Bastos (✉)
Instituto de Química, Universidade de São Paulo, São Paulo, SP, Brazil
e-mail: baptista@iq.usp.br; elbastos@iq.usp.br

© Springer Nature Switzerland AG 2019   39
B. Pedras (ed.), *Fluorescence in Industry*, Springer Ser Fluoresc (2019) 18: 39–102,
https://doi.org/10.1007/4243_2018_1, Published online: 8 February 2019

Keywords Cosmetics · Drug discovery · Dyes · Fluorescence · Hair · Labels · Microscopy · Nanomaterials · Pharmaceuticals · Photoprotection · Probes · Skin

Abbreviations

ε	Molar absorption coefficient
λ^{ABS}	Absorption wavelength
λ^{EM}	Emission wavelength
λ^{EX}	Excitation wavelength
τ	Lifetime of the singlet excited state
ϕ_{FL}	Fluorescence quantum yield
AFP	*Aequorea*-derived fluorescent proteins
API	Active pharmaceutical ingredients
C-dots	Carbon nanodots
CNTs	Carbon nanotubes
CT	Charge transfer
DSNPs	Fluorescent dye-doped silica nanoparticles
FCCS	Fluorescence cross-correlation spectroscopy
FCS	Fluorescence correlation spectroscopy
FLM	Fluorescence lifetime measurements
FP	Fluorescent protein
FPR	Fluorescence photobleaching recovery
FRET	Förster resonance energy transfer
GFP	Green fluorescent protein
GM	Goeppert-Mayer
GO	Graphene oxide
MB	Molecular beacons
NCE	New chemical entities
NDs	Nanodiamonds
NIR	Near infrared
PeT	Photoinduced electron transfer
Ph. Eur.	*European Pharmacopoeia*
QD	Semiconductor quantum dots
spFRET	Solution-phase single-pair FRET
TRF	Time-resolved fluorescence
UCNPs	Lanthanide-based upconversion nanoparticles
USP	*US Pharmacopeia*

Fundamental Aspects and Scope

A species promoted to a singlet excited state can undergo several different processes to return to the ground state, one of them being the emission of light. When there is no change in the spin state during emission, this process is called fluorescence. In this chapter, we describe the application of fluorescent molecules, proteins, and nanostructured materials, hereafter called *fluorophores*, in pharmacology and cosmetology. Steady-state and time-resolved spectroscopic platforms and bioimaging techniques that have been used for the investigation of complex biological and biochemical problems are presented in detail. These methods rely on changes in fluorescence intensity, polarization, or lifetime caused by intermolecular interactions that may involve energy transfer, electron transfer, and coupling with plasmonic and photonic structures. Therefore, a brief description of some of the fundamental concepts in photoscience is included in the text to stimulate the interest of the novice in the field.

In the first two sections, photophysical and toxicological properties of well-known fluorophores and emerging fluorescent materials are presented focusing their use in the study of metabolic processes and monitoring of drug delivery. The following section reviews the use of fluorescence in pharmaceutical applications. Next, we describe the use of fluorescence-based methods for the characterization of human hair and skin structures and fluorescent colors as additives in cosmetic formulations. Finally, some perspectives and challenges associated to the use of fluorescence technology in these fields are discussed.

1 Fluorophores

The use of fluorescence in biosensing and bioimaging depends on the excitation of a given fluorophore and on the detection of the signal that ultimately results from this process. The intrinsic properties of a fluorophore determine its practical utility [1–16]. The following section briefly presents some of the structural, physicochemical, and photophysical properties of fluorophores widely used in biological and biochemical applications. A brief explanation of the concepts related to the application of fluorophores is given in Box 1; detailed information can be found elsewhere [17–21].

> **Box 1 Fundamental Photophysical Concepts and Keywords [22]**
> Fundamental parameters of light *absorption* in solution are the maximum absorption wavelength (λ^{ABS}), which corresponds to the energy difference between the ground state and higher energy levels, and the *molar absorption coefficient* (ε) that is related to the one-photon cross section (σ_1) at a given wavelength and correlates the absorption of the solution with the solute concentration through the Beer-Lambert-Bouguer law. Some fluorophores

(continued)

Box 1 (continued)

can be excited by absorbing one- or two-photons; in *two-photon absorption the cross section*, σ_2, is often expressed in units of Goeppert-Mayer (GM). Several fluorophores are *anisotropic*, meaning that their excitation with polarized light is orientation-dependent. For typical fluorescent molecules such as fluorescein, the polarization is not observed if the molecule is rotating rapidly in solution under conditions typically used in biological assays. Molecular *fluorescence* in solution implies light emission resulting from the deactivation of a singlet excited state, usually S_1, which was previously populated upon light absorption by a given species. The wavelength of emission (λ^{EM}) is lower in energy (i.e., longer, red-shifted, bathochromically shifted) than the excitation wavelength (λ^{EX}) due to energy losses by intra- and intermolecular processes such as solvent and internal reorganization. The difference between the spectral positions of the absorption band maxima and fluorescence arising from the same electronic transition is called *Stokes shift*. When the Stokes shift is small, the degree of superposition of the absorption and emission spectra is high; therefore, the fluorophore is more susceptible to self-quenching *via* energy transfer. In solids, *excitons* are generated upon the absorption of light, and electron-hole recombination leads to fluorescence.

The *lifetime of the singlet excited state* (τ, also called *fluorescence lifetime*) depends on internal and environmental factors, ranging from roughly 0.1 to 100 ns. Physical and chemical processes occurring within this time range (or even faster, down to the sub-femtosecond regime) can be monitored by *time-resolved measurements* using modern pulsed lasers. Shorter lifetimes of singlet states are often related to non-radiative deactivation processes, including the population of *triplet states* via *intersystem crossing*. The formation of triplet excited states may have important biological implications besides their well-known role converting molecular oxygen into highly reactive singlet oxygen ($^1O_2$). The ratio between the number of photons absorbed and emitted is a measure of the *fluorescence quantum yield* (ϕ_{FL}) of a given species. The ϕ_{FL} is related to the fluorescence lifetime, and the $\phi_{FL} \times \varepsilon$ product is proportional to the *brightness* of a fluorophore; the latter has been used as a figure of merit for the applicability of *fluorescent probes*. *Energy transfer* between an excited species and a ground-state species can occur even when they are separated by distances considerably exceeding the sum of their van der Waals radii when, among other factors, the emission spectra of the excited species (energy donor) superimpose the absorption spectra of the energy acceptor in the ground state. This process is known as *Förster resonance energy transfer (FRET)* and has been used to determine the relative position and dynamics of two fluorophores in complex systems. Although several texts refer to this phenomenon as *fluorescence* resonance energy transfer, this use is discouraged by IUPAC because no fluorescence is involved in the energy transfer. Singlet excited states can be deactivated non-radiatively by electron transfer, a phenomenon called *photoinduced electron transfer* (PeT).

1.1 Small-Molecule Fluorophores

1.1.1 Endogenous Fluorophores

Living organisms produce several fluorescent metabolites [5]. Collectively, these endogenous fluorophores (also referred to as *intrinsic fluorophores*) can give rise to *autofluorescence*, a phenomenon that can obfuscate the fluorescence of exogenous fluorophores in cellulo and in vivo but also can be used for bioimaging and in diagnostic applications [12, 23–26]. Examples of naturally occurring fluorophores include aromatic amino acids, enzyme cofactors, vitamins, porphyrins, and proteins (Fig. 1). Normal skin exhibits substantial autofluorescence, which is originated from fluorophores distributed throughout its different layers [27]. The most intense emission originates from the dermis and subcutaneous fat epidermis, but the epidermis and the *stratum corneum* are also fluorescent [10, 12, 14, 28, 29]. Dyes absorbing red and near-infrared (NIR) light circumvent background problems resulting from autofluorescence while benefiting from deeper tissue penetration compared to UV light [30]. Furthermore, long-wavelength excitation prevents photoinduced DNA damage (λ^{EX} ~260 nm; ε from 7000 to 15,000 L mol^{-1} cm^{-1}) [5, 12, 14, 30–37].

1.1.2 Exogenous Fluorophores

Fluorescent dyes are small in size and can be tailored to exhibit high brightness and photostability, as well as narrow bandwidth relative to fluorescent proteins [1, 2, 11, 15, 38]. They can be designed to be membrane permeable to illuminate intracellular milieu or membrane impermeable to report on extracellular species, although controlling localization is not necessarily simple [15]. The utility of these fluorophores is determined by their physicochemical and photophysical properties such as acidity and basicity (pK_a and/or pK_{aH}), lipophilicity (e.g., log P and log D), chemical stability, size, ease of functionalization and complexation, fluorescence quantum yield, lifetime of the singlet excited state, absorption and emission wavelengths, photostability, and nonlinear optical properties. A didactic description of the properties of cationic organic dyes was presented by Scaiano and coauthors [39]. Extensive description of the use of fluorophores as labels for biomolecules (Fig. 2a), enzyme substrates (Fig. 2b), environmental indicators (Fig. 2c), and cellular stains (Fig. 2d) is available in the literature [5].

The study of endogenous fluorophores led to the development of nonnatural fluorescent amino acids that can be conveniently incorporated into proteins, enabling their labeling in live cells [15]. This strategy addresses the concern that the fusion of bulky fluorescent proteins to other biological structures causes functional perturbation [40]. Several fluorescent amino acids, e.g., L-(7-hydroxycoumarin-4-yl)-ethylglycine [41, 42], were incorporated at amber codons by evolving orthogonal aminoacyl-tRNA synthetases to specifically recognize the fluorescent amino acid [43]. This approach has also been used to introduce a new fluorophore into cyan

Fig. 1 Excitation (**a**) and emission spectra (**b**) of the principal endogenous fluorophores. The best relative excitation/emission conditions have been considered. Reproduced from Ref. [12]. Copyright © 2005 Elsevier Inc.

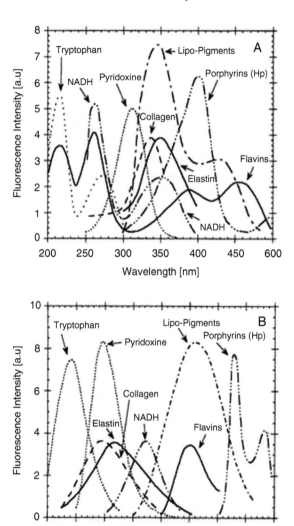

fluorescent protein, altering its spectral properties by inducing FRET between the endogenous and exogenous fluorophores [44].

Polycyclic aromatic compounds have been widely used as exogenous fluorescent dyes. For example, the naphthalene derivative 5-((2-aminoethyl)amino)naphthalene-1-sulfonic acid (EDANS; Lucifer yellow, $\lambda^{Abs} = 336$ nm; $\lambda^{EM} = 520$ nm, $\varepsilon = 6100$ L mol^{-1} cm^{-1}, $\phi_{FL} = 0.27$ in water) [45] is used in biochemical applications, particularly in FRET-based experiments [46] including molecular beacons [47]. Pyrene-derived molecules also find use as probes, and their environmental sensitivity can be used to report on RNA folding [48]. Pyrene exhibits a long-lived excited state ($\tau \sim 100$ ns), allowing excimer formation ($\lambda^{EM} = 490$ nm) that can be used to measure biomolecular processes, such as protein conformation [49].

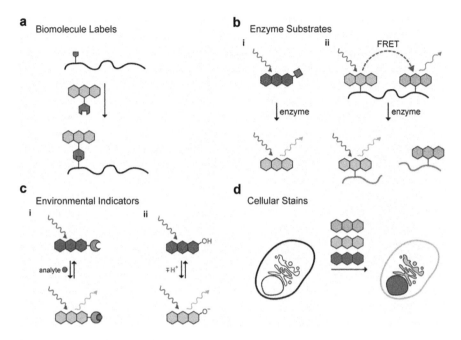

Fig. 2 Use of fluorophores as labels for biomolecules (**a**), enzyme substrates (**b**), environmental indicators (**c**), and cellular stains (**d**). (**a**) Site-specific labeling of a biomolecule by an orthogonal reaction between two functional groups (red). (**b**) Enzyme substrates: (i) enzyme-catalyzed removal of a blocking group (red) elicits a change in fluorescence; (ii) enzyme catalyzes the cleavage of a labeled biomolecule (red) and concomitant decrease in FRET. (**c**) Environmental indicators: (i) binding of an analyte (red) elicits a change in fluorescence; (ii) protonation of a fluorophore elicits a change in fluorescence. (**d**) Staining of subcellular domains by distinct fluorophores. Adapted from reference [5]. Copyright © 2008 American Chemical Society

Molecular probes built on the coumarin scaffold include useful biomolecular labels. Various derivatives of 7-hydroxycoumarin can be used to assay an assortment of hydrolases [50, 51] and dealkylases [52]. Peptidyl derivatives of 7-amino-4-methylcoumarin (AMC) are widely used to measure protease activity [53]. Microarrays of coumarin substrates have been built to examine protease specificities [54]. AMC has also been elaborated to prepare substrates for other enzymes including deacetylases [55] and esterases [56].

Fluorescein and its derivatives remains the most widely utilized fluorophore class in modern biochemical, biological, and medicinal research. Fluorescein can exist in seven prototropic forms, serving as a scaffold for preparing indicator molecules. The most biologically relevant forms are the monoanion and the dianion ($pK_a = 6.4$) [57], the latter form being more fluorescent ($\lambda^{Abs} = 490$ nm, $\lambda^{EM} = 514$ nm, $\varepsilon = 93,000$ L mol^{-1} cm^{-1}, $\phi_{FL} = 0.95$) [17]. The structure of fluorescein can be modified further to tune properties such as pK_a or excitation/emission wavelengths. For example, 2,7-difluoro-fluorescein (i.e., Oregon green) is more acidic ($pK_a = 4.6$) and more photostable than fluorescein but maintains fluorescein-like wavelengths [58]. 2,4,7,7-Tetrachlorofluorescein (TET) exhibits a $\lambda^{Abs}/\lambda^{EM}$ of 521/536 nm [15] and has been used as a traditional automated

DNA sequencing dye. Appending fluorescein with various chelating moieties affords sensors for biologically relevant ions. For example, the calcium indicator Fluo-3 developed by Tsien and coworkers [59] can be used to measure calcium ion fluxes in live cells and is employed widely in high-throughput screening [60].

Rhodamines are isologues of fluorescein. Rhodamine 110 (Rh110) has been used to build useful caspase substrates to assay apoptosis [61]. Rh110-based substrates have also been developed for phosphatases [62], esterases [63], and metal-ion catalysis in a cellular context [64]. Other rhodamine derivatives have been assembled to detect reactive oxygen species in cells [65]. Rhodamines containing rigid julolidine ring systems show higher quantum yields than more flexible analogs [66] and exhibit longer excitation and emission wavelengths [5]. Sulforhodamine 101 (SRh101) is a julolidine-based dye that is common in bioresearch. Amine-reactive sulfonyl chloride derivatives of SRh101 are sold under the trademark Texas red [15]. Rhodamine labels are often paired with fluorescein derivatives for FRET-based experiments [15] and have proven useful for DNA sequencing. Hybrid structures between fluorescein and rhodamine are called "rhodols" and have been used to build ion indicators [67]. A noteworthy addition to the suite of small-molecule fluorophores is the NIR silicon-rhodamine dyes [68] and their extension to DNA stains [69] and cytoskeletal stains [70] with improved spectral properties for long-term and in vivo imaging. Another class of fluorogenic molecules consists of the carbofluoresceins and carborhodamines [71], which have been used for super-resolution microscopy applications [5].

The boron difluoride dipyrromethene (BODIPY) dye structure has been used to build a variety of useful fluorescent labels and probes [72, 73]. This dye class features insensitivity of the spectral properties to environment, small Stokes shift, and considerable lipophilicity [74]. The simplest BODIPY scaffold shows fluorescein-like parameters with $\lambda^{Abs} = 505$ nm, $\lambda^{EM} = 511$ nm, $\varepsilon = 91{,}000$ L mol^{-1} cm^{-1}, and $\phi_{FL} = 0.94$ and is commonly called BODIPY-FL [75]. BODIPYs are particularly useful labels for fluorescence polarization techniques [76] and their small Stokes shift causes efficient self-quenching of overlabeled biomolecules. Therefore, BODIPY dyes have been used to create protease substrates, because proteolysis of densely labeled proteins leads to an increase in fluorescence intensity [77].

Cyanine dyes having associated polymethine structures are useful as labels [78], DNA stains [79], and membrane potential sensors [80–82]. CyDyes have a sulfoindocyanine structure and are useful biomolecular labels being the standard fluorophores for microarrays [83–85]. These compounds are given common names according to the number of carbon atoms between the dihydroindole units, e.g., Cy3 ($\lambda^{Abs} = 554$, $\lambda^{EM} = 568$ nm, $\varepsilon = 1.3 \times 10^5$ L mol^{-1} cm^{-1}, and $\phi_{FL} = 0.14$ in water), Cy5 ($\lambda^{Abs} = 652$ nm, $\lambda^{EM} = 672$ nm, $\varepsilon = 2.0 \times 10^5$ L mol^{-1} cm^{-1}, and $\phi_{FL} = 0.18$ in water), and Cy7 ($\lambda^{Abs} = 755$ nm, $\lambda^{EM} = 788$ nm, and ϕ_{FL} 0.02 in water) [5, 15, 84]. The introduction of a fused benzo ring in the dihydroindole moieties elicits a bathochromic shift of 20–30 nm [83], and the resulting probes are designated with a ".5" suffix (e.g., Cy5.5). The CyDye pairs are also often used for FRET experiments [86] and can be utilized as photoswitchable probes for ultrahigh-resolution imaging [87]. Another example of cyanine dye is indocyanine green (ICG), which is used in medical diagnostics [88–90]. The structure and brightness of the most important classes of fluorescent dyes are presented in Fig. 3 [5].

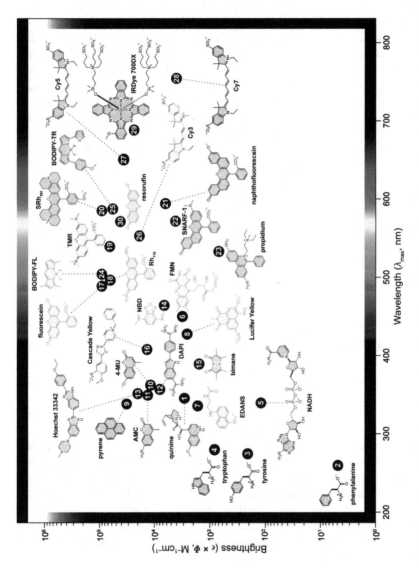

Fig. 3 Structure, brightness, and wavelength of maximum absorption of the major classes of fluorophores. The color of the structure indicates its wavelength of maximum emission. For clarity, only the fluorophoric moiety of some molecules is shown. Reprinted from Ref. [5]. Copyright © 2008 American Chemical Society

1.2   Fluorescent Proteins

The two most frequently cited limitations of organic fluorophores for live-cell imaging are the lack of molecular specificity, which requires sophisticated approaches for attaching fluorophores to biomolecules, and the challenge of ensuring cell permeability [91]. Due to their genetic encodability, the use of fluorescent proteins (FPs) as fluorophores circumvents both these limitations. Since the 1990s, FPs have been subjected to extensive protein engineering to tune their optical properties (wavelength range, brightness, and photostability), as well as biochemical properties (rate of protein folding and chromophore maturation, pK_a, oligomerization state, robustness in fusions and in oxidizing environments) [15]. Among fluorescent proteins, the most notorious example is the green fluorescent protein (GFP), which was first isolated from the jellyfish *Aequorea victoria* and became a landmark in the development of fluorescent tools for cell and molecular biology [4]. The cyclization of the tripeptide Ser65-Tyr66-Gly67, in this 238-amino acid protein (26.9 kDa), produces a 4-(p-hydroxybenzylidene)imidazolidin-5-one (HBI) fluorophore that emits green light ($\lambda^{EM} = 509$ nm) upon excitation at 395/475 nm ($\phi_{FL} = 0.79$) [4, 92]. The internal fluorophore is formed without requiring any accessory cofactors, gene products, or enzymes/substrates other than molecular oxygen and is formed in a tightly packed proteic barrel that excludes solvent molecules, preventing fluorescence quenching by water [4]. The GFP gene is frequently used as a reporter of expression and has been extensively modified to make biosensors [15].

1.3   Fluorescent Nanostructured Materials

Fluorescent nanostructured materials have emerged as important fluorophores that often circumvent some limitations of small-molecule fluorophores such as low photostability [14, 93–108]. The first class of widely used fluorescent nanostructured materials was the semiconductor quantum dots (QDs), which are colloidal semiconductor nanocrystals with dimensions between about 1 and 10 nm [3, 9, 14, 100, 109–111]. Excitons are generated in the nanocrystals upon the absorption of light, and electron-hole recombination leads to fluorescence. Since then, several new materials have been developed; examples include nanodiamonds (NDs), carbon nanodots (C-dots), graphene oxide (GO), carbon nanotubes (CNTs), lanthanide-based upconversion nanoparticles (UCNPs), and fluorescent dye-doped silica nanoparticles (DSNPs) [29]. Each class of fluorophore has its own benefits and liabilities and may fit an application's need better than the other.

Depending on its size, each QD can comprise hundreds to thousands of atoms, a large fraction (10%) of which are located at the nanocrystal surface, i.e., QDs have high surface area-to-volume ratio. Most of the QDs used in analytical applications are synthesized as core/shell structures, where the core nanocrystal is overcoated with another semiconductor material to protect and improve its optical properties. The most studied QD material is core/shell CdSe/ZnS, which first generated excitement

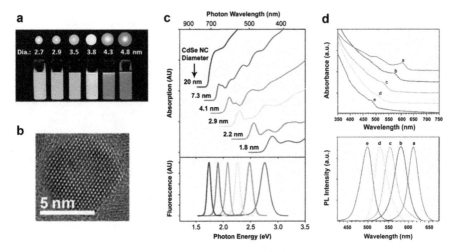

Fig. 4 (**a**) Size-tunable PL of CdSe QDs. The photograph was taken under UV illumination (365 nm). (**b**) Transmission electron microscopy image of a CdSe/ZnS QD. Reprinted with permission from Ref. [119]. Copyright © 2011, American Chemical Society (**c**) Size-dependent absorption and fluorescence spectra of CdSe QDs. Reprinted with permission from Ref. [120]. Copyright © 2010, American Chemical Society. (**d**) Absorption and PL spectra of $Zn_xCd_{1-x}Se$ QDs with Zn mole fractions of (a) $x = 0$, (b) 0.28, (c) 0.44, (d) 0.55, and (e) 0.67. Reprinted with permission from Ref. [121]. Copyright © 2003, American Chemical Society

for biological imaging and analysis. Under UV illumination, differentially sized QDs show a bright rainbow of photoluminescence (Fig. 4a, b) due to quantum confinement. The bright photoluminescence is the result of high quantum yields ($\phi = 0.1$–0.9) combined with large molar extinction coefficients (ε between 10^5 and 10^7 L mol^{-1} cm^{-1}) [3, 14, 93, 112]. As shown in Fig. 4c, d, QDs have broad absorption spectra that continuously increase in magnitude from their first exciton peak to shorter wavelengths in the near-UV. The excited-state lifetimes of QD tend to be longer than those of fluorescent dyes (10 ns), and these nanocrystals usually show superior resistance to photobleaching and chemical degradation and higher two-photon absorption cross sections (10^3–10^4 GM) compared to molecular fluorescent dyes [113]. Consequently, QDs have been used as probes for tracking dynamic processes over time and for two-photon imaging of tissues or other complex biological specimens where NIR excitation mitigates challenges associated with autofluorescence and attenuation of excitation light by strong protein absorbance (e.g., hemoglobin) in the visible region [114, 115]. The *blinking* or fluorescence intermittency of QDs can be observed at the single particle level. Although detrimental in some applications of QDs, the observation of blinking is useful to confirm tracking of a single QD [116, 117], and it has enabled super-resolution imaging [14, 118]. In the following sections, we denote the peak photoluminescence wavelength of QDs by using a subscript number (e.g., QD$_{525}$), as presented elsewhere [14]; if no material is mentioned explicitly, the reader should assume the QD material is CdSe/ZnS.

Proper fluorophore choice allows the imaging of single molecules, biomolecules, organelles, subcellular structures, cells, and tissues in vivo. The properties of several

Organic fluorophores	
Material	Molecular dyes, proteins
Size	< 4.0 nm
Functionalizable	Yes
Quantum yield	up to 1.0
FWHM	Broad, red-tailed
Photostability	Variable, generally good
Lifetime	< 10 ns
Multiplexing	Yes
2PE	Yes

Dye-doped NPs	
Material	Silica
Size	2 – 200 nm
Functionalizable	Yes
Quantum yield	up to 1.0
FWHM	Broad, red-tailed
Photostability	Very good
Lifetime	< 10 ns
Multiplexing	Yes
2PE	Yes

Graphene oxide	
Material	Carbon
Size	ca. 0.6 nm vs. length
Functionalizable	Yes
Quantum yield	up to 0.1
FWHM	> 80 nm
Photostability	Excellent
Lifetime	< 10 ns
Multiplexing	Yes
2PE	Yes

Carbon nanotubes	
Material	Carbon
Size	diameter 0.3 nm vs. length
Functionalizable	Yes
Quantum yield	< 0.25
FWHM	> 60 nm
Photostability	Excellent
Lifetime	< 5 ns
Multiplexing	Yes
2PE	Yes

C-dots	
Material	Carbon
Size	< 10 nm
Functionalizable	Yes
Quantum yield	up to 0.6
FWHM	> 60 nm
Photostability	Excellent
Lifetime	< 10 ns
Multiplexing	Yes
2PE	Yes

Nanodiamonds	
Material	Carbon
Size	5 – 20 nm
Functionalizable	Yes
Quantum yield	up to 1.0
FWHM	> 60 nm
Photostability	Excellent
Lifetime	10 – 20 ns
Multiplexing	Yes
2PE	Yes

UCNPs	
Material	Lantanide-doped matrix
Size	20 – 50 nm
Functionalizable	Yes
Quantum yield	up to 1.0
FWHM	< 15 nm
Photostability	Excellent
Lifetime	> 100 ms
Multiplexing	Yes
2PE	Yes

Quantum dots	
Material	Semiconductor
Size	2 – 10 nm
Functionalizable	Yes
Quantum yield	up to 0.9
FWHM	25 – 35 nm
Photostability	Very good
Lifetime	> 10 ns
Multiplexing	Yes
2PE	Yes

Metal nanoclusters	
Material	Usually gold and silver
Size	< 2.0 nm
Functionalizable	Yes
Quantum yield	< 0.2
FWHM	> 60 nm
Photostability	Excellent
Lifetime	> 100 ms
Multiplexing	Yes
2PE	Yes

Fig. 5 Comparison of the physicochemical and photophysical properties of several kinds of fluorophores

classes of fluorophores are compared in Fig. 5 [14]. Organic fluorophores are usually mono-reactive, but more prone to photobleaching and less facile multiplexing, i.e., simultaneous measure of multiple analytes [15, 93, 122–125]. Fluorescent dye-doped silica nanoparticles are very bright and have good resistance to photobleaching, despite being much larger in size and less facile multiplexing compared to QDs [126]. Nanodiamonds have excellent resistance to photobleaching and very high quantum yield, but its emission wavelength is not easily tunable, and its molar absorption coefficient is low [127]. Graphene oxide that is resistant to photobleaching has intrinsic aqueous solubility, but its broad photoluminescence emission is not easily tuned [128]. C-dots are non-blinking, photostable small nanoparticles; interestingly, the emission wavelength depends on the excitation wavelength for reasons that are still poorly understood [129]. Carbon nanotubes show excellent resistance to photobleaching and NIR emission, but the obtaining of pure samples is challenging, and their intensity of photoluminescence is weak [130]. Lanthanide-based upconversion nanoparticles show multiple narrow emission lines and are larger in size than other nanostructured fluorophores [131]. Noble metal clusters that contain between a few and a hundred atoms, e.g., Au, Ag, and Pt, are smaller than 2 nm, exhibit no apparent plasmonic properties, and have excitation and emission bands similar to those of molecular dyes [132]. NCs have been used for biological imaging and analysis [14, 131, 132].

Castellano and coauthors developed a promising sensitizer based on the combination of quantum dots and polyaromatic molecular fluorophores [133]. Cadmium selenide semiconductor nanoparticles, selectively excited by green light, engage in interfacial Dexter-like triplet-triplet energy transfer with surface-anchored polyaromatic carboxylic acid acceptors, extending the excited-state lifetime by six orders of magnitude. Net triplet energy transfer also occurs from surface acceptors to freely diffusing molecular solutes, further extending the lifetime while sensitizing singlet oxygen in an aerated solution. The successful translation of triplet excitons from semiconductor nanoparticles to the bulk solution implies that such materials are generally effective surrogates for molecular triplets, a phenomenon relevant to optoelectronics, solar energy conversion, and photobiology [133].

2 Toxicology of Selected Fluorophores

Fluorophores have been extensively used to study biological and biochemical processes in vivo. Examples include the staining of tumors for the diagnoses and surgical treatment of cancer [3, 12, 14–16, 33, 34, 134–137]. However, fluorophores can act as toxicants, and the factors influencing their chemical toxicity must be investigated [138]. The most effective and appropriate fluorophores for pharmacological, clinical, and medical use will be determined by their fluorescent characteristics and safety. Fluorophores used for these applications include Alexa Fluor dyes (e.g., Alexa Fluor 488 and 514), BODIPY dyes (e.g., BODIPY-FL, 8-phenyl BODIPY, and BODIPY R6G), Cy dyes (e.g., Cy 5.5 and Cy 7), cypate, fluorescein, indocyanine green, Oregon green, rhodamine dyes (rhodamine 110, rhodamine 6G, and rhodamine X), rhodol, TAMRA, Texas red, and Tokyo green (Fig. 3). Although the photophysical properties of most conventional fluorescent dyes and pigments are readily available [5], detailed information on the cytotoxicity, tissue toxicity, in vivo toxicity, carcinogenicity, and mutagenicity of non-FDA-approved fluorophores is limited [138]. Fortunately, in most cases, the amounts of fluorophore required for the in vivo bioanalysis and bioimaging are usually much lower than the toxic doses reported in the literature.

The FDA approved the use of indocyanine green (ICG) as a contrast agent for retinal angiography in the late 1950s. ICG toxicity in cells, tissues, animals, and humans is well studied; however, specific information on the mutagenicity of ICG is still not available. The typical clinical dose of ICG is 3.5 μg kg^{-1}, and the maximum intravenous dose for humans is 5 mg kg^{-1}; these doses are at least 10,000 times lower than the lowest corresponding LC_{50} values reported [139]. At a concentration of 2.5 mg mL^{-1}, ICG induces apoptosis in the MIO-M1 cultured human retinal glial cells through both reduction of thymidine incorporation and induction of the caspase cascade [140]. At lower concentration, ICG (5 mg mL^{-1}) is toxic to cultured lens epithelial cells [139]. In liver cells, ICG accumulates intracellularly, decreases dehydrogenase activity in the mitochondria [141–143], and inhibits mitochondrial oxygen consumption [144, 145]. ICG (0.25 mg mL^{-1}) causes the upregulation of the cell cycle arrest protein p21 as well as cell cycle arrest genes Bax and p53 in cultured human retinal pigment epithelial cells, ARPE19, exposed to visible light for 15 min [146].

ICG has photodynamic toxicities on the inner retina when exposed to light [139, 147]. It has been shown that the toxicity of ICG increases with the duration of light exposure, increasing the rate of DNA synthesis [143]. A dose of 0.25 mg of ICG administered into rabbit eyes causes the loss of photoreceptor outer segments after 10 min of endoillumination [148]. Photodynamic toxicity has also been attributed to the production of a low level of reactive oxygen species [149] and has been considered for use in laser-induced photodynamic therapy against infectious agents [150, 151] and cancer cells [152]. The use of ICG during macular surgery is not recommended because of its toxicity to the retina. There have been at least two reported deaths owing to ICG injection resulting from anaphylactic reactions and cardiorespiratory arrest, and the estimated death rate is 0.0003% [153].

Fluorescein was approved by the FDA in 1976 as a contrast agent for angiography. In excised embryonic chicken retina, a concentration of 0.2 mg mL^{-1} fluorescein inhibits neuron growth [154]. Fluorescein may irritate the nerve roots when injected intrathecally (within the spinal cord) and affects the central nervous system [155, 156]. The typical clinical dose ranges from 7.1 to 10 mg kg^{-1}, with 30 mg kg^{-1} used for the examination of the anterior segment of the eye [138, 157, 158]. Fluorescein may be a developmental toxicant, which has been shown to cross the placenta in rats and humans. Adverse reactions to intravenous fluorescein occur in ca. 5% of patients [159]. The primary route of excretion is the kidney [138]. Cardiac arrest and death are very rare; the mortality rate is 0.0005% [153, 160].

Rhodamine 6G (Rh6G) is a non-FDA-approved fluorophore that has demonstrated mutagenicity and toxicity in cells, tissues, and organisms. Rh6G is cytotoxic to Friend leukemia cells and doxorubicin-resistant variant cells, ARN15, in culture [161]. Rh6G is also lethal to *Salmonella* TA98 and TA100 [162]. In mice, Rh6G accumulates in mitochondria and disrupts oxidative phosphorylation and ATP synthesis by inhibiting the function of ATPase and the processing of the mitochondrial matrix peptides [163]. At an intravenous dose of 0.5 mg/kg/day for 4 days, ATP synthesis in pregnant mice was reduced by almost half. At a concentration of 10 μg dye mg^{-1} protein, 86% of ATPase production in the mice was inhibited [164]. Studies suggest that Rh6G is a mutagen, carcinogen and teratogen at high concentrations [138, 165]. Limited toxicity information is available for rhodamine 110 (Rh110), also commonly known as rhodamine green. Rh110 accumulates in the mitochondria in human lymphoblastoid cells but shows no cytotoxic effects at concentrations up to 10 μmol L^{-1} [166].

Texas red alters the physicochemical characteristics of bovine serum albumin (BSA) [167]. Conjugates of Texas red/BSA have shown lower phototoxic effects in rat erythrocytes exposed to epiillumination compared to BSA conjugates of FITC and BODIPY-FL [168]. A BODIPY-verapamil conjugate (>10 μmol L^{-1}) is more toxic to multidrug resistant human KB carcinoma cell lines in culture than verapamil alone [169].

3   Pharmaceutics

Pharmaceutics deals with new chemical entities, active pharmaceutical ingredients, and drugs. Fluorescence has been used in combination with other analytical methods to characterize the quality of bulk drug materials by setting limits for their active ingredient content [33, 170]. However, since spectrophotometric methods are, a priori, non-specific, they have been less prescribed for the assay of bulk drug materials in pharmacopeial monographs (e.g., *US Pharmacopeia* (USP) and *European Pharmacopoeia* (Ph. Eur.)) than chromatographic and titration methods (Table 1) [33].

This situation is slowly changing, because fluorescent biosensor technology reached a point where many useful labels and reporters are commercially available and powerful equipments are now more affordable. Fluorescence is used for labeling specific cell compartments and components and to report on biological activities

Table 1 Proportion of various analytical methods prescribed for the assay of bulk drug materials in Ph. Eur. 4 and USP XXVII [33]

Method	Ph. Eur. 4 (%)	USP 27 (%)
HPLC	15.5	44
GC	2.0	2.5
Titration	69.5	40.5
Acid-base	57.5	29.5
Aqueous mixtures	21	5.5
Indicator	6.5	4.5
Potentiometric	14.5	1.0
Nonaqueous	36.5	24
Indicator	9.5	14
Potentiometric	27	10
Redox (iodometry, nitritometry, etc.)	6.5	5.5
Other (complexometry, argentometry, etc.)	5.5	5.5
UV-Vis spectrophotometry	9.5	8.5
Microbiological assay (antibiotics)	3.0	2.5
Other (IR, NMR, polarimetry, *fluorimetry*, atomic absorption spectroscopy, polarography, gravimetry, etc.)	0.5	2.0

or environmental changes (e.g., pH, ion concentration, electrical potential) [136]. Immunofluorescent probes can be directed to bind not only to specific proteins but also to specific conformations, cleavage products, or site modifications, such as phosphorylation and SUMOylation [171]. Individual peptides and proteins can be engineered to autofluorescence by expressing them as *Aequorea*-derived fluorescent protein (AFP) chimeras inside cells resulting in fluorescent proteins [14, 172, 173] and transcriptional reporters [174–178]. RNA riboswitches have inspired a new class of small-molecule and metabolite sensors that rely on fluorescence modulation upon interaction with an RNA aptamer, often based on the design of the Spinach aptamer [179–184].

Such high levels of specificity enable the use of several different fluorophores to investigate the complex interactions that occur among and between subcellular constituents using miniaturized [185, 186] and batch platforms. Modern fluorescent-based imaging techniques allow one to peer inside cells to visualize, track, and quantify molecules, ions, proteins, nucleic acids, and biochemical reactions [15]. The choice of the specific technique and fluorophore (or even class of fluorophore) depends on several factors, including the physicochemical and photophysical properties of the fluorophore, the nature of the experiment, and the potential for perturbation of the dye and the cellular state of interest [14, 15, 31, 34, 187].

3.1 Biosensing

In bioanalytical applications, the great advantage of FRET is the ability to turn fluorescence "on" or "off" in response to biorecognition events (e.g., ligand-receptor binding, enzyme activity, DNA hybridization) or other physicochemical stimuli (e.g., pH change) [15, 75, 122, 188]. Since measured signals are not strictly based on the accumulation of fluorophores, FRET methods can be applied in the ensemble and down to the level of single particles. Numerous configurations using different classes of fluorophores and FRET have been reported for the detection of metal ions [14, 189–191], small molecules [3, 122, 186, 192, 193], toxins and toxicants [3, 14, 190, 194, 195], drugs [73, 122, 135, 137, 196], proteases [192, 195, 197–199] and nucleases [192, 197, 198, 200], hybridization assays [189, 190, 192, 201, 202], immunoassays [122, 198, 203], and pH sensing [75, 187, 204, 205]. Examples are summarized in Table 2.

The oldest and most used family of analyte sensors encompasses small-molecule and genetically encoded calcium indicators (GECIs). Ca^{2+} is a ubiquitous second messenger that serves as a key node in many signaling pathways. Its intricate spatial distribution and exquisitely regulated dynamics have long lured researchers into developing ever more sophisticated tools for monitoring the spatial and temporal patterns of Ca^{2+} signals from living cells to whole organisms. Since the introduction of synthetic Ca^{2+} indicators in the 1980s and genetically encoded indicators in the

Table 2 Selected examples of tools to monitor cellular dynamics by FRET-based detection

Monitored item	Name/description of tool	Type of detection	GE	Key advantages/ disadvantages
Ca^{2+}	GCamPs/GECIs	Intensity change (single FP sensor) or FRET-based	Yes	Plus: permits long-term measurement of Ca^{2+} transients in transgenic organisms
				Minus: may perturb endogenous calcium dynamics
	Small-molecule Ca^{2+} indicators	Intensity change or spectral shift in dye	No	Plus: fast response time
				Minus: intracellular concentration of dye can be very high
Kinase activity	KTRs	Change in localization (nucleus versus cytoplasm) of single FP reporter	Yes	Plus: multiplexing for detection of up to 4 different kinase activities possible
Voltage change across a membrane	Genetically encoded voltage sensors (based on conformational or photophysical change of sensing domain)	Intensity change (single FP sensor) or FRET-based	Yes	Plus: targeting to small pool of neurons in live animals possible
				Minus: slow response times compared to small-molecule probes
	Small-molecule-based voltage sensors	Intensity change or FRET-based	No	Minus: delivery to membrane is difficult; partitioning in other membranes likely
Cell cycle stages	Fucci (fluorescent, ubiquitination-based cell cycle indicator)	Cell cycle state-dependent degradation of FP-reporters (green/red color change at M–G1 transition, yellow at G1–S transition)	Yes	Minus: requires delivery of two reporters
	CDK2-based localization change	Change in localization (nucleus versus cytoplasm) of single FP reporter	Yes	Plus: single color allowing it to be multiplexed with other probes
NADH/ NAD^+	SoNar	Conformational change of single FP fused to NADH/NAD^+-sensing domain	Yes	Plus: insensitive to changes in pH; ratiometric
Molecular crowding	GimRET	FRET-based	Yes	Plus: can be targeted to different organelles to monitor crowding

Reproduced from Ref. [15]
GE genetic encoding

1990s, probes have been subjected to intense iterative optimization. Although synthetic dyes remained the gold standard for many years with respect to dynamic range, response kinetics, and indicator linearity, GECIs have recently come of age and surpass synthetic dyes for many applications [194]. However, a recent comparison of small-molecule Ca^{2+} indicators and GCamP6 sensors revealed the superiority of small-molecule dyes, particularly Cal-520 and Rhod-4, for monitoring local Ca^{2+} signals such as Ca^{2+} puffs induced by IP3-mediated release of Ca^{2+} from the endoplasmic reticulum [206].

The two most common classes of GECIs are the single FP-based indicators (GCamP platform), which are intensity-based and rely on collection of a single fluorescence channel, and FRET-based indicators, which are ratiometric and require dual-channel recording [194, 207–209]. Both platforms were subjected to extensive iterative optimization to improve performance metrics [194, 208, 210]. A major breakthrough in GECI performance came from insights derived from the analysis of crystal structures [211, 212] and large-scale mutagenesis and screening strategies [213–215]. Screening and functional assessment upon electrical stimulation in neurons was particularly valuable for identifying sensors with improved response kinetics and amplitudes [213]. A cell-free biomimetic sensing platform using virus-like particles (VLPs) with intact ligand-gated Ca^{2+} channels and GECI was reported [216]. The best in class GECIs are now sufficiently sensitive to permit in vivo imaging in a wide range of model systems, including worm, zebrafish, fly, mouse, and nonhuman primates [194, 207, 208, 217]. Although state-of-the-art green GCamPs permit robust measurement of Ca^{2+} transients, most notably for monitoring neuronal activity, there are some limitations for its use in vivo. In particular, poor tissue penetration of blue excitation light and spectral overlap with optogenetic tools for controlling neuronal signals have led to concerted efforts to expand the color palette of single FP sensors into the red. Whereas early generation red calcium indicators suffered from diminished performance compared to their green counterparts, sensitive indicators based on R-CamP (derived from mRuby) and R-GECO (derived from mApple) were developed [15, 209].

Single QD detection has been used to measure biomolecular interactions (e.g., DNA hybridization, DNA-protein binding, ligand-receptor binding) and to detect small molecules via two-color co-localization, FRET, and, more recently, charge transfer [218–221]. In the latter, CdSe/ZnS QD_{560} were functionalized with ferrocene-maltose binding protein (MBP) conjugates, resulting in charge transfer (CT) quenching [221]. MBP undergoes a conformational change (scissoring) upon binding maltose, resulting in an increase in the distance between the ferrocene and the QD with a corresponding increase in the QD PL. By monitoring individual QDs, it was possible to detect maltose with a dynamic range spanning a remarkable six orders of magnitude (100 pmol L^{-1} to 100 μmol L^{-1}). Interestingly, unlike single-pair FRET systems that show on-off switching of QD photoluminescence, the single QD-CT system exhibited constant emission with an increase in intensity upon binding maltose [221]. This behavior was attributed to a "gray" state associated with CT. Solution-phase single-pair FRET (spFRET) with QDs has also been used as a platform for sensitive assays. For example, Zhang and coauthors conjugated streptavidin-QD$_{605}$ with biotinylated capture probe oligonucleotides for a sandwich DNA hybridization assay, where the reporter oligonucleotides were labeled with Cy5 acceptors [222, 223]. spFRET was monitored by flowing the sample solution (mixed with QDs, capture and reporter oligonucleotides) through a glass microcapillary with laser excitation at 488 nm in a small observation volume. A fluorescence burst coincidence analysis between the donor and acceptor detection channels was used to observe FRET and measure DNA hybridization. The LOD was 4.8 fmol L^{-1} target.

Several studies were carried out using QD-fluorescence cross-correlation spectroscopy (FCCS) to monitor ligand-receptor interactions. A pair of early studies investigated the streptavidin-biotin interaction in association with QDs (i.e., QD-streptavidin and QD-biotin conjugates) [224, 225]. It was found that the streptavidin-biotin dissociation constant ($K_d = k_{off}/k_{on}$) increased from 10^{-15} mol L^{-1} with the native system to 10^{-10} mol L^{-1} in association with QDs, where the association rate (k_{on}) decreased by six orders of magnitude and the dissociation rate increased by one order of magnitude [225].

Plasmon-coupled photoluminescence provided a sevenfold enhancement in the measured photoluminescence compared with farfield excitation of PL, and this enhancement was attributed to directional emission at the surface plasmon resonance (SPR) angle (47°, 405 nm laser excitation). SPR-based DNA hybridization assays and immunoassays have been developed by using NIR-emitting QDs as reporters [226]. However, in this work, the reflectivity was used as the analytical parameter (as in traditional SPR) rather than photoluminescence. The QDs provided a 25-fold signal amplification due to a putative combination of its optical mass (large size and high dielectric constant) and coupling of its emission into propagating surface plasmons. The minimum detectable DNA concentration was 100 fmol L^{-1}. This type of experiment must rely on spatial registration (i.e., discrete spots) and SPR imaging for multiplexing and thus does not take full advantage of the optical properties of QDs, but it is interesting from the perspective of multimodal detection (SPR and PL) [14].

QDs have been found to be an excellent platform for optical encoding and multiplexed analysis due to their spectrally narrow photoluminescence and the ability to excite multiple colors of QDs by using a single laser line [227, 228]. QDs also provided excellent long-term stability (photoluminescence remained after 2–4 weeks) in addition to better sensitivity for IHC labeling of prostate stem cell antigen in specimens from prostate resections or prostatectomies [229]. Immunocytochemical (ICC) and immunohistochemical (IHC) labeling have been two of the most widespread applications of QDs in microscopy [14]. Primary or secondary antibody-QD conjugates can be used for direct and indirect staining, respectively, albeit the latter is often preferred due to the high cost of primary antibodies and the potential loss of activity upon conjugation. QD-IHC has been used for five-color molecular profiling of human prostate cancer cells [230]. Finally, high-throughput multiplexed screening assays have been developed around flow cytometry instrumentation by using optically encoded bead technology [231]. A QD-barcode-based DNA hybridization assay was used to identify optimal probe oligonucleotide and target sequence lengths for fast hybridization kinetics and good hybridization efficiency [232].

3.1.1 Allergens, Toxins, and Toxicants

Cholera, ricin, Shiga-like toxin 1, and staphylococcal enterotoxin B were detected simultaneously by using a sandwich immunoassay with QD labels [233]. The corresponding reporter antibodies were conjugated with QD_{510}, QD_{555}, QD_{590}, and QD_{610} and offered limits of detection (LODs) in the range of 3–300 ng mL^{-1}. The assay was carried out in a single microtiter plate well with excitation at 330 nm. Many similar examples of spectral multiplexing based on QD photoluminescence intensity can be found in the literature [14, 125, 199, 234, 235].

The sensitization capacity for allergic reaction was tested on mice by examining the T-cell growth with regional lymph or on guinea pigs by examining their inflammatory reaction, but new testing methods have been developed, such as the direct peptide reactivity assay (DPRA) method that examines the test substance's ability to directly bind to protein or the potency of sensitization (hCLAT) method where THP-1, a dendritic cell line, is used to determine molecule expression required for antigen presentation. Human monocyte cell line, THP-1, is treated with test compounds and stained with fluorescence-labeled antibodies for activation markers of Langerhans cells (CD40, CD86) and then analyses by flow cytometry for expression of activation markers. These new testing methods are expected to evolve with basic dermatological immunology. Stimulation response against the permeability of chemical agents was previously tested on human and animal organs, but development and production of skin models have allowed us to test on materials that are more similar to skin [29].

3.1.2 Cancer Cells and Tumors

Most NIR dyes used as contrast agents have the capability of polarity sensing; their distribution in tumor-bearing rodents has been recovered by lifetime gating. In diseases such as cancer, these probes accumulate in pathological tissues in response to physiological alterations such as enhanced blood flow and vascular permeability. While conventional molecular probes often lack specificity in pathology, contrast agents accumulate in diseased tissues upon interaction with disease-specific conditions or molecular events, thus enhancing contrast. For example, noninvasive fluorescence lifetime imaging of tumors has been achieved with receptor-targeted NIR fluorescent probes [236]. In addition, monitoring fluorescence lifetime of encapsulated NIR dyes [237] is employed to identify the biodegradability of the drug polymer carrier in vivo or the delivery of the therapeutic vehicle toward the target disease [238]. Indocyanine green (ICG) analogs are important NIR dyes, which can be attached to biomolecules, such as short peptides at designed positions, to create optical bioprobes particularly useful in cancer diagnostics [21]. In vivo and ex vivo canine mammary gland tissue was imaged by NIR with lifetime sensitive detection and localization of exogenous fluorescent contrast agents, concerning the ability of this technique to detect spontaneous mammary tumors and regional lymph nodes [21, 239].

A fluorescent protein-based competitive immunoassay was used for the simultaneous detection of two tumor markers in human serum, carcinoembryonic antigen, and α-fetoprotein, by using QD_{520} and QD_{620} [172]. Binding of the corresponding antibodies increased the effective size of the QD conjugates and increased the emission polarization. Several groups have reported labeling of fixed cells and tissue with QD-antibody conjugates. For example, Wu and coauthors demonstrated the use of QD probes for ICC labeling of different subcellular targets, including cell surface receptors (HER2), cytoskeletal components (actin and microtubules), and nuclear antigens associated with SK-BR-3 (human breast cancer) or 3T3 cells (mouse fibroblast) [240]. QD-secondary antibody conjugates, or a combination of QD-streptavidin conjugates with biotinylated secondary antibody (or phalloidin for actin), were used for labeling. Turning to IHC, caveolin-1 and proliferating cell nuclear agent were detected in a lung cancer tissue microarray [241]. Compared with conventional IHC, QD-based labeling improved detection rates from 47% to 57% for caveolin-1 and from 77% to 86% for PCNA. Similarly, QD-IHC has been shown to improve the sensitivity of quantitative detection of the HER2 breast cancer marker in formalin-fixed paraffin-embedded tissue specimens [242].

Many studies have highlighted the utility of QD-based spectral imaging for studying human cancers. Multispectral imaging and IHC labeling with QD_{605}-streptavidin conjugates (via biotinylated secondary antibodies) were used to identify three molecular markers in breast cancer tissue: HER2, estrogen receptor, and progesterone receptor [243, 244]. Five different molecular subtypes of breast cancer cell heterogeneity were identified and corresponded to different 5-year patient prognoses [244]. Spectral imaging was used to separate out tissue autofluorescence, and this separation was further aided by the brightness of QDs. In contrast, multispectral imaging with multiple colors of QD was used to reliably detect and characterize tumor cells in complex tissue microenvironments [124, 245]. For example, four protein biomarkers—CD15, CD30, CD45, and Pax5—were simultaneously mapped in lymph node biopsy specimens by using four colors of QD-secondary antibody conjugates (QD_{525}, QD_{565}, QD_{605}, and QD_{655}) [124]. This method permitted reliable identification of low-abundance (~1%) Hodgkin's and Reed-Sternberg cells; identifying such cells is essential for differentiating Hodgkin's lymphoma from non-Hodgkin's lymphoma and benign lymphoid hyperplasia [124]. In another study, four protein markers associated with prostate cancer—E-cadherin, high-molecular-weight cytokeratin (CK HMW), p63, and α-methylacyl CoA racemase (AMACR)—were quantitatively mapped in biopsy specimens [245]. As in the previous study, the biomarkers were detected using four different colors of QD-secondary antibody conjugate (QD_{565}, QD_{605}, QD_{655}, QD_{705}), revealing molecular and morphological details that are unseen with traditional staining methods [14]. One of the most important observations was that progressive changes in benign prostate glands start with a single malignant cell and ultimately lead to a malignant gland [245]. Overall, spectral imaging with QDs is an ideal methodology for tackling the challenge of tumor heterogeneity that exists at the molecular, cellular, and tissue-architecture levels, as well as between individuals. A greater understanding of tumor growth mechanisms and more rigorous classification schemes are expected to lead to more effective stage-specific and personalized treatments of cancer. Spectral imaging of fixed mouse spleen tissue sections IHC were labeled with five colors of QD-antibody conjugate (QD_{525}, QD_{565}, QD_{605}, QD_{625}, QD_{650}) [246]. B cells, T cells, leukocytes, thymocytes, and macrophages were labeled with the QDs via antibodies targeting CD45 (type C protein tyrosine phosphatase receptor), CD11c (integrin αX), CD31 (platelet endothelial cell adhesion molecule), CD4 (a glycoprotein), and CD11b (integrin αM) antigens, respectively [246].

3.1.3 Infectious Diseases and Pathogens

A flow cytometric method for rapid (10 min), sensitive (femtomole level), and parallel detection of genetic markers for the infectious diseases HIV, hepatitis B, hepatitis C, syphilis, and malaria was reported [247]. In another related study, a flow cytometric QD-barcode method was used for the simultaneous analysis of ten single-nucleotide polymorphisms [248]. Continued development of QD-barcode assays is expected to result in improved methods for sensitive, high-throughput assays for broad panels of pathogen and disease markers.

Although naturally occurring betalains are only weakly or nonfluorescent, betalain analogs with much higher fluorescence efficiencies can be obtained by conjugating betalamic acid with a highly emissive chromophore such as 7-amino-4-methylcoumarin. The resultant coumarin-containing betalain (cBeet120, Fig. 6) not only exhibits good fluorescence and high photostability but also was found to be promising as a dye for the selective live-cell fluorescence imaging of malaria parasites inside *Plasmodium*-infected erythrocytes [249]. The betalain moiety in cBeet120 extends the π-conjugation of the coumarin chromophore, resulting in a spectroscopically convenient bathochromic spectral shift of the absorption and emission. In addition, compared to the free or unconjugated coumarin, cBeet120 is quite water soluble, which facilitates the accumulation of the dye in the cells.

The tight binding of trophozoites of *Giardia lamblia* to host cells occurs by means of the ventral adhesive disc, a spiral array of microtubules and associated proteins such as giardins. It has been shown that knock down of the small ubiquitin-like modifier (SUMO) results in less adhesive trophozoites, decreased cell proliferation and deep morphological alterations, including at the ventral disc [171]. Confocal immunofluorescence microscopy was used to characterize the localization of SUMO proteins in the parasite (Fig. 7).

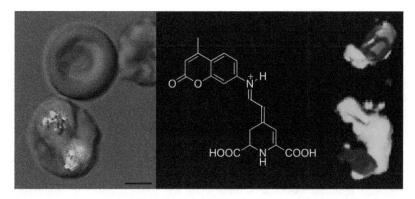

Fig. 6 An example of live-cell imaging of *Plasmodium*-infected red blood cells with cBeet120 (molecular structure indicated in the figure). The blue and green fluorescence emissions are from Hoechst 33342 (a selective stain for DNA) and cBeet120, respectively. Scale bar = 2 μm. Reproduced from Ref. [249]. Copyright © 2013 Gonçalves et al. Creative Commons Attribution License

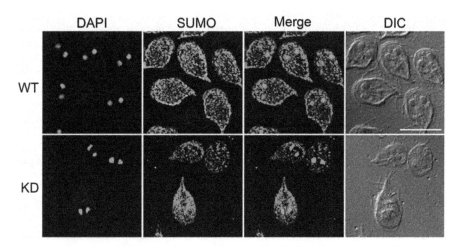

Fig. 7 Immunofluorescence analysis of wild-type (WT) and SUMO knocked-down *Giardia lamblia* using anti-GlSUMO (green). DAPI staining (blue) and the merged fluorescence images of the same field are also shown, as well as the corresponding DIC images. Scale bar = 10 μm. Reproduced from Ref. [171]. Copyright 2016 John Wiley and Sons

3.2   High-Throughput Systems for Drug Discovery

Fluorescent tracers are the method of choice for tagging purposes in life science research and development, where it is continuously replacing radio-labeling [122]. Due to its superior sensitivity to environmental properties and also its multidimensionality, i.e., its ability to provide various simultaneous readouts such as intensity, lifetime, anisotropy, and spectral characteristics, fluorescence has gained major attraction in life science applications. By attaching a dye to the biological molecule of interest, even minor biological activities will lead to environmental changes of the dye and can be monitored by changes in at least one of the fluorescence parameters [17, 250].

However, not all fluorescence techniques are fulfilling the prerequisites for high-throughput systems (HTS). Using bulk fluorescence techniques such as total fluorescence emission, robust assay systems which are compatible with HTS can be miniaturized down to volumes of 10 μL, but not further [251]. In contrast, applying confocal fluorescence microscopy, the measurement volume is as small as one femtoliter and assay volumes of 1 μL are established [122]. Besides minute assay volumes, confocal techniques offer the key advantage of applying new efficient fluorescence read-out methods, which reach the highest possible sensitivity by tracking biological activities down to the single-molecule level.

The most detailed information on biomolecular processes can be obtained by detecting fluorescence from single molecules. Therefore, fluorescence-based assays using single-molecule detection (SMD) techniques are evolving as a very important tool in science [3, 8, 87, 116, 127, 239]. These techniques use samples of highly diluted fluorophores and include not only the direct detection and analysis of single-molecule events but also spectroscopic analysis by means of applying fluctuation methods such as fluorescence correlation spectroscopy (FCS) or fluorescence intensity distribution analysis (FIDA) [122, 253, 254]. They enable to distinguish and quantify different molecules of a sample (e.g., bound and unbound) and, furthermore, open up the way to combine a number of different readouts, which are inherent to the fluorescence signal. The use of multiple, simultaneously acquired parameters improves the resolution of distinct species significantly, and hence it opens the door to study complex assemblies of biomolecular interactions. Thus, the major advantages of these techniques toward HTS are their high statistical accuracy, even at measurement times of about one second, and their very low consumption of precious biological material. As a consequence, they are increasingly applied in HTS.

Jäger and coauthors reviewed fluorescence techniques for high-throughput drug discovery [122]. In particular, the family of fluorescence techniques, FIDA, which is based on confocal single-molecule detection, has opened up a new field of HTS applications. Haupts and collaborators compared macroscopic versus microscopic fluorescence techniques in (ultra)-HTS [255]. Rüdiger and coauthors reviewed single-molecule detection technologies in miniaturized HTS and studied binding assays for G-protein-coupled receptors using FIDA and fluorescence anisotropy [254]. As a confocal detection technology, FIDA inherently allows reduction of the assay volume to the microliter range and below without any loss of signal. Palo and coauthors introduced a method that combines the benefits of both FIDA and fluorescence lifetime analysis [256]. It is based on fitting the two-dimensional histogram of the number of photons detected in counting time intervals of given width and the sum of excitation to detection delay times of these photons. Referred to as fluorescence intensity and lifetime distribution analysis (FILDA), the technique distinguishes fluorescence species on the basis of both their specific molecular brightness and the lifetime of the excited state and is also able to determine absolute fluorophore concentrations.

Parker and coauthors used fluorescence polarization to develop HTS assays for nuclear receptor-displacement and kinase inhibition [257]. This method is a solution-based, homogeneous technique, requiring no immobilization or separation of reaction components. As an example, the authors described the fluorescence polarization-based estrogen receptor assay based on the competition of fluorescein-labeled estradiol and estrogen-like compounds for binding to the estrogen receptor. Banks and coauthors described fluorescence polarization assays for HTS of G-protein-coupled receptors in 384-well microtiter plates [258]. Li and coauthors developed an assay for measuring the activity of an enzyme that transfers multiple adenine-containing groups to an acceptor protein, based on fluorescence polarization technology in a 1536-well plate format [259]. Texas red (rhodamine) was covalently conjugated to the adenine of the donor substrate through a C6 spacer arm. As a result of the transfer of the adenine-containing moieties to the acceptor protein substrate, the rotational correlation time of the Texas red conjugate increased, hence increasing the degree of fluorescence polarization. Owicki compared the perspectives of fluorescence polarization and anisotropy in HTS [260]. Qian and coauthors used fluorescence polarization to discover novel inhibitors of Bcl-xL, a factor preventing apoptosis [261]. Lu and collaborators developed a fluorescence polarization bead-based coupled assay to target activity/conformation states of a protein kinase [262]. This assay is based on the principle of the Immobilized Metal Assay for Phosphochemicals (IMAPTM) (MoleTMcular Devices, Sunnyvale, Calif., USA), a fluorescence polarization assay based on the affinity capture of phosphorylated peptides [263, 264].

Auer and coauthors recommended fluorescence correlation spectroscopy for lead discovery by miniaturized HTS, i.e., femtoliter volumes [265]. Schwille and coauthors reported that fluorescence correlation spectroscopy reveals fast optical excitation-driven intramolecular dynamics of yellow fluorescent proteins [266]. Koltermann and coauthors proposed rapid assay processing by integration of dual-color fluorescence cross-correlation spectroscopy (RAPID FCS) as an ideal tool for ultra-HTS when combined with nanotechnology, which can probe up to 105 samples per day [267].

Beasley and coauthors described the evaluation of compound interference in immobilized metal-ion affinity-based fluorescence polarization (IMAP) detection with a four million member compound collection [268]. IMAP is a non-separation-based, antibody-independent FP assay that can be applied to many types of protein kinases and phosphatases. In this technology, a fluorescently labeled peptide substrate is phosphorylated and then captured on immobilized metal nanoparticles. The binding of the phosphorylated peptide to the nanoparticles is detected using FP. Loomans and coauthors reported HTS with immobilized metal-ion affinity-based fluorescence polarization detection as a homogeneous assay for protein kinases [269]. Klumpp and coauthors discussed the development of homogeneous, miniaturized assays for the identification of novel kinase inhibitors from very large compound collections [270]. In particular, the suitability of TR-FRET based on phospho-specific antibodies, an antibody-independent fluorescence polarization approach using metal-coated beads (IMAP technology), and the determination of adenosine triphosphate consumption through chemiluminescence was evaluated.

Europium-labeled recombinant protein G as a fast and sensitive immunoreagent for time-resolved immunofluorometry had been described [271]. Using luminescent lanthanides, instead of conventional fluorophores, as donor molecules in resonance energy transfer measurements offers many technical advantages and opens up a wide range of applications including farther measurable distances with greater accuracy, insensitivity to incomplete labeling, and the ability to use generic relatively large labels. The principles and biophysical applications of lanthanide-based probes are discussed elsewhere [272–274].

Genetically encoded fluorescent indicators based on FRET technology were used to monitor various cell functions [275], miniaturization of cell-based β-lactamase-dependent assays to ultrahigh-throughput formats to identify agonists of human liver X receptors [276]. Eggeling and coauthors recommended the combination of FRET and two-color global fluorescence correlation spectroscopy (2CG-FCS) for high-throughput applications in drug screening [277]. Homogeneous time-resolved fluorescence resonance energy transfer (TR-FRET) assays represent a highly sensitive and robust HTS method for the quantification of kinase activity [278]. A widely applicable high-throughput TR-FRET assay for the measurement of kinase autophosphorylation was reported [279]. Inhibition of the vascular endothelial growth factor receptor 2 (VEGRF-2) was used as prototype. Glickman and coauthors published a comparison of ALPHAScreen, TR-FRET, and TRF as assay methods for FXR nuclear receptors [280, 281].

Two-photon excitation fluorescence cross-correlation spectroscopy (2PE-FCCS) was also used to interrogate the binding kinetics between the agonist Leu-enkephalin (BLEK) and human 1 opioid receptor (hMOR) in cell membrane nanopatches [282]. QD_{605} were conjugated with BLEK to bind the hMOR receptors, and the FCCS data were converted into fractional receptor occupancy for Hill plot analysis and extraction of K_d. Interestingly, the conjugation of BLEK to comparatively large QDs did not affect its binding properties. Cramb's group has developed 3C-FCCS that is greatly facilitated by the multiplexing advantages of QDs [282–285]. Polymer spheres barcoded with QD_{525}, QD_{605}, and QD_{655} could be successfully identified and quantitated in the presence of more than an 800-fold excess of free QDs, and the size of the triply labeled spheres could be determined [282]. 3C-FCCS was also able to identify and determine the size of DNA trimers (~80 nm) labeled with the three colors of QD [285]. The technique requires short acquisition times (1 min) and has been useful for measuring molecular exchanges in signal transduction pathways or the assembly of tripartite biomolecular complexes.

3.3   Bioimaging

3.3.1   Epifluorescence Microscopy

Epifluorescence microscopy and QDs have been widely applied for live-cell labeling and imaging [286]. Spatiotemporal multicolor labeling of live A594 cells (human alveolar adenocarcinoma) was demonstrated by using mixed delivery techniques over several days [287]. Initially, QD_{520} were delivered to cells by using a cationic polymer, and the cells were cultured for 3 days so that the QD_{520} were largely within late endosomes. QD_{635}-cell-penetrating peptide conjugates were then delivered to cells so as to label early endosomes, followed by cytosolic microinjection of QD_{550}-Cy 3 conjugates and subsequent incubation with Cy5-QD_{635}-Arg-Gly-Asp (RGD) peptide conjugates to label aVb3 integrins on the cell membrane. When combined with four narrow bandpass filters, these QD probes provided four distinct spectral windows for imaging each of the aforementioned cellular components with excitation at 457 nm. The QD_{550}-Cy3 and Cy5-QD_{635}-RDG conjugates were FRET pairs where the QDs were used as antennae to enable observation of Cy3/Cy5 photoluminescence with excitation at 457 nm (nonresonant with the dye absorption) and a reduction of dye photobleaching rates (indirect excitation via FRET) [287]. QDs with emission wavelengths analogous to those of Cy3/Cy5 can also be used directly, as demonstrated with QD_{580} instead of Cy3. Additional examples of live-cell imaging are discussed below in the context of other microscopy techniques. Chan and Nie demonstrated live-cell imaging with receptor-mediated endocytosis of transferrin-QD conjugates [288].

3.3.2   Confocal Microscopy

Confocal microscopy is a high-resolution imaging technique that is capable of axial sectioning. It particularly benefits from high brightness fluorophores, such as QDs, due to its reduced light throughput compared with epifluorescence microscopy (the trade-off for higher resolution). Matsuno and coauthors used QDs and confocal laser scanning microscopy to generate three-dimensional (3D) images of the intracellular localization of either growth hormone or prolactin, along with its corresponding mRNA, by using combined immunohistochemical labeling (QD_{655}-antibody conjugates) and in situ hybridization (QD_{605}-oligonucleotide conjugates) [289]. The spatial distributions of mRNA and its encoded protein provide insight into cellular protein synthesis. Chan and coauthors used similar methodology to simultaneously visualize vesicular monoamine transporter 2 mRNA and tyrosine hydroxylase protein within the cytoplasm of dopaminergic neurons [290]. Lee and coworkers used confocal imaging with QDs to observe agonist-induced endocytosis of two types of tagged G-protein-coupled receptors (GPCRs): (1) influenza hemagglutinin (HA) peptide-tagged kOR and (2) green fluorescent protein (GFP)-tagged A3AR (adenosine receptors) [291]. These GPCRs are overexpressed on the membrane of human osteosarcoma cells. The kOR receptor was labeled with QD-anti-HA (antibody) conjugates, and the QD and GFP photoluminescence signals allowed real-time parallel visualization of the internalization of both GPCR types when stimulated with agonists. This format is potentially useful for high-throughput screening of agonists for drug discovery since GPCRs are involved in several major diseases [291].

Winkler and coauthors presented confocal fluorescence coincidence analysis as an alternative to HTS. Confocal fluorescence coincidence analysis extracts fluorescence fluctuations that occur coincidentally in two different spectral ranges from a tiny observation volume of less than 1 fL. This procedure makes it possible to monitor whether an association between molecular fragments labeled with different fluorophores is established or broken, providing access to the characterization of a variety of cleavage and ligation reactions in biochemistry.

3.3.3   Spectral Imaging

Spectral imaging allows the use of spectral unmixing (i.e., image deconvolution), and a library of reference spectra permits quantitative resolution of the unique fluorescence contribution from each emitter and discrimination of background autofluorescence, even with emission overlap between different fluorophores [292–295]. This technical capability is ideal for pairing with the multiplexing advantages of some fluorophores to facilitate detection and visualization of multiple biomarkers in complex biological systems. Spectral imaging coupled with linear unmixing overcomes the limitations imposed by spectral overlap and bandpass filters. However, the linear unmixing algorithm has itself a fundamental limit to the number of labels it can distinguish. The maximum number of different labels that can be unambiguously identified in an image is limited by the number of spectral channels used to record the digital image. Biological labeling, image acquisition, and image analysis technique that combined both combinatorial labeling and spectral imaging were used to distinguish 28 different kinds of objects in a single fluorescence image [296].

Another prospective application of spectral imaging is intracellular thermometry. QDs exhibit a bathochromic shift with increasing temperature due to alteration of electron lattice interactions [297]. Yang et al. used this effect to measure intracellular heat generation in NIH/3T3 murine fibroblast cells after Ca^{2+} stress and cold shock [298]. QD photoluminescence collected from the specimen through an inverted microscope was directed to a spectrograph to measure temperature-dependent spectral shifts, revealing an inhomogeneous intracellular temperature response.

3.3.4   Two-Photon Fluorescence Microscopy

Two-photon fluorescence microscopy (2PE) is widely used for biological imaging. It offers high resolution, comparable to that of confocal microscopy, by limiting excitation of fluorescence to a small focal volume where there is sufficient flux of NIR photons for 2PE. NIR excitation penetrates more deeply into tissues and generates less autofluorescence compared to the visible excitation used in conventional (one-photon excitation [1PE]) confocal microscopy. Multiple fluorescent dyes can also be simultaneously interrogated due to frequent overlap in their 2PE spectra, even when their 1PE spectra have minimal overlap [252]. 2PE fluorescence microscopy, however, is still prone to photobleaching of dyes [252], and such dyes already have limited brightness in 2PE due to their small two-photon absorption cross sections (typically 300 GM) [299]. Despite that fact, small water-soluble molecules with σ_2 around 300 GM have several potential applications in 2PE microscopy [300–303].

QDs are less prone to photobleaching and show remarkably large two-photon absorption cross sections (10^3–10^4 GM) [113, 304]. One application of QDs in 2PE microscopy is high-resolution cellular imaging with minimal autofluorescence background and photodamage. For example, CdTe QDs was imaged in highly autofluorescent BY-2-T (tobacco) cells and found that the signal-to-noise (S/N) ratio, or QD PL-to-autofluorescence ratio, increased from ~3 to 11 comparing 1PE at 405 nm to 2PE at 800 nm. Most importantly, unlike 1PE at 405 or 488 nm (2 mW), 2PE at 800 nm (20 mW) caused no discernible photodamage to QGY human hepatocellular carcinoma cells [305]. Folate-conjugated InP/ZnS QDs were delivered to KB cells (human epithelial) via receptor-mediated uptake and imaged their accumulation in multivesicular bodies with 2PE at 800 nm [306]. Similarly, folate conjugation and 2PE were used to deliver and image ZnS:Mn/ZnS QDs in T47D and MCF-7 breast cancer cells [307]. Paramagnetic Mn-doped Si QDs were shown to function as a non-cytotoxic, multimodal contrast agent for magnetic resonance imaging and 2PE fluorescence imaging of macrophage cells [308]. Another useful application for QDs in 2PE microscopy is in vivo imaging. In vivo 2PE imaging of mouse vasculature, viz., angiography, in skin and adipose tissue was demonstrated by the injection of amphiphilic polymer-coated CdSe/ZnS QD$_{550}$ and 2PE at 880 nm [113].

2PEF and second-harmonic generation (SHG) imaging were used to monitor the progress of hMSC cultures stimulated with adipogenic and osteogenic differentiation factors and maintained under atmospheric or hypoxic (5%) oxygen concentrations. Based on multispectral analysis approaches, it was demonstrated that the major endogenous chromophores that contribute to the observed TPEF signal from hMSCs include NAD(P)H, flavoproteins, and lipofuscin (Fig. 8) [296].

3.3.5 Super-Resolution Imaging

Optical microscopy is an unrivaled tool for obtaining structural and molecular information at the micrometer scale. However, the opportunity for insight at the nanometer scale, where complex and important biochemistry remains to be elucidated, is limited by the Abbe's diffraction limit (typically 200–300 nm). To address this shortcoming, a variety of super-resolution imaging techniques have emerged.

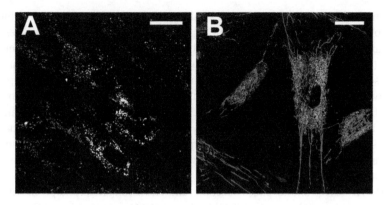

Fig. 8 Lysosomal localization of lipofuscin. Lipofuscin autofluorescence (green channel) is co-localized with (**a**) LysoTracker Red (red channel) and not (**b**) MitoTracker Orange (red channel). Co-localization is indicated by the yellow color in panel (**a**). Note that not all lysosomes contain lipofuscin. Bar = 30 µm. Reproduced from Ref. [296]. Copyright © 2010 William L. Rice et al., Creative Commons Attribution License

One group of these techniques, referred to as single-molecule localization methods, includes stochastic optical reconstruction microscopy (STORM), photoactivated localization microscopy, and super-resolution optical fluctuation imaging (SOFI).

An independent component analysis method was used to resolve two closely spaced (23 nm) QD emitters, based on their blinking, with a standard wide-field microscope [309]. Chien et al. similarly used the blinking of QDs to image microtubules in fixed CHO (Chinese hamster ovary) cells with 30 nm resolution and an acquisition time that was less than 1 min [310]. QDs were also used to develop the SOFI imaging technique that records a movie to capture fluorescence intermittency (i.e., blinking) over time and statistically analyzes the signal fluctuations [118]. Since the blinking of QDs follows power law statistics and occurs over all time scales, arbitrary frame rates can be used. The advantages of the technique include a fivefold improvement in resolution and reduced background, as demonstrated by labeling the a-tubulin network of fixed 3T3 fibroblast cells with QD_{625}-secondary antibody conjugates [118].

Super-resolution microscopy methods such as stochastic optical reconstruction microscopy (STORM) have enabled visualization of subcellular structures below the optical resolution limit. However, due to the poor temporal resolution, these methods have mostly been used to image fixed cells or slow dynamic processes. Fast dynamic processes and their relationship to the underlying ultrastructure or nanoscale protein organization cannot be discerned. To overcome this limitation, Tam and coauthors developed a correlative and sequential imaging method that combines live-cell and super-resolution microscopy and use a microfluidic platform for sample preparation [311]. For mitochondrial imaging, cells were incubated with MitoTracker Orange and stably transfected cell line for GFP-tubulin was used, which were generated by transfecting BS-C-1 cells with a plasmid encoding GFP-tubulin (Fig. 9) [311].

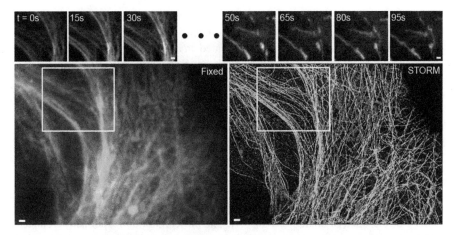

Fig. 9 Correlative microscopy using microfluidics. Microtubules and mitochondria were imaged sequentially using epifluorescence microscopy and STORM. The live imaging examples from various time points selected from a sequence of live-cell images acquired at 100 ms per frame (top) are from the region inside the white boxes (bottom). The experimental sequence was as follows: live-cell imaging of microtubules (top left), followed by live-cell imaging of mitochondria (top right); onstage automated fixation, automated immunostaining of mitochondria, and epifluorescence imaging of mitochondria (bottom left, red); STORM of mitochondria (bottom right, red); automated immunostaining of microtubules and epifluorescence imaging of microtubules (bottom left, green); and STORM imaging of microtubules (bottom right, green). Since the STORM images were acquired sequentially, fiduciary beads were used to align the two images. Scale bars, 1 μm. Reproduced from Ref. [311]; Copyright © 2014 Johnny Tam et al., Creative Commons CC0

3.3.6 Near-Field Scanning Optical Microscopy

Near-field scanning optical microscopy (NSOM) is a scanning probe imaging technique that provides optical resolution below the diffraction limit, with simultaneous topographic information. QDs and NSOM were combined for high-resolution (~50–100 nm) imaging of cell surface receptors in several studies [312–314]. For example, one study focused on imaging antigen-specific T-cell receptor (TCR) response to activation and expansion [312]. Vc2Vd2 T-cell surface receptors specifically bind phosphoantigens, resulting in T-cell expansion (i.e., rapid proliferation of clonal T cells) as part of an immune response. QD_{655} were conjugated with anti-TCR antibodies and used to label the TCRs on T cells obtained from macaques. The NSOM-measured distribution of QD photoluminescence across individual cells was correlated to the cell surface distribution of TCRs. Before activation and expansion, a larger number of smaller TCR sites were associated with Vc2Vd2 T cells (primarily 50 nm in size) and more uniformly distributed compared with a nonengaging ab T-cell phenotype (90 nm in size). After activation (i.e., infection of the macaques), the TCRs aggregated on the Vc2Vd2 T-cell membrane, and this pattern was sustained in daughter cells. The aggregated receptors recognized phosphoantigen and imbued the Vc2Vd2 T cells with much more potent effector function, suggesting an important role for TCR aggregation in immune response [312]. In these experiments, the QD_{655} were found to be a much more effective label than several common fluorescent dyes, permitting repeated imaging of cells by virtue of their resistance to both photobleaching and chemical degradation. Building on this approach, these authors subsequently used two-color NSOM to visualize the clustering and interactions of membrane receptors during TCR/CD3-mediated signaling in T cells [315]. Images of QD_{605} and QD_{650} photoluminescence were combined with topographical information and indicated co-clustering of CD3 with CD4 or CD8 co-receptors during T-cell activation. The authors also noted that co-stimulation of CD28 receptors significantly enhanced the clustering of TCR/CD3 with CD4 co-receptors, but not CD8 co-receptors. QD-antibody conjugates and NSOM were combined to investigate the distribution of CD4 receptor proteins on T-helper cell surfaces [313] and hyaluronan receptor CD44 on the surface of mesenchymal stem cells [314]. The enrichment corresponded to clusters of CD44 domains (200–600 nm in size) that were primarily observed at the peaks and ridges of cell surface topography plots would not have been discerned without the dual topographical and optical capability of the NSOM technique, facilitated by the use of QDs.

PNT2 samples were examined using SNOM (Fig. 10). Topography information can be seen in Fig. 10a, which was acquired to demonstrate that the extensive dual-labeling protocol did not result in changes to the structure of the sample. As can be seen in Fig. 10a, the cellular structure remained well preserved, and features can clearly be distinguished. Corresponding optical images were acquired for the topography image shown in Fig. 10a to reveal the location of ZO-1 and E-cadherin (shown in Fig. 10b, c, respectively). The resulting topography image (Fig. 10e) reveals a network of filopodia from neighboring cells that have established contact. The simultaneously acquired fluorescence image is shown in Fig. 10d. A row of fluorescence corresponding to E-cadherin location can be seen running parallel to the edge of one cell (see arrow on Fig. 10d). Additional rows of fluorescence are also present along the length of the large filopodium in the center of the image (identified by arrowhead). These results confirm that the localization of E-cadherin can be successfully examined in samples that have been dual stained [316].

3.3.7 Fluorescence Lifetime Imaging Microscopy

Although bioprobe fluorescence and QD photoluminescence intensity measurements are straightforward and versatile, reliable quantitative measurements are very challenging when the concentration of fluorophore is not controlled. Such is almost always the case when, for example, QDs are delivered to cells, immobilized at an interface, or subject to dilution in flow systems. Although ratiometric photoluminescence intensity methods address this challenge, ratiometric QD probes are still not available for every application. Fluorescence (or photoluminescence) lifetime analysis is independent of the concentration of the fluorophore and is also very well suited to quantitative measurements when concentration is uncertain.

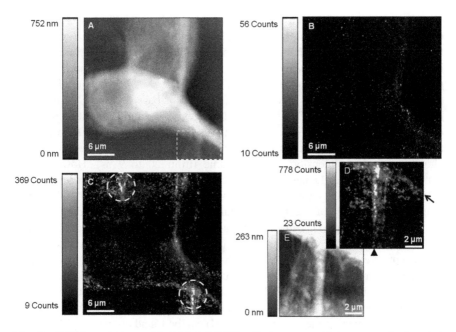

Fig. 10 SNOM acquisitions of dual-labeled PNT2 cells. Images reveal the topography (**a** and **e**) and fluorescence (**b**, **c**, and **d**) response. Images (**b**) and (**c**) were obtained with different filters to spectrally distinguish between red and green fluorescence, identifying ZO-1 and E-cadherin locations, respectively. The boxed region in (**a**) was scanned at higher resolution to generate the detailed E-cadherin fluorescence image (**d**) and the accompanying topography is shown in (**e**). Circled regions in (**c**) and the arrow/arrowhead in (**d**) highlight E-cadherin clusters. Reproduced from Ref. [316] Copyright © 2012 Kelly-Ann D. Walker et al., Creative Commons Attribution License

In addition to being a measure of FRET and CT quenching efficiency, changes in emission properties, such as lifetime, can reflect physicochemical changes in the local microenvironment (e.g., temperature and pH) by virtue of changes in radiative and non-radiative rates [317, 318]. Jaque's group has investigated this phenomenon for thermometry applications and found that optimum temperature sensitivity can be obtained with smaller-sized QDs and the use of CdTe over CdSe [319]. Spectral shifts are accompanied by changes in photoluminescence lifetime where, for example, CdTe QD_{515} (1 nm in diameter) exhibited a thermal sensitivity of $-0.017\ °C^{-1}$ (measured as the relative change in lifetime per degree Celsius) between 27 and 50 °C. This sensitivity was comparable to rhodamine B-doped microspheres ($0.0016\ °C^{-1}$) and Kiton red dye ($0.011\ °C^{-1}$) [319], suggesting that QDs might be useful probes for FLIM-based measurement of intracellular temperature. Such measurements are important for identifying "hot" malignant cells (higher metabolic activity compared with healthy cells) and monitoring hyperthermic treatment of those cells.

FLIM can also be used for visualization and quantitative sensing with QD probes based on FRET or CT quenching. For example, Pai and Cotlet modified vaterite

(CaCO$_3$) microparticles with CdSe/ZnS QD$_{525}$-TDTomato (fluorescent protein) conjugates and measured the QD-TDTomato FRET efficiency (~67%) via two-color FLIM [320]. Although only a proof-of-concept study, this format could be readily extended to sensing proteolytic activity by expressing the fluorescent protein with a peptide linker that was a substrate for a protease of interest [321, 322]. A FLIM-based chloride ion (Cl$^-$) sensor based on the CT quenching interaction between CdSe/ZnS QD$_{610}$ and lucigenin, a chloride-sensitive indicator dye, was developed [323]. The lucigenin was conjugated to the QDs to provide the proximity for CT quenching, resulting in a marked decrease in the QD photoluminescence intensity and its lifetime (from 20 to 9 ns). However, Cl$^-$ dynamically quenches lucigenin, and the collisional lucigenin-Cl$^-$ interaction competed with the QD-lucigenin CT interaction. Increasing concentrations of Cl$^-$ led to a progressive recovery of the QD photoluminescence lifetime, including a linear response for 0.5–50 mM Cl$^-$ in a complex solution mimicking an intracellular matrix [323]. FLIM-based sensing of Cl$^-$ was demonstrated, but once again only as a proof-of-concept. Nevertheless, as FLIM with fluorescent proteins and organic dyes continues to grow in importance for cellular imaging and sensing [324–326], the utility of combining QDs with FLIM will also grow. More information on imaging of protein molecules using Förster resonance energy transfer (FRET) and fluorescence lifetime imaging measurements (FLIM) can be found elsewhere [327–329].

4 Cosmetics

Cosmetic science is a field focused on human perception of aesthetic beauty. Due to the subjective aspect of physical beauty, cosmetic formulations are designed to change the appearance of the human body, e.g., hair, skin, lips, and nails, to improve self- and, eventually, external perception. The human skin is essential for health maintenance, and skin care delays the deterioration of the skin functions, viz., has an antiaging effect [29]. In addition to its visual appeal, hair has an important physiological role, for example, protecting the body against excessive sun exposition. Consequently, most research carried out by the cosmetic industry pursues the development of products for skin and hair care [330].

The study of the structure, characteristics, and functions of both hair and skin tissues contribute for the development of cosmetics that match (and possibly exceed) consumers' expectations. Fluorescence allowed the investigation of the internal structures of the skin and hair at the molecular level. For example, the combination of techniques, such as optical coherent tomography (OCT) [331, 332], second-harmonic generation microscopy [333–335], and fluorescence molecular imaging (FMI) [336–339], led to observation of the deep internals of the skin such as collagen fiber, making it possible to extract information on complex biological and biochemical processes.

The visual perception of cosmetics is equally important, stimulating the search for new safe ingredients. Although the importance of color in cosmetic formulation is clear for the consumer, the importance of fluorescence in cosmetic formulations is somewhat more obscure. However, fluorescent colors have been used to enhance brightness and to create new visual effects that cannot be obtained with normal color additives.

This section aims to present the use of fluorescence-based techniques to investigate the properties of human skin and hair and presents some applications of fluorescent dyes in hair coloring, makeup formulation, and development of photoprotective agents.

4.1 Skin

The skin accounts for about 15% of the total body weight and can be considered the largest organ of the body. In an adult, it is a living tissue system with an area of roughly 2 m^2. Unlike most other organs, the skin is accessible for visual inspection that allows the recognition of pigmentation, texture, and cosmetic appearance. It is formed by three major tissue layers; the epidermis, the dermis, and the subcutaneous tissue. The thin outermost epidermis can be considered a sheet of epithelial cells called keratinocytes stacked 10–15 high. When keratinocytes migrate toward the skin surface, they flatten, die, and cement together to form a proteinaceous, thin, and tough outer layer called the *stratum corneum*, which attenuates the absorption of UV radiation before it reaches living cells due to the presence of melanin. The dermis is much thicker than the epidermis, has fewer cells, and is connective tissue formed by three major types of fibers: collagen, reticulum, and elastin. Collagen makes up about 70% of the dry weight of the dermis, and is optically birefringent. The scattering of light by collagen is an important feature of the optics of the dermis. Comprehensive information on skin structure, physiology, and aging can be found elsewhere [340, 341].

The autofluorescence of normal skin is distributed throughout its different layers [27]. The main chromophores and fluorophores found in skin and their photophysical properties are presented in Tables 3 and 4, respectively [342]. The level of epidermal fluorescence depends mainly upon the amount of melanin, while dermis fluorescence originates from elastin and collagen. Collagen fluorescence (because of cross-links) has different characteristics depending on the type of skin tissue investigated. At least one collagen-related peak could be identified for all of the investigated skin samples; however, in most cases, two peaks were discernable. Hair and other hard keratins are distinctly different from whole sections of skin. The kynurenines, which are metabolites and degradation products of tryptophan, were identified in hair and nail. There were also striking similarities between fluorescence excitation-emission matrices of *stratum corneum* in hair and nail [343].

With FLIM, keratinocytes exhibit long-medium values, visually corresponding to the blue-green range, whereas melanocytes have a medium to short fluorescence

Table 3 Principal chromophores of skin

Skin chromophore	Abs[a]	FL[b]	λ^{ABS} (nm)
Oxyhemoglobin	UV-Vis	NO	412, 542, 577
Deoxyhemoglobin	UV-Vis	NO	430, 555, 760
Melanin	UV-Vis	Low	Monotonic increase to short wavelengths
Porphyrins	VIS	Yes	λ^{EX}: ~405; λ^{EM}: 600
Bilirubin	VIS	NO	460
NAD/NADH	UV	Yes	λ^{EX}: ~350; λ^{EM}: 460
DNA/RNA	UV	NO	260
Tryptophan	UV	Yes	λ^{EX}: 295; λ^{EM}: 340–350
Urocanic acid	UV	NO	280
Collagen x-links	UV	Yes	λ^{EX}: 335, 370; λ^{EM}: 380, 460
Elastin x-links	UV-Vis	Yes	λ^{EX}: 420, 460; λ^{EM}: 500, 540
Keratin (dry), horn	UV	Yes	λ^{EX}: 370; λ^{EM}: 460

Adapted from Ref. [342]
NO not observed
[a]Spectral range of absorption
[b]Occurrence of fluorescence

Table 4 The principal fluorophores of skin

Fluorophore	λ^{EX} (nm)	λ^{EM} (nm)	Fl intensity
Tryptophan	295	340–350	Strong
Pepsin digestible collagen cross-links	335–340	380–390	Secondary
Collagenase digestible collagen cross-links	365–380	420–440	Strong
Elastin cross-links	420	500	Weak
NADH	350	460	Weak

Adapted from Ref. [342]

decay time coded in the yellow-red range (Fig. 11) [344, 345]. Interestingly, the change in structure of collagen and elastin caused by photoaging correlates with a decline in the intensity of the green skin fluorescence [346]. This change in fluorescence intensity is also observed in chronoaged skin, which in middle age begins to lose its green fluorescence and, in later years, loses its blue fluorescence. It is very likely that the decline in the vigorous "glow" common to young, healthy skin is related at least in part to this observed loss of fluorescence.

Many noninvasive devices to investigate the biomechanical properties of the skin have been developed, but the measurements of skin deformation after suction or torsion are still widely used techniques in cosmetic research. Video microscopes have made it possible to enlarge the surface of the skin, but the internal structure could only be roughly observed with ultrasonic diagnosis. It was also challenging to improve the resolution with thin tissue like the skin. Under these situations, a competition of research has begun using intrinsic fluorescence methods such as multiphoton confocal microscopes to observe the internal structures. Genetic vectors and congenic methods have developed, and there are alternate methods being developed.

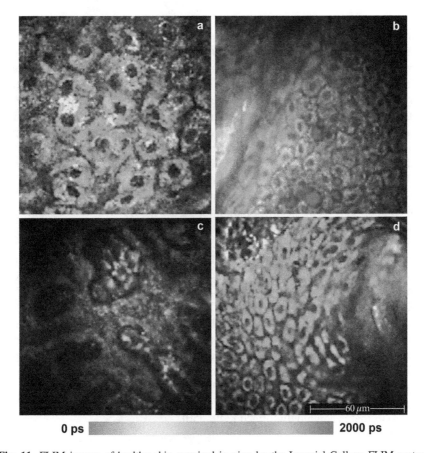

0 ps   2000 ps

Fig. 11 FLIM images of healthy skin acquired in vivo by the Imperial College FLIM system incorporated in the DermaInspect. A false-color scale, in this case corresponding to 0–2000 ps, is used to define the fluorescence lifetime decay rate of each fluorophore: for each color, there is a corresponding value of lifetime decay. (**a**) 760 nm excitation wavelength. In the false-color images, keratinocytes appear green-blue, showing a longer fluorescence lifetime with respect to melanin aggregates corresponding to yellow-red areas with a shorter lifetime value. (**b**) 760 nm excitation wavelength. Keratinocytes of the basal layer appearing yellow-red because of the presence of pigmentation. (**c**) 760 nm excitation wavelength. Dermal papillae surrounded by pigmented keratinocytes and melanocytes with low FLIM values. (**d**) 760 nm excitation wavelength. A hair follicle with bluish perifollicular cells showing a shorter lifetime. Adapted from Ref. [344]. Copyright © 2012 Stefania Seidenari et al., Creative Commons Attribution License

These techniques are being applied and used in academic research, with various fluorescent pigments used to label the immune cells of mice ears so they can be observed alive to discover new immune reaction pathways [29].

Skin's optical properties play an important role as well, and devices measuring these characteristics assess reflected light after illumination of the skin surface. Skin autofluorescence is intense enough to compromise the measurement of epidermal transmission spectra in the UV-A region. The portion of absorbed 260–290 nm radiation which is emitted as fluorescence at 330–360 nm is falsely recorded by the broadband detector as transmittance at 260–290 nm. The comparatively high epidermal transmittance at the 330–360 nm fluorescence emission band can cause order-of-magnitude errors in the measurement of 260–290 nm epidermal transmittance, especially if the samples are heavily pigmented. Different noninvasive methods have been proposed for evaluating skin complexion in vivo. These include quantitative measurements of skin color, using colorimetry—i.e., L*a*b* and individual typological angle (ITA°)—or of the intensity of specular reflection and the back-scattering of light from the skin [343].

Tryptophan fluorescence is a good marker for noninvasive evaluation of the epidermal cell proliferation rate. Since this fluorescence is associated with the cell proliferation, its decline with age may indicate the reduced replicative capacity of epidermal cells, i.e., replicative senescence [347, 348]. Photoaging as well as intrinsic aging reduced cell renewal. In contrast, the fluorescence of collagen and elastin cross-links increased with age on the forearms and face (temples), indicating an accumulation of advanced glycation end products (AGE), aldimine adducts formed sugars and proteins [349, 350].

Optical properties including color and brightness were strongly linked to age, as was the formation of AGEs, as measured by spectrofluorimetry. Due to the variability in the types of sugars and the types of amino acids with free amine groups, there are a number of possible types of AGE. Interestingly, quite a few of these products have fluorescent properties and can be measured noninvasively in the skin. AGE formation can also be assayed by measuring the formation of specific AGE species using immunological-based techniques. For example, carboxymethyl lysine (CML) is an AGE that is commonly formed [349, 350].

4.2 Hair

Hair is a protein filament that grows from follicles found in the dermis. The microscopic structure of the hair is formed by the cuticle, the medulla, and the cortex. The cuticle consists of overlapping scales that cover the hair shaft. The medulla is the canal of air or liquid-filled cells in the center of the cortex. The cortex is the main body of the hair which consists of keratinized fibers oriented parallel to the long axis of the hair [351–354]. The dominant human hair characteristic, both macroscopically and microscopically, is hair color. The color of the hair depends on the pigment present, the surface transparency, and the reflectivity of the hair. There are two observed pigments that account for the color of human hair. Melanin is the brown pigment, and phaeomelanin is the red pigment. Pigment granules of phaeomelanin are not observable with light microscopy owing to the small size of the granules [29, 342, 343, 352, 355].

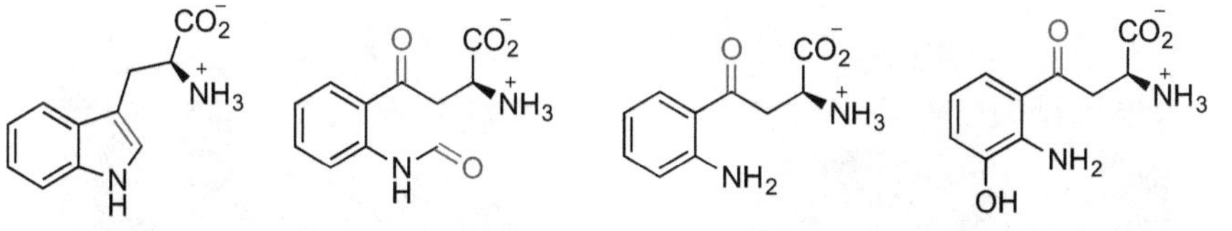

Fig. 12 Structures of L-tryptophan and its photooxidation products

Fluorescence of gray hair, in both unpigmented and melanin-pigmented fibers, is caused by tryptophan and its metabolic or photooxidation conversion products including chromophores such as *N*-formylkynurenine, kynurenine, and 3-hydroxykynurenine (Fig. 12). Unpigmented hair is characterized by higher intensity of fluorescence than pigmented hair. Melanin is also selectively quenching longer wavelength fluorescence originating from L-Trp conversion products. Yellow coloration of gray hair is related to L-Trp-derived metabolic or photooxidation chromophores, especially 3-hydroxykynurenine [356]. Finally, melanin oxidation products have been found to emit fluorescence at 540 nm [357].

Several studies have been published on the penetration of fluorescein isothiocyanate (FITC)-labeled proteins and protein hydrolysates into human hair using fluorescence microscopy or fluorescence laser scanning confocal microscopy. The latter method provides a new dimension of hair fiber observation and diagnostic perspective, as will be shown, due to its significantly higher resolution than classical microscopy, which locates penetrated labeled proteins more precisely [358]. Fluorescence laser scanning confocal microscopy represents an excellent tool to demonstrate the substantive effect for hair care active ingredients labeled with a fluorescent marker. This technique makes it possible not only to observe the absorption of FITC-labeled protein into the hair cuticle but also to detect the penetration into hair fibers by creating optical sections and reconstructing them as a three-dimensional image. This noninvasive technique offers possibilities for the exploration of the linkage and penetration of hair care active ingredients into hair fibers, thus enabling product developers to optimize hair care formulas of the future [358].

The binding of charged substances from external aqueous media to hair has been investigated through the use of fluorescence microscopy (Fig. 13). Eleven hair samples, reflecting various ethnic groups and cosmetic treatments, were tested. Rhodamine 6G, a cationic dye representative of drugs such as cocaine and opiates, showed incorporation throughout the hair of all samples except one. In contrast, fluorescein, an anionic dye representative of drugs such as THC carboxylic acid, was not readily incorporated. The incorporation of rhodamine 6G was faster for chemically "straightened" and bleached African-American female hair than for untreated hair. Incorporation of rhodamine 6G followed a pH dependence, but an ionic strength dependence could not be established. These studies support that (1) electrostatic interactions explain the preferential binding of cationic drugs of abuse to hair; (2) the hair matrix, or the non-helical portion of hair, is accessible to

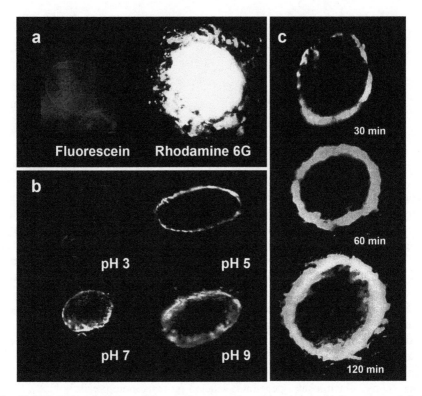

Fig. 13 Effect of incubation conditions on the penetration of dyes in hair. (**a**) Caucasian male hair was exposed to a concentration of 10 µg mL^{-1} dye in 10 mmol L^{-1} phosphate buffer, pH 5.6 and 37 °C, for 2 h; Note the difference in dye incorporation of rhodamine 6G, cationic, vs. fluorescein, anionic. (**b**) Effect of pH on the incorporation of rhodamine 6G into African female untreated hair. The hair was incubated for 2 h in 5 µg mL^{-1} of rhodamine 6G following the procedure described in the text. Incorporation was most pronounced at pH 7.0. (**c**) Caucasian male hair1 µg mL^{-1} of rhodamine 6G in 10 mM phosphate buffer, pH 5.6, and 37 °C, for 30, 60, and 120 min. Rhodamine 6G and fluorescein were digitally colored in orange and green, respectively. Adapted from Ref. [359]. Copyright © 2000 Elsevier Science Ireland Ltd. All rights reserved

external solutions and thus subject to contamination; and (3) cosmetic treatments may alter the helical portion of hair thereby increasing its accessibility to external contamination [359].

4.3 Fluorescent Colors

4.3.1 Hair Coloring

Materials that fluoresce are referred to as optical brighteners and are used in consumer products. Optical brighteners are dyes that absorb light in the ultraviolet and violet region (usually 340–370 nm) of the electromagnetic spectrum and re-emit light in the blue region (typically 420–470 nm). These additives often enhance fabrics or paper by producing a whitening effect to make these materials look less yellow through an increase of the amount of blue light reflected. Unfortunately, optical brighteners are not approved color additives for cosmetics in the United States. However, ingredients that provide visual effects that can alter the performance and visual properties of a product are certainly not prohibited. FDA only approves the following fluorescent colors for use in cosmetics, and there are limits on their intended uses: D&C Orange No. 5, No. 10, and No. 11; and D&C Red No. 21, No. 22, No. 27, and No. 28 [21 CFR 74.2254, 74.2260, 74.2261, 74.2321, 74.2322, 74.2327, and 74.2328] [360, 361].

Temporary hair dyes consist of acidic textile dyes that are water soluble and possess a high molecular weight. They are named for their ability to be removed after a single shampooing. These large molecules are unable to penetrate the cuticle and are instead temporarily deposited on the surface layers until they are washed off. Temporary dyes are formulated as rinses, gels, mousses, and sprays and are primarily used to remove unwanted tones, add highlights, or subtly color the hair [362–364]. Fluorescent pigments or colorants are important to the cosmetics industry. They can produce brilliant, very lively colors that cannot be produced with conventional coloring substances. When they do not absorb in the visible region, they can also lighten and significantly enhance the reflectance of substrates onto which they are deposited, such as the skin and hair. Hair coloring has an impressive visual effect and usually involve temporary hair dyes (Fig. 14). However, it is important to consider the characteristics of the fluorescent dye (e.g., toxicity, phototoxicity, and excitation wavelength) before its use in cosmetic formulations. There are examples of FDA-approved fluorescent dyes whose photoexcitation followed by ISC populates the triplet excited state and produces singlet oxygen via energy transfer to molecular oxygen, i.e., act as photosensitizers [365]. There is general concern on the occurrence of deleterious photochemical processes promoted by exogenous pigments at the surfaces of skin and hair, e.g., phototoxicity of pigments used in tattoos [366].

4.3.2 Makeup

Preserving both biomechanical and optical properties of the skin during aging and/or photoaging is one of the main goals of cosmetic research. Nonetheless, cosmetics and skin care products have traditionally focused on the camouflaging of the most easily characterized signs of aging, such as wrinkles since addressing of the seemingly more intangible problem of renewing the glow of youth in more mature skin impose several difficulties that are related to the eventual occurrence of photochemical reactions. While these applications do not actually add optical brighteners as color additives, they add other ingredients that visually perform this task, as in light-diffusing makeup.

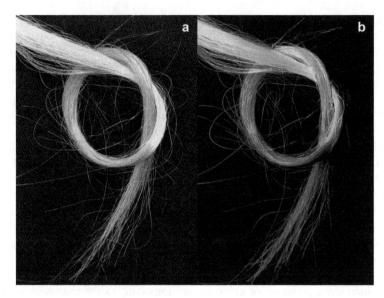

Fig. 14 Discolored human hair dyed with fluorescent pigments under visible light (**a**) and irradiated with black light (**b**). Copyright © 2018 Erick Leite Bastos, Creative Commons Attribution License

L'Oréal patented a cosmetic composition comprising flakes containing fluorescent agents entrapped in a hybrid matrix. Fluorescent flakes were prepared from a tetramethoxysilane and methyltrimethoxysilane polymer, dicocoylethyl hydroxyethylmonium methosulfate (and) propylene glycola, glycereth-25 PCA isostearate, hexylene glycol, and D&C Red 28 or D&C Orange 5. The fluorescent flakes were incorporated into a nail enamel that imparted a red-orange fluorescent color after application to the nails, into a powder foundation and a lip gloss [358]. Additionally, fluorescent proteins (emission at 630 nm) were used to cosmetically revert the age-related decline in skin fluorescence [358].

4.4 Fluorescence Quenchers and Photoprotection

Tryptophan decomposition can lead to the formation of kynurenines that are believed to be the culprit behind a photo-yellowing process that can be of particular concern for those with gray hair [356]. Cystine is well-recognized as an effective quencher molecule. Quenching of L-Trp excitation energy by cystine results in the subsequent breakdown of the latter into cysteic acid (Fig. 15). Singlet oxygen has been shown to be involved in the oxidation of cystine in the presence of protonated L-Trp [367].

Hairs have intrinsic fluorescence that depends on the excitation wavelength, an indication of the presence of a range of fluorophores [368]. When excited in visible

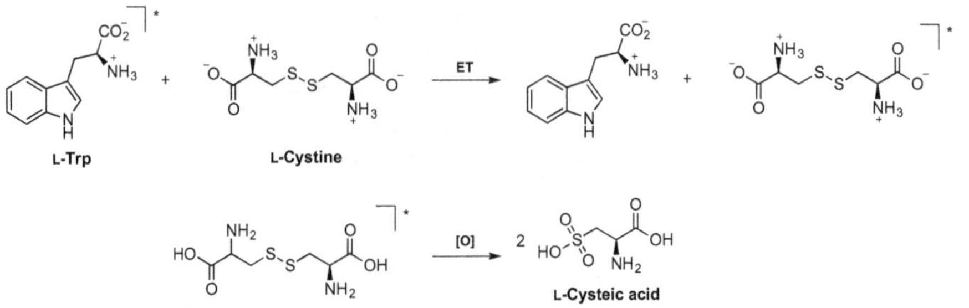

Fig. 15 Excitation energy transfer from L-tryptophan to L-cystine and subsequent oxidation to give L-cysteic acid

region, dark hair presents lower fluorescence than light hair. The weaker fluorescence of black hair results from the inner filter effect and light re-absorption due to the presence of larger amounts of melanin [369]. Furthermore, the ϕ_{FL} of synthetic eumelanin was found to be very low ($<7 \times 10^{-4}$) [355], and oxidized melanin has been shown to fluoresce at 540 nm [357].

Melanin is the sunscreen of nature; but perhaps surprisingly, it does not have especially strong UV absorptive capability [355]. Heavy pigmentation associated with high melanin concentrations provides some attenuation of light penetration into hair fibers; but melanin too is known as a quencher. Therefore, melanin molecules can accept absorbed energy to become excited and subsequently undergo sacrificial decomposition in place of more important proteins. Melanin granules present weak light emission themselves. However, melanin precursors and derivatives are photoactive and contribute to the photoinduced processes generated in the visible region, including the light emission. As any chromophore, melanin can form oxidant species (excited states and radicals) after light absorption and photosensitization reactions are prone to occur [369]. Nofsinger and colleagues observed the formation of ROS from eumelanin irradiated by UV-A [370]. Yet, melanin is also a good absorber in the visible part of the spectra. Besides being a visible light chromophore, melanin is able to engage in photosensitization reactions triggered by visible light. Chiarelli-Neto and coworkers characterized the excited-state reactions of melanin and proposed that a possible explanation for the damage induced by visible light in hair and skin is the formation of $^1O_2$ after excitation of melanin [369, 371].

Ethylhexyl methoxycrylene was developed as a photostabilizer that quenches the singlet excited state of UV filters [372]. A photostabilizer that quenches the singlet excited state reduces the number of molecules that enter the triplet excited state, thus further reducing the possibilities for destructive photochemical reactions. Theoretically, a singlet quencher such as ethylhexyl methoxycrylene would be effective at lower concentrations than a triplet quencher in all formulations and would improve the performance of formulations that are difficult for triplet quenchers to photostabilize, such as those that combine avobenzone and cetyl methoxycinnamate.

The scientific knowledge available today allows one to conclude that further developments in the field of sun protection both to hair and skin are on high demand.

The recently developed hair products that have sunscreen included in its formulation will not help much, because hair bleaching is mainly caused by visible light and the visible part of the spectra is responsible to cause photooxidative damage similar to the UV-A. New sunscreens operate in a wider spectral range (including visible light), so that people can get a better protection of skin and hair [373].

5 Perspectives

Fluorescent tools have launched biological research into a new realm of understanding of cellular processes and dynamics at the single-cell level. These tools are enabling the characterization of stochasticity and heterogeneity exhibited by biological systems, which could not adequately be probed by techniques that rely on bulk analysis of populations of cells. Early applications of fluorescent tools entailed monitoring protein dynamics, which continues to be a field of active development. Recent advances in photoactivatable and photoswitchable probes enable more sophisticated measurements of protein stability and turnover, and multiplexing capabilities are pushing the field toward high-throughput proteomics using data-rich fluorescent techniques. New tools for nucleic acid imaging are surging, and single-molecule detection used in combination with highly multiplexed labeling is enabling unprecedented quantitative global analysis of cellular responses. Fluorescent sensors are also increasingly providing insight into the molecular components of the cell: small molecules, secondary metabolites, metals, and ions. Tools can now probe global shifts in cellular state that are not well captured by any single molecular indicator, such as membrane potential, cellular division, and tissue differentiation.

With rapid advances in fluorescent tool development and improvements in microscopy platforms, the most significant challenges currently facing the field are new methods for processing and interpreting the vast amounts of data being generated. There is an increasing need for strategies to incorporate complex data into systems-level modeling and user-friendly software for automated data analysis. Fluorescent tools have now been developed to monitor a vast array of cellular constituents in live cells. Now, a true systems-level biological understanding of cellular dynamics is conceivable by measuring these constituents in parallel, with temporal and intracellular resolution within single cells. What is needed, then, is a multiplexed approach that integrates information from several monitoring techniques into a systems-level model of cellular dynamics. Thus far, studies incorporating simultaneous monitoring of multiple cellular constituents are fairly limited, but recent work is beginning to reveal some of the parameters contributing to what was previously considered to be biological noise.

"Omics" studies have historically relied on methods that separate the biomolecules of interest from a bulk population of cells, such as mass spectrometry, microarrays, and next-generation DNA and RNA sequencing. Although these methods are extremely powerful, cells have evolved complex and redundant mechanisms for regulating cellular functions in response to stress and other stimuli.

Moreover, there is growing recognition of both heterogeneity and dynamics in cellular states and cellular signaling pathways that call for single-cell, time-lapse measurements to elucidate mechanisms of regulation. Simultaneous interrogation of multiple levels of cellular dynamics (e.g., RNA level, protein level, and metabolite level) may reveal novel regulatory mechanisms that govern cellular dynamics. In this respect, one limitation of fluorescence microscopy is the small number of parameters that can be imaged simultaneously due to spectral overlap of optical probes. A recent approach for circumventing this limitation is the development of a phenotypic screen to identify the ideal reporter cell lines whose phenotypic profile captures the effect of different classes of drugs. Alternatively, a common approach involves tracking dynamics of biomarkers of interest using appropriate probes, followed by fixing and staining for a broader repertoire of biomarkers that help define features of the cellular state, albeit at a fixed point in time. Using immunofluorescence, cells can be reprobed with a variety of different markers to build a high-dimensional data set.

Finally, fluorescence dyes used for cosmetic purposes must be safe for human use and environmentally friendly. This is a challenging task not only for fluorescent colors but also for conventional dyes and pigments. Several natural products have been found to fluoresce, but their stabilization and tinctorial strength must be improved to allow industrial application.

Acknowledgments We thank the São Paulo Research Foundation (FAPESP), the Brazilian National Council for Scientific and Technological Development (CNPq), the Coordination for the Improvement of Higher Education Personnel (CAPES), and Natura Cosmetics for the financial support.

Conflict of Interest Statement The authors declare that there are no conflicts of interest associated with this publication.

References

1. Merola F, Levy B, Demachy I, Pasquier H (2010) Photophysics and spectroscopy of fluorophores in the green fluorescent protein family. Springer Ser Fluoresc 8:347–384. https://doi.org/10.1007/978-3-642-04702-2_11
2. Liu J, Liu C, He W (2013) Fluorophores and their applications as molecular probes in living cells. Curr Org Chem 17:564–579
3. Matea CT, Mocan T, Tabaran F, Pop T, Mosteanu O, Puia C, Iancu C, Mocan L (2017) Quantum dots in imaging, drug delivery and sensor applications. Int J Nanomedicine 12:5421–5431
4. Tsien RY (1998) The green fluorescent protein. Annu Rev Biochem 67:509–544. https://doi.org/10.1146/annurev.biochem.67.1.509
5. Lavis LD, Raines RT (2008) Bright ideas for chemical biology. ACS Chem Biol 3:142–155. https://doi.org/10.1021/cb700248m
6. Hanson GT, Hanson BJ (2008) Fluorescent probes for cellular assays. Comb Chem High Throughput Screen 11:505–513. https://doi.org/10.2174/138620708785204090
7. Warrier S, Kharkar PS (2014) Fluorescent probes for biomedical applications (2009-2014). Pharm Pat Anal 3:543–560. https://doi.org/10.4155/ppa.14.41

8. Zheng Q, Lavis LD (2017) Development of photostable fluorophores for molecular imaging. Curr Opin Chem Biol 39:32–38. https://doi.org/10.1016/j.cbpa.2017.04.017

9. Chandan HR, Schiffman JD, Balakrishna RG (2018) Quantum dots as fluorescent probes: synthesis, surface chemistry, energy transfer mechanisms, and applications. Sens Actuators B 258:1191–1214. https://doi.org/10.1016/j.snb.2017.11.189

10. Haque A, Faizi MSH, Rather JA, Khan MS (2017) Next generation NIR fluorophores for tumor imaging and fluorescence-guided surgery: a review. Bioorg Med Chem 25:2017–2034. https://doi.org/10.1016/j.bmc.2017.02.061

11. Zhang X, Liu J, Wang W et al (2013) Applications of fluorescent molecular probes in cell biology. In: Fluorophores. Nova Science Publishers, Hauppauge, NY, pp 29–52

12. Monici M (2005) Cell and tissue autofluorescence research and diagnostic applications. Biotechnol Annu Rev 11:227–256

13. Lavis LD (2017) Teaching old dyes new tricks: biological probes built from fluoresceins and rhodamines. Annu Rev Biochem 86:825–843. https://doi.org/10.1146/annurev-biochem-061516-044839

14. Petryayeva E, Algar WR, Medintz IL (2013) Quantum dots in bioanalysis: a review of applications across various platforms for fluorescence spectroscopy and imaging. Appl Spectrosc 67:215–252. https://doi.org/10.1366/12-06948

15. Specht EA, Braselmann E, Palmer AE (2017) A critical and comparative review of fluorescent tools for live-cell imaging. Annu Rev Physiol 79:93–117. https://doi.org/10.1146/annurev-physiol-022516-034055

16. Peng Z, Han X, Li S et al (2017) Carbon dots: biomacromolecule interaction, bioimaging and nanomedicine. Coord Chem Rev 343:256–277. https://doi.org/10.1016/j.ccr.2017.06.001

17. Lakowicz JR (2006) Principles of fluorescence spectroscopy. Springer US, Boston, MA

18. Valeur B, Berberan-Santos, MN, Martin MM, Plaza P (2012) Photophysics and photochemistry of supramolecular systems. In: Schalley CA (ed) Analytical methods in supramolecular chemistry, 2nd edn. Wiley-VCH Verlag & Co., Weinheim, pp 287–336

19. Turro NJ, Ramamurthy V, Scaiano JC (2012) Modern molecular photochemistry of organic molecules. Photochem Photobiol 88:1033. https://doi.org/10.1111/j.1751-1097.2012.01178.x

20. Klán P, Wirz J (2009) Photochemistry of organic compounds: from concepts to practice. John Wiley & Sons Ltd, Chichester, p 582

21. Marcu L, French PMW, Elson DS (2015) Fluorescence lifetime spectroscopy and imaging: principles and applications in biomedical diagnostics. CRC Press, Boca Raton, p 570

22. Braslavsky SE (2007) Glossary of terms used in photochemistry, 3rd edition (IUPAC Recommendations 2006). Pure Appl Chem 79:293–465. https://doi.org/10.1351/pac200779030293

23. Yang C, Hou VW, Girard EJ et al (2014) Target-to-background enhancement in multispectral endoscopy with background autofluorescence mitigation for quantitative molecular imaging. J Biomed Opt 19:76014

24. Roshchina VV (2012) Vital autofluorescence: application to the study of plant living cells. Int J Spectrosc 2012:1–14. https://doi.org/10.1155/2012/124672

25. Lu H-H, Wu Y-M, Chang W-T et al (2014) Molecular imaging of ischemia and reperfusion in vivo with mitochondrial autofluorescence. Anal Chem 86:5024–5031. https://doi.org/10.1021/ac5006469

26. Wagnieres GA, Star WM, Wilson BC (1998) In vivo fluorescence spectroscopy and imaging for oncological applications. Photochem Photobiol 68:603–632. https://doi.org/10.1111/j.1751-1097.1998.tb02521.x

27. Fellner MJ (1976) Green autofluorescence in human epidermal cells. Arch Dermatol 112:667. https://doi.org/10.1001/archderm.1976.01630290017003

28. Zeng H, MacAulay C, Palcic B, McLean DI (1995) Spectroscopic and microscopic characteristics of human skin autofluorescence emission. Photochem Photobiol 61:639–645. https://doi.org/10.1111/j.1751-1097.1995.tb09881.x

29. Hosoi J, Koyama J, Ozawa T (2017) New aspects of cosmetics and cosmetic science. Elsevier, New York

30. Martinic I, Eliseeva SV, Petoud S (2017) Near-infrared emitting probes for biological imaging: organic fluorophores, quantum dots, fluorescent proteins, lanthanide(III) complexes and nanomaterials. J Lumin 189:19–43. https://doi.org/10.1016/j.jlumin.2016.09.058
31. Ye Y, Bloch S, Kao J, Achilefu S (2005) Multivalent carbocyanine molecular probes: synthesis and applications. Bioconjug Chem 16:51–61. https://doi.org/10.1021/bc049790i
32. He J, Yang L, Yi W et al (2017) Combination of fluorescence-guided surgery with photodynamic therapy for the treatment of cancer. Mol Imaging 16:153601211772291. https://doi.org/10.1177/1536012117722911
33. Siddiqui MR, AlOthman ZA, Rahman N (2017) Analytical techniques in pharmaceutical analysis: a review. Arab J Chem 10:S1409–S1421. https://doi.org/10.1016/j.arabjc.2013.04.016
34. Rudin M, Weissleder R (2003) Molecular imaging in drug discovery and development. Nat Rev Drug Discov 2:123–131. https://doi.org/10.1038/nrd1007
35. Thapaliya ER, Zhang Y, Dhakal P et al (2017) Bioimaging with macromolecular probes incorporating multiple BODIPY fluorophores. Bioconjug Chem 28:1519–1528. https://doi.org/10.1021/acs.bioconjchem.7b00166
36. Chrzanowski SM, Vohra RS, Batra A et al (2016) Near-infrared optical imaging noninvasively detects acutely damaged muscle. Am J Pathol 186:2692–2700
37. Ravoori MK, Singh S, Bhavane R et al (2016) Multimodal magnetic resonance and near-infrared-fluorescent imaging of intraperitoneal ovarian cancer using a dual-mode-dual-gadolinium liposomal contrast agent. Sci Rep 6:38991. https://doi.org/10.1038/srep38991
38. Wolff M, Kredel S, Wiedenmann J et al (2008) Cell-based assays in practice: cell markers from autofluorescent proteins of the GFP-family. Comb Chem High Throughput Screen 11:602–609. https://doi.org/10.2174/138620708785739880
39. Pitre SP, McTiernan CD, Scaiano JC (2016) Library of cationic organic dyes for visible-light-driven photoredox transformations. ACS Omega 1:66–76. https://doi.org/10.1021/acsomega.6b00058
40. Chen X, Zaro JL, Shen WC (2013) Fusion protein linkers: property, design and functionality. Adv Drug Deliv Rev 65:1357–1369
41. Wang J, Xie J, Schultz PG (2006) A genetically encoded fluorescent amino acid. J Am Chem Soc 128:8738–8739. https://doi.org/10.1021/ja062666k
42. Koopmans T, van Haren M, van Ufford LQ et al (2013) A concise preparation of the fluorescent amino acid L-(7-hydroxycoumarin-4-yl) ethylglycine and extension of its utility in solid phase peptide synthesis. Bioorg Med Chem 21:553–559. https://doi.org/10.1016/j.bmc.2012.10.055
43. Bryson DI, Fan C, Guo LT et al (2017) Continuous directed evolution of aminoacyl-tRNA synthetases. Nat Chem Biol 13:1253–1260. https://doi.org/10.1038/nchembio.2474
44. Kuhn SM, Rubini M, Müller MA, Skerra A (2011) Biosynthesis of a fluorescent protein with extreme Pseudo-Stokes shift by introducing a genetically encoded non-natural amino acid outside the fluorophore. J Am Chem Soc 133:3708–3711. https://doi.org/10.1021/ja1099787
45. Hudson EN, Weber G (1973) Synthesis and characterization of 2 fluorescent sulfhydryl reagents. Biochemistry 12:4154–4161. https://doi.org/10.1021/bi00745a019
46. Maggiora LL, Smith CW, Zhang ZY (1992) A general-method for the preparation of internally quenched fluorogenic protease substrates using solid-phase peptide-synthesis. J Med Chem 35:3727–3730. https://doi.org/10.1021/jm00099a001
47. Tyagi S, Kramer FR (1996) Molecular beacons: probes that fluoresce upon hybridization. Nat Biotechnol 14:303–308. https://doi.org/10.1038/nbt0396-303
48. Smalley MK, Silverman SK (2006) Fluorescence of covalently attached pyrene as a general RNA folding probe. Nucleic Acids Res 34:152–166. https://doi.org/10.1093/nar/gkj420
49. Sahoo D, Narayanaswami V, Kay CM, Ryan RO (2000) Pyrene excimer fluorescence: a spatially sensitive probe to monitor lipid-induced helical rearrangement of apolipophorin III. Biochemistry 39:6594–6601. https://doi.org/10.1021/bi992609m

50. Gee KR, Sun WC, Bhalgat MK et al (1999) Fluorogenic substrates based on fluorinated umbelliferones for continuous assays of phosphatases and beta-galactosidases. Anal Biochem 273:41–48. https://doi.org/10.1006/abio.1999.4202
51. Babiak P, Reymond JL (2005) A high-throughput, low-volume enzyme assay on solid support. Anal Chem 77:373–377. https://doi.org/10.1021/ac048611n
52. Yamazaki H, Inoue K, Mimura M et al (1996) 7-ethoxycoumarin O-deethylation catalyzed by cytochromes P450 1A2 and 2E1 in human liver microsomes. Biochem Pharmacol 51:313–319. https://doi.org/10.1016/0006-2952(95)02178-7
53. Zimmerman M, Ashe B, Yurewicz EC, Patel G (1977) Sensitive assays for trypsin, elastase, and chymotrypsin using new fluoroenic substrates. Anal Biochem 78:47–51. https://doi.org/10.1016/0003-2697(77)90006-9
54. Salisbury CM, Maly DJ, Ellman JA (2002) Peptide microarrays for the determination of protease substrate specificity. J Am Chem Soc 124:14868–14870. https://doi.org/10.1021/ja027477q
55. Wegener D, Wirsching F, Riester D, Schwienhorst A (2003) A fluorogenic histone deacetylase assay well suited for high-throughput activity screening. Chem Biol 10:61–68. https://doi.org/10.1016/s1074-5521(02)00305-8
56. Lavis LD, Chao T-Y, Raines RT (2006) Latent blue and red fluorophores based on the trimethyl lock. Chembiochem 7:1151–1154. https://doi.org/10.1002/cbic.200500559
57. Lavis LD, Rutkoski TJ, Raines RT (2007) Tuning the pK(a) of fluorescein to optimize binding assays. Anal Chem 79:6775–6782. https://doi.org/10.1021/ac070907g
58. Sun WC, Gee KR, Klaubert DH, Haugland RP (1997) Synthesis of fluorinated fluoresceins. J Org Chem 62:6469–6475. https://doi.org/10.1021/jo9706178
59. Minta A, Kao JPY, Tsien RY (1989) Fluorescent indicators for cytosolic calcium based on rhodamine and fluorescein chromophores. J Biol Chem 264:8171–8178
60. Inglese J, Johnson RL, Simeonov A et al (2007) High-throughput screening assays for the identification of chemical probes. Nat Chem Biol 3:466–479. https://doi.org/10.1038/nchembio.2007.17
61. Liu JX, Bhalgat M, Zhang CL et al (1999) Fluorescent molecular probes V: a sensitive caspase-3 substrate for fluorometric assays. Bioorg Med Chem Lett 9:3231–3236
62. Kupcho K, Hsiao K, Bulleit B, Goueli SA (2004) A homogeneous, nonradioactive high-throughput fluorogenic protein phosphatase assay. J Biomol Screen 9:223–231. https://doi.org/10.1177/1087057103262840
63. Lavis LD, Chao T-Y, Raines RT (2006) Fluorogenic label for biomolecular imaging. ACS Chem Biol 1:252–260. https://doi.org/10.1021/cb600132m
64. Streu C, Meggers E (2006) Ruthenium-induced allylcarbamate cleavage in living cells. Angew Chem Int Ed 45:5645–5648. https://doi.org/10.1002/anie.200601752
65. Koide Y, Urano Y, Kenmoku S et al (2007) Design and synthesis of fluorescent probes for selective detection of highly reactive oxygen species in mitochondria of living cells. J Am Chem Soc 129:10324–10325. https://doi.org/10.1021/ja073220m
66. Karstens T, Kobs K (1980) Rhodamine-B and rhodamine-101 as reference substances for fluorescence quantum yield measurements. J Phys Chem 84:1871–1872. https://doi.org/10.1021/j100451a030
67. Whitaker JE, Haugland RP, Ryan D et al (1992) Fluorescent rhodol derivatives—versatile, photostable labels and tracers. Anal Biochem 207:267–279. https://doi.org/10.1016/0003-2697(92)90011-u
68. Lukinavičius G, Umezawa K, Olivier N et al (2013) A near-infrared fluorophore for live-cell super-resolution microscopy of cellular proteins. Nat Chem 5:132–139. https://doi.org/10.1038/nchem.1546
69. Lukinavičius G, Blaukopf C, Pershagen E et al (2015) SiR-Hoechst is a far-red DNA stain for live-cell nanoscopy. Nat Commun 6. https://doi.org/10.1038/ncomms9497
70. Lukinavičius G, Reymond L, D'Este E et al (2014) Fluorogenic probes for live-cell imaging of the cytoskeleton. Nat Methods 11:731–733. https://doi.org/10.1038/nmeth.2972

71. Grimm JB, Sung AJ, Legant WR et al (2013) Carbofluoresceins and carborhodamines as scaffolds for high-contrast fluorogenic probes. ACS Chem Biol 8:1303–1310. https://doi.org/10.1021/cb4000822
72. Loudet A, Burgess K (2007) BODIPY dyes and their derivatives: syntheses and spectroscopic properties. Chem Rev 107:4891–4932. https://doi.org/10.1021/cr078381n
73. Marfin YS, Solomonov AV, Timin AS, Rumyantsev EV (2017) Recent advances of individual BODIPY and BODIPY-based functional materials in medical diagnostics and treatment. Curr Med Chem 24:1–28. https://doi.org/10.2174/0929867324666170601092327
74. Karolin J, Johansson LBA, Strandberg L, Ny T (1994) Fluorescence and absorption spectroscopic properties of dipyrrometheneboron difluoride (BODIPY) derivatives in liquids, lipid-membranes, and proteins. J Am Chem Soc 116:7801–7806. https://doi.org/10.1021/ja00096a042
75. Johnson I (1998) Fluorescent probes for living cells. Histochem J 30:123–140. https://doi.org/10.1023/A:1003287101868
76. Banks P, Gosselin M, Prystay L (2000) Impact of a red-shifted dye label for high throughput fluorescence polarization assays of G protein-coupled receptors. J Biomol Screen 5:329–334. https://doi.org/10.1177/108705710000500504
77. Thompson VF, Saldana S, Cong JY, Goll DE (2000) A BODIPY fluorescent microplate assay for measuring activity of calpains and other proteases. Anal Biochem 279:170–178. https://doi.org/10.1006/abio.1999.4475
78. Buschmann V, Weston KD, Sauer M (2003) Spectroscopic study and evaluation of red-absorbing fluorescent dyes. Bioconjug Chem 14:195–204. https://doi.org/10.1021/bc025600x
79. Cosa G, Focsaneanu KS, McLean JRN et al (2001) Photophysical properties of fluorescent DNA-dyes bound to single- and double-stranded DNA in aqueous buffered solution. Photochem Photobiol 73:585–599. https://doi.org/10.1562/0031-8655(2001)073<0585:ppofdd>2.0.co;2
80. Smith JC (1990) Potential-sensitive molecular probes in membranes of bioenergetic relevance. Biochim Biophys Acta 1016:1–28. https://doi.org/10.1016/0005-2728(90)90002-1
81. Plasek J, Sigler K (1996) Slow fluorescent indicators of membrane potential: a survey of different approaches to probe response analysis. J Photochem Photobiol B 33:101–124. https://doi.org/10.1016/1011-1344(96)07283-1
82. Zhou W-L, Yan P, Wuskell JP et al (2007) Intracellular long-wavelength voltage-sensitive dyes dynamics of action potentials in axons and thin for studying the dendrites. J Neurosci Methods 164:225–239. https://doi.org/10.1016/j.jneumeth.2007.05.002
83. Mujumdar SR, Mujumdar RB, Grant CM, Waggoner AS (1996) Cyanine-labeling reagents: sulfobenzindocyanine succinimidyl esters. Bioconjug Chem 7:356–362. https://doi.org/10.1021/bc960021b
84. Waggoner A (1995) Covalent labeling of proteins and nucleic-acids with fluorophores. Biochem Spectrosc 246:362–373
85. Waggoner A (2006) Fluorescent labels for proteomics and genomics. Curr Opin Chem Biol 10:62–66. https://doi.org/10.1016/j.cbpa.2006.01.005
86. Schobel U, Egelhaaf HJ, Brecht A et al (1999) New-donor-acceptor pair for fluorescent immunoassays by energy transfer. Bioconjug Chem 10:1107–1114. https://doi.org/10.1021/bc990073b
87. Bates M, Huang B, Dempsey GT, Zhuang X (2007) Multicolor super-resolution imaging with photo-switchable fluorescent probes. Science 317:1749–1753. https://doi.org/10.1126/science.1146598
88. Grosenick D, Wabnitz H, Ebert B (2012) Recent advances in contrast-enhanced near infrared diffuse optical imaging of diseases using indocyanine green. J Near Infrared Spectrosc 20:203–221. https://doi.org/10.1255/jnirs.964

89. Toyota T, Fujito H, Suganami A et al (2014) Near-infrared-fluorescence imaging of lymph nodes by using liposomally formulated indocyanine green derivatives. Bioorg Med Chem 22:721–727. https://doi.org/10.1016/j.bmc.2013.12.026

90. Alander JT, Kaartinen I, Laakso A et al (2012) A review of indocyanine green fluorescent imaging in surgery. Int J Biomed Imaging 2012:940585

91. Giepmans BNG, Adams SR, Ellisman MH, Tsien RY (2006) Review—the fluorescent toolbox for assessing protein location and function. Science 312:217–224. https://doi.org/10.1126/science.1124618

92. Heim R, Prasher DC, Tsien RY (1994) Wavelength mutations and posttranslational autoxidation of green fluorescent protein. Proc Natl Acad Sci 91:12501–12504. https://doi.org/10.1073/pnas.91.26.12501

93. Kairdolf BA, Qian X, Nie S (2017) Bioconjugated nanoparticles for biosensing, in vivo imaging, and medical diagnostics. Anal Chem 89:1015–1031. https://doi.org/10.1021/acs.analchem.6b04873

94. Khomein P, Swaminathan S, Young ER, Thayumanavan S (2017) Fluorescence enhancement through incorporation of chromophores in polymeric nanoparticles. J Inorg Organomet Polym Mater 28(2):1–7. https://doi.org/10.1007/s10904-017-0670-1

95. Chen W (2008) Nanoparticle fluorescence based technology for biological applications. J Nanosci Nanotechnol 8:1019–1051. https://doi.org/10.1166/jnn.2008.301

96. Fu A, Wilson RJ, Smith BR et al (2012) Fluorescent magnetic nanoparticles for magnetically enhanced cancer imaging and targeting in living subjects. ACS Nano 6:6862–6869. https://doi.org/10.1021/nn301670a

97. Huang H-C, Barua S, Sharma G et al (2011) Inorganic nanoparticles for cancer imaging and therapy. J Control Release 155:344–357. https://doi.org/10.1016/j.jconrel.2011.06.004

98. Cheraghi M, Negahdari B, Daraee H, Eatemadi A (2017) Heart targeted nanoliposomal/nanoparticles drug delivery: an updated review. Biomed Pharmacother 86:316–323. https://doi.org/10.1016/j.biopha.2016.12.009

99. Gao X, Du C, Zhuang Z, Chen W (2016) Carbon quantum dot-based nanoprobes for metal ion detection. J Mater Chem C Mater Opt Electron Devices 4:6927–6945. https://doi.org/10.1039/C6TC02055K

100. Wegner KD, Hildebrandt N (2015) Quantum dots: bright and versatile in vitro and in vivo fluorescence imaging biosensors. Chem Soc Rev 44:4792–4834. https://doi.org/10.1039/C4CS00532E

101. Hou Y, Cao S, Wang L et al (2015) Morphology-controlled dual clickable nanoparticles via ultrasonic-assisted click polymerization. Polym Chem 6:223–227. https://doi.org/10.1039/C4PY01045K

102. Feng Y, Panwar N, Tng DJH et al (2016) The application of mesoporous silica nanoparticle family in cancer theranostics. Coord Chem Rev 319:86–109. https://doi.org/10.1016/j.ccr.2016.04.019

103. Syamchand SS, Sony G (2015) Multifunctional hydroxyapatite nanoparticles for drug delivery and multimodal molecular imaging. Microchim Acta 182:1567–1589. https://doi.org/10.1007/s00604-015-1504-x

104. Xu X, Liu R, Li L (2015) Nanoparticles made of π-conjugated compounds targeted for chemical and biological applications. Chem Commun 51:16733–16749. https://doi.org/10.1039/C5CC06439B

105. Kim K, Lee M, Park H et al (2006) Cell-permeable and biocompatible polymeric nanoparticles for apoptosis imaging. J Am Chem Soc 128:3490–3491. https://doi.org/10.1021/ja057712f

106. Sun M, Li Z-J, Liu C-L et al (2014) Persistent luminescent nanoparticles for super-long time in vivo and in situ imaging with repeatable excitation. J Lumin 145:838–842. https://doi.org/10.1016/j.jlumin.2013.08.070

107. Kang KA, Wang J (2014) Smart dual-mode fluorescent gold nanoparticle agents. Wiley Interdiscip Rev Nanomed Nanobiotechnol 6:398–409. https://doi.org/10.1002/wnan.1267

108. Nedosekin DA, Foster S, Nima ZA et al (2015) Photothermal confocal multicolor microscopy of nanoparticles and nanodrugs in live cells. Drug Metab Rev 47:346–355. https://doi.org/10.3109/03602532.2015.1058818
109. Rosenthal SJ, Chang JC, Kovtun O et al (2011) Biocompatible quantum dots for biological applications. Chem Biol 18:10–24. https://doi.org/10.1016/j.chembiol.2010.11.013
110. Delehanty JB, Mattoussi H, Medintz IL (2009) Delivering quantum dots into cells: strategies, progress and remaining issues. Anal Bioanal Chem 393:1091–1105. https://doi.org/10.1007/s00216-008-2410-4
111. Lin G, Yin F, Yong K-T (2014) The future of quantum dots in drug discovery. Expert Opin Drug Discov 9:991–994. https://doi.org/10.1517/17460441.2014.928280
112. Hoshino A, Fujioka K, Oku T et al (2004) Physicochemical properties and cellular toxicity of nanocrystal quantum dots depend on their surface modification. Nano Lett 4:2163–2169. https://doi.org/10.1021/nl048715d
113. Larson DR, Zipfel WR, Williams RM et al (2003) Water-soluble quantum dots for multiphoton fluorescence imaging in vivo. Science 300:1434–1436. https://doi.org/10.1126/science.1083780
114. Helmchen F, Denk W (2005) Deep tissue two-photon microscopy. Nat Methods 2:932–940. https://doi.org/10.1038/nmeth818
115. Na RH, Stender IM, Ma LX, Wulf HC (2000) Autofluorescence spectrum of skin: component bands and body site variations. Skin Res Technol 6:112–117. https://doi.org/10.1034/j.1600-0846.2000.006003112.x
116. Clarke S, Pinaud F, Beutel O et al (2010) Covalent monofunctionalization of peptide-coated quantum dots for single-molecule assays. Nano Lett 10:2147–2154. https://doi.org/10.1021/nl100825n
117. Ehrensperger M-V, Hanus C, Vannier C et al (2007) Multiple association states between glycine receptors and gephyrin identified by SPT analysis. Biophys J 92:3706–3718. https://doi.org/10.1529/biophysj.106.095596
118. Dertinger T, Colyer R, Iyer G et al (2009) Fast, background-free, 3D super-resolution optical fluctuation imaging (SOFI). Proc Natl Acad Sci U S A 106:22287–22292. https://doi.org/10.1073/pnas.0907866106
119. Algar WR, Susumu K, Delehanty JB, Medintz IL (2011) Semiconductor quantum dots in bioanalysis: crossing the valley of death. Anal Chem 83:8826–8837. https://doi.org/10.1021/ac201331r
120. Smith AM, Nie S (2010) Semiconductor nanocrystals: structure, properties, and band gap engineering. Acc Chem Res 43:190–200. https://doi.org/10.1021/ar9001069
121. Zhong XH, Han MY, Dong ZL et al (2003) Composition-tunable ZnxCd1-xSe nanocrystals with high luminescence and stability. J Am Chem Soc 125:8589–8594. https://doi.org/10.1021/ja035096m
122. Jäger S, Brand L, Eggeling C (2003) New fluorescence techniques for high-throughput drug discovery. Curr Pharm Biotechnol 4:463–476. https://doi.org/10.2174/1389201033377382
123. Xiao L, Guo J (2015) Multiplexed single-cell in situ RNA analysis by reiterative hybridization. Anal Methods 7:7290–7295. https://doi.org/10.1039/c5ay00500k
124. Liu J, Lau SK, Varma VA et al (2010) Multiplexed detection and characterization of rare tumor cells in Hodgkin's lymphoma with multicolor quantum dots. Anal Chem 82:6237–6243. https://doi.org/10.1021/ac101065b
125. Medintz IL, Farrell D, Susumu K et al (2009) Multiplex charge-transfer interactions between quantum dots and peptide-bridged ruthenium complexes. Anal Chem 81:4831–4839. https://doi.org/10.1021/ac900412j
126. Bae SW, Tan W, Hong J-I (2012) Fluorescent dye-doped silica nanoparticles: new tools for bioapplications. Chem Commun 48:2270–2282. https://doi.org/10.1039/c2cc16306c
127. Mochalin VN, Shenderova O, Ho D, Gogotsi Y (2012) The properties and applications of nanodiamonds. Nat Nanotechnol 7:11–23. https://doi.org/10.1038/nnano.2011.209

128. Loh KP, Bao Q, Eda G, Chhowalla M (2010) Graphene oxide as a chemically tunable platform for optical applications. Nat Chem 2:1015–1024. https://doi.org/10.1038/nchem.907
129. Baker SN, Baker GA (2010) Luminescent carbon nanodots: emergent nanolights. Angew Chem Int Ed Engl 49:6726–6744. https://doi.org/10.1002/anie.200906623
130. Wu H-C, Chang X, Liu L et al (2010) Chemistry of carbon nanotubes in biomedical applications. J Mater Chem 20:1036–1052. https://doi.org/10.1039/b911099m
131. Wang F, Banerjee D, Liu Y et al (2010) Upconversion nanoparticles in biological labeling, imaging, and therapy. Analyst 135:1839–1854. https://doi.org/10.1039/c0an00144a
132. Shiang Y-C, Huang C-C, Chen W-Y et al (2012) Fluorescent gold and silver nanoclusters for the analysis of biopolymers and cell imaging. J Mater Chem 22:12972–12982. https://doi.org/10.1039/c2jm30563a
133. Mongin C, Garakyaraghi S, Razgoniaeva N et al (2016) Direct observation of triplet energy transfer from semiconductor nanocrystals. Science 351:369–372. https://doi.org/10.1126/science.aad6378
134. Yang I, Lee JW, Hwang S et al (2017) Live bio-imaging with fully bio-compatible organic fluorophores. J Photochem Photobiol B Biol 166:52–57. https://doi.org/10.1016/j.jphotobiol.2016.11.009
135. Chakraborty C, Hsu C-H, Wen Z-H, Lin C-S (2009) Recent advances of fluorescent technologies for drug discovery and development. Curr Pharm Des 15:3552–3570. https://doi.org/10.2174/138161209789207006
136. Vogel HG (2008) Drug discovery and evaluation: pharmacological assays. Springer, Berlin
137. Shashkova S, Leake MC (2017) Single-molecule fluorescence microscopy review: shedding new light on old problems. Biosci Rep 37. pii: BSR20170031. https://doi.org/10.1042/BSR20170031
138. Choyke PL, Alford R, Simpson HM et al (2009) Toxicity of organic fluorophores used in molecular imaging: literature review. Mol Imaging 8:341–354. https://doi.org/10.2310/7290.2009.00031
139. Gandorfer A, Haritoglou C, Gandorfer A, Kampik A (2003) Retinal damage from indocyanine green in experimental macular surgery. Invest Ophthalmol Vis Sci 44:316–323. https://doi.org/10.1167/iovs.02-0545
140. Kawahara S, Hata Y, Miura M et al (2007) Intracellular events in retinal glial cells exposed to ICG and BBG. Invest Ophthalmol Vis Sci 48:4426–4432. https://doi.org/10.1167/iovs.07-0358
141. Frangioni JV (2003) In vivo near-infrared fluorescence imaging. Curr Opin Chem Biol 7:626–634. https://doi.org/10.1016/j.cbpa.2003.08.007
142. Lai CC, Wu WC, Chuang LH et al (2005) Prevention of indocyanine green toxicity on retinal pigment epithelium with whole blood in stain-assisted macular hole surgery. Ophthalmology 112:1409–1414. https://doi.org/10.1016/j.ophtha.2005.02.025
143. Narayanan R, Kenney M, Kamjoo S et al (2005) Toxicity of indocyanine green (ICG) in combination with light on retinal pigment epithelial cells and neurosensory retinal cells. Curr Eye Res 30:471–478. https://doi.org/10.1080/02713680590959312
144. Enaida H, Sakamoto T, Hisatomi T et al (2002) Morphological and functional damage of the retina caused by intravitreous indocyanine green in rat eyes. Graefes Arch Clin Exp Ophthalmol 240:209–213. https://doi.org/10.1007/s00417-002-0433-7
145. Laperche Y, Oudea MC, Lostanlen D (1977) Toxic effects of indocyanine green on rat liver mitochondria. Toxicol Appl Pharmacol 41:377–387. https://doi.org/10.1016/0041-008X(77)90039-4
146. Yam HF, Kwok AKH, Chan KP et al (2003) Effect of indocyanine green and illumination on gene expression in human retinal pigment epithelial cells. Invest Ophthalmol Vis Sci 44:370–377. https://doi.org/10.1167/iovs.01-1113
147. Haritoglou C, Yu A, Freyer W et al (2005) An evaluation of novel vital dyes for intraocular surgery. Invest Ophthalmol Vis Sci 46:3315–3322. https://doi.org/10.1167/iovs.04-1142

148. Kwok AKH, Lai TYY, Yeung C-K et al (2005) The effects of indocyanine green and endoillumination on rabbit retina: an electroretinographic and histological study. Br J Ophthalmol 89:897–900

149. Abels C, Fickweiler S, Weiderer P et al (2000) Indocyanine green (ICG) and laser irradiation induce photooxidation. Arch Dermatol Res 292:404–411. https://doi.org/10.1007/s004030000147

150. Omar GS, Wilson M, Nair SP (2008) Lethal photosensitization of wound-associated microbes using indocyanine green and near-infrared light. BMC Microbiol 8. https://doi.org/10.1186/1471-2180-8-111

151. Jori G, Brown SB (2004) Photosensitized inactivation of microorganisms. Photochem Photobiol Sci 3:403–405. https://doi.org/10.1039/b311904c

152. Baumler W, Abels C, Karrer S et al (1999) Photo-oxidative killing of human colonic cancer cells using indocyanine green and infrared light. Br J Cancer 80:360–363. https://doi.org/10.1038/sj.bjc.6690363

153. Hope-Ross M, Yannuzzi LA, Gragoudas ES et al (1994) Adverse reactions due to indocyanine green. Ophthalmology 101:529–533

154. Kato S, Madachi-Yamamoto S, Hayashi Y et al (1983) Effect of sodium fluorescein on neurite outgrowth from the retinal explant culture: an in vitro model for retinal toxicity. Dev Brain Res 11:143–147. https://doi.org/10.1016/0165-3806(83)90211-0

155. Placantonakis DG, Tabaee A, Anand VK et al (2007) Safety of low-dose intrathecal fluorescein in endoscopic cranial base surgery. Neurosurgery 61:161–165. https://doi.org/10.1227/01.neu.0000279993.65459.7b

156. Pouliquen H, Algoet M, Buchet V, Le BH (1995) Acute toxicity of fluorescein to turbot (Scophthalmus maximus). Vet Hum Toxicol 37:527–529

157. Susumu K, Oh E, Delehanty JB et al (2011) Multifunctional compact zwitterionic ligands for preparing robust biocompatible semiconductor quantum dots and gold nanoparticles. J Am Chem Soc 133:9480–9496. https://doi.org/10.1021/ja201919s

158. O'Goshi KI, Serup J (2006) Safety of sodium fluorescein for in vivo study of skin. Skin Res Technol 12:155–161. https://doi.org/10.1111/j.0909-752X.2006.00147.x

159. Butner RW, McPherson AR (1983) Adverse reactions in intravenous fluorescein angiography. Ann Ophthalmol 15:1084–1086

160. Lentner A, Boehler U, Bohler U (1995) Photosensitivity reaction to intravenously administered fluorescein. Photodermatol Photoimmunol Photomed 11:178–179. https://doi.org/10.1111/j.1600-0781.1995.tb00163.x

161. Lampidis TJ, Castello C, Del Giglio A et al (1989) Relevance of the chemical charge of rhodamine dyes to multiple drug resistance. Biochem Pharmacol 38:4267–4271. https://doi.org/10.1016/0006-2952(89)90525-X

162. Elliott GS, Mason RW, Edwards IR (1990) Studies on the pharmacokinetics and mutagenic potential of rhodamine b. Clin Toxicol 28:45–59. https://doi.org/10.3109/15563659008993475

163. Hood RD, Jones CL, Ranganathan S (1989) Comparative developmental toxicity of cationic and neutral rhodamines in mice. Teratology 40:143–150. https://doi.org/10.1002/tera.1420400207

164. Ranganathan S, Hood RD (1989) Effects of in vivo and in vitro exposure to rhodamine dyes on mitochondrial function of mouse embryos. Teratog Carcinog Mutagen 9:29–37

165. Nestmann ER, Douglas GR, Matula TI et al (1979) Mutagenic activity of rhodamine dyes and their impurities as detected by mutation induction in Salmonella and DNA damage in Chinese hamster ovary cells. Cancer Res 39:4412–4417

166. Jeannot V, Salmon JM, Deumie M, Viallet P (1997) Intracellular accumulation of rhodamine 110 in single living cells. J Histochem Cytochem 45:403–412

167. Bingaman S, Huxley VH, Rumbaut RE (2003) Fluorescent dyes modify properties of proteins used in microvascular research. Microcirculation 10:221–231. https://doi.org/10.1080/mic.10.2.221.231

168. Rumbaut RE, Sial AJ (1999) Differential phototoxicity of fluorescent dye-labeled albumin conjugates. Microcirculation 6:205–213

169. Lelong IH, Guzikowski AP, Haugland RP et al (1991) Fluorescent verapamil derivative for monitoring activity of the multidrug transporter. Mol Pharmacol 40:490–494

170. Offermanns S, Rosenthal W (2008) Encyclopedia of molecular pharmacology. Springer, Berlin

171. Di Genova BM, da Silva RC, da Cunha JPC et al (2017) Protein SUMOylation is involved in cell-cycle progression and cell morphology in Giardia lamblia. J Eukaryot Microbiol 64:491–503. https://doi.org/10.1111/jeu.12386

172. Tian J, Zhou L, Zhao Y et al (2012) Multiplexed detection of tumor markers with multicolor quantum dots based on fluorescence polarization immunoassay. Talanta 92:72–77. https://doi.org/10.1016/j.talanta.2012.01.051

173. Wu M-S, Shi H-W, He L-J et al (2012) Microchip device with 64-site electrode array for multiplexed immunoassay of cell surface antigens based on electrochemiluminescence resonance energy transfer. Anal Chem 84:4207–4213. https://doi.org/10.1021/ac300551e

174. Mukherjee A, Walker J, Weyant KB, Schroeder CM (2013) Characterization of flavin-based fluorescent proteins: an emerging class of fluorescent reporters. PLoS One 8. https://doi.org/10.1371/journal.pone.0064753

175. Zaslaver A, Bren A, Ronen M et al (2006) A comprehensive library of fluorescent transcriptional reporters for Escherichia coli. Nat Methods 3:623–628. https://doi.org/10.1038/nmeth895

176. Hackett EA, Esch RK, Maleri S, Errede B (2006) A family of destabilized cyan fluorescent proteins as transcriptional reporters in S. cerevisiae. Yeast 23:333–349. https://doi.org/10.1002/yea.1358

177. Wang X, Errede B, Elston TC (2008) Mathematical analysis and quantification of fluorescent proteins as transcriptional reporters. Biophys J 94:2017–2026. https://doi.org/10.1529/biophysj.107.122200

178. Gross S, Piwnica-Worms D (2005) Spying on cancer: molecular imaging in vivo with genetically encoded reporters. Cancer Cell 7:5–15

179. Han KY, Leslie BJ, Fei J et al (2013) Understanding the photophysics of the Spinach-DFHBI RNA aptamer-fluorogen complex to improve live-cell RNA imaging. J Am Chem Soc 135:19033–19038. https://doi.org/10.1021/ja411060p

180. Paige JS, Nguyen-Duc T, Song W, Jaffrey SR (2012) Fluorescence imaging of cellular metabolites with RNA. Science 335:1194

181. Zhang J, Fei J, Leslie BJ et al (2015) Tandem spinach array for mRNA imaging in living bacterial cells. Sci Rep 5. https://doi.org/10.1038/srep17295

182. Pothoulakis G, Ellis T (2015) Using spinach aptamer to correlate mRNA and protein levels in escherichia coli. Methods Enzymol 550:173–185. https://doi.org/10.1016/bs.mie.2014.10.047

183. Kellenberger CA, Hammond MC (2015) In vitro analysis of riboswitch-spinach aptamer fusions as metabolite-sensing fluorescent biosensors. Methods Enzymol 550:147–172. https://doi.org/10.1016/bs.mie.2014.10.045

184. Pothoulakis G, Ceroni F, Reeve B, Ellis T (2014) The Spinach RNA aptamer as a characterization tool for synthetic biology. ACS Synth Biol 3:182–187. https://doi.org/10.1021/sb400089c

185. Pope AJ, Haupts UM, Moore KJ (1999) Homogeneous fluorescence readouts for miniaturized high-throughput screening: theory and practice. Drug Discov Today 4:350–362. https://doi.org/10.1016/S1359-6446(99)01340-9

186. Wölcke J, Ullmann D (2001) Miniaturized HTS technologies—uHTS. Drug Discov Today 6:637–646. https://doi.org/10.1016/S1359-6446(01)01807-4

187. Albani JR (2004) Structure and dynamics of macromolecules: absorption and fluorescence studies. Elsevier B.V, Amsterdam, p 426

188. Milligan G (2004) Applications of bioluminescence- and fluorescence resonance energy transfer to drug discovery at G protein-coupled receptors. Eur J Pharm Sci 21:397–405. https://doi.org/10.1016/j.ejps.2003.11.010

189. Lai W-F, Rogach AL, Wong W-T (2017) Chemistry and engineering of cyclodextrins for molecular imaging. Chem Soc Rev 46:6379–6419. https://doi.org/10.1039/C7CS00040E

190. Zheng P, Wu N (2017) Fluorescence and sensing applications of graphene oxide and graphene quantum dots: a review. Chem Asian J 12:2343–2353. https://doi.org/10.1002/asia.201700814

191. Prasuhn DE, Feltz A, Blanco-Canosa JB et al (2010) Quantum dot peptide biosensors for monitoring caspase 3 proteolysis and calcium ions. ACS Nano 4:5487–5497. https://doi.org/10.1021/nn1016132

192. Algar WR, Tavares AJ, Krull UJ (2010) Beyond labels: a review of the application of quantum dots as integrated components of assays, bioprobes, and biosensors utilizing optical transduction. Anal Chim Acta 673:1–25. https://doi.org/10.1016/j.aca.2010.05.026

193. Melo CV, Okumoto S, Gomes JR et al (2013) Spatiotemporal resolution of Bdnf neuroprotection against glutamate excitotoxicity in cultured hippocampal neurons. Neuroscience 237:66–86. https://doi.org/10.1016/j.neuroscience.2013.01.054

194. Rose T, Goltstein PM, Portugues R, Griesbeck O (2014) Putting a finishing touch on GEC's. Front Mol Neurosci 7. https://doi.org/10.3389/fnmol.2014.00088

195. Sapsford KE, Granek J, Deschamps JR et al (2011) Monitoring botulinum neurotoxin a activity with peptide-functionalized quantum dot resonance energy transfer sensors. ACS Nano 5:2687–2699. https://doi.org/10.1021/nn102997b

196. Butler MS (2004) The role of natural product chemistry in drug discovery. J Nat Prod 67:2141–2153. https://doi.org/10.1021/np040106y

197. Lin M, Gao Y, Diefenbach TJ et al (2017) Facial layer-by-layer engineering of upconversion nanoparticles for gene delivery: near-infrared-initiated fluorescence resonance energy transfer tracking and overcoming drug resistance in ovarian cancer. ACS Appl Mater Interfaces 9:7941–7949. https://doi.org/10.1021/acsami.6b15321

198. Hall MD, Yasgar A, Peryea T et al (2016) Fluorescence polarization assays in high-throughput screening and drug discovery: a review. Methods Appl Fluoresc 4. https://doi.org/10.1088/2050-6120/4/2/022001

199. Algar WR, Malanoski AP, Susumu K et al (2012) Multiplexed tracking of protease activity using a single color of quantum dot vector and a time-gated forster resonance energy transfer relay. Anal Chem 84:10136–10146. https://doi.org/10.1021/ac3028068

200. Huang S, Xiao Q, He ZK et al (2008) A high sensitive and specific QDs FRET bioprobe for MNase. Chem Commun:5990–5992. https://doi.org/10.1039/b815061c

201. Algar WR, Wegner D, Huston AL et al (2012) Quantum dots as simultaneous acceptors and donors in time-gated forster resonance energy transfer relays: characterization and biosensing. J Am Chem Soc 134:1876–1891. https://doi.org/10.1021/ja210162f

202. Algar WR, Krull UJ (2007) Towards multi-colour strategies for the detection of oligonucleotide hybridization using quantum dots as energy donors in fluorescence resonance energy transfer (FRET). Anal Chim Acta 581:193–201. https://doi.org/10.1016/j.aca.2006.08.026

203. Kattke MD, Gao EJ, Sapsford KE et al (2011) FRET-based quantum dot immunoassay for rapid and sensitive detection of Aspergillus amstelodami. Sensors 11:6396–6410. https://doi.org/10.3390/s110606396

204. Dennis AM, Rhee WJ, Sotto D et al (2012) Quantum dot-fluorescent protein FRET probes for sensing intracellular pH. ACS Nano 6:2917–2924. https://doi.org/10.1021/nn2038077

205. Morikawa TJ, Fujita H, Kitamura A et al (2016) Dependence of fluorescent protein brightness on protein concentration in solution and enhancement of it. Sci Rep 6. https://doi.org/10.1038/srep22342

206. Lock JT, Parker I, Smith IF (2015) A comparison of fluorescent Ca2+ indicators for imaging local Ca2+ signals in cultured cells. Cell Calcium 58:638–648. https://doi.org/10.1016/j.ceca.2015.10.003

207. Rodriguez EA, Campbell RE, Lin JY et al (2017) The growing and glowing toolbox of fluorescent and photoactive proteins. Trends Biochem Sci 42:111–129. https://doi.org/10.1016/j.tibs.2016.09.010
208. Broussard GJ, Liang R, Tian L (2014) Monitoring activity in neural circuits with genetically encoded indicators. Front Mol Neurosci 7. https://doi.org/10.3389/fnmol.2014.00097
209. Dana H, Mohar B, Sun Y et al (2016) Sensitive red protein calcium indicators for imaging neural activity. Elife 5. https://doi.org/10.7554/eLife.12727
210. Palmer AE, Qin Y, Park JG, McCombs JE (2011) Design and application of genetically encoded biosensors. Trends Biotechnol 29:144–152. https://doi.org/10.1016/j.tibtech.2010.12.004
211. Akerboom J, Rivera JDV, Guilbe MMR et al (2009) Crystal structures of the GCaMP calcium sensor reveal the mechanism of fluorescence signal change and aid rational design. J Biol Chem 284:6455–6464. https://doi.org/10.1074/jbc.M807657200
212. Wang Q, Shui B, Kotlikoff MI, Sondermann H (2008) Structural basis for calcium sensing by GCaMP2. Structure 16:1817–1827. https://doi.org/10.1016/j.str.2008.10.008
213. Wardill TJ, Chen T-W, Schreiter ER et al (2013) A neuron-based screening platform for optimizing genetically-encoded calcium indicators. PLoS One 8. https://doi.org/10.1371/journal.pone.0077728
214. Chen T-W, Wardill TJ, Sun Y et al (2013) Ultrasensitive fluorescent proteins for imaging neuronal activity. Nature 499:295–300. https://doi.org/10.1038/nature12354
215. Thestrup T, Litzlbauer J, Bartholomaeus I et al (2014) Optimized ratiometric calcium sensors for functional in vivo imaging of neurons and T lymphocytes. Nat Methods 11:175–182. https://doi.org/10.1038/nmeth.2773
216. Kushida Y, Arai Y, Shimono K, Nagai T (2017) Biomimetic chemical sensing by fluorescence signals using a virus-like particle-based platform. ACS Sens. https://doi.org/10.1021/acssensors.7b00537
217. Ni Q, Mehta S, Zhang J (2017) Live-cell imaging of cell signaling using genetically encoded fluorescent reporters. FEBS J:1–17. https://doi.org/10.1111/febs.14134
218. Bruchez M, Moronne M, Gin P et al (1998) Semiconductor nanocrystals as fluorescent biological labels. Science 281:2013–2016. https://doi.org/10.1126/science.281.5385.2013
219. Bruchez MP (2011) Quantum dots find their stride in single molecule tracking. Curr Opin Chem Biol 15:775–780. https://doi.org/10.1016/j.cbpa.2011.10.011
220. Pinaud F, Clarke S, Sittner A, Dahan M (2010) Probing cellular events, one quantum dot at a time. Nat Methods 7:275–285. https://doi.org/10.1038/nmeth.1444
221. Opperwall SR, Divakaran A, Porter EG et al (2012) Wide dynamic range sensing with single quantum dot biosensors. ACS Nano 6:8078–8086. https://doi.org/10.1021/nn303347k
222. Zhang C, Hu J (2010) Single quantum dot-based nanosensor for multiple DNA detection. Anal Chem 82:1921–1927. https://doi.org/10.1021/ac9026675
223. Zhang CY, Yeh HC, Kuroki MT, Wang TH (2005) Single-quantum-dot-based DNA nanosensor. Nat Mater 4:826–831. https://doi.org/10.1038/nmat1508
224. Swift JL, Heuff R, Cramb DT (2006) A two-photon excitation fluorescence cross-correlation assay for a model ligand-receptor binding system using quantum dots. Biophys J 90:1396–1410. https://doi.org/10.1529/biophysj.105.069526
225. Swift JL, Cramb DT (2008) Nanoparticles as fluorescence labels: is size all that matters? Biophys J 95:865–876. https://doi.org/10.1529/biophysj.107.127688
226. Malic L, Sandros MG, Tabrizian M (2011) Designed biointerface using near-infrared quantum dots for ultrasensitive surface plasmon resonance imaging biosensors. Anal Chem 83:5222–5229. https://doi.org/10.1021/ac200465m
227. Han MY, Gao XH, Su JZ, Nie S (2001) Quantum-dot-tagged microbeads for multiplexed optical coding of biomolecules. Nat Biotechnol 19:631–635. https://doi.org/10.1038/90228
228. Agrawal A, Zhang CY, Byassee T et al (2006) Counting single native biomolecules and intact viruses with color-coded nanoparticles. Anal Chem 78:1061–1070. https://doi.org/10.1021/ac051801t

229. Ruan Y, Yu W, Cheng F et al (2012) Detection of prostate stem cell antigen expression in human prostate cancer using quantum-dot-based technology. Sensors 12:5461–5470. https://doi.org/10.3390/s120505461
230. Xing Y, Chaudry Q, Shen C et al (2007) Bioconjugated quantum dots for multiplexed and quantitative immunohistochemistry. Nat Protoc 2:1152–1165. https://doi.org/10.1038/nprot.2007.107
231. Krutzik PO, Nolan GP (2006) Fluorescent cell barcoding in flow cytometry allows high-throughput drug screening and signaling profiling. Nat Methods 3:361–368. https://doi.org/10.1038/nmeth872
232. Gao Y, Stanford WL, Chan WCW (2011) Quantum-dot-encoded microbeads for multiplexed genetic detection of non-amplified DNA samples. Small 7:137–146. https://doi.org/10.1002/smll.201000909
233. Goldman ER, Clapp AR, Anderson GP et al (2004) Multiplexed toxin analysis using four colors of quantum dot fluororeagents. Anal Chem 76:684–688. https://doi.org/10.1021/ac035083r
234. Suzuki M, Husimi Y, Komatsu H et al (2008) Quantum dot FRET biosensors that respond to pH, to proteolytic or nucleolytic cleavage, to DNA synthesis, or to a multiplexing combination. J Am Chem Soc 130:5720–5725. https://doi.org/10.1021/ja710870e
235. Xia Z, Rao J (2009) Biosensing and imaging based on bioluminescence resonance energy transfer. Curr Opin Biotechnol 20:37–44. https://doi.org/10.1016/j.copbio.2009.01.001
236. Bloch S, Lesage F, McIntosh L et al (2005) Whole-body fluorescence lifetime imaging of a tumor-targeted near-infrared molecular probe in mice. J Biomed Opt 10:054003. https://doi.org/10.1117/1.2070148
237. Almutairi A, Akers WJ, Berezin MY et al (2008) Monitoring the biodegradation of dendritic near-infrared nanoprobes by in vivo fluorescence imaging. Mol Pharm 5:1103–1110. https://doi.org/10.1021/mp8000952
238. Tarte K, Klein B (1999) Dendritic cell-based vaccine: a promising approach for cancer immunotherapy. Leukemia 13:653–663
239. Reynolds J, Troy T, Mayer R et al (1999) Imaging of spontaneous canine mammary tumors using fluorescent contrast agents. Photochem Photobiol 70:87–94. https://doi.org/10.1111/j.1751-1097.1999.tb01953.x
240. Wu XY, Liu HJ, Liu JQ et al (2003) Immunofluorescent labeling of cancer marker Her2 and other cellular targets with semiconductor quantum dots. Nat Biotechnol 21:41–46. https://doi.org/10.1038/nbt764
241. Chen H, Xue J, Zhang Y et al (2009) Comparison of quantum dots immunofluorescence histochemistry and conventional immunohistochemistry for the detection of caveolin-1 and PCNA in the lung cancer tissue microarray. J Mol Histol 40:261–268. https://doi.org/10.1007/s10735-009-9237-y
242. Chen C, Peng J, Xia H-S et al (2009) Quantum dots-based immunofluorescence technology for the quantitative determination of HER2 expression in breast cancer. Biomaterials 30:2912–2918. https://doi.org/10.1016/j.biomaterials.2009.02.010
243. Chen C, Sun S-R, Gong Y-P et al (2011) Quantum dots-based molecular classification of breast cancer by quantitative spectroanalysis of hormone receptors and HER2. Biomaterials 32:7592–7599. https://doi.org/10.1016/j.biomaterials.2011.06.029
244. Yang X-Q, Chen C, Peng C-W et al (2011) Quantum dot-based quantitative immunofluorescence detection and spectrum analysis of epidermal growth factor receptor in breast cancer tissue arrays. Int J Nanomedicine 6:2265–2273. https://doi.org/10.2147/ijn.s24161
245. Liu J, Lau SK, Varma VA et al (2010) Molecular mapping of tumor heterogeneity on clinical tissue specimens with multiplexed quantum dots. ACS Nano 4:2755–2765. https://doi.org/10.1021/nn100213v
246. Jennings TL, Becker-Catania SG, Triulzi RC et al (2011) Reactive semiconductor nanocrystals for chemoselective biolabeling and multiplexed analysis. ACS Nano 5:5579–5593. https://doi.org/10.1021/nn201050g

247. Giri S, Sykes EA, Jennings TL, Chan WCW (2011) Rapid screening of genetic biomarkers of infectious agents using quantum dot barcodes. ACS Nano 5:1580–1587. https://doi.org/10.1021/nn102873w

248. Xu HX, Sha MY, Wong EY et al (2003) Multiplexed SNP genotyping using the Qbead (TM) system: a quantum dot-encoded microsphere-based assay. Nucleic Acids Res:31. https://doi.org/10.1093/nar/gng043

249. Gonçalves LCP, Tonelli RR, Bagnaresi P et al (2013) A nature-inspired betalainic probe for live-cell imaging of plasmodium-infected erythrocytes. PLoS One 8. https://doi.org/10.1371/journal.pone.0053874

250. Eigen M, Rigler R (1994) Sorting single molecules: application to diagnostics and evolutionary biotechnology. Proc Natl Acad Sci U S A 91:5740–5747. https://doi.org/10.1073/pnas.91.13.5740

251. Lavery P, Brown MJB, Pope AJ (2001) Simple absorbance-based assays for ultra-high throughput screening. J Biomol Screen 6:3–9. https://doi.org/10.1177/108705710100600102

252. Diaspro A, Chirico G, Collini M (2005) Two-photon fluorescence excitation and related techniques in biological microscopy. Q Rev Biophys 38:97–166. https://doi.org/10.1017/S0033583505004129

253. Wachsmuth M, Conrad C, Bulkescher J et al (2015) High-throughput fluorescence correlation spectroscopy enables analysis of proteome dynamics in living cells. Nat Biotechnol 33:384–389. https://doi.org/10.1038/nbt.3146

254. Rudiger M, Haupts U, Moore KJ, Pope a J (2001) Single-molecule detection technologies in miniaturized high throughput screening: binding assays for G protein-coupled receptors using fluorescence intensity distribution analysis and fluorescence anisotropy. J Biomol Screen 6:29–37. https://doi.org/10.1177/108705710100600105

255. Haupts U, Maiti S, Schwille P, Webb WW (1998) Dynamics of fluorescence fluctuations in green fluorescent protein observed by fluorescence correlation spectroscopy. Proc Natl Acad Sci U S A 95:13573–13578. https://doi.org/10.1073/pnas.95.23.13573

256. Palo K, Brand L, Eggeling C et al (2002) Fluorescence intensity and lifetime distribution analysis: toward higher accuracy in fluorescence fluctuation spectroscopy. Biophys J 83:605–618. https://doi.org/10.1016/S0006-3495(02)75195-3

257. Parker GJ (2000) Development of high throughput screening assays using fluorescence polarization: nuclear receptor-ligand-binding and kinase/phosphatase assays. J Biomol Screen 5:77–88. https://doi.org/10.1177/108705710000500204

258. Banks P, Gosselin M, Prystay L (2000) Fluorescence polarization assays for high throughput screening of G protein-coupled receptors. J Biomol Screen 5:159–167

259. Li Z, Mehdi S, Patel I et al (2000) An ultra-high throughput screening approach for an adenine transferase using fluorescence polarization. J Biomol Screen 5:31–37. https://doi.org/10.1177/108705710000500107

260. Owicki JC (2000) Fluorescence polarization and anisotropy in high throughput screening: perspectives and primer. J Biomol Screen 5:297–306. https://doi.org/10.1177/108705710000500501

261. Qian J, Voorbach MJ, Huth JR et al (2004) Discovery of novel inhibitors of Bcl-xL using multiple high-throughput screening platforms. Anal Biochem 328:131–138. https://doi.org/10.1016/j.ab.2003.12.034

262. Lu Z, Yin Z, James L et al (2004) Development of a fluorescence polarization bead-based coupled assay to target different activity/conformation states of a protein kinase. J Biomol Screen 9:309–321. https://doi.org/10.1177/1087057104263506

263. Gaudet EA, Sen HK, Zhang Y et al (2003) A homogeneous fluorescence polarization assay adaptable for a range of protein serine/threonine and tyrosine kinases. J Biomol Screen 8:164–175. https://doi.org/10.1177/1087057103252309

264. Sportsman JR, Gaudet JD and EA (2003) Fluorescence polarization assays in signal transduction discovery. Comb Chem High Throughput Screen 6:195–200

265. Auer M, Moore KJ, Meyer-Almes FJ et al (1998) Fluorescence correlation spectroscopy: lead discovery by miniaturized HTS. Drug Discov Today 3:457–465

266. Schwille P, Kummer S, Heikal AA et al (2000) Fluorescence correlation spectroscopy reveals fast optical excitation-driven intramolecular dynamics of yellow fluorescent proteins. Proc Natl Acad Sci U S A 97:151–156. https://doi.org/10.1073/pnas.97.1.151

267. Koltermann A, Kettling U, Bieschke J et al (1998) Rapid assay processing by integration of dual-color fluorescence cross-correlation spectroscopy: high throughput screening for enzyme activity. Proc Natl Acad Sci U S A 95:1421–1426. https://doi.org/10.1073/pnas.95.4.1421

268. Beasley JR, Dunn DA, Walker TL et al (2003) Evaluation of compound interference in immobilized metal ion affinity-based fluorescence polarization detection with a four million member compound collection. Assay Drug Dev Technol 1:455–459. https://doi.org/10.1089/154065803322163768

269. Loomans EEMG, van Doornmalen AM, Wat JWY, Zaman GJR (2003) High-throughput screening with immobilized metal ion affinity-based fluorescence polarization detection, a homogeneous assay for protein kinases. Assay Drug Dev Technol 1:445–453. https://doi.org/10.1089/154065803322163759

270. Klumpp M, Boettcher A, Becker D et al (2006) Readout technologies for highly miniaturized kinase assays applicable to high-throughput screening in a 1536-well format. J Biomol Screen 11:617–633. https://doi.org/10.1177/1087057106288444

271. Markela E, Ståhlberg TH, Hemmilä I (1993) Europium-labelled recombinant protein G. A fast and sensitive universal immunoreagent for time-resolved immunofluorometry. J Immunol Methods 161:1–6. https://doi.org/10.1016/0022-1759(93)90192-A

272. Thibon A, Pierre VC (2009) Principles of responsive lanthanide-based luminescent probes for cellular imaging. Anal Bioanal Chem 394:107–120

273. Wang X, Chang H, Xie J et al (2014) Recent developments in lanthanide-based luminescent probes. Coord Chem Rev 273–274:201–212

274. Selvin PR (2002) Principles and biophysical applications of lanthanide-based probes. Annu Rev Biophys Biomol Struct 31:275–302. https://doi.org/10.1146/annurev.biophys.31.101101.140927

275. Nagai T, Yamada S, Tominaga T et al (2004) Expanded dynamic range of fluorescent indicators for Ca2+ by circularly permuted yellow fluorescent proteins. Proc Natl Acad Sci 101:10554–10559. https://doi.org/10.1073/pnas.0400417101

276. Chin J, Adams AD, Bouffard A et al (2003) Miniaturization of cell-based beta-lactamase-dependent FRET assays to ultra-high throughput formats to identify agonists of human liver X receptors. Assay Drug Dev Technol 1:777–787. https://doi.org/10.1089/154065803772613417

277. Eggeling C, Kask P, Winkler D, Jäger S (2005) Rapid analysis of Förster resonance energy transfer by two-color global fluorescence correlation spectroscopy: trypsin proteinase reaction. Biophys J 89:605–618. https://doi.org/10.1529/biophysj.104.052753

278. Lundin K, Blomberg K, Nordström T, Lindqvist C (2001) Development of a time-resolved fluorescence resonance energy transfer assay (cell TR-FRET) for protein detection on intact cells. Anal Biochem 299:92–97. https://doi.org/10.1006/abio.2001.5370

279. Moshinsky DJ, Ruslim L, Blake RA, Tang F (2003) A widely applicable, high-throughput TR-FRET assay for the measurement of kinase autophosphorylation: VEGFR-2 as a proto-type. J Biomol Screen 8:447–452. https://doi.org/10.1177/1087057103255282

280. Glickman JF, Wu X, Mercuri R et al (2002) A comparison of ALPHAScreen, TR-FRET, and TRF as assay methods for FXR nuclear receptors. J Biomol Screen 7:3–10. https://doi.org/10.1177/108705710200700102

281. Newman M, Josiah S (2004) Utilization of fluorescence polarization and time resolved fluorescence resonance energy transfer assay formats for SAR studies: Src kinase as a model system. J Biomol Screen 9:525–532. https://doi.org/10.1177/1087057104264597

282. Swift JL, Burger MC, Massotte D et al (2007) Two-photon excitation fluorescence cross-correlation assay for ligand-receptor binding: cell membrane nanopatches containing the human mu-opioid receptor. Anal Chem 79:6783–6791. https://doi.org/10.1021/ac0709495

283. Blades ML, Grekova E, Wobma HM et al (2012) Three-color fluorescence cross-correlation spectroscopy for analyzing complex nanoparticle mixtures. Anal Chem 84:9623–9631. https://doi.org/10.1021/ac302572k

284. Heuff RF, Swift JL, Cramb DT (2007) Fluorescence correlation spectroscopy using quantum dots: advances, challenges and opportunities. Phys Chem Chem Phys 9:1870–1880. https://doi.org/10.1039/b617115j

285. Wobma HM, Blades ML, Grekova E et al (2012) The development of direct multicolour fluorescence cross-correlation spectroscopy: towards a new tool for tracking complex biomolecular events in real-time. Phys Chem Chem Phys 14:3290–3294. https://doi.org/10.1039/c2cp23278b

286. Jaiswal JK, Mattoussi H, Mauro JM, Simon SM (2003) Long-term multiple color imaging of live cells using quantum dot bioconjugates. Nat Biotechnol 21:47–51. https://doi.org/10.1038/nbt.767

287. Delehanty JB, Bradburne CE, Susumu K et al (2011) Spatiotemporal multicolor labeling of individual cells using peptide-functionalized quantum dots and mixed delivery techniques. J Am Chem Soc 133:10482–10489. https://doi.org/10.1021/ja200555z

288. Chan WCW, Nie SM (1998) Quantum dot bioconjugates for ultrasensitive nonisotopic detection. Science 281:2016–2018. https://doi.org/10.1126/science.281.5385.2016

289. Matsuno A, Itoh J, Takekoshi S et al (2005) Three-dimensional imaging of the intracellular localization of growth hormone and prolactin and their mRNA using nanocrystal (Quantum dot) and confocal laser scanning microscopy techniques. J Histochem Cytochem 53:833–838. https://doi.org/10.1369/jhc.4A6577.2005

290. Chan PM, Yuen T, Ruf F et al (2005) Method for multiplex cellular detection of mRNAs using quantum dot fluorescent in situ hybridization. Nucleic Acids Res 33. https://doi.org/10.1093/nar/gni162

291. Lee J, Kwon Y-J, Choi Y et al (2012) Quantum dot-based screening system for discovery of G protein-coupled receptor agonists. Chembiochem 13:1503–1508. https://doi.org/10.1002/cbic.201200128

292. DaCosta RS, Wilson BC, Marcon NE (2007) Fluorescence and spectral imaging. ScientificWorldJournal 7:2046–2071. https://doi.org/10.1100/tsw.2007.308

293. Hagen N, Kudenov MW (2013) Review of snapshot spectral imaging technologies. Opt Eng 52:090901. https://doi.org/10.1117/1.OE.52.9.090901

294. Haraguchi T, Shimi T, Koujin T et al (2002) Spectral imaging fluorescence microscopy. Genes Cells 7:881–887

295. Garini Y, Young IT, McNamara G (2006) Spectral imaging: principles and applications. Cytom A 69:735–747

296. Rice WL, Kaplan DL, Georgakoudi I (2010) Two-photon microscopy for non-invasive, quantitative monitoring of stem cell differentiation. PLoS One 5. https://doi.org/10.1371/journal.pone.0010075

297. Varshni YP (1967) Temperature dependence of energy gap in semiconductors. Physica 34:149–154. https://doi.org/10.1016/0031-8914(67)90062-6

298. Yang J-M, Yang H, Lin L (2011) Quantum dot nano thermometers reveal heterogeneous local thermogenesis in living cells. ACS Nano 5:5067–5071. https://doi.org/10.1021/nn201142f

299. He GS, Tan L-S, Zheng Q, Prasad PN (2008) Multiphoton absorbing materials: molecular designs, characterizations, and applications. Chem Rev 108:1245–1330. https://doi.org/10.1021/cr050054x

300. Mariz IFA, Pinto S, Lavrado J et al (2017) Cryptolepine and quindoline: understanding their photophysics. Phys Chem Chem Phys 19:10255–10263. https://doi.org/10.1039/C7CP00455A

301. Rodrigues ACB, Mariz I de FA, Maçoas EMS et al (2018) Bioinspired water-soluble two-photon fluorophores. Dye Pigment 150:105–111. https://doi.org/10.1016/j.dyepig.2017.11.020

302. Rodrigues CAB, Mariz IFA, MaçÔas EMS et al (2012) Two-photon absorption properties of push-pull oxazolones derivatives. Dye Pigment 95:713–722. https://doi.org/10.1016/j.dyepig.2012.06.005

303. Pawlicki M, Collins HA, Denning RG, Anderson HL (2009) Two-photon absorption and the design of two-photon dyes. Angew Chem Int Ed 48:3244–3266

304. Resch-Genger U, Grabolle M, Cavaliere-Jaricot S et al (2008) Quantum dots versus organic dyes as fluorescent labels. Nat Methods 5:763–775. https://doi.org/10.1038/nmeth.1248

305. Wang T, Chen J-Y, Zhen S et al (2009) Thiol-capped CdTe quantum dots with two-photon excitation for imaging high autofluorescence background living cells. J Fluoresc 19:615–621. https://doi.org/10.1007/s10895-008-0452-9

306. Bharali DJ, Lucey DW, Jayakumar H et al (2005) Folate-receptor-mediated delivery of InP quantum dots for bioimaging using confocal and two-photon microscopy. J Am Chem Soc 127:11364–11371. https://doi.org/10.1021/ja051455x

307. Geszke M, Murias M, Balan L et al (2011) Folic acid-conjugated core/shell ZnS:Mn/ZnS quantum dots as targeted probes for two photon fluorescence imaging of cancer cells. Acta Biomater 7:1327–1338. https://doi.org/10.1016/j.actbio.2010.10.012

308. Tu C, Ma X, Pantazis P et al (2010) Paramagnetic, silicon quantum dots for magnetic resonance and two-photon imaging of macrophages. J Am Chem Soc 132:2016–2023. https://doi.org/10.1021/ja909303g

309. Lidke KA, Rieger B, Jovin TM, Heintzmann R (2005) Superresolution by localization of quantum dots using blinking statistics. Opt Express 13:7052–7062. https://doi.org/10.1364/opex.13.007052

310. Chien F-C, Kuo CW, Chen P (2011) Localization imaging using blinking quantum dots. Analyst 136:1608–1613. https://doi.org/10.1039/c0an00859a

311. Tam J, Cordier GA, Bálint Š et al (2014) A microfluidic platform for correlative live-cell and super-resolution microscopy. PLoS One 9:1–20. https://doi.org/10.1371/journal.pone.0115512

312. Chen Y, Shao L, Ali Z et al (2008) NSOM/QD-based nanoscale immunofluorescence imaging of antigen-specific T-cell receptor responses during an in vivo clonal V gamma 2V delta 2 T-cell expansion. Blood 111:4220–4232. https://doi.org/10.1182/blood-2007-07-101691

313. Chen J, Wu Y, Wang C, Cai J (2008) Nanoscale organization of CD4 molecules of human T helper cell mapped by NSOM and quantum dots. Scanning 30:448–451. https://doi.org/10.1002/sca.20128

314. Chen J, Pei Y, Chen Z, Cai J (2010) Quantum dot labeling based on near-field optical imaging of CD44 molecules. Micron 41:198–202. https://doi.org/10.1016/j.micron.2009.11.002

315. Zhong L, Zeng G, Lu X et al (2009) NSOM/QD-based direct visualization of CD3-induced and CD28-enhanced nanospatial coclustering of TCR and coreceptor in nanodomains in T Cell activation. PLoS One 4. https://doi.org/10.1371/journal.pone.0005945

316. Walker K-AD, Morgan C, Doak SH, Dunstan PR (2012) Quantum dots for multiplexed detection and characterisation of prostate cancer cells using a scanning near-field optical microscope. PLoS One 7:e31592. https://doi.org/10.1371/journal.pone.0031592

317. Walker GW, Sundar VC, Rudzinski CM et al (2003) Quantum-dot optical temperature probes. Appl Phys Lett 83:3555–3557. https://doi.org/10.1063/1.1620686

318. Maestro LM, Jacinto C, Silva UR et al (2011) CdTe quantum dots as nanothermometers: towards highly sensitive thermal imaging. Small 7:1774–1778. https://doi.org/10.1002/smll.201002377

319. Haro-Gonzalez P, Martinez-Maestro L, Martin IR et al (2012) High-sensitivity fluorescence lifetime thermal sensing based on CdTe quantum dots. Small 8:2652–2658. https://doi.org/10.1002/smll.201102736

320. Pai RK, Cotlet M (2011) Highly stable, water-soluble, intrinsic fluorescent hybrid scaffolds for imaging and biosensing. J Phys Chem C 115:1674–1681. https://doi.org/10.1021/jp109589h

321. Boeneman K, Mei BC, Dennis AM et al (2009) Sensing caspase 3 activity with quantum dot-fluorescent protein assemblies. J Am Chem Soc 131:3828–3829. https://doi.org/10.1021/ja809721j
322. Lowe SB, Dick JAG, Cohen BE, Stevens MM (2012) Multiplex sensing of protease and kinase enzyme activity via orthogonal coupling of quantum dot peptide conjugates. ACS Nano 6:851–857. https://doi.org/10.1021/nn204361s
323. Ruedas-Rama MJ, Orte A, Hall EAH et al (2012) A chloride ion nanosensor for time-resolved fluorimetry and fluorescence lifetime imaging. Analyst 137:1500–1508. https://doi.org/10.1039/c2an15851e
324. Provenzano PP, Eliceiri KW, Keely PJ (2009) Multiphoton microscopy and fluorescence lifetime imaging microscopy (FLIM) to monitor metastasis and the tumor microenvironment. Clin Exp Metastasis 26:357–370. https://doi.org/10.1007/s10585-008-9204-0
325. Sun Y, Phipps J, Elson DS et al (2009) Fluorescence lifetime imaging microscopy: in vivo application to diagnosis of oral carcinoma. Opt Lett 34:2081–2083. https://doi.org/10.1364/ol.34.002081
326. Shcherbo D, Souslova EA, Goedhart J et al (2009) Practical and reliable FRET/FLIM pair of fluorescent proteins. BMC Biotechnol 9. https://doi.org/10.1186/1472-6750-9-24
327. Barroso M, Sun Y, Wallrabe H, Periasamy A (2013) Nanometer-scale measurements using FRET and FLIM microscopy. In: Gilmore AM (ed) Luminescence: the instrumental key to the future of nanotechnology. CRC Press, Boca Raton, pp 259–290
328. Rudkouskaya A, Sinsuebphon N, Intes X et al (2017) Fluorescence lifetime FRET imaging of receptor-ligand complexes in tumor cells in vitro and in vivo. Proc SPIE 10049:1006917. https://doi.org/10.1117/12.2258231
329. Chandler A, Chandler A, Wallrabe H (2017) Differential levels of metabolic activity in isolated versus confluent/partially confluent HeLa cells are analyzed by autofluorescent NAD (P) H using multi-photon FLIM microscopy. Proc SPIE 10069:1–6. https://doi.org/10.1117/12.2267657
330. Alpert A, Altenburg M, Bailey D, Barnes L, Barnes L (2009) Milady's standard cosmetology. Cengage Learning, Clifton Park, p 897
331. Zimnyakov DA, Tuchin VV (2002) Optical tomography of tissues. Quantum Electron 32:849–867. https://doi.org/10.1070/QE2002v032n10ABEH00
332. Huang D, Swanson EA, Lin CP et al (1991) Optical coherence tomography. Science 254:1178–1181
333. Campagnola P (2011) Second harmonic generation imaging microscopy: applications to diseases diagnostics. Anal Chem 83:3224–3231. https://doi.org/10.1021/ac1032325
334. Gibb L, Matthews D (2002) Two photon microscopy and second harmonic generation. Report
335. Stoller P, Celliers PM, Reiser KM, Rubenchik AM (2003) Quantitative second-harmonic generation microscopy in collagen. Appl Optics 42:5209. https://doi.org/10.1364/AO.42.005209
336. Ntziachristos V (2006) Fluorescence molecular imaging. Annu Rev Biomed Eng 8:1–33. https://doi.org/10.1146/annurev.bioeng.8.061505.095831
337. Hilderbrand SA, Weissleder R (2010) Near-infrared fluorescence: application to in vivo molecular imaging. Curr Opin Chem Biol 14:71–79
338. Wang Y, Shyy JY-J, Chien S (2008) Fluorescence proteins, live-cell imaging, and mechanobiology: seeing is believing. Annu Rev Biomed Eng 10:1–38. https://doi.org/10.1146/annurev.bioeng.010308.161731
339. Ntziachristos V, Bremer C, Weissleder R (2003) Fluorescence imaging with near-infrared light: new technological advances that enable in vivo molecular imaging. Eur Radiol 13:195–208. https://doi.org/10.1007/s00330-002-1524-x
340. Venus M, Waterman J, McNab I (2011) Basic physiology of the skin. Surgery 29:471–474. https://doi.org/10.1016/j.mpsur.2011.06.010
341. Farage MA, Miller KW, Maibach HI (2017) Textbook of aging skin, 2nd edn. Springer-Verlag, Berlin, p 616

342. Kollias N, Zonios G, Stamatas GN (2002) Fluorescence spectroscopy of skin. Vib Spectrosc 28:17–23. https://doi.org/10.1016/S0924-2031(01)00142-4
343. McMullen RL, Chen S, Moore DJ (2012) Spectrofluorescence of skin and hair. Int J Cosmet Sci 34:246–256. https://doi.org/10.1111/j.1468-2494.2012.00709.x
344. Seidenari S, Arginelli F, Bassoli S et al (2012) Multiphoton laser microscopy and fluorescence lifetime imaging for the evaluation of the skin. Dermatol Res Pract 2012:1–9. https://doi.org/10.1155/2012/810749
345. Benati E, Bellini V, Borsari S et al (2011) Quantitative evaluation of healthy epidermis by means of multiphoton microscopy and fluorescence lifetime imaging microscopy. Skin Res Technol 17:295–303. https://doi.org/10.1111/j.1600-0846.2011.00496.x
346. Leffell DJ, Stetz ML, Milstone LM, Deckelbaum LI (1988) In vivo fluorescence of human skin: a potential marker of photoaging. Arch Dermatol 124:1514–1518
347. Campisi J (1997) The biology of replicative senescence. Eur J Cancer A 33:703–709
348. Cristofalo VJ, Lorenzini A, Allen RG et al (2004) Replicative senescence: a critical review. Mech Ageing Dev 125:827–848
349. Singh R, Barden A, Mori T, Beilin L (2001) Advanced glycation end-products: a review. Diabetologia 44:129–146
350. Ott C, Jacobs K, Haucke E et al (2014) Role of advanced glycation end products in cellular signaling. Redox Biol 2:411–429
351. Daniel CR, Piraccini BM, Tosti A (2004) The nail and hair in forensic science. J Am Acad Dermatol 50:258–261
352. Tobin DJ (2008) Human hair pigmentation—biological aspects. Int J Cosmet Sci 30:233–257
353. Yang CC, Cotsarelis G (2010) Review of hair follicle dermal cells. J Dermatol Sci 57:2–11
354. Westgate GE, Botchkareva NV, Tobin DJ (2013) The biology of hair diversity. Int J Cosmet Sci 35:329–336
355. Meredith P, Sarna T (2006) The physical and chemical properties of eumelanin. Pigment Cell Res 19:572–594
356. Daly S, Bianchini R, Polefka T et al (2009) Fluorescence and coloration of grey hair. Int J Cosmet Sci 31:347–359. https://doi.org/10.1111/j.1468-2494.2009.00500.x
357. Kayatz P, Thumann G, Luther TT et al (2001) Oxidation causes melanin fluorescence. Invest Ophthalmol Vis Sci 42:241–246
358. Freis O, Gauché D, Griesbach U, Haake H-M (2010) Fluorescence laser scanning confocal microscopy to assess the penetration of low molecular protein hydrolyzates into hair. Cosmet Toilet 125:30–35
359. DeLauder SF, Kidwell DA (2000) The incorporation of dyes into hair as a model for drug binding. Forensic Sci Int 107:93–104. https://doi.org/10.1016/S0379-0738(99)00153-X
360. FDA approved fluorescent colors. https://www.fda.gov/ForIndustry/ColorAdditives/ColorAdditiveInventories/ucm115641.htm
361. FDA 2. https://www.fda.gov/forindustry/coloradditives/coloradditivesinspecificproducts/incosmetics/ucm110032.htm
362. Madhusudan Rao Y, Shayeda, Sujatha P (2008) Formulation and evaluation of commonly used natural hair colorants. Indian J Nat Prod Resour 7:45–48
363. Bechtold T (2009) Natural colorants in hair dyeing. In: Bechtold T, Mussak R (eds) Handbook of natural colorants. John Wiley & Sons Ltd, Chichester, pp 339–350
364. Allam KV, Kumar GP (2011) Colorants—the cosmetics for the pharmaceutical dosage forms. Int J Pharm Pharm Sci 3:13–21
365. Rosenthal I, Yang GC, Bell SJ, Scher AL (1988) The chemical photosensitizing ability of certified colour additives. Food Addit Contam 5:563–571. https://doi.org/10.1080/02652038809373719
366. Laux P, Tralau T, Tentschert J et al (2016) A medical-toxicological view of tattooing. Lancet 387:395–402. https://doi.org/10.1016/S0140-6736(15)60215-X

367. Liu F, Fang Y, Chen Y, Liu J (2011) Dissociative excitation energy transfer in the reactions of protonated cysteine and tryptophan with electronically excited singlet molecular oxygen (a1Δg). J Phys Chem B 115:9898–9909. https://doi.org/10.1021/jp205235d
368. Jachowicz J, McMullen RL (2011) Tryptophan fluorescence in hair-examination of contributing factors. J Cosmet Sci 62:291–304
369. Chiarelli-Neto O, Pavani C, Ferreira AS et al (2011) Generation and suppression of singlet oxygen in hair by photosensitization of melanin. Free Radic Biol Med 51:1195–1202. https://doi.org/10.1016/j.freeradbiomed.2011.06.013
370. Nofsinger JB, Liu Y, Simon JD (2002) Aggregation of eumelanin mitigates photogeneration of reactive oxygen species. Free Radic Biol Med 32:720–730. https://doi.org/10.1016/S0891-5849(02)00763-3
371. Chiarelli-Neto O, Ferreira AS, Martins WK et al (2014) Melanin photosensitization and the effect of visible light on epithelial cells. PLoS One 9. https://doi.org/10.1371/journal.pone.0113266
372. Syed IA, Najdek L, Ionita-Manzatu MC, Susak M (2014) Topical compositions comprising inorganic particulates and an alkoxylated diphenylacrylate compound. United States ELC Management LLC, New York. Patent #8765156
373. Tonolli PN, Neto OC, Santacruz-Perez C et al (2017) Lipofuscin generated by UVA exposure makes human skin keratinocytes sensitive to visible light. Free Radic Biol Med 112:65–66. https://doi.org/10.1016/j.freeradbiomed.2017.10.093

Instrumentation for Fluorescence Lifetime Measurement Using Photon Counting

David J. S. Birch, Graham Hungerford, David McLoskey, Kulwinder Sagoo, and Philip Yip

Contents

Abstract We describe the evolution of HORIBA Jobin Yvon IBH Ltd, and its time-correlated single-photon counting (TCSPC) products, from university research beginnings through to its present place as a market leader in fluorescence lifetime spectroscopy. The company philosophy is to ensure leading-edge research capabilities continue to be incorporated into instruments in order to meet the needs of the diverse range of customer applications, which span a multitude of scientific and engineering disciplines. We illustrate some of the range of activities of a scientific instrument company in meeting this goal and highlight by way of an exemplar the performance of the versatile *DeltaFlex* instrument in measuring fluorescence lifetimes. This includes resolving fluorescence lifetimes down to 5 ps, as frequently observed in energy transfer, nanoparticle metrology with sub-nanometre resolution

D. J. S. Birch (✉) and P. Yip
Photophysics Research Group, Department of Physics, Scottish Universities Physics Alliance, University of Strathclyde, Glasgow, Scotland, UK

HORIBA Jobin Yvon IBH Ltd, Glasgow, Scotland, UK
e-mail: Djs.birch@strath.ac.uk

G. Hungerford, D. McLoskey, and K. Sagoo
HORIBA Jobin Yvon IBH Ltd, Glasgow, Scotland, UK

© Springer Nature Switzerland AG 2019
B. Pedras (ed.), *Fluorescence in Industry*, Springer Ser Fluoresc (2019) 18: 103–134,
https://doi.org/10.1007/4243_2018_2, Published online: 2 February 2019

and measuring a fluorescence lifetime in as little as 60 μs for the study of transient species and kinetics.

Keywords *DeltaFlex* · Fluorescence lifetime · HORIBA Jobin Yvon IBH · Photon counting · TCSPC

1 Introduction

Fluorescence is a phenomenon that finds application across numerous disciplines, from life sciences and biomedicine to nanotechnology and materials. Fluorescence is not only a powerful molecular research tool but also underpins critically important breakthroughs in techniques [1, 2]. This is perhaps most notable in healthcare, where fluorescence is ubiquitous in disease diagnostics and provided the basis for sequencing the human genome. Understanding, controlling and designing fluorescence assays relies on being able to accurately and precisely measure the properties of fluorescence. This quickly comes down to the design and functionality of instrumentation. Our company, HORIBA Jobin Yvon IBH Ltd, specializes in the design, development and manufacture of fluorescence lifetime systems and associated components and software, the fluorescence lifetime τ being the exponential constant describing the decay of fluorescence following δ-function impulse excitation.

The company is one of the very earliest spin-out companies from a Scottish University, celebrating the 40th anniversary of its founding in 2017. Now, as part of HORIBA, it plays a key role in HORIBA's present-day market-leading position in sales of fluorescence spectrometers [3]. The global market for analytical instrumentation exceeds \$50Bn and within this figure fluorescence spectroscopy and microscopy total \$500 m with fluorescence lifetime spectroscopy the most rapidly growing part of the whole fluorescence market, increasing at 8% compound annual growth rate [3]. The reason for this is because fluorescence lifetime measurements bring with them not only the salient properties of fluorescence such as a high sensitivity (down to the single-molecule limit), spectral specificity and non-destructive nature but additional advantages which include:

1. Time-resolved capability for revealing dynamic information and kinetic rate parameters
2. Increased specificity by means of temporal discrimination against background and unwanted fluorescence
3. Independence from fluorophore concentration changes such as those caused by photobleaching
4. Ease of calibration and comparison between samples, which contrasts with the complexity of absolute fluorescence intensity and quantum yield measurements

Put simply the benefits of fluorescence lifetimes compared to steady-state fluorescence are analogous to the motion picture compared to the still photograph, the former providing everything the latter does and much more besides.

It was with these opportunities in mind that in 1978, co-author David Birch moved to a lectureship in physics at the University of Strathclyde in Glasgow hoping to bridge between what was then a quite large separation globally between the university and industry sectors. He joined former colleagues from the Physics Department at the University of Manchester, Bob Imhof and Tony Hallam, having together formed IBH Consultants Ltd in 1977. At the time, two techniques for fluorescence lifetime measurement were starting to emerge from early research [4, 5]: time-correlated single-photon counting (TCSPC) and phase modulation. IBH worked on developing TCSPC [6, 7] to the level where today it is now firmly established as the world's favourite. However, in 1977, it was not clear that either of the techniques would "stand the test of time" and support a business as back then fluorescence lifetime measurement was still the preserve of the specialist. Certainly, there was not the established market there is today. A NATO ASI meeting in St Andrews in 1980 [8] brought many of the major exponents of the two techniques together without resolving the best way forward for the field. This was partly because few applications at the time were exposing any deficiencies in either technique by stretching them to their limits and in particular single-molecule detection had yet to be demonstrated and become dominated by photon counting. Originally applications were clustered around chemistry, but these expanded a long time ago to embrace areas as diverse as the life sciences, medicine, nanotechnology, materials and energy, thus widening the customer base and doing much to support our business growth.

The fact that there was no established TCSPC market back in the late 1970s and early 1980s also had advantages for us in that we were in the enviable and unusual position of being able to build the market the way we thought it should look and naturally this reflected our beliefs and experience. The fact that our founders were also active researchers meant we had a good knowledge and empathy with the needs of many of the applications of interest to our customers. With PhD theses in excited state decay time measurement, and experience of working in industry behind them, our founders set about looking for a manufacturer to turn their ideas into products. Initially Edinburgh-based Nuclear Enterprises, who manufactured exactly the NIM (Nuclear Instrument Modules) needed for TCSPC, showed interest. However, it was by attracting Edinburgh Instruments to the market that the early IBH designs and decay analysis software became available to customers for the first time in the *Model 199*. Other companies producing TCSPC instruments around that time, and who subsequently departed the TCSPC market, included Photochemical Research Associates Inc., a spin-out from the University of Western Ontario, and Applied Photophysics Ltd., a spin-out from the Royal Institution in London. Around 1990 our company became a manufacturer of complete systems in its own right and then in 2003 joined HORIBA to form HORIBA Jobin Yvon IBH Ltd. This brought a TCSPC range to HORIBA to complement its steady-state fluorimeters and provided increased global exposure for our products.

The philosophy which drove the company to success in the early days still inspires it today. This is based on satisfying the need for the latest research capabilities to be made widely available to customers. The most important thing for any company is to survive as without survival nothing is possible. However, good staff, products, marketing, customer support, etc. can then become the means, not only for survival but for growth. In the following sections, we try to bring out the meaning of "good products" from the standpoint of choosing a fluorescence lifetime spectrometer and aspects to look out for. We will cover the major considerations involved in this decision-making and illustrate them with examples in fluorescence lifetime spectroscopy while keeping in mind that they can be easily adapted across many other areas of scientific instrumentation, particularly those in other areas of spectroscopy. Decision-making criteria naturally include performance specifications but of comparable standing are ease of use, functionality, versatility, reliability and in some cases modularity and the flexibility to upgrade so as to keep up with new developments. The latter in particular align perfectly with our company philosophy of making the latest cutting-edge techniques available so that customers can then "explore the future" with confidence.

2 Fluorescence Lifetime Systems

2.1 From Our Instrument Origins to Today

Like many instrument companies, our origins lie in University research, initially in the Schuster Laboratories at the University of Manchester and subsequently in the Department of Physics at the University of Strathclyde in Glasgow. From these origins, our products continue to have a close relationship with research requirements. Figure 1 shows a photograph of the TCSPC fluorescence lifetime spectrometer system which started us out in the 1970s on this "journey of a lifetime" [9, 10]. Features that were then novel included the use of optical monochromators to select both the excitation and emission wavelengths and an all-metal thyratrongated coaxial nanosecond flashlamp, the latter designed and constructed in-house. The use of an emission monochromator, rather than cut-off filters used previously to select fluorescence, opened the door to improved spectroscopic capabilities, such as time-resolved emission spectra and reduced stray light detection. A bespoke cryostat included a temperature-regulated, manually controlled 4-sample turret, which was kept under vacuum to eliminate oxygen quenching and frost formation at low temperatures. The all-metal flashlamp generated a spark, was usually filled with nitrogen or hydrogen, operated stably at 30 kHz and had the salient advantage over earlier designs based on glass or ceramic enclosures in containing the bugbear of spark-induced radio-frequency oscillations in the decay curves. Subsequent versions of the flashlamp [11] were commercialized and became the affordable pulsed source of choice for over two decades until semiconductor diodes emerged around 2000. General purpose NIM timing electronics, originally developed for nuclear physics and combined with a multichannel analyser, were used for timing and decay acquisition, respectively. These were precise and fairly linear, but not at all user-friendly owing to the many cables and manually selected settings.

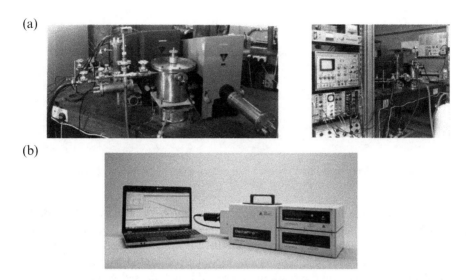

Fig. 1 (**a**) TCSPC fluorescence lifetime spectrometer from the early 1970s developed in John Birk's Photophysics Group in Manchester University [9, 10] and incorporating the all-metal coaxial flashlamp, two optical monochromators and NIM timing electronics. (**b**) The clean-cut lines of one of its descendants, the *DeltaPro*, with its on-line data analysis, epitomizes the ease of use of present-day versions

A lifetime determination then, as now, involved recording the fluorescence decay and the excitation pulse in separate measurements in order to perform reconvolution analysis to extract lifetime values that would otherwise be corrupted by the finite temporal instrumental pulse of typical full width at half maximum (FWHM) of ~2 ns. The non-linear least squares data analysis algorithm written by Bob Imhof had a number of rigorous features [12]. These included termination of the iterations to find the lifetime only when the chi-squared goodness of fit criterion had found a true minimum rather than when its rate of descent had slowed, as is common in the widely used Marquardt algorithm. The algorithm also included the capability to iteratively shift the function fitted to the decay data in order to correct for the photoelectron transit time dependence on wavelength in the photomultiplier and the finite width of the discrete timing channels. Needless to say in the early years, data was stored on paper tape and analysed on a central institutional mainframe computer. The subsequent introduction of the microprocessor changed all this by allowing, for the first time, on-line computation and almost immediate answers. How times change!

In terms of pulsed sources, mode-locked lasers offered an expensive alternative to flashlamps in the early days, but they were clearly a tour de force, and both types were subsequently replaced for general use by the advent of low-cost semiconductor sources. However, trade-offs and optimization of the source repetition rate, dead time and the inherent time resolution of the overall system are still sometimes required. This is largely governed by the application and discussed further in Sect. 2.2.

(a)   (b)

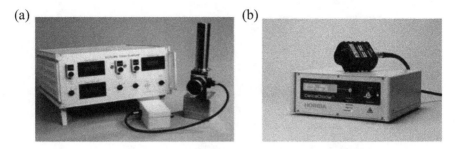

Fig. 2 (a) *Model 5000F* coaxial nanosecond flashlamp and (b) the picosecond *DeltaDiode*

The mainstay of our success in the early manufacturing years was the IBH *Model 5000F* version of the all-metal coaxial flashlamp shown in Fig. 2a. Typically operating at up to 50 kHz repetition rate, this design was gold plated as an additional measure to further reduce distortion on decay curves caused by radio-frequency oscillations generated by the spark discharge. Although the pulse energy was low, typically $\sim<1$ pJ, and average power <1 μW, single-photon timing detection sensitivity made flashlamps workable. When filled with nitrogen, the 337 nm line proved to be quite intense. With hydrogen (or deuterium for extra intensity), the flashlamp came into its own as a very workable and affordable alternative to frequency-doubled mode-locked lasers for UV excitation of protein intrinsic fluorescence. However, it was in the visible where flashlamps often proved inadequate. The introduction of semiconductor light-emitting diodes and lasers spanning the UV to near infrared changed all this. Our *NanoLED* range provided the first UV LED excitation of protein fluorescence decay [13], and more recently, the *DeltaDiode* range (Fig. 2b), operating at repetition rates up to 100 MHz, has a unique place in the field having set a new benchmark in data acquisition rates [14]. The combined advantages of the concomitant $\sim1000\times$ higher data collection rate, narrower pulses, maintenance-free operation, higher stability and monochromatic nature of their laser variants quickly led to semiconductor sources superseding flashlamps in TCSPC.

Quite early on in the company's development, we recognized the need and opportunities in accessing as much as possible of the fluorescence fingerprint (Eq. 1) in a single measurement and in the 1980s set about this task by developing TCSPC signal multiplexing and routing techniques for the first time. The approach aims at retrieving simultaneously the maximum information content available in the multidimensional contour defining the fluorescence signature of a sample, which can be conveniently expressed as a function F:

$$\text{Fluorescence} = \text{F}(I, \lambda_{\text{exc}}, \lambda_{\text{em}}, p, r, t) \tag{1}$$

where I describes the fluorescence intensity, λ_{exc}, λ_{em} the excitation and emission wavelengths that give rise to their respective spectra, p the polarization that can be used in anisotropy measurements, r the position as used in microscopy and t the time described by the fluorescence lifetime τ.

Initially we worked on a two-channel approach [15] to correct for any temporal change in the excitation pulse by means of simultaneous acquisition of fluorescence and excitation (SAFE). Coincidentally around the same time, Kiyoaki Hara was developing the *NAES* lifetime system at HORIBA to achieve a similar goal. The main difference between our approaches was the *NAES* system used multiple time-to-amplitude converters (TACs) for timing, whereas we used a single TAC, which ensured a common time calibration. Our two-channel implementation was also ideal for simultaneously recording the two orthogonal planes of polarization needed for anisotropy decay measurements [16]. Soon we extended the approach to four channels by using fibre optic coupling to a spectrograph to demonstrate for the first time simultaneous decay measurements at multiple fluorescence wavelengths [17]. This pointed the way towards much faster measurement of time-resolved emission spectra (TRES) than the sequential wavelength-stepping approach used previously. Nevertheless, measurement times were still measured in hours because of the low repetition rate limitation of the flashlamp, and TCSPC took a while to shake off its reputation as a slow method even after semiconductor diode excitation became widespread.

Our research and development in multiplexing led to us to combine the SAFE and dual fluorescence channels in the IBH *Model 5000W SAFE* (Fig. 3a). This instrument was an expansion of the popular space-saving folded geometry of the *Model 5000U* which had a single fluorescence channel. The *5000W SAFE* rapidly became the most versatile fluorescence lifetime system yet produced as it included not only three channels but the capability to multiplex up 16 TCSPC channels simultaneously [18]. This was achieved by virtue of the first application-specific integrated circuit (ASIC) designed for fluorescence. Based on complementary metal-oxide-semiconductor (CMOS) technology, the ASIC is illustrated in Fig. 3b in a NIM module which were still commonly used at the time. The ASIC provided the first commercial readout system for multi-anode microchannel plate photomulitpliers, thus opening the door to various forms of multiplexed imaging. Moreover, the device broke through the 1–2% "stop" to "start" rate barrier of conventional TCSPC and permitted count rates up to 37% [19]. Joining the company in 1995, co-author David McLoskey led this research out of his PhD [20] and into the marketplace.

(a)   (b)

Fig. 3 (a) Model *5000W SAFE* and (b) its 16 channel TCSPC multiplexing module

Together these developments provided a pointer for future technologies aimed at accessing more of the fluorescence signature by means of simultaneous measurement. This goal has still to be fully achieved and is a fertile area of research today. CMOS single-photon avalanche diode (SPAD) array detectors combined with individual diode timing offer significant potential for rapid TCSPC multiplexing, particularly in fluorescence lifetime imaging microscopy (FLIM) where the micron size SPAD pixels are less of a limit on sensitivity [21] as they are in TRES [22]. Conventional photomultiplier variants still have advantages over SPADs in spectroscopy where the highest time resolution and sensitivity are required. This is because photomultipliers have a more constant temporal response with photon wavelength, much larger detection area offering several orders of magnitude, higher sensitivity and wider wavelength spectral response including below 400 nm [7].

As can be seen from the foregoing, much of our work has been in close collaboration with university research, the application awareness of which helping to shape the innovation and specifications which our customers require. The UK's present market leadership in TCSPC was initiated by members of the University of Strathclyde's Photophysics Research Group in forming its spin-out IBH back in 1977, and this successful collaboration is ongoing. Our Glasgow facility also works closely with our sister company HORIBA Instruments Inc. based in New Jersey who are pursuing the goal of "total spectroscopy" in steady-state fluorimeters. Widely used for liquid analysis, the HORIBA *Aqualog* combines a CCD detector with a spectrograph to give absorption, transmission and excitation-emission matrix (A-TEEM™) in a single measurement at up to 4000× the rate of conventional single-channel fluorimeters. The HORIBA *Duetta* is a more general purpose version with extended response up to 1100 nm.

2.2 *The* DeltaFlex

Moving rapidly to today, our instruments have evolved from the early days to be much more user-friendly while still retaining the cutting-edge research capabilities required by our customers. For example, the recently released *EzTime* software, with its optimized touchscreen interface, is unique for the field and dramatically simplifies the process of fluorescence lifetime measurement via script automation and "one-click" automated fitting and presentation of results. In addition, the instrument auto-identification and configuration provided by our F-Link bus technology ensures easy plug-and-play when swapping components, auto-calibration, retrofit and upgrades with excitation sources and sample compartment accessories. These features are greatly facilitating the broadening of the user base away from the TCSPC specialist and opening up new applications across the disciplines. Our TCSPC journey has come a very long way from the confines of NIM electronics.

Monolithic time-to-digital converters (TDCs) have largely replaced TACs and ADCs in recent systems such as the *DeltaFlex*. While these do not offer the extreme electrical time resolution achievable with TACs (a few hundred femtoseconds is readily achievable with a TAC nowadays), all-digital TDCs in combination with 100 MHz diode light sources can attain previously undreamt of measurement times of less than 1 ms—a long way from the several hours that we endured in the early days. Such short measurement times allow TCSPC measurements to be continuously streamed to disk; rates over 20,000 histograms per second are opening up whole new applications for TCSPC in 3D imaging, light detection and ranging (LIDAR). Personal computers are now so powerful that photons need not be counted and logged in fixed size histograms but may now be "tagged" with information about conditions existing at the instant of their detection and subsequently streamed to disk for sorting and manipulation—each retaining their own "individuality" within resultant files that range from a few hundred megabytes to potentially terabytes in size.

The *DeltaFlex* (Fig. 4) offers all-round spectroscopic tuning and fully automated operation under software control with flexibility for ease of upgrade. In the *DeltaFlex*, the fluorescence emission is selected using a monochromator as opposed to a long pass optical filter in the *DeltaPro*. This offers the capability for TRES with fluorescence decays measured sequentially at incrementing fluorescence wavelengths which can then be temporally sliced in software to reveal the TRES. Global analysis of the resultant datasets can also determine the decay-associated spectra. When required, an excitation monochromator can be added. The Seya-Namioka monochromator used in the *DeltaFlex* is ideal for fluorescence decay studies as it contains only two mirrors and a single concave holographic diffraction grating in a compact construction. This minimizes the temporal dispersion due to differences in the light path to a level where it is not the limiting factor in time resolution when exciting with LED and diode lasers combined with photomultiplier detection. Hence there is no need to incur the sensitivity loss of subtractive dispersive monochromators caused by them having an additional grating and more mirrors (see Sect. 3.1).

In terms of pulsed sources for the UV, 250–370 nm sub-nanosecond LED sources, which can be pulsed up to a repetition rate of 25 MHz, are used. It should be noted that LED sources are not polarized, and if maximum sensitivity is required, there is no need to use polarizers with these sources. In the visible to NIR regime, 375–1310 nm picosecond *DeltaDiode* laser sources are used with repetition rates of up to 100 MHz depending on the time range required (10 ns–100 μs). The selected repetition rate depends on the lifetime to be measured and is typically equivalent to the reciprocal of the measurement time range, which is selected to avoid re-excitation during the fluorescence decay. High-throughput *DeltaHub* timing electronics with their short (<10 ns) dead time are optimum in utilizing the 100 MHz repetition rate of the *DeltaDiodes*. These laser sources are vertically polarized so should be used with at least a polarizer in the emission arm at an angle of the magic angle 54.7° to prevent polarization artefacts from distorting the measured fluorescence decay [1].

For the much longer time ranges needed for phosphorescence decay measurements (100 μs–10 s), *DeltaDiode* lasers may be operated in a "burst mode" by fast gating of their 100 MHz pulse train. Hence the same excitation source can be used to cover picosecond to second time scales and with no operator intervention required. Longer-pulse LEDs, *SpectraLEDs*, are also available from 265 to 1275 nm for phosphorescence decay measurement using multichannel scaling rather than

(a)

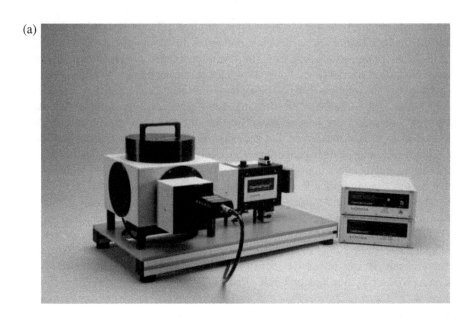

(b)

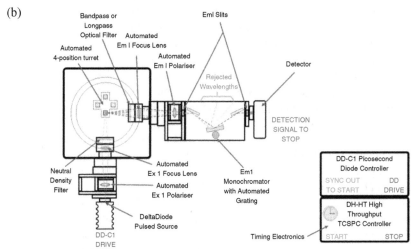

Fig. 4 The *DeltaFlex*—a truly modular instrument which satisfies the wide range of fluorescence lifetime applications and overcomes the fear of obsolescence by facilitating upgrade. Photo (**a**) and schematic (**b**)

TCSPC. The UV capability of LEDs, as first demonstrated in protein fluorescence decay [13], has also made protein phosphorescence decay more accessible as a structural tool [23]. *SpectraLEDs* are an alternative to the xenon flashlamp commonly used for phosphorescence excitation and offer numerous advantages such as a sharper turn-off (near absence of tail), software control of duration (and thus average power) and operation at a wider range of repetition rates—the combination of which allows the excitation parameters to be optimized to suit the sample in order to minimize the time required for measurement.

In the *DeltaFlex*, all the optical components and sample turret can be motorized and directly controlled with the *EzTime* software, which is particularly useful when used in conjunction with the *EzTime*'s scripting capabilities. Detector options include the near infrared (NIR), and a T-format can be easily added if required. The spare ports on the sides and underneath the sample chamber help ensure optimum optical alignment for a wide range of sample environments including cryostats and front face illumination. Figure 5 shows some of the optional configurations available with the *DeltaFlex* and also its compact version, the *DeltaPro*. The high-throughput *DeltaHub* timing electronics and *PPD* detection module are recommended for 95% of fluorescence lifetime applications. However should higher timing resolution be required, a microchannel plate photomultiplier (MCP-PM) or hybrid (*HPPD*) detector and (TAC-based) *FluoroHub A+* timing electronics can be selected. With a narrow optical pulse excitation source such as the *DD-375L* or *DD-405L*, the temporal FWHM can be routinely reduced down to below 50 ps enabling lifetimes down to 5 ps to be recovered. The *DeltaDiode* lasers implement novel synchronization circuits that adapt to further optimize performance, when used

(a)   (b)

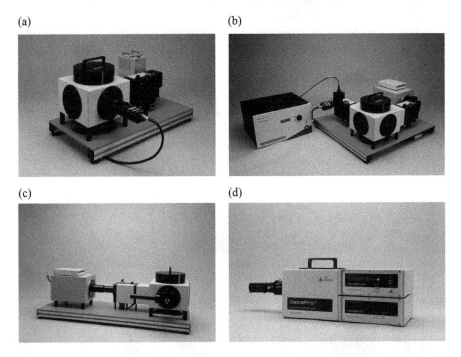

(c)   (d)

Fig. 5 The *DeltaFlex* fluorescence lifetime system incorporating (**a**) ps diode laser excitation and near-infrared detection up to 1700 nm, (**b**) continuum laser excitation, excitation monochromator and microchannel plate photomultiplier detection, (**c**) coupling optics for an ultra-fast Ti/Sapphire laser for use in such as multiphoton excitation, (**d**) with filter rather than optical monochromator fluorescence wavelength selection in the *DeltaPro*

in such high-resolution measurements, by automatically correcting for electronic jitter in the signal path.

The trade-off incurred by the higher timing resolution offered by a MCP-PM is that the system data acquisition rate is reduced owing to the increased dead time, firstly, by the MCP-PM, which can typically only support count rates of 20 kHz as compared to the standard *PPD-850* detector where the upper limit is ~10 MHz; secondly by the larger dead time of the matching timing electronics (~1 μs for the *FluoroHub A+* as compared to 10 ns for the *DeltaHub*) and concomitant matching source repetition rate (20 MHz *FluoroHub A+* vs 100 MHz *DeltaHub*); and finally, it is a common practice to increase the number of data channels when seeking the highest time-resolution measurements, which again increases the measurement time. Conversely, at the other end of the scale, runtimes with the *DeltaFlex* can be programmed to record 1–26,000 decays measured sequentially in as little as 1 ms to 1 min per decay.

Historically MCP-PMs were the only alternative offering a faster impulse response than conventional photomultipliers. However hybrid photomultipliers, with the metal dynode stages of a traditional photomultiplier replaced with a silicon avalanche photodiode, are finding increasing use. These can also be used in conjunction with low timing jitter electronics, such as the *FluoroHub A+*, with a time resolution of a few hundred fs per bin.

Although the transit time spread of the MCP-PM is shorter, the fact that the *HPPD* hybrid version of our *PPD* detector range exhibits a reduced after-pulse, as illustrated in Fig. 6a, and can sustain higher count rates (MHz compared to the 20,000 cps for the MCP-PM), is making these devices more popular. Certain versions have transit time spreads that enable lifetimes up to the limit of the TCSPC technique to be recovered, while others have high quantum efficiencies. The reduction in after-pulse helps reconvolution (Eq. 2) when determining shorter-lived decays, as shown in Fig. 6b, c, which compare a typical MCP-PM with a *HPPD* when determining a 60 ps decay time.

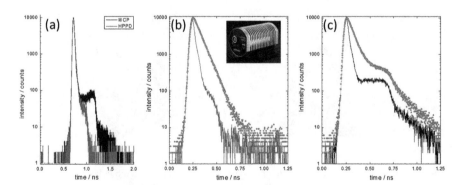

Fig. 6 Comparison of (**a**) MCP-PM and *HPPD* detector responses measured directly using a *DeltaDiode-405L*. A FWHM of 38 ps was obtained for both detectors. Measured decay of DASPI (lifetime ~60 ps) with a (**b**) *HPPD* (Inset) and (**c**) MCP-PM detector

Fig. 7 The *DeltaTime* hybrid steady-state and TCSPC lifetime system

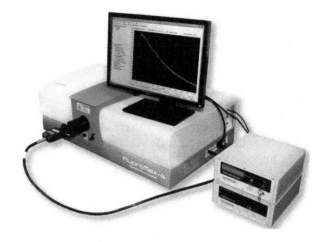

Finally, it should be noted that the flexibility of our TCSPC range extends beyond the stand-alone capabilities of the *DeltaFlex* and *DeltaPro*. Key components, including diode sources, detectors and timing electronics, are readily available separately for upgrade of existing instruments in the field such as the *FluoroLog* and *FluoroMax*. Indeed hybrid TCSPC-steady-state systems are available as factory orders to give an all-round spectral-temporal performance in a space-saving system (Fig. 7). These instruments provide a full research capability for accessing the fluorescence contour offered by Eq. 1 including quantum yield, steady-state polarization and the excitation-emission matrix (EEM) in addition to offering the measurement of lifetimes. TCSPC components such as the *DeltaDiode* can also be added to microscopes for FLIM applications.

2.3 *Software*

Software for fluorescence lifetime spectroscopy fulfils three roles: instrument control, decay analysis and data display. Along with the hardware required to make the measurements, major advances have occurred in the associated software. The purpose of the software has expanded to control and optimize the hardware for data acquisition, as well as its original use in analysing the resultant data. It should assist the user by being intuitive and simplify the measurement process, enabling lifetime data to be obtained with ease. However, although the aim is the smooth collection and determination of fluorescence parameters, the interpretation of these data is very dependent on the sample under study, and specific knowledge relating to this is still needed for the user to put the outcome in context and draw sensible conclusions.

The fact that microprocessors and memory can be added to almost all of the individual components making up a TCSPC system means that a degree of "intelligence" can be incorporated. The TCSPC system can then be thought of as a collection of intelligent nodes. The software can interrogate the system to ascertain the status of these nodes, such as the wavelength of the monochromator, lens and polarizer positions, as well as the time range and repetition rate of the excitation source. This "knowledge" enables the use of scripting (a language akin to a computer programme) to be used to automate data collection and fitting. The software is now able to exert a high degree of control over the hardware and is integral to the measurement system. Overall this means that software can be employed to assist the novice user from the simple matching of the excitation source repetition rate to the measurement time range (to avoid more than one optical excitation pulse arriving in the histogram time window) to estimating the data range over which to perform the data analysis. On the other hand, for more experienced users, it allows access to equipment settings and control over the data analysis parameters to tailor the measurement process to the sample.

Advances in computer operating systems and hardware mean that the software should take account of the latest versions (at the point of writing, Windows 10) and be optimized for use with touchscreen monitors allowing for equipment settings and data collection to be performed with a simple "tap" on the screen. An example of such software is the new *EzTime* package. Screenshots showing aspects of the *EzTime* user interface are shown in Fig. 8. This software package controls the hardware and performs data collection and analysis. With just one "tap" on the screen, it is possible to collect data and analyse it as the sum of exponentials and other models (1, 2 and 3 exponential models can be automatically fitted as soon as data acquisition stops). The output of the analysis (as well as any selected graph data) can be output as a spreadsheet that can be opened by most popular spreadsheet/ graphing programmes. A simple scripting language combined with the "intelligent nodes" in the *DeltaFlex* optical system enables the tailoring of the measurement to the sample under study, whether it is the scheduling of repetitive measurements, the control of temperature, polarization or wavelength throughout the data acquisition process. An autoscript assistant also enables users unfamiliar with scripting to obtain scripts for common measurement types without realizing that they are writing one. These can be further optimized using an "on-board" script editor if required.

2.4   The Role of the Manufacturer

The role of a spectroscopy instrument manufacturer overlaps with the research of the customer in satisfying the latter's requirements, but in fact the manufacturer's role is much broader. This is true even if we compare it to a researcher designing and developing an instrument for laboratory use. Design and development are on the face of it common to both, but in reality, the manufacturer has to take on-board many additional considerations. These include testing prior to CE marking and compliance with other regulatory requirements such as WEEE and RoHS, scaling up the number of systems to satisfy orders, maintenance of stock levels, factoring into design the continuity of component supply, implementing quality control systems, ease of use

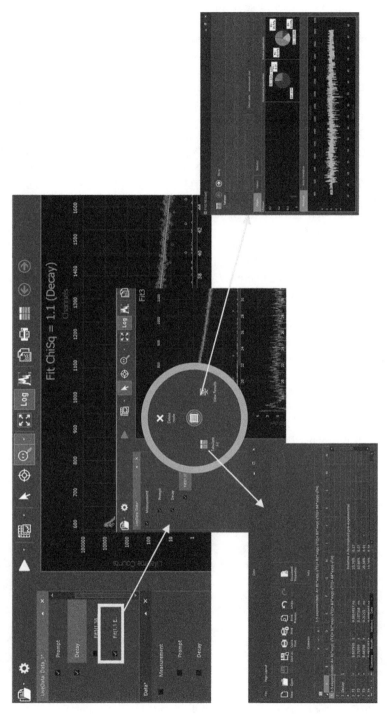

Fig. 8 Screenshots of the user interface for the *EzTime* software package

in order to support a user base with varying levels of skill, meeting delivery deadlines, installation, attaining test specifications and customer support, which includes manuals, training, applications advice and ensuring the availability and reproducibility of replacement parts for the anticipated lifespan of the product. Hand in hand with manufacturing goes the marketing and global sales network and all the usual functions familiar to most companies such as human resource management, ensuring a safe and attractive working environment, and finance operations such as payroll and accounting. In order to achieve all of this, the company has to ensure the funds for financing the work required are always in place. Last, but by no means least, the company needs to sell the products at prices that are attractive yet still make a profit for the business to continue, invest in new developments, and grow! The necessity for products to be affordable and yet of high performance inevitably brings with it compromises at times. As such, it is equally inappropriate for products to be overdesigned as well as under-designed. The term "horses for courses" comes to mind in such a context!

The manufacturer-customer landscape has changed considerably, and very much for the better, in recent years. The relationship is now much more of a partnership which strives to serve mutual interest. Companies are increasingly aware of the need to invest in the market and give something back to the community they serve. With this in mind, we initiated the FluoroFest series of international workshops in Prague in 2009, and the 12th in the series was held in Glasgow in April 2017 as part of our 40th anniversary celebrations (Fig. 9). These didactic workshops offer a programme designed to be of interest to multidisciplinary delegates and span an introduction to the basics of fluorescence to more advanced applications presented by world-leading experts. All delegates receive copies of the lectures and support material. FluoroFest differs from the usual academic research conference because of the opportunity for hands-on time and small group training on a wide range of fluorescence instruments covering steady-state spectroscopy, lifetimes, EEM and microscopy. These all support themes taken from some of the hottest topics in fluorescence worldwide, such as instrumentation/techniques, life sciences/biomedical and nanotechnology/materials. In addition to plenary and invited lectures, contributed talks, student flash presentations and posters are included. For the Glasgow event, generic skills for

(a)   (b)

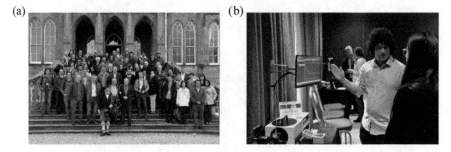

Fig. 9 (**a**) Delegates at the 2017 12th FluoroFest dinner at Ross Priory on the banks of Loch Lomond. (**b**) Co-author Philip Yip demonstrating the *DeltaPro* at the Glasgow FluoroFest

postgraduate and postdoctoral delegates were included with publication advice given in a lecture by the Institute of Physics Publishing, who also received papers from the meeting for a special issue of *Methods and Applications in Fluorescence*. In keeping with the partnership between the community and vendor, FluoroFest always includes a forum on shaping the future of fluorescence wherein industry listens to delegates outlining their needs and aspirations for new technology and techniques.

3   Demonstrating Time Resolution Through Applications

In this section, we illustrate some of the principles of fluorescence lifetime measurement by focusing on the high time resolution of the *DeltaFlex* in a few of its many applications. In an application-focused scientific instrument company, it is important to have application specialists with diverse user experience from a customer perspective. Co-author Kulwinder Sagoo joined our company in 2002, with a chemistry background and a Master's degree from Brock University. Her research making use of time-resolved spectroscopy for biomolecule characterization then continued at McMaster University prior to joining IBH. Co-author Graham Hungerford obtained his PhD in the Photophysics Group at the University of Strathclyde in 1991 for work on near-infrared TCSPC and then continued in research involving fluorescence at the Universities of Strathclyde, Leuven and Minho (Chemistry and Physics Departments) before joining the company in 2008. Together they work with the rest of the team to ensure that the company continues to deliver on getting the ultimate in performance into customers' laboratories.

The push for reliable limits of time resolution embodies our corporate philosophy. Although time resolution lies at the heart of fluorescence lifetime spectroscopy, it can in fact take on several interpretations. For example:

1. The fastest fluorescence decay time which can be resolved
2. The smallest difference in fluorescence lifetime which can be resolved
3. The minimum time needed to measure a fluorescence decay

3.1   Fastest Decay Times

Interpretation number 1 concerns measuring the decay constant associated with the difference between the fluorescence decay $F(t)$ of a sample and the instrumental pulse (response) profile $P(t')$ recorded by tuning the emission monochromator to the excitation wavelength and replacing the sample with a scattering medium of negligible temporal broadening such as a silica colloid. $P(t')$ incorporates the temporal broadening associated with the excitation pulse, spread in optical path length, the detector and the timing electronics. In principle the narrower $P(t')$, the better the ultimate time resolution which can be achieved. The two measurements are then

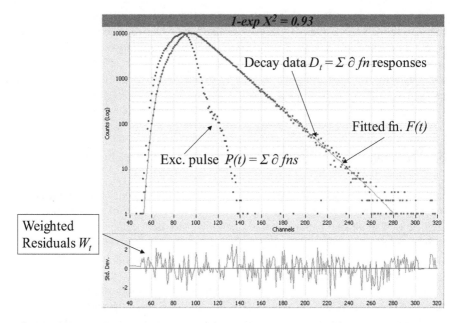

Fig. 10 Fitting to a fluorescence decay using reconvolution analysis [2]. Copyright © 2015 by John Wiley & Sons, Inc. Reproduced with permission

handled in software using iterative convolution analysis whereby the excitation pulse is considered to be composed of a series of δ-functions and the fluorescence decay a linear superposition of the corresponding fluorescence impulse responses [2], i.e.

$$F(t) = \int_0^t P(t')i(t-t')dt' = P(t) \otimes i(t) \tag{2}$$

where the sample's fluorescence δ-function impulse response $i(t-t')$ contains the fluorescence lifetime τ_f according to

$$i(t) = i(0)\exp\left(-\frac{t}{\tau_f}\right) \tag{3}$$

The variable t' is a moving time delay that defines the instant at which each δ-function component of the instrumental pulse generates the start of a fluorescence response. On iteration, the best fit of $F(t)$ from Eqs. 2 and 3 to the measured fluorescence decay data D_t defines τ_f by means of minimizing the sum of the squares of the weighted residuals W_t using the chi-squared (χ^2) goodness of fit criterion as illustrated in Fig. 10, where

$$W_t = \frac{D_t - F(t)}{\sqrt{D_t}} \qquad (4)$$

and

$$\chi^2 = \sum_{\substack{\text{Data} \\ \text{channels}}} W_t^2 \qquad (5)$$

A χ^2 value from Eq. 5, which is normalized according to the number of detection channels and the number of fitted parameters, should approach 1 if the correct model is being applied.

Clearly, many systematic errors can limit the time resolution, for example, the detection of stray excitation light which mirrors $P(t')$, temporal instabilities in the light source and timing electronics, non-linearity in the time base, etc. These have an increasingly detrimental effect on the goodness of fit as the number of counts in the data (i.e. the statistical precision) increases. If we take, for example, time dispersion in optical monochromators, this can be eliminated by the use of two monochromators configured such that the time dispersion is subtracted. However, this comes at a heavy price in terms of loss in throughput due to reflections from all the additional gratings and mirrors required. We use the proprietary *TDM* time domain monochromator optimized for fluorescence lifetime measurements and based on a Seya-Namioka geometry with only one grating and two mirrors in order to maximize throughput and minimize stray light. Moreover, other sources of temporal broadening, such as that of the detector, exceed any optical transit time dispersion, and, given such contributions add in quadrature as the sum of the square of their values, the monochromator time dispersion is a second-order effect [7]. The minimal change in instrumental pulse for the *TDM* is illustrated in Fig. 11 at zero-, first- and second-order grating positions for a *DeltaDiode* wavelength of 378 nm and MCP-PM detection.

A generally accepted "rule of thumb" in reconvolution analysis using Eqs. 2–5 is that it is possible to extract a decay component ~1/10th of the instrumental FWHM. Hence from Fig. 11, it should be possible to extract ~5 ps decay time. Indeed this was confirmed in a recent study of lycopene, a carotenoid commonly found in tomatoes, other red fruit and some vegetables [24]. Commercially available lycopene (LYC) was compared with a control lycopene sample extracted from tomato and lycopene extracted from tomato following ultrasonic treatment at 584 kHz for 15 min (T15) and 60 min (T60) [25]. Figure 12 shows the raw data obtained for these four samples and a typical fit.

Table 1 shows the results of decay analysis for the four lycopene samples according to

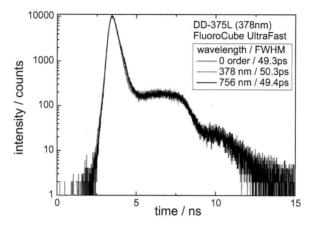

Fig. 11 Instrumental pulse profiles demonstrating minimum wavelength transit time dispersion of ~1 ps recorded using the *TDM* time domain monochromator and a *DeltaDiode* laser at 378 nm

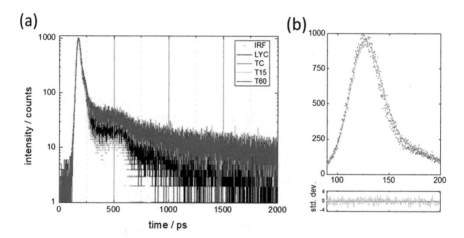

Fig. 12 (**a**) Time-resolved fluorescence decay of various lycopene samples [25] monitored at 550 nm, with excitation at 409 nm using a *DeltaDiode*. These are for commercially available lycopene (LYC) extracted from tomato without ultrasonic treatment (TC) and with ultrasonic treatment for 15 min (T15) and 60 min (T60). The instrumental response function (IRF) is also shown and has a FWHM = 36 ps. (**b**) Decay, IRF and fitted function, along with weighted residuals for LYC

$$i(t) = \sum f_n \exp\left(-\frac{t}{\tau_n}\right), \quad \text{for } n = 3 \tag{6}$$

The results in Table 1 confirm a dominant decay component of around 5 ps, and this is consistent with work previously reported from a calculated value of 5 ps [26] and measured value of 4.7 ps obtained using ultra-fast absorption spectroscopy

Table 1 Results of the three-exponential fluorescence decay analysis of the four lycopene samples according to Eq. 6 and including average decay time $<\tau>$ [25]

Sample	Decay time/ps			Fraction/%			$\langle\tau\rangle$	χ^2
	τ_1	τ_2	τ_3	f_1	f_2	f_3		
LYC	5.1 ± 0.8	807 ± 20		84	16		6.1	1.15
TC	4.4 ± 0.4	313 ± 69	2014 ± 183	62	8	30	7.1	1.21
T15	4.8 ± 0.9	322 ± 69	1764 ± 123	60	8	32	8.0	1.18
T60	5.1 ± 1.1	277 ± 57	1758 ± 123	62	7	31	8.2	1.19

Details are provided in the caption to Fig. 12

[27]. Given that light which travels 0.3 mm in 1 ps, it can be seen that care is needed to maintain geometric consistency between recording the instrumental response function and the fluorescence decay.

3.2 Resolving Differences in Decay Curves

In the previous example, time resolution was demonstrated through finding the minimum decay component which could be measured. Equally important is resolving the difference between decay curves. This problem is manifest in quenching, energy transfer, binding, etc. and is particularly demanding in the case of fluorescence polarization anisotropy decay, where it is highly dependent on the precision of the instrument.

It is informative to consider an analogy between resolving power in optical spectroscopy and mass spectrometry in resolving small difference in measurands, the equivalent expressed here as:

$$\text{Resolving Power} = \frac{\tau_f}{\Delta_f} = \frac{1}{\text{Resolution}} \tag{7}$$

For a fluorescence lifetime, $\Delta\tau_f$ is essentially described by the statistical error calculated from the reconvolution analysis. This is influenced by the number of counts (statistical precision) recorded and systematic errors such as non-linearity in the time base or the detection of stray excitation light. All such factors are reflected in the goodness of fit.

We can similarly think about the resolution of parameters derived from measuring the fluorescence lifetime, such as distance resolution between donors and acceptors undergoing fluorescence resonance energy transfer (FRET) or the radius resolution from the fluorescence anisotropy decay of a rotating fluorophore. Together these approaches offer complementary distance measurement and resolution. FRET typically offers an effective range of ~1–10 nm and fluorescence anisotropy decay ~0.1 to ≥ 10 nm [1]. This complementarity comes about because FRET falls off rapidly as $\sim r^{-6}$, and the rotational correlation time of anisotropy decay of a rotating fluorophore at an equivalent rate of $\sim r^3$. Both offer sub-nanometre resolution, even down to 0.1 nm, and this is well-illustrated by the application of anisotropy decay to nanoparticle metrology.

Synthetic nanoparticles have joined with nature's indigenous varieties to create an urgent need for their size measurement. This is particularly true for the 1–10 nm range which can easily traverse cellular membranes and are of increasing concern in respect of the associated implications for human health. It is perhaps surprising that, although there are a wide range of available methods for this task, such as electron microscopy and different scattering approaches [28], at this point in time, there are no universal standards with which to compare nanoparticles. Here we describe how fluorescence decay measurements offer a comparatively low-cost and portable solution to address this need. Although the underlying theory of fluorescence polarization anisotropy decay is relatively straightforward and has been extensively reviewed for fluorophores in solution [29–31], from a measurement point of view, it does present some additional demands on measurement precision and accuracy to that of determining a simple fluorescence lifetime from a decay. This is because the latter effectively compares a fluorescence decay with a zero baseline, whereas anisotropy decay, as can be seem from Eq. 8, depends on, and compares, the difference between two decay curves. Here is a brief summary of the theory of the simplest case of relevance. This relates to a nanoparticle labelled with a dye, and the linear addition of other rotational kinetics, such as free dye rotation or dye wobbling on the nanoparticle, which can also be successfully analysed.

By recording vertically (V) and horizontally (H) polarized fluorescence decay curves, $I_{VV}(t)$ and $I_{VH}(t)$, orthogonal to vertically (V) polarized excitation, a time-resolved anisotropy function $r(t)$ is generated, i.e.

$$r(t) = \frac{[I_{VV}(t) - GI_{VH}(t)]}{[I_{VV}(t) + 2GI_{VH}(t)]} \tag{8}$$

where G is a factor ($G = I_{HV}(t)/I_{HH}(t)$) which corrects for transmission efficiencies and

$$I_{VV}(t) = \exp\left(-\frac{t}{\tau_f}\right)\left[1 + 2r_0 \exp\left(-\frac{t}{\tau_c}\right)\right] \tag{9}$$

$$I_{VH}(t) = \exp\left(-\frac{t}{\tau_f}\right)\left[1 - r_0 \exp\left(-\frac{t}{\tau_c}\right)\right] \tag{10}$$

Leading to

$$r(t) = r_0 \exp\left(-\frac{t}{\tau_c}\right) \tag{11}$$

where r_0 is the initial anisotropy at $t = 0$, which has a maximum value of 0.4, and τ_c the rotational correlation time describing the rate of depolarization due to isotropic rotation.

The simplest depolarization occurs for a fluorophore that can be treated as a spherical rigid rotor undergoing Brownian rotation in an isotropic medium such as a solvent. In this case, τ_c can be expressed by the Stokes-Einstein equation:

$$\tau_c = \frac{\eta V}{kT} = \frac{1}{6D} \tag{12}$$

where η is the microviscosity, V the hydrodynamic volume $= 4\pi R^3/3$ prescribed by the rotor of hydrodynamic radius R, T the temperature, k the Boltzmann constant and D the rotational diffusion coefficient such that

$$R_p = \left(\frac{3kT\tau_c}{4\pi\eta}\right)^{1/3} \tag{13}$$

The adaptation of this simple theory to a dye attached electrostatically or covalently to a nanoparticle is straightforward as it is then the nanoparticle hydrodynamic radius which is prescribed as depicted in Fig. 13a. Ideally, the fluorescence lifetime τ_f is $\sim \tau_c$ [32]. If $\tau_f \gg \tau_c$, the depolarization is too fast to measure accurately during the fluorescence lifetime and if $\tau_f \ll \tau_c$ too little depolarization occurs during the fluorescence lifetime. However, these theoretical constraints have to a large extent been circumvented by modern high-repetition rate sources such as the *DeltaDiode* combined with high-throughput *DeltaHub* timing electronics. This is because they facilitate the accumulation of significant number of counts, with even several exponents down from the peak of a fluorescence decay still providing a useful

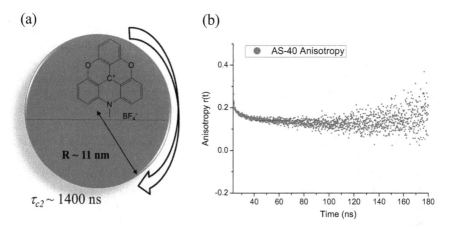

Fig. 13 Rotating 11 nm radius LUDOX AS-40 nanoparticles (**a**) labelled with Me-ADOTA and (**b**) their anisotropy decay measured with a *DeltaFlex* using *DeltaDiode* excitation at 503 nm, fluorescence detected at 570 nm [38]

monitor of the particle rotation such that $\tau_f \ll \tau_c$ can still produce a viable measurement, as we will illustrate here.

In the case where the dye partitions between being bound to the nanoparticle and free rotation

$$r(t) = (1 - f)r_0 \exp\left(-\frac{t}{\tau_{c1}}\right) + f r_0 \exp\left(-\frac{t}{\tau_{c2}}\right) \qquad (14)$$

where f is interpreted as the fraction of fluorescence due to dye bound to silica nanoparticles rotating with a correlation time τ_{c2} and $1 - f$ due to dye molecules unbound in the colloid and rotating faster with a correlation time τ_{c1}.

Silica colloids of well-defined size and produced by the Stöber process [33, 34] are readily available under the LUDOX tradename (Grace). Indeed they are widely used in fluorescence lifetime spectroscopy to scatter the excitation pulse when recording the instrumental response for reconvolution analysis using Eq. 2.

Early work [35] with fluorescence anisotropy for the metrology of LUDOX nanoparticles used 6-Methoxy quinolinium labels that require excitation in the ultraviolet. Unfortunately, these suffered from the likelihood of data contamination caused by the excitation of background fluorescence in what is not a spectroscopically pure sample. Most dyes emitting in the visible part of the spectrum (thus circumventing background fluorescence problems) have too short a fluorescence lifetime (i.e. $\tau_f \ll \tau_c$) to be optimum for use with nanoparticles. Recently, a range of triangulenium dyes with ~20 ns fluorescence lifetime have been synthesized [36, 37], which seem to be ideal for the task and bring the spectral range within that of the *DeltaDiodes* with their high-repetition rate ps pulses. Me(thyl)-ADOTA (Fig. 13a) is one such example [38]. Its anisotropy decay (Eq. 8) when electrostatically bound to silica particles in LUDOX AS-40 at pH 9–10 is shown in Fig. 13b.

Table 2 shows that over the range studied, the errors determined demonstrate a resolving power ($R_p/\Delta R_p$) for these kinds of measurements of at least 10 (Eq. 7), and it is clear that sub-nanometre resolution is being obtained on the smallest particles. In fact increasing D_p by ~26× gives sub-nanometre resolution for all the particles and ~0.1 nm for the two smallest [38].

The ability to acquire high data rates efficiently becomes essential where samples are changing rapidly such as in a flow cell or during polymerization. Indeed the formation of silica gel provides a good example of this. In the case of growing silica nanoparticles, an average radius of a much larger size distribution as compared to LUDOX is obtained. Both hydrogels and alcogels can be studied. Table 3 illustrates this for an alcogel of tetraethyl orthosilicate (TEOS).

What is interesting about Table 3 as compared to Table 2 is that a resolving power better than 10 is obtained even for such small (~1 nm) nanoparticles, which give <20× lower rotational times. This reflects the instrument stability, linearity of time base, stray light rejection and reliability of data analysis. Moreover, the small sizes obtained for TEOS with the *DeltaDiode* are consistent with those previously reported for tetramethyl orthosilicate (TMOS) using fs two-photon excitation [39].

Table 2 Fluorescence anisotropy decay analysis of various LUDOX colloids labelled with Me-ADOTA using Eqs. 13 and 14 [38]

	With Me-ADOTA [38]					With 6-MQ [35]		
LUDOX	$D_p/10^3$ (cts)	τ_{c1} (ns)	τ_{c2} (ns)	χ^2	$R_p \pm \Delta R_p$ (nm)	R_m (nm)	$R_p \pm \Delta R_p$ (nm)	$D_p/10^5$ (cts)
SM-AS	3.0	3.48 ± 0.93	100.0 ± 18.0	1.18	4.6 ± 0.3	3.5	4.0 ± 0.4 (SM-30)	1.0
AM	3.5	4.08 ± 1.30	210.0 ± 19.5	1.09	5.9 ± 0.2	6.0	6.4 ± 0.5 (AM-30)	5.0
AS-40	12.0	12.1 ± 2.06	1424 ± 471	1.18	11.1 ± 1.1	11.0	11.0 ± 1.6 (AS-40)	10.0

The anisotropy data was acquired in typically ~20 min. The microviscosity was taken to be 10^{-3} Pa s (1 cp). The measured average particle radius R_p, the manufacturer's values R_m and the particle radii obtained previously [35] using 6-methoxyquinoline (6-MQ) are shown. The latter were measured in ~10 h with 10^5, 5×10^5 and 10^6 counts in the peak D_p of the difference curve $I_{VV}(t) - GI_{VH}(t)$ for the 3.5 nm, 6 nm and 11 nm particles, respectively, at a channel width of 28 ps, whereas only $\leq 10^4$ counts were required in D_p at a channel width of 104 ps for a better precision in R_p to be obtained with Me-ADOTA excited by a *DeltaDiode* at 503 nm. τ_{c2} defines the particle radii and τ_{c1} most probably describes the ADOTA dye wobbling on the nanoparticle. The data were analysed using impulse reconvolution analysis [7] analogous to that described in Sect. 3.1 for fluorescence lifetimes

Table 3 The results of the anisotropy analysis of Me-ADOTA in TEOS under acidic conditions [38]

Time (h)	τ_{c1} (ns)	$(1-f)r_0$ (%)	τ_{c2} (ns)	fr_0 (%)	χ^2	$<R_p>$ (nm)
19.0	0.20 ± 0.19	16.01	3.00 ± 0.70	83.99	1.11	1.42 ± 0.10
21.0	0.34 ± 0.09	18.54	3.14 ± 0.83	81.46	1.02	1.45 ± 0.12
23.0	0.43 ± 0.11	21.51	4.03 ± 1.45	78.49	1.04	1.57 ± 0.17
43.5	0.30 ± 0.12	12.93	4.07 ± 1.14	87.07	1.03	1.58 ± 0.14
46.5	0.32 ± 0.11	14.27	4.50 ± 1.10	85.73	1.04	1.63 ± 0.12
48.5	0.83 ± 0.19	23.40	6.00 ± 2.26	76.60	1.06	1.79 ± 0.20
67.0	0.50 ± 0.14	14.95	6.13 ± 1.59	85.05	0.98	1.81 ± 0.14

Over 48 h a monotonically increasing average nanoparticle radius $<R_p>$ of ~1.4–1.8 nm, as determined from τ_{c2}, is observed. Such measurements provide a stern test of the resolution of the technique and reveal close to 0.1 nm precision with $D_p \sim 3500$ cts acquired in ~20 min. The τ_{c1} value of ~0.3 ns up to 46.5 h is generally consistent with free dye rotation in water before the gel time t_g (~50 h) is reached, further complicating the kinetics

Interestingly, although the measurements were performed using the *DeltaFlex*, the versatility and spectral capabilities of this instrument are not always essential for these kinds of measurements, and indeed, the *DeltaPro* (Fig. 5d), with its lower cost, fluorescence selection using cut-off filters and dichroic polarizers, would be adequate for many such applications in nanoparticle metrology. The ultraprecision afforded by the research capabilities of the *DeltaFlex* in analysing complex fluorescence decay kinetics is not necessarily always needed as the fluorescence decay just provides a marker with which to track particle rotation. All that is required is for an adequate representation of the kinetics to be as good as the statistical precision of the data. This can be achieved by first fitting the fluorescence decay (the denominator in Eq. 8) using a series of exponentials and then using impulse reconvolution to iterate the rotational terms (e.g. Eq. 14) to give the best fit to the numerator in Eq. 8 which contains the rotational information [2].

3.3 Minimum Measurement Times

The final interpretation of time resolution concerns the need to detect transient species as occurs in many areas of analysis such as chromatography and flow cytometry. Here, we highlight the application of the *DeltaFlex* in bringing some of the advantages of fluorescence lifetime spectroscopy to the study of stopped-flow kinetics. In stopped-flow reactants are expelled from syringes, mixed and injected into a flow cell. The flow is then stopped and the ensuing reaction/interaction monitored by fluorescent labelling of one of the reactants or by using its intrinsic fluorescence.

Typically reactant syringes expel the reactants, which are rapidly mixed and enter a flow cell to be detected. Until relatively recently, lifetime instrumentation was not efficient enough to collect sufficient data on the critical ms timescales on which many reactions occur. However, the recent introduction of very low dead time electronics, coupled with high repetition rate excitation sources, has made this possible. Moreover, the advantages of lifetime measurements listed in Sect. 1 can be brought to bear. Specifically, the obviation of photochemical bleaching and determination of the absolute measurand of decay time represent demonstrable advantages for stopped-flow. In the TCSPC histogram streaming measurement mode, available in both *DeltaFlex* and *DeltaPro* systems, up to 26,000 time-resolved fluorescence decays can be obtained, with each histogram seamlessly collected in as little as 1 ms. The 100 MHz capability of the *DeltaDiode* is ideally matched to the very low dead time of the *DeltaHub* (10 ns), and together they have pushed data acquisition rates in TCSPC to new limits. When short data collection times are employed, this efficiency is required in order to obtain data of sufficient statistical precision for accurate decay time determination. The stopped flow accessory fits easily into the standard cuvette (10 mm path length) holder via modification of the sample chamber lid shown in Fig. 4. Using the control software, data collection can be started either using an external TTL signal from the stopped-flow accessory or manually, within the software, in the histogram streaming measurement mode.

In order to illustrate this capability, we consider the interaction of curcumin with protein [40]. Curcumin is commonly found in turmeric and is of wider interest as it is thought to have potential health benefits due to its antioxidant properties. The role of serum albumin in blood makes it an ideal model protein with which to observe this interaction. For stopped-flow, a solution of curcumin in DMSO was placed in one syringe, with the other reactant syringe filled with bovine serum albumin in buffer. In this solvent mixture, curcumin is weakly fluorescent and exhibits a short fluorescence lifetime. Upon interaction with the protein, both the fluorescence quantum yield and lifetime increase. The data acquisition was started to provide a "background," and the reactants expelled manually. After decay analysis, the change in average lifetime during the course of the experiment was plotted, and this is displayed in Fig. 14 along with the relative change in fluorescence intensity.

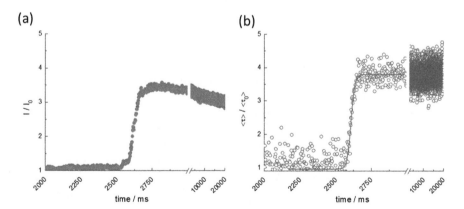

Fig. 14 Comparison between (**a**) relative fluorescence intensity (I/I_0) and (**b**) relative average lifetime data ($<\tau>/<\tau_0>$) obtained from a stopped-flow measurement of curcumin binding to bovine serum albumin. Data was collected every 5 ms. The excitation source was a *DeltaDiode-395L* operating at 100 MHz with fluorescence detected at 500 nm

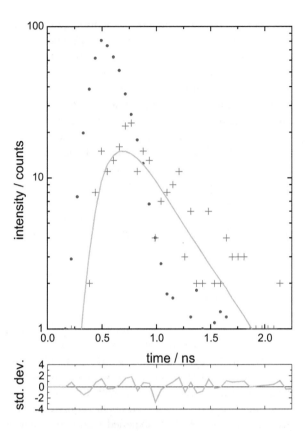

Fig. 15 Fluorescence decay (plus sign) of a BODIPY derivative [42] measured in 60 μs, along with the instrumental response (circle), weighted residuals (solid line) and fitted function (solid line). The lifetime obtained of 366 ps compares favourably with that measured for the same sample to a higher precision (~14,000 counts) with an MCP-PM. For the latter a value of 376 ± 12 ps was obtained [14]

Reaction rates can be determined from such plots. However, it is interesting to see how the constancy of average lifetime change is preserved in the face of the gradual photobleaching, bringing to bear a major advantage of lifetime determination (see Sect. 1) and eliminating a systematic error which would otherwise affect kinetic calculations based on relative intensity.

Hand in hand with such considerations goes the question of how many counts are required to gain the precision needed for a fluorescence lifetime measurement? Of course, if it is a static and stable sample, TCSPC offers the opportunity to count as many photons as possible in order to obtain the most stringent test of the kinetic model and maximize precision. However, for a transient sample, such as in a stopped-flow experiment, the question is very apposite. Previous work suggested fewer than 200 counts in a decay were sufficient to obtain 10% precision for a 2.5 ns lifetime [41]. Figure 15 supports this finding as a total of 233 counts recorded in the fluorescence decay of a BODIPY derivative [42] gave a lifetime of 366 ps, which compares well with a lifetime of 376 ± 12 ps obtained from 14,000 counts [14].

4 Conclusions

Fluorescence lifetime spectroscopy is now a mature technique, but innovation to improve performance and usability is ongoing. Perhaps the main developments over the past 20 years have been the introduction of semiconductor sources and the faster data acquisition to take advantage of their multi-MHz repetition rate. Together these are enabling more of the multidimensional characteristics of fluorescence to be measured. Our company's early innovations in multiplexed TCSPC helped point the way, but there is still much more to be done before the full spectroscopic contour afforded by Eq. 1 can be realized. Even when this is achieved, conventional simplex steady-state and lifetime spectrometers are likely to be just as important as they are today as there are nearly always compromises to be made in new developments, and the search for higher performance implies dedicated and specific, rather than general capabilities. We watch and work with keen interest to help the future unfold.

Acknowledgement PY wishes to thank the QuantIC Technology Hub for a research fellowship.

References

1. Lakowicz JR (2006) Principles of fluorescence spectroscopy3rd edn. Springer, New York
2. Birch DJS, Chen Y, Rolinski OJ (2015) Fluorescence. In: Andrews DL (ed) Photonics. Biological and medical photonics, spectroscopy and microscopy, vol 4. Wiley, Hoboken, p 1
3. Strategic Directions International Inc. (2014) Global assessment report 13th edition and market forecast 2014-18. The laboratory analytical and life science instrumentation industry
4. Bollinger LM, Thomas GE (1961) Measurement of the time-dependence of scintillation intensity by a delayed coincidence method. Rev Sci Instrum 32:1044
5. Gaviola Z (1926) Ein Fluorometer, apparat zur messing von Fluoreszenzabklingungszeiten. Z Phys 42:853
6. O'Connor DV, Phillips D (1984) Time-correlated single photon counting. Academic Press, London
7. Birch DJS, Imhof RE (1991) Time-domain fluorescence spectroscopy using time-correlated single-photon counting. In: Lakowicz JR (ed) Techniques. Topics in fluorescence spectroscopy, vol 1. Plenum, New York, p 1
8. Cundall RB, Dale RE (eds) (1983) Time resolved fluorescence spectroscopy in biochemistry and biology. NATO advanced science institutes series a: life sciences, vol 69. Plenum, New York
9. Birch DJS (1975) Delayed coincidence fluorescence studies of 9,10-diphenylanthracene and the first four all-trans diphenyl polyenes. PhD thesis, University of Manchester
10. Birch DJS, Imhof RE (1977) A single-photon counting fluorescence decay-time spectrometer. J Phys E Sci Instrum 10:1044
11. Birch DJS, Imhof RE (1981) Coaxial nanosecond flashlamp. Rev Sci Instrum 52:1206
12. Imhof RE, Birch DJS (1982) Kinetic modelling and time resolution in reconvolution analysis. In: Bouchy M (ed) Deconvolution and reconvolution of analytical signals: application to fluorescence spectroscopy. ENSIC-INPL, Nancy, p 411
13. McGuiness CD, Sagoo K, McLoskey D, Birch DJS (2004) A new sub-nanosecond LED at 280 nm: application to protein fluorescence. Meas Sci Technol 15:L19

14. McLoskey D, Campbell D, Allison A, Hungerford G (2011) Fast time-correlated single-photon counting fluorescence lifetime acquisition using a 100 MHz semiconductor excitation source. Meas Sci Technol 22:067001
15. Birch DJS, Imhof RE, Dutch A (1984) Pulse fluorometry using simultaneous acquisition of fluorescence and excitation. Rev Sci Instrum 55:1255
16. Birch DJS, Holmes AS, Gilchrist JR, Imhof RE, Al-Alawi SM, Nadolski BZ (1987) A Multiplexed single photon instrument for routine measurement of time-resolved fluorescence anisotropy. J Phys E Sci Instrum 20:471
17. Birch DJS, Holmes AS, Imhof RE, Nadolski BZ, Suhling K (1988) Multiplexed array fluorometry. J Phys E Sci Instrum 21:415
18. McLoskey D, Birch DJS, Sanderson A, Suhling K, Welch E, Hicks PJ (1996) Multiplexed single-photon counting 1: a time-correlated fluorescence lifetime camera. Rev Sci Instrum 67:2228
19. Suhling K, McLoskey D, Birch DJS (1996) Multiplexed single-photon counting 2: the statistical theory of time-correlated measurements. Rev Sci Instrum 67:2238
20. McLoskey D (1995) Multiplexed single-photon timing array fluorometry: application to tryptophan fluorescence in microemulsions. PhD thesis, University of Strathclyde
21. Poland SP, Krstajić N, Monypenny J, Coelho S, Tyndall D, Walker RJ, Devauges V, Richardson J, Dutton N, Barber P, Day-Uei Li D, Suhling K, Ng T, Henderson RK, Ameer-Beg SM (2015) A high speed multifocal multiphoton fluorescence lifetime imaging microscope for live-cell FRET imaging. Biomed Opt Express 6:277. https://doi.org/10.1364/BOE.6.000277
22. Krstajić N, Levitt J, Poland S, Ameer-Beg S, Henderson R (2015) 256 × 2 SPAD line sensor for time resolved fluorescence spectroscopy. Opt Express 23:5653. https://doi.org/10.1364/OE.23.005653
23. Sagoo K, Hirsch R, Johnston P, McLoskey D, Hungerford G (2014) Pre-denaturing transitions in human serum albumin probed using time-resolved phosphorescence. Spectrochim Acta A 124:611
24. Polívka T, Sundstrom V (2005) Carotenoid excited states photophysics, ultrafast dynamics and photosynthetic functions. In: Di Bartolo B, Forte O (eds) Frontiers of optical spectroscopy: investigating extreme physical conditions with advanced optical techniques. Kluwer Academic, Dordrecht
25. Bot F, Anese M, Lemos MA, Hungerford G (2015) Use of time-resolved spectroscopy as a method to monitor carotenoids present in tomato extract obtained using ultrasound treatment. Phytochem Anal 27:32
26. Fujii R, Onaka K, Nagae H, Koyama Y, Watanabe Y (2001) Fluorescence spectroscopy of all-trans-lycopene: comparison of the energy and the potential displacements of its 2 Ag state with those of neurosporene and spheroidene. J Lumin 92:213
27. Zhang J-P, Fujii R, Qian P, Inaba T, Mizoguchi T, Koyama Y, Onaka K, Watanabe Y (2000) Mechanism of the carotenoid-to-bacteriochlorophyll energy transfer via the S1 state in the LH2 complexes from purple bacteria. J Phys Chem B 104:3683
28. Linsinger TPJ, Roebben G, Gilliland D, Calzolai L, Rossi F, Gibson N, Klein C (2012) Requirements on measurements for the implementation of the European Commission definition of the term "nanomaterial". European Commission Joint Research Centre Report. https://doi.org/10.22787/63490
29. Steiner RF (1991) Fluorescence anisotropy: theory and applications. In: Lakowicz JR (ed) Principles. Topics in fluorescence spectroscopy, vol 2. Plenum, New York, p 1
30. Kawski A (1993) Fluorescence anisotropy: theory and applications of rotational depolarization. Crit Rev Anal Chem 23:459
31. Smith TA, Ghiggino KP (2015) A review of the analysis of complex time-resolved fluorescence anisotropy data. Methods Appl Fluoresc 3:022001
32. Weber G (1952) Polarization of the fluorescence of macromolecules: I. Theory and experimental method. Biochem J 52:145

33. Iler RK (1979) The chemistry of silica: solubility, polymerization, colloid and surface properties, and biochemistry. Wiley, New York
34. Brinker CJ, Scherer GW (1990) Sol-gel science. Academic Press, London
35. Apperson K, Karolin J, Martin RW, Birch DJS (2009) Nanoparticle metrology standards based on the time resolved fluorescence anisotropy of silica colloids. Meas Sci Technol 20:025310
36. Laursen BW, Krebs FC (2001) Synthesis, structure, and properties of azatriangulenium salts. Chem Eur J 7:1773
37. Sorensen TJ, Thyrhaug E, Szabelski M, Luchowski R, Gryczynski I, Gryczynski Z, Laursen BW (2013) Azadioxatriangulenium: a long fluorescence lifetime fluorophore for large biomolecule binding assay. Methods Appl Fluoresc 1:025001
38. Stewart HL, Yip P, Rosenberg M, Sørensen TJ, Laursen BW, Knight AE, Birch DJS (2016) Nanoparticle metrology of silica colloids and super-resolution studies using the ADOTA fluorophore. Meas Sci Technol 27:045007
39. Karolin J, Geddes CD, Wynne K, Birch DJS (2002) Nanoparticle metrology in sol-gels using multiphoton excited fluorescence. Meas Sci Technol 13:21
40. Lemos MA, Hungerford G (2013) The binding of curcuma longa extract with bovine serum albumin monitored via time-resolved fluorescence. Photochem Photobiol 89:1071
41. Köllner M, Wolfrum J (1992) How many photons are necessary for fluorescence lifetime measurements? Chem Phys Lett 200:199
42. Hungerford G, Allison A, McLoskey D, Kuimova MK, Yahioglu G, Suhling K (2009) Monitoring sol-to-gel transitions via fluorescence lifetime determination using viscosity sensitive fluorescent probes. J Phys Chem B 113:12067

Applications of Submersible Fluorescence Sensors for Monitoring Hydrocarbons in Treated and Untreated Waters

Vadim B. Malkov and Jeremy J. Lowe

Contents

Abstract With development of new methods for oil and gas exploration, and in an effort to increase efficiency of petroleum production, rapid analysis of oil in water (OIW) has become increasingly important. Cost-effective, real-time analysis of refined oil products in water—whether their presence is as products or contaminants—enables nimble response to changing concentrations of analytes, which is not only of interest within the energy industry but also useful to ensure compliance within other industrial, municipal, and environmental applications. As a result of lab and field experimentation with a UV-fluorescence sensor responding to polycyclic aromatic hydrocarbons (PAH) present in any oil derived from mineral sources, it was determined that OIW could be successfully measured through correlation of PAH and OIW concentration in industrial process water, wastewater, municipal water treatment, natural seawater, and source water samples containing stable oil content. In samples with repeatable and sustainable content, the concentration of hydrocarbons can be quantified, while the most useful application of the examined instrumentation was found for event detection and trending of oil in water.

V. B. Malkov (✉) and J. J. Lowe
Hach Company, Loveland, CO, USA
e-mail: vmalkov@hach.com

© Springer Nature Switzerland AG 2019   135
B. Pedras (ed.), *Fluorescence in Industry*, Springer Ser Fluoresc (2019) 18: 135–172,
https://doi.org/10.1007/4243_2018_6, Published online: 9 February 2019

Keywords Cooling water · Energy industry · Industrial wastewater · Municipal drinking water · Oil and grease · Oil-in-water · OIW · PAH · Petroleum production · Polycyclic aromatic hydrocarbons · Process water · Refinery · Seawater · Source water · UV fluorescence · Wastewater

Abbreviations

AOX	Adsorbable organic halides
API	American Petroleum Institute
ATR	Attenuated total reflectance
BOD	Biochemical oxygen demand
BTEX	Benzene, toluene, ethylbenzene, xylenes
CDOM	Colored dissolved organic matter
CHC	Chlorinated hydrocarbons
COD	Chemical oxygen demand
DAF	Dissolved air flotation
DDT	Dichlorodiphenyltrichloroethane
DOC	Dissolved organic carbon
DOM	Dissolved organic matter
EEM	Excitation-emission matrices
FEWS	Fiber optic evanescent wave spectroscopy
FTIR	Fourier transformation infrared spectroscopy
GC	Gas chromatography
GC/MS	Gas chromatography/mass spectrometry
GPRS	Global packet radio service
HOAB	High output air-blast system
HPLC	High-performance liquid chromatography
IAF	Induced air flotation
LOD	Limit of detection
LR	Low range
MAH	Monocyclic aromatic hydrocarbons (BTEX)
MCA	Multicomponent analysis
OIW	Oil in water
PAH	Polycyclic aromatic hydrocarbons
PCB	Polychlorinated biphenyls
PPB	Parts per billion
QMB	Quartz microbalance
RIfS	Reflectometric Interference Spectrometry
SEC	Standard error of calibration
SEP	Standard error of prediction
UN	United Nations
UV	Ultraviolet

UVMA   UV multiwavelength absorptiometry
WHO   World Health Organization
WW   Wastewater

1   Introduction

Environmental monitoring is a crucial element in the implementation of regulatory initiatives formed by nation states. In particular, regulatory oversight of water quality governs industry activities, which must maintain certain standards and terms that affect water resources across multiple locations and seasonal cycles.

Protecting all water sources—oceanic, groundwater, and surface water—is of critical importance to sustaining the wide variety of fauna and flora that depend upon a resource free from contaminants and pollutants to sustain life at acceptable levels of quality.

Given the ubiquitous use of hydrocarbons in the modern world, it may be argued that the proximity of hydrocarbons and hydrocarbon derivatives to potable or treatable water sources is an indicator of acceptable levels of pollutants at water sources; at a minimum their presence offers a point at which a measurable gauge of pollution may be established. Data derived from such empirical research may then be used to influence stakeholders in their adoption of strategies that would proactively set an agreed standard of protection.

To date, environmental monitoring has utilized both qualitative and quantitative techniques to assess offending articles. Yet the demand for technologies able to analyze or detect the presence of known and unknown substances continues to rise among various stakeholders tasked with providing remedial solutions to complex sources of pollution and contamination.

Development of technologies suitable for challenging environments has been a priority in recent years. Industries, such as water treatment, chemical, and maritime, present conditions in which complex chemical processes often produce by-products that interfere with the accuracy of installed instruments. While many advancements have been made over the last few decades to technologies that measure hydrocarbons in water, research must continue to keep pace with environmental regulations and industry demand to ensure water quality even in the most harsh and uncompromising situations.

2   Literature Review

The instrumentation sector has directed a substantial amount of effort toward the development of spectroscopic solutions for a wide range of applications, with a core focus on the ability of substances to interact with light. Light absorbance, fluorescence, and luminescence are a means of detection and measurement; therefore, technologies that utilize these methods offer the required data for intervention sought by a highly demanding and diverse market.

The enormous progress in this sector over the last 20 years has been attributed to the curiosities formulated in laboratory research projects throughout the 1980s and demonstrated in a multiplicity of applications driven by powerful algorithms and data processing software throughout the 1990s and beyond. Published research and innovations increased in the 2000s, with many key ingredients added to the knowledge base; yet many researchers were prone to reinventing the subject within areas of specialization based on the assumption that software can solve all problems [1].

This period of innovation advanced a wide range of analytical technologies used in environmental monitoring, with many applications, pollutants, and contaminants being of interest across numerous industry and stakeholder groups. The synopsis of significant research is presented below, with focus on those pertaining to the optical measurement of hydrocarbons in water.

2.1   Contaminants and Pollutants: What to Measure and How

Because of the impractical nature of monitoring every substance, there is an increasing need to integrate our understanding and model a wide range of processes, structures, and scales in the context of environmental monitoring [2] as a way to prioritize the values that are most responsible for protecting our natural resources. Based on information provided by the World Health Organization (WHO), a comprehensive list has been compiled of the most concerning pollutants and contaminants found in a variety of water sources [3]. Among other potentially harmful substances or indicators of harmful chemical reactions, the list includes:

- Adsorbable organic halides (AOX)
- Biochemical oxygen demand (BOD)
- Chemical oxygen demand (COD)
- Dichlorodiphenyltrichloroethane (DDT)
- Dissolved organic carbon (DOC)
- Polycyclic aromatic hydrocarbons (PAH)
- Polychlorinated biphenyls (PCBs)

Additionally, certain organic pollutants, such as chlorinated aliphatic hydrocarbons and chlorinated aromatic hydrocarbons (CHCs), are cited as carcinogenic contaminants and persistent toxic environmental pollutants that present a substantial threat to ecological systems and human life [4, 5]. Examples of important CHCs include the following:

- Monochlorobenzene
- 1,2-dichlorobenzene
- 1,3-dichlorobenzene

- Trichloroethylene
- Perchloroethylene
- Chloroform

CHCs present an additional challenge to industry in that they are problematic to detect in aqueous environments [6–8]; however, effective detection is essential to monitor and control regimes that are fundamental components in meeting stringent regulatory requirements that can result in highly punitive consequences if not met.

A variety of detection techniques exist [4, 6, 8] that involve:

- High-performance liquid chromatography (HPLC)
- Gas chromatography (GC)
- Gas chromatography and mass spectrometry (GC/MS)
- Absorption spectroscopy (e.g., UV)
- Raman spectroscopy
- Fluorescence spectroscopy

These techniques were considered as the most appropriate form of detection able to measure at low concentrations, yet upon careful evaluation of the existing market were found to be lacking: the prevailing methods engage complex and time-consuming processes involving solvents and other potentially harmful reagents generally suited to laboratory conditions [4, 8]. In addition, the application of sensitive online detection was less prevalent for in situ and remote monitoring, further demonstrating the need for technological advancement to rapidly detect the concerning hydrocarbons in aqueous environments.

2.2 Compliance: Why Effectiveness Matters

Industry, municipal, and environmental stakeholders are faced with a wide range of substances and a complex array of issues in their quest to ensure compliance with the policies of local, national, and international regulatory bodies. Four key categories of water quality monitoring have been contextualized and grouped according to existing sensor capabilities, using regulatory compliance as the main driver for effective environmental monitoring [9]:

1. Metals
2. Radioisotopes
3. Biological contaminants
4. Volatile organic compounds

An example of the stringent standards set by a government body may be found in the US Environment Protection Agency's legally enforceable regulatory requirements, standards, and policies; see Table 1. This brief sample of contaminants drawn from the wider list offers an insight into the low level of tolerance specified by the EPA that is echoed by similar regulatory bodies around the world.

Table 1 A sample of EPA national primary drinking water standards for microorganisms, solvents, and hydrocarbons [9]

Contaminant	Maximum contaminant level goal (mg/L)	Maximum contaminant level (mg/L)
Cryptosporidium	Zero	1%
Giardia lamblia	Zero	0.1%
Heterotrophic plate count (HPC)	n/a	a
Legionella	Zero	a
Total coliforms (including fecal coliform and *E. coli*)	Zero	5.0%[b]
Turbidity	0.3	1 NTU
Viruses (enteric)	Zero	a
Benzo(a)pyrene (PAHs)	Zero	0.002
Total trihalomethanes (TTHMs)	n/a	0.08

[a]EPA's surface water treatment rules require systems using surface water or groundwater under the direct influence of surface water to (1) disinfect their water and (2) filter their water or meet criteria for avoiding filtration
[b]More than 5.0% samples total coliform-positive in a month

The private sector has played a key role in the development of innovative technologies and techniques aimed at monitoring the impact of the issues raised by Chapman and Ho et al. [3, 9], while the legally enforceable regulatory thresholds have served as a mutually important driver for the commercial interests of private organizations and the need for them to maintain compliance.

As a result, a wide range of technologies has been developed to provide the measurements required by international, national, state, local, and industry-specific regulators. Key attributes of acceptable monitoring technologies include:

– Levels of accuracy
– Timely response
– In situ or infield capability

The suitability of technology to a given issue is directly related to the environment within which the technology is expected to perform. Water and wastewater offer an interesting contrast that presents incongruent conditions that require adjustments to the appropriate technology.

Continuous and discrete sampling techniques have historically found a place within the sampling regimes and strategic decision-making of industrial organizations as a means of satisfying increasingly stringent compliance targets. Meanwhile innovators are presented with new opportunities to meet the demand for better online monitoring of water systems created by the inability of laboratories to provide the responsiveness required by industry for the effective deployment of risk management strategies and early warning systems [10].

> There is a clear need to be able to rapidly detect (and respond) to instances of accidental (or deliberate) contamination, due to the potentially severe consequences to human health. Detecting in real time is the most optimal way to ensure an appropriate and timely response [10].

2.2.1 Detection of Heavy Metals in Water

A number of innovative techniques have developed in recent years aimed at the detection and analysis of heavy metal ions: (1) inductively coupled plasma mass spectrometry, (2) anodic stripping voltammetry, (3) X-ray fluorescence, and (4) microprobes. In general, these techniques involve expensive equipment and require sample pre-treatment and/or analyte preconcentration steps [11].

The current approach is based on previous advances in recognizing elements based on organic chelators, organic polymers, proteins, peptides, and DNA/RNA; peptides offer combinatorial chemistry that allows for optimal amino acid sequencing for specific metal-ion recognition [11].

2.2.2 Detection of Radioisotopes in Water

The need for effective detection of radioisotopes stems from the natural occurrence of radioactivity in groundwater and seawater with the consequential effective dose received by consumers [12].

Specific priority is given to radiometric detection of uranium in the hexavalent oxidation state via several methods that measure concentration in both groundwater and seawater: spectrophotometry, X-ray fluorescence spectroscopy, inductive coupled plasma mass spectroscopy, and alpha liquid scintillation [12]. Effective measurement in seawater may be achieved through the application of robust radioanalytical methods using high-resolution α-spectroscopy as a means of detecting activity and concentration of ^{238}U and ^{234}U [12].

2.2.3 Detection of Biological Contaminants in Water

The conventional approach to detecting harmful microbial contamination has relied on time-consuming enrichment steps followed by biochemical identification over an assay period of up to 1 week [13].

Biological sensors able to detect and identify microorganisms in real time have since been developed, using a combination of biological receptor compounds (antibody, enzyme, nucleic acid, etc.) and a physical or physical-chemical transducer that directs "real-time" observation of a specific biological event. Refer to Fig. 1.

Sensors that employ optically based transduction methods to achieve a more robust, easy-to-use, portable, economical system are evaluated in Table 2 [13].

When suitably specific antibodies can be aligned with the target analyte, biosensors offer an exciting alternative to traditional methods of detecting microorganisms in water, providing an effective solution for rapid, real-time, and multiple analyses to be carried out in the field and laboratory [13].

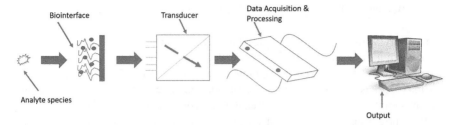

Fig. 1 Schematic diagram is showing the main components of a biosensor. The biological event, e.g., antibody-antigen interaction, elicits a physicochemical change at the biointerface, e.g., change of mass, heat change, or change in electrical potential, which is converted by the transducer to an electrical signal. The output from the transducer is then amplified, processed, and, finally, displayed as a measurable signal (Adapted from [13])

Table 2 Comparison of several commercially available SPR-based biosensors

Principle	BIAcore 3000 (prism-based SPR)	IBIS (vibrating mirror SPR)	Plasmon (broad-range SPR)	SPREETA (prism-based SPR)	IASys (resonant mirror)
Flow injection analysis (FIA) system	Yes	Yes	No	Yes	No
Temperature control	Yes	Yes	Yes	No[a]	Yes
Autosampler	Yes	No	Yes	No	Yes
Microfluidics	Yes	No	No	No	No
Disposable sensing element	Yes	Yes	Yes	Optimal	Yes
Refractive index range	1.33–1.40	1.33–1.43	1.33–1.48	1.33–1.40	–
Limit of detection (RIU)	3×10^{-7}	2×10^{-6}	6×10^{-6}	3×10^{-7}	$>1 \times 10^{-6}$

[a]Does not offer temperature control but offers temperature compensation by correcting the signal for temperature fluctuations

2.2.4 Detection of Organic Compounds in Water

A modular fiber-optic system [14] offers a cost-effective, rugged, field-based solution for the detection of hydrocarbons in aquatic solutions. In trials a wide range of benzene-based pollutants was detected using a PVC polymer coating and quantified at 500 ppm. While the results offered significant improvement in detection level and demonstrated the importance of the polymer coating, the technology proved unable to extend beyond the ppm range toward the ppb range required by industry [14].

When comparing the solid crystal approach used in the conventional attenuated total reflectance (ATR) approach to the optical fiber light-guide of the fiber optic evanescent wave (FEWS) approach, the latter has shown more favorable results by demonstrating an increased number of internal reflections, which leads to a higher absorption rate and thereby an increase in sensitivity [8]. However, to improve the detection capability toward the identification of pollutants in the presence of numerous interferents, additional efforts may be required to further increase the signal-to-noise ratio.

While the FEWS approach advocated by Lu et al. is cited as offering a viable solution, the technique is limited to ppm and is therefore considered unsatisfactory by industrial applications that routinely demand measurement of constituent elements in ppb. This problem is further compounded by background spectral noise that is said to derive from interfering absorbance of water and other interfering molecules and thereby reduce the reliability of retrieved data [14].

Chemical Sensor Arrays

Spectroscopic methods based on reflection of light and specifically internal reflection spectroscopy are strongly dependent on the nature of the surface and can therefore be used to study surfaces [15].

Shear horizontal surface acoustic wave (SH-SAW) and estimation theory (ET) have been used to illustrate the identification of chemical arrays in aquatic solutions [16]. This approach is used as a basis to derive a higher degree of accuracy through the processing of signals received from a sensor. The selection of benzene, toluene, and ethylbenzene/xylenes array (BTEX) was presented in a solution containing interferents, configured to speciate and quantify BTEX in the presence of what is said to be noisy data.

A study team used a polymer-coated SH-SAW device and signal processing methodology with estimation theory for the detection and quantification of BTEX compounds in concentrations in the range of 10–2000 ppb, the results of which are summarized in Table 3 [16].

Sensor arrays based on random forests analysis offer a quantitative approach with a focus on the accuracy of algorithms rather than the selected mode of the instrument utilized. While sensor selection is important, the multivariate assessment of unary, binary, ternary, quaternary, and quinary combinations of BTEX solutions, including the added naphthalene, offers the most accurate means by which the properties of a solution may be determined [17].

Table 3 Estimated concentrations of BTEX compounds obtained using the measurement data of an LNAPL sample in DI water compared to actual concentrations [16]

Analyte of interest[a]	Concentrations (ppb)		
	Actual	Estimated	% difference
Benzene	610	597	2
Toluene	874	919	5
Ethylbenzene and xylenes	150	154	3

[a]Collected using an SH-SAW device coated with 0.8 µg PIB

Thirteen sensors obtained from various manufacturers were selected for assessment. While the measurement of simple binary mixtures can be analyzed within a reasonable degree of accuracy, through the application of simple statistical models, mixtures having several components require a higher degree of predictive analysis.

The accuracy of predictions was found to be directly related to the experimental error of the sensor responses measurement. For each sensor array, an average coefficient of variation of 1.8% was obtained for sensor responses to mixtures having a nominal component concentration of 10 mg/L and an average coefficient of 3.5% to sensor responses to the mixture with a nominal component concentration of 1 mg/L [17].

These measurements and the random forests model were used to predict the mixture concentrations. The root-mean-square error between the measured and predicted concentrations were found to be 0.2–1.5 mg/L in the samples with a nominal concentration of 10 mg/L and 0.06–0.23 mg/L in the samples with a nominal concentration of 1 mg/L. It was found that the accuracy of the random forests predictions correlated with that of the sensitivity and selectivity of the sensor array. The level of accuracy was also found not to be unduly affected by an increase in mixture complexity. As a result, the random forests statistical technique offers promise for analyzing the output from partially selective sensors to quantifying the concentration of multiple components in mixtures [17].

Smart Sensors and OIW Detection

The issue for the accuracy of detection for hydrocarbons within aquatic solutions as evaluated by Cooper et al. [17] is based on the earlier work of Dickert et al. [18] in that new concepts were used to explore the key focus of carcinogenic polycyclic aromatic hydrocarbons (PAH). PAH may form through incomplete combustion processes of diesel engines and released to the atmosphere, absorbed into soot particles, deposited in the ground, and then washed into the surface and oceanic waters.

A common trait of the PAH-type compounds is fluorescence—the ability to emit light after initial irradiation. Efficient detectability of PAH is of critical importance and is directly related and proportional to the selected transducer and its capacity to measure the optimized signal emitted by the chosen sensor [18].

The signals derived from these transducers form the basis upon which measurement can be made, data produced and evaluated, and appropriate action taken, as shown in Fig. 2.

UV-fluorescence spectroscopy has been cited as the ideal method of determining PAH concentration levels in a target sample [19]. A reference method based upon prescribed regulatory thresholds of known PAH species seeks to set an achievable photometric determination metric derived from empirical data. UV spectra of PAH species are often difficult to differentiate due to size; therefore, grouping of PAH offers an opportunity for multicomponent analysis (MCA). The correlation of the grouped spectra using a PLS mathematical algorithm is ideally suited to the task [20].

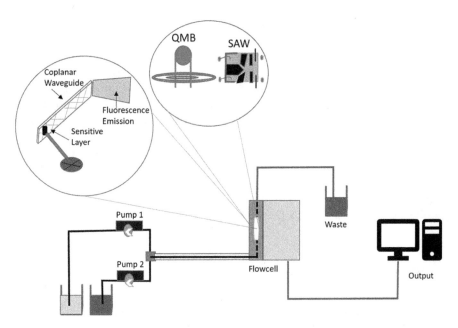

Fig. 2 Fluorescence sensor and mass-sensitive devices and their integration into a flow system [13]

The results of the analysis may be tested through a standard error of calibration (SEC) and the standard error of prediction (SEP) calculated for each component [20].

The issue of interferent background noise as cited by [8] may be eliminated by mathematical or experimental means: a UV multiwavelength absorptiometry (UVMA) based on the K-matrix algorithm [20]. The coefficient matrix K is extended by a polynomial of q degrees which models the indifferent background signal:

$$K = \begin{vmatrix} k_1(\lambda_1) \dots k_P(\lambda_1)\ 1\ \lambda_1 \dots \lambda_1 q \\ \vdots \qquad\qquad \vdots\ \ \vdots\ \ \vdots \qquad \vdots \\ k_1(\lambda_n) \dots k_p(\lambda_n)\ 1\ \lambda_{n\dots}\lambda_n q \end{vmatrix}$$

where k_j (λ_i) is the regression coefficient k of component p_j at wavelength λ_i. The latter will be recorded at n wavelengths. Mathematically, this method is based on the fact that any background function can be approached by a polynomial of degree q. The result of that modified K-matrix approach is said to be a vector which contains the concentrations and the polynomial coefficients [20].

These advancements enabled the development of portable or submersible fluorometers suitable for real-time processes and field applications [21]. The established efficacy of excitation-emission matrices (EEMs) and online laboratory grade spectrofluorometers able to investigate the composition and dynamics of dissolved organic matter (DOM) in aquatic environments that are produced over a range of excitation (λ_{Ex}) and emission (λ_{Em}) wavelengths from ~200 to 300 nm [deep ultraviolet (UV) domain] is considered. The methodology may use xenon as a light source and a PMT or silicon photodiode as the reference detector and is ideally suited to the detection of DOM fluorophores found in natural waters, particularly phenanthrene, a typical petrogenic hydrocarbon.

Recent advancements in semiconductor LED technology are now able to offer a possibility of using deep UV LED, in the range of 250–300 nm—comparable to laboratory grade functionality. These developments are based on aluminum gallium nitride (AlGaN) multiple-quantum-well active layer designs having a continuous-wave output power in the 0.1–1 mW range at 20 mA driving current [21].

The main advantages of these portable instruments are spectral selectivity, small size and weight, low power consumption, stability, and low cost. Deep UV LEDs present a number of opportunities to the sensor marketplace, particularly for the measurement of phenanthrene, a known carcinogen identifiable through an analysis of oil in aquatic environments [21].

3 Practical Applications

Detecting oil and grease in industrial and environmental waters presents a challenge due to the variability of analytes and the matrix, which is dependent on the application in which the water is used. This challenge is even more pronounced where online analysis is implemented. Direct OIW measurement can be accurately accomplished in a lab setting; however, because the process is long, complicated, and costly (and therefore relatively infrequently performed), lab analysis of OIW isn't particularly useful for event detection, trending, or quantification in situ.

Instead, if lab analysis can be used to correlate OIW concentration with another parameter that is easily measured by online instrumentation, monitoring becomes simpler, more efficient, and more useful as an operational decision-making tool. Several methods of indirectly monitoring OIW concentration have been established; these methods are summarized in Table 4 and will be briefly discussed in Sect. 3.

Correct correlation depends on selecting a model compound that is found in a broad range of oils (providing no limitations for the type of product to be analyzed), but not present in the water matrix. The model compound should generate a strong signal, free of interference, that can be detected and correctly interpreted by online instrumentation.

Two major groups of aromatic hydrocarbons are consistently used as a model for optical methods: monocyclic aromatic hydrocarbons (benzene, toluene, ethyl-benzene, and xylenes, known collectively as BTEX) and polycyclic aromatic hydrocarbons (PAH). Of the two groups, PAH compounds derived from naphtha-lene, anthracene, and phenanthrene and present in oils feature a system of conjugated bonds (Fig. 3) that contribute to the substance's fluorescent properties.

Practical applications of OIW monitoring using UV fluorescence may be grouped in three main categories—event detection, trending, and quantification—with one common limitation: the oils must be derived from mineral source. Natural oils based on vegetable or animal sources or synthetic oils cannot be detected with this method as they do not contain PAH.

Table 4 Comparison of different methods for OIW measurements with their respective advantages and shortcomings

Method	Advantages	Disadvantages
Lab analysis of a grab sample	Ultimate method for direct determination of oil and grease in water	Long and complex analysis, special equipment required, representative sample required
Nephelometry (light surface scattering)	Cost-effective online instrumentation	Difficult to distinguish between turbidity caused by oil and other particulate matters
UV absorbance	Robust, well-known technology	Interference from non-oil compounds, biological matter, and suspended solids
VIS fluorescence	LPP (Low Price Point) online analyzers and submersible probes	Low sensitivity to PAH, interference from natural organic matter
UV fluorescence	High sensitivity and selectivity toward PAH, wide range of measurements, online analyzers, and submersible probes available	Relatively high price, requires calibration per each matrix/application

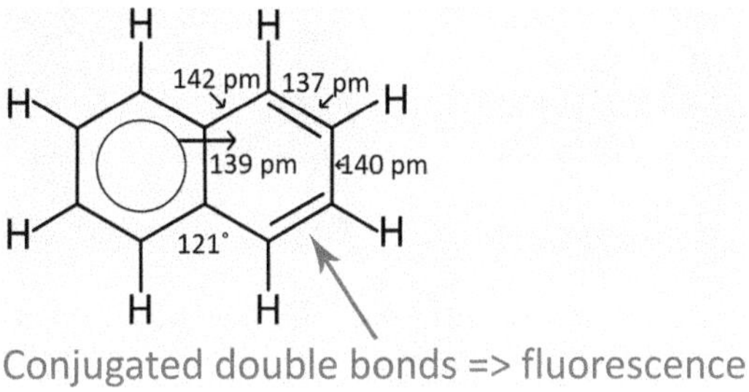

Fig. 3 Structural formula of naphthalene as an example of PAH

Applications that have changing oil content, variety, and concentrations are well suited to monitoring PAH for event detection, while applications that have water matrices with constant oil content (e.g., various oils in a constant proportion) are well suited to trending. Finally, applications that have water containing a single oil are well suited to quantitative measurement of concentration. In any case, process calibration is preferred over standard calibration.

With the goal of providing accurate, real-time measurements of oil content using a moderately priced, compact, reagentless sensor, the use cases described in Sect. 3 implemented a UV-fluorescence probe (Fig. 4) having a standard Xenon flash lamp with interference filter producing light at 254 nm (excitation) and collecting feedback at 360 nm (emission). For this study, the probe generated a 4–20 mA signal that was processed by the Hach SC1000 controller, and readings were displayed in raw format (mA), in relative (% of scale), or in absolute concentration of OIW based on calibrations.

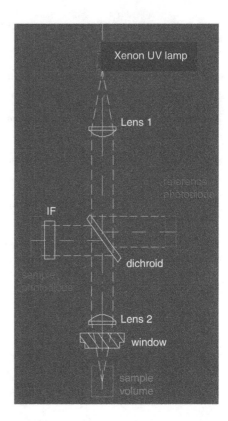

Fig. 4 UV-fluorescence OIW sensor and principle of its operation (photo credit: Hach Company)

This probe was chosen based on its potential applicability to all refined and crude oils because the method is highly sensitive and quite specific to PAH-type compounds, which are considered to be constituents of all such products.

Main technical specifications for the tested UV-fluorescence sensor are presented in Table 5. The measurement range limits are set at the factory and the calibration standards are also available for both low- and high-range sensors. Optional equipment, such as automated cleaning system, was found to be very helpful during field studies.

Depending on the intended analyte, a smaller, less-expensive alternative to the UV-fluorescence sensor is a visible light-based (LED) fluorescence probe (VIS-fluorescence) producing light at 370–460 nm (excitation) and measuring feedback at 520–715 nm (emission). Both types of sensors are available in either stainless steel or titanium body and produce an analog signal that can be registered by a standard controller.

Prior to field testing, a laboratory performance evaluation was conducted to establish correlation between the model compound (phenanthrene) and total OIW concentration to confirm the probe's ability to effectively measure OIW in the field [22]. Results of the laboratory calibrations are presented in Figs. 5, 6, 7, and 8.

Table 5 UV-fluorescence sensor technical specifications

Parameter	Specification
Detection parameter (PAH)	Phenanthrene (model compound)
Measuring principle	UV fluorescence (excitation 254 nm/emission 360 nm)
Measuring range	Related to phenanthrene
Low-range probe	0–50 ppb and 0–500 ppb, corresponding to 0–1.5 ppm and 0–15 ppm oil calibration standards
High-range probe	0–500 ppb and 0–5000 ppb, corresponding to 0–15 ppm and 0–150 ppm oil calibration standards
Limit of detection	1.2 ppb (phenanthrene)
Probe housing material	Stainless steel, titanium
Mounting options	Chain (submersible) In-line (in-pipe mounting hardware) Bypass (flow cell)

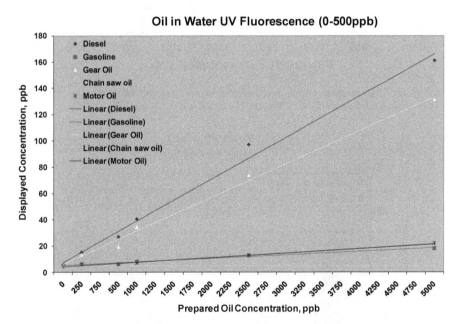

Fig. 5 Calibration curves for series of available oil products in water calibrated to phenanthrene [22]. The calibration coefficients were established based on the linear regression for several oil products including diesel fuel, gasoline, gear oil, and motor oil; however, chain saw oil did not show the presence of the model compound as it is a natural product derived from a biological raw material

From analysis of the literature and through extensive professional experience, a list of potential applications has been identified for the OIW sensors in municipal waters (drinking water and wastewater) as well as numerous opportunities in industrial waters (Table 6).

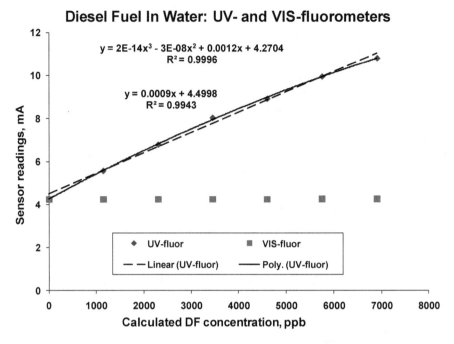

Fig. 6 UV fluorescence: diesel fuel calibration test results. In comparing UV- and VIS-fluorescence sensors, only the UV-based probe detected diesel fuel [22]

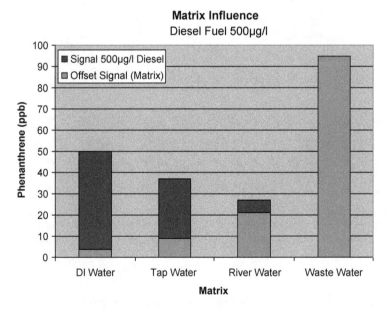

Fig. 7 Water matrix influence at 500 ppb of diesel fuel concentration. A set of experiments was conducted in DI water, tap water, river water, and wastewater to define the matrix effect on the measurement of OIW concentration with a UV-based sensor (Fig. 8, Y-axis = ppb). The UV-fluorescence probe provided a limit of detection (LOD) for diesel fuel significantly lower than 1 ppm [22]

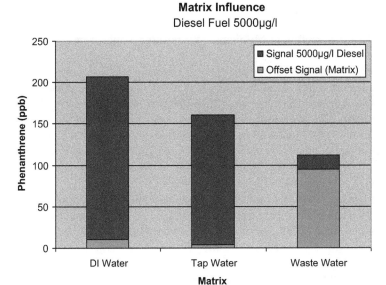

Matrix Influence
Diesel Fuel 5000µg/l

Fig. 8 Water matrix influence at 5000 ppb of diesel fuel concentration. The comparison of performance of the UV-fluorescence and UV-spectrophotometric (Hach UVAS, 254 nm) sensors indicates that the UV-fluorescence sensor was ten times more sensitive to phenanthrene than the UV-spectrophotometer [22]

Table 6 Potential applications for OIW sensors

Industry	Application
Drinking water	Environmental—early detection in natural sources Detection of oil in artificial reservoirs, as well as in source water (intake) Desalination plants water intake monitoring
Wastewater	Municipal Wastewater Treatment Plant (WWTP) inlet Monitoring direct and indirect discharge Storm water runoff Membrane plants (water reuse) Groundwater reclamation sites
Industrial water	Fuel storage tank area drainage systems Cooling water Condensate return Leaks from heat exchangers Turbine oil in power plant process water Effluent monitoring after oil-water separators (refineries) Aircraft and truck washdown facilities Petrochemical: detection of OIW separated from the crude oil Maritime applications (e.g., holding tanks wash down: <15 ppm OIW discharge limit)

3.1 Use Cases of UV-Fluorescence for Event Detection

3.1.1 Event Detection at Municipal Wastewater Applications

A municipal wastewater treatment plant in Germany was concerned with possible oil contamination from local factories and several small towns and villages nearby. A UV-fluorescence high-range sensor (0–5000 ppb) was submersed in the tail water at the bottom of a weir located between the screen and sand trap, allowing strong mixing of the sample and a fairly even distribution of the oil contamination. To test the performance of the probe, a mixture of 2 L of diesel fuel in 20 L water was spilled into the wastewater several meters upstream of the weir. An immediate response from the sensor was observed as shown in Fig. 9.

After several days of testing with readings close to zero, the sensor alerted plant staff to an actual oil spillage accident (Fig. 10), which produced what a plant operator described as "an emulsion visible on the water surface."

A large water reclamation facility in St. Petersburg, Russia, that treated water from a commercial marine port area was concerned with possible excessive discharge of oil products from the ships and port infrastructure. A UV-fluorescence LR sensor (0–500 ppb) was installed at an inlet collector before the central aeration station in the open well (chain mounted sensor) and the test lasted for several months.

A grab sample analysis was conducted in laboratory at the beginning of the test to establish a baseline for normal OIW concentration (rectangular dot on

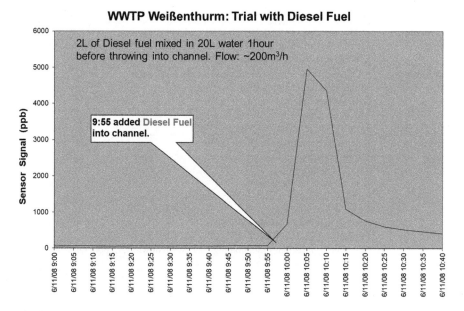

Fig. 9 Experiment at a wastewater treatment plant, high-range (HR) sensor [23]

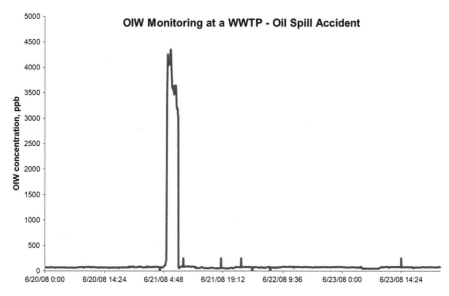

Fig. 10 Real-life oil accident at a wastewater treatment plant, HR sensor [23]

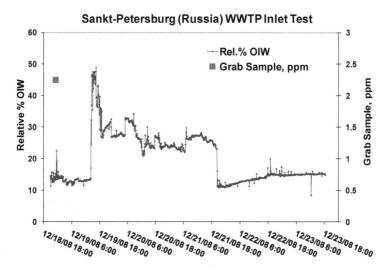

Fig. 11 Wastewater treatment plant inlet collector test (Russia), first response [23]

the chart, Fig. 11). Shortly after, an event was registered by the instrument (Fig. 11).

Because the event happened right before the weekend and the oil concentration was elevated just for couple days, the violation could have gone unnoticed; however, the OIW monitor registered and recorded it (Fig. 11). Therefore, based on this finding, a decision was made to equip the monitoring point with an autosampler connected to the same controller and driven by the OIW instrument.

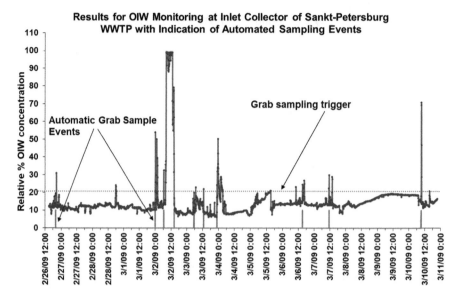

Fig. 12 Wastewater treatment plant test (Russia)—long-term trial with autosampler [23]

The utility elected to combine the OIW with an autosampler for additional coverage of a contamination event. Exceeding a pre-programmed concentration threshold of 3.5 ppm OIW (~25% of the scale shown in Fig. 12) would trigger an automatic grab sampling routine, providing the opportunity to analyze the stored water samples later in a laboratory (within 4 h from the sampling time) to determine absolute concentration and nature of oil products at the time of event, which would help to pinpoint the source of contamination.

3.1.2 Event Detection at Municipal Drinking Water Applications

A municipal drinking water plant in the town of Rifle, Colorado, was concerned with potential contamination of the main water source for the city in light of active oil and gas exploration activities in the area. To mitigate the concerns, a full-scale OIW analyzer was purchased by the municipality and installed at the water intake (Fig. 13), and later a UV-fluorescence OIW sensor was added to the panel for testing [24].

The sensor was calibrated versus the analyzer output, and the ongoing readings from both instruments were recorded and analyzed (Fig. 14).

The UV-fluorescence sensor was submersed in a trough-type flow cell as a part of the Hach Source Water Monitoring Panel and connected to a Hach SC1000 controller serving the panel. As seen from the chart in Fig. 15, the readings from both OIW instruments were similar throughout the test. Also, in several instances, the sensor followed the analyzer closely when some excursions from the baseline

Fig. 13 Installation at a drinking water intake. Right—inside the shed (photo credit: Hach Company)

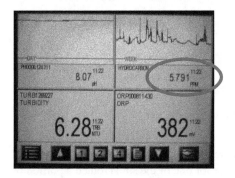 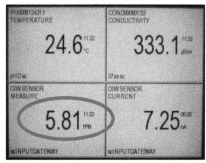

Fig. 14 Readings of both instruments: OIW analyzer (left screen, circled) and OIW sensor (right screen, circled) after calibration (photo credit: Hach Company)

were detected (see circled area on the graph in Fig. 15). In other cases, the deviations registered by the analyzer were due to its patterned performance and/or maintenance events.

The results recorded by the low-range UV-fluorescence sensor reflected the baseline and events very well against the full-scale OIW analyzer. Per the utility operator, the analyzer required weekly maintenance at a minimum, while the sensor sustained no maintenance during the entire test (more than a month of operation).

This test results allowed Parachute, Colorado, a town of approximately 1100 residents, to implement a complete solution to monitor drinking water intake for contamination related to oil and gas exploration.

The town's concerns of contamination were well founded: at least four spills occurred northwest of Parachute along Parachute Creek in late 2007 and 2008. Presumably because Parachute's water treatment plant is located above the confluence of Parachute Creek and the Colorado River, the town and others potentially impacted were not immediately notified of the incidents. Later, fines were imposed

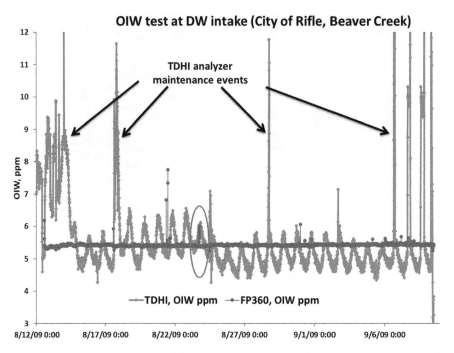

Fig. 15 Analysis results collected by both instruments side by side [24]

on the firms responsible. Fortunately, an agreement was reached between the State of Colorado, the town, and the firm responsible for one of the spills routed a portion of the fines toward purchase of monitoring equipment for the Town of Parachute's water treatment plant, rather than filling state coffers.

A Hach Source Water Monitoring Panel was purchased and installed at the water treatment plant to continuously monitor the source water for the community. An autosampler was also installed (Fig. 16) and programmed to begin sample collection if OIW levels exceeded a certain set point, which was an important feature for the utility's small, four-person staff that perform a variety of other functions in the community.

The source water from the Colorado River is initially pumped into a small storage tank. The water is then treated by the membrane system in the treatment plant. Should the instrument array detect a problem, the raw water pump is stopped until the cause of the problem can be investigated. In the meantime, the treatment plant can continue to operate using water in the storage tank. Thus, the monitoring system including an OIW sensor provides peace of mind for all residents of Parachute.

Fig. 16 Town of Parachute Source Water Monitoring Panel and Automatic Sampler (photo credit: Hach Company)

3.1.3 Event Detection in Industrial Applications

Cooling Water

A gas production facility in Southwestern Colorado was concerned with repeated oil leaks from the heat exchanger serving the compressors to produce liquid carbon dioxide and needed a system that would detect a problem before damage became severe.

A UV-fluorescence high-range sensor (0–5000 ppb, titanium) was installed in a shed with the chemical feed system for corrosion control, near the cooling tower. The sample feed was provided with the flow cell as shown in Fig. 17.

After several days of normal operation with readings close to zero, an oil leak from the heat exchanger was detected (Fig. 18). The contamination costs the utility more than $1000 USD in oil loss. Besides the direct loss of oil, the company had to complete a series of cleanup procedures which required a biocide, oil-absorbing mats, higher chlorine injection, draining of entire cooling tower water, and re-establishing chlorine and polymer feed to the cooling water, requiring approximately 8 h of work time and an additional $1000 USD in materials.

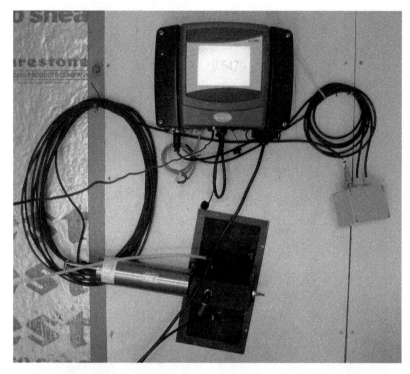

Fig. 17 Cooling tower installation with a flow-through cell (photo credit: Hach Company)

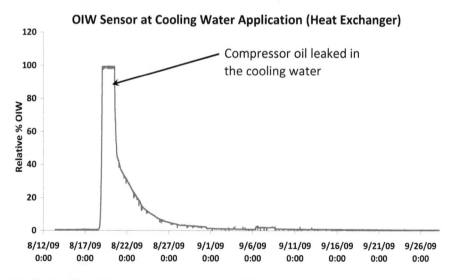

Fig. 18 Real-life oil leak from the heat exchanger [24]

While the cost of the accident was significant, it could have been exponentially worse, had there been no sensor installed. According to facility personnel, without such early notification, the leak would have been discovered only after the shutdown of the compressor due to low oil levels. This could have led to possible plant shutdown or damage to the compressor and extensive cleanup of the cooling tower.

As seen in Fig. 18, the accident went unnoticed for about 2 days, because communication between the instrument and the plant's monitoring system had not been established, despite the controller's capability for such communication. Remote access and control via Global Packet Radio Service (GPRS) were also capabilities that had gone unused, as the test site was extremely remote and lacked cellular service from any of the major US wireless carriers during the time of testing.

Nevertheless, the sensor proved very useful by saving the company significant amount of money at a nominal cost, which was not exacerbated by the maintenance requirements, which only required cleaning of the sensor window at intervals depending on the application. For example, when the sensor is submersed in a slow-moving dirty sample, the cleaning may occur daily depending on the sample conditions and its origin (application). The cleaning frequency can be significantly minimized by using an air-blast auto-cleaning system managed from the same controller (for submersible installations) or the flow-through cell. In the latter case, the maintenance may be nothing other than occasional cleaning of the strainer installed in the sample line feeding the flow cell.

In fact, plant staff had not conducted any maintenance on the sensor during the 45-day test period. The flow cell and sensor window fouling found during the posttest inspection, shown in Fig. 19, did not prevent the probe from producing correct readings or affect sensor performance in any other ways throughout the duration of the test.

Steam Condensate

A metal work plant uses steam in a closed-loop system for process heating purposes. The steam heats up a mineral oil-based heat transfer fluid. Any leakage of the steam/heat transfer fluid at heat exchanger surfaces causes contamination of the

Sample Inlet   Sample Outlet   Sensor Window

Fig. 19 After-test inspection of the flow cell and sensor window (photo credit: Hach Company)

steam condensate with mineral oil. Problems occur at low ppm oil levels in the condensate water when it is used for a new steam generation cycle.

In the past, a turbidimeter utilizing the light scattering technology with a sample cooler was used to monitor the condensate for oil contamination; however, a turbidimeter can only detect oil droplets, not dissolved mineral oils, which do show some solubility in water at low ppm range, especially at higher temperatures. Consequently, using a turbidimeter as an early warning for OIW is impossible by definition, because turbidimetry cannot detect dissolved matter. In addition, any other particles in the condensate will increase readings and can cause false alarms.

A direct side-by-side comparison of the turbidimeter and a UV-fluorescence sensor was made possible by installing the probe in flow cell at the outlet of the turbidimeter. For performance checks, an oil/water mixture was pumped from a separate tank through the installed instruments.

The UV-fluorescence sensor reacted quickly to the test sample, which was comprised of water with dissolved oil, while the turbidimeter responded with a significant delay. Only after some mixing time in the tank did oil droplets form in the sample to the point where the turbidimeter could detect contamination. After the test, the turbidimeter was removed, and the tested UV-fluorescence probe was purchased and installed at this location.

Over a 2-year testing period, significant advantages were shown in measuring OIW in industrial water and wastewater using an online sensor in comparison to the weekly grab sample analysis method. Maintenance was minimized by using the automatic air-blast cleaning system in submersible installations. Continuous readings and correlation to lab results were achieved, allowing for consistent real-time monitoring and event detection.

Although the end users were primarily concerned with raw data indicating a leakage event, it would have been possible to calibrate the OIW readings to real oil concentration since only one type of oil was present and constant measuring conditions were provided. This would be accomplished by taking grab samples for lab testing and then calculating calibration factors from the probe readings and lab results. Such approach could provide reliable trending between lab and process results and serve as a step toward quantification of the sensor readings.

3.2 Use Cases of UV-Fluorescence for Trending

3.2.1 Trending at Crude Oil Refineries

Conversations with petrochemical industry representatives revealed three potential applications for UV-fluorescence testing at refineries: wastewater after American Petroleum Institute (API) separators (prior to bioreactor), in desalters and in cokers. The desalter and coker applications normally involve high sample temperatures and high concentrations of chloride (desalter); therefore, they require special body materials (titanium), at a minimum.

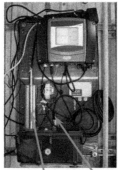 

UV-sensor   VIS-sensor   UV-sensor measuring window

Fig. 20 Refinery installation—wastewater stream after IAF and before the bioreactor (photo credit: Hach Company)

Because the wastewater application doesn't require special precautions, a test was devised for a refinery in Wyoming, USA, where the plant had implemented induced air flotation (IAF) to minimize OIW content and tried several approaches to monitor oil concentration in wastewater discharge prior to bio-treatment, as excess oil can wear out bacteria, which are costly to replace.

This refinery tried a full-scale online OIW analyzer based on chemical method; however, they abandoned the instrument due to its high maintenance requirements and unreliable results. Personnel conducted several different lab analyses of samples, along with online turbidity monitoring (surface scattering) to evaluate the OIW content and keep it under control.

Both a UV- (low range [LR], 0–500 ppb) and VIS-fluorescence (attuned to crude oil) sensors were installed on the same sample line with the surface scatter after the IAF device and the data were collected with the single multiparameter controller (Fig. 20, left). The sensors and controller were mounted on a panel along with a flow-through type chamber made from available materials.

Several challenges were encountered in this application. First, sensors experienced premature fouling (Fig. 20, right) due to the slow sample flow through the chamber. Unfortunately, the flow could not be increased without interfering with the surface scatter sharing the sample and hence the results provided by that instrument. Also, the automated air-blast cleaning system was not available at the time of testing.

Another challenge was with the nature of the oils in the sample. Per plant management, there was never only a single type of crude oil being treated at the refinery at a time. Therefore, the sample content was never consistent in terms of the oil type and consequently the PAH content in the sample. This provided the most challenging conditions and resulted in some inconsistency between the readings received from both sensors and the grab sample analyses (Fig. 21). All attempts to calibrate the sensors using grab sample data failed due to the aforementioned limitations.

On a positive side, it should be noted that a similar trend was found between the results collected from both sensors. However, the VIS-fluorescence sensor provided a significantly narrower range of response (Fig. 21) and therefore was excluded from all further experiments.

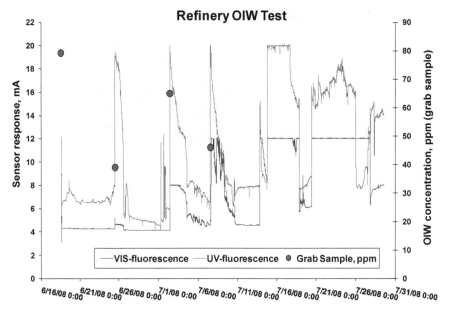

Fig. 21 Refinery wastewater application test results [24]

Fig. 22 UV-fluorescence sensor equipped with HOAB installed in stream of industrial wastewater effluent after API and DAF prior to filtration (photo credit: Hach Company)

A subsequent test was conducted at a different refinery with lower to no variation in oil types in the wastewater discharge. A UV-fluorescence sensor was installed in submersible mode using a chain mounting kit at the facility's wastewater treatment system including API separator, dissolved air flotation (DAF) system, and then conventional sand-gravel filtration of the discharge (Fig. 22). The purpose of the test was to monitor relatively clean water after the separation and DAF cleaning and immediately before the filters, with allowed OIW content of up to 10 ppm. The threshold was determined by plant staff based on their treatment realities and filter capabilities.

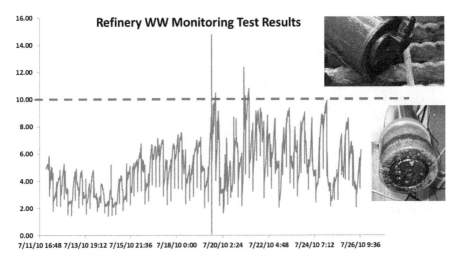

Fig. 23 Results of the effluent monitoring prior to filtration [24] (photo credit: Hach Company)

The probe was installed right after the DAF unit in a trough collecting the sample right before it was pumped into the filters. Plant staff was concerned with potentially too high concentration of oil in water at this location, because it could cause excessive fouling of the filters, and most importantly, it could jeopardize the discharge limit of oil into the city sewer system creating a permit violation.

The sample temperature at installation was ~25 °C. The PTFE 1/4″ tubing was connected to the probe and a high output air-blast system (HOAB), which was installed inside the control room (~5–7 m from the sampling point). The probe was connected to a Hach SC1000 controller also located in the control room, which was not classified as a hazardous location. The entire installation did not require Class 1 Division 2 certification.

Plant staff conducted grab sample analysis every 4 h and was very concerned with possible OIW fluctuations within this time frame. The operators were trained on the probe calibration by slope correction using a factor calculated based on the grab sample analysis results, and verified the sensor readings against lab analyses after the initial calibration. An alarm was set at 10 ppm OIW to be able to divert the sample for additional treatment once the threshold is exceeded (Fig. 23).

After a successful 2-week test period, results demonstrated that the sensor with air-blast auto-cleaning worked longer (several weeks with HOAB versus a couple days without) before manual cleaning was required. The probe displayed expected performance during the trial and delivered continuous results without additional interaction.

3.2.2   Trending in Natural Seawater

Several hydrocarbon sensors, including the UV-fluorescence probes used in fresh water and wastewater testing, have been thoroughly tested in laboratory and field settings to verify performance in in-situ monitoring of OIW in coastal moored and profiled deployments [25].

During these tests, OIW sensors (PAH, crude oil, and refined fuels) were evaluated for response time, linearity, accuracy/precision, and reliability. The testing, at all test sites and modes, was conducted on regular production units that allowed functions such as calibration, auto-ranging, data logging, and reporting of data as PAH or oil concentration.

Of all sensors, the UV-fluorescence technology most consistently performed to specifications, displaying remarkable specificity to the petroleum products without reacting to interfering factors such as turbidity, CDOM, chlorophyll, etc.

Accuracy of this type of sensors employing indirect measurement usually depends on the accuracy of the reference method used for calibration, as well as stability of the measured sample from the standpoint of the analyte nature and content. However, the precision expressed as standard deviation of the series of measurements describes both stability of the matrix and the instrument's optical and electrical systems. The UV-fluorescence sensor demonstrated very good precision in all measurement conditions and sample matrices.

3.3 Use Cases of UV-Fluorescence for Quantification of OIW

3.3.1 Quantification at a Metal Works Plant

A Chrysler auto parts plant in Michigan collects wastewater from all plant operations—containing a variety of lubricants from the metal works—in a holding tank and chemically treats it prior to discharging it into a municipal collector. Given the consistency of the plant operations and the used lubricant, the OIW content was considered to be stable that allowed for an attempt to calibrate the sensor and quantify readings during the test.

A UV-fluorescence probe (LR, 0–500 ppb) was installed in an open well, chain-mounted at the inlet to the wastewater treatment facility (Fig. 24).

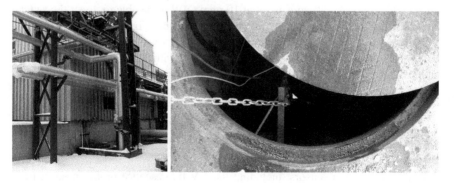

Fig. 24 Installation of the submersible probe (photo credit: Hach Company)

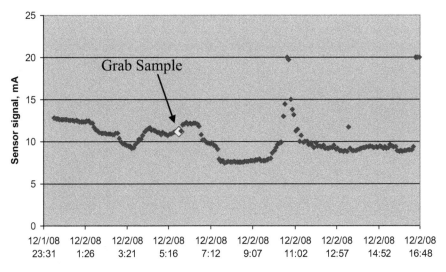

Fig. 25 Process calibration procedure [23]

Over the course of a 2-month test, data were logged by a controller equipped with remote wireless access: Process calibration was performed at predetermined intervals using the process grab sample technique shown in Figs. 25 and 26.

Following laboratory analysis, the obtained OIW concentration was entered in a simple linear relationship, which yielded the calibration equation shown in Fig. 26. The calibration coefficients (slope and offset) generated by the equation were used to calculate the concentration range in units of oil, and then the probe was calibrated to display the OIW concentration by introducing a calculated factor (calibration coefficient) via the controller.

After the calibration, facility personnel were able to adjust their process to optimize the chemical treatment (Fig. 27). Positive spikes were discussed with facility management, to ascertain whether they were attributable to a process change or an instrumentation error; in all cases there was an explanation providing legitimacy to the probe's response.

As seen from Fig. 27 and according to the facility personnel, the process optimization resulted in smoother wastewater treatment operation.

3.3.2 Quantification at a Glass Production Plant

A bottle production plant blends wastewater from different sources within the plant—including glass recycling—into a single stream before sending it out to a municipal wastewater treatment plant. More than ten different types of oil are used during the production process (e.g., lubricants for pumps) and can easily come into contact with process and wash waters. In addition, rainwater that contacts truck loading docks and fuel oil tank areas adds to the plant's normal wastewater stream.

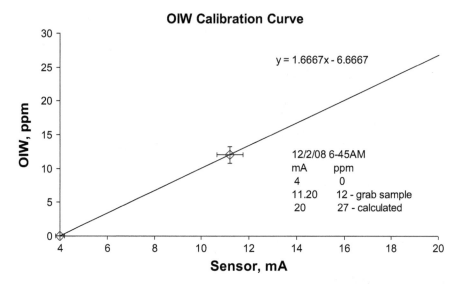

Fig. 26 Industrial wastewater test—calibration [23]

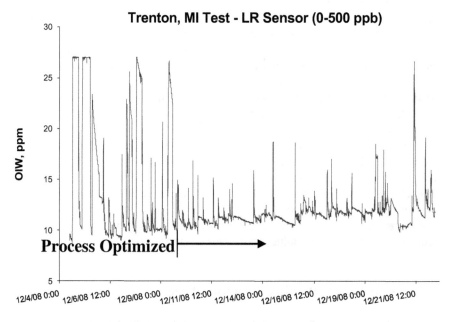

Fig. 27 Auto parts plant in Michigan (industrial wastewater)—test results [23]

High oil concentrations in the wastewater must be avoided in order to protect the sewer system and the biological stages of the municipal wastewater treatment plant. Therefore, the final wastewater stream is passed through an oil/water separator.

Water authorities set a threshold value of 20 ppm (mg/L) total oil in water after the separator, verified by a weekly grab sample analysis. Typical lab results found oil in the low ppm range (<5 ppm).

In order to reduce the number of lab analyses and to allow for continuous monitoring of excessive mineral oil contamination of the wastewater stream, a UV-fluorescence probe was installed in an inspection chamber close to the final wastewater discharge outlet. The system was expected to provide an early warning with low interference from particles and/or other components in the wastewater. The probe was installed by using a stainless steel chain mounting set (Fig. 28).

During the first 4 weeks, multiple grab samples were taken and compared to the sensor readings. A correlation factor was derived from the corresponding sensor readings and the lab results, allowing sensor calibration and the readings to display "ppm OIL" directly (e.g., with a factor of 34, a reading of 100 ppb PAH is displayed as 3.4 ppm OIL). Because the compositions of the oils coming from different sources in the plant are not predictable, and variations of the factor may be significant, a warning level was set to 15 ppm OIL. Typical readings were below 4 ppm OIL as shown in the diagram (Fig. 29).

Fig. 28 Oil-in-water probe installation in inspection chamber after oil/water separator (photo credit: Hach Company)

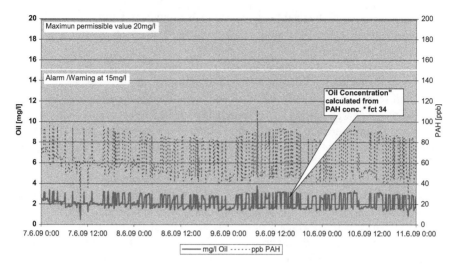

Fig. 29 Sensor's readings displayed as ppb of PAH and mg/L of oil. Reprinted with permission of Hach Company

Fig. 30 Head of the OIW probe with air-blast cleaning system—area of the measuring window is cleaned automatically (photo credit: Hach Company)

A HOAB unit was soon added to the unit, as shown in Fig. 30, to eliminate the need to manually clean the measuring window every 3–5 days. Automatic cleaning was set to perform 1 min per hour, and no additional manual cleaning was necessary during the subsequent 10-week test period.

The sensor readings were frequently verified with lab analyses, and the quantification was considered to be successful, especially after implementation of the automatic cleaning system.

4 Conclusions

Analysis of the literature along with laboratory and field testing demonstrated that the best results can be expected when UV-fluorescence sensors are used for qualitative continuous monitoring of water samples, providing an excellent early warning device that instantaneously reacts to contamination of water by hydrocarbons. Important observations include:

- The tested UV-fluorescence sensor adequately and specifically responded to petrochemical products containing PAH, including crude oil.
- UV-fluorescence technology provides better sensitivity and selectivity in OIW analysis than UV-absorbance and/or VIS-fluorescence.
- UV-fluorescence technology demonstrated higher sensitivity to dissolved and dispersed mineral oil versus a turbidimeter, as it is more specific to oils and not affected by other components in the sample.
- Sensor performance can be improved by implementing a specially designed flow cell, which provides faster flow and/or automated cleaning system (air-blast) for open channel mounting.
- Oil spill detection in the low ppm range is the most common use case for the sensor. It was greatly simulated by the wave tank seawater test where the probe showed excellent performance for all types of oil used.
- The signal produced by the UV-fluorescence sensor could be accurately correlated to OIW concentration. Sensors can be calibrated by using either purchased standards or based on the process grab sample analysis; however, inconsistent oil content may prohibit quantification of the results. A good amount of research should be done when selecting a calibration standard or method to obtain quantifiable results.
- Testing demonstrated linear response to concentrations of various types of petrochemical products when there was no variation in the product nature. The linearity is supported by a very low sensitivity of the probe to interfering compounds found in natural or treated waters.

References

1. Blais F (2004) Review of 20 years of range sensor development. J Electron Imaging 13(1):23–235
2. Macleod CJA (2010) What can we learn from systems based approaches: from systems biology to earth systems science? In: 5th international congress on environmental modelling and software, 162. Brigham Young University BYU ScholarsArchive

3. Chapman DV, World Health Organization (1996) Water quality assessments: a guide to the use of biota, sediments and water in environmental monitoring, 2nd edn. University Press, Cambridge
4. Lu R, Mizaikoff B, Li WW, Qian C, Katzir A, Raichlin Y, Sheng GP, Yu HQ (2013) Determination of chlorinated hydrocarbons in water using highly sensitive mid-infrared sensor technology. Sci Rep 3:2525
5. Bürck J, Mensch M, Krämer K (1998) Field experiments with a portable fiber-optic sensor system for monitoring hydrocarbons in water. Field Anal Chem Technol 2(4):205–219
6. Rosenberg E, Krska R, Kellner R (1994) Theoretical and practical response evaluation of a fibre optic sensor for chlorinated hydrocarbons in water. Fresenius J Anal Chem 348(8-9):560–562
7. Krska R, Taga K, Kellner R (1993) New IR fibre-optic chemical sensor for in-situ measurements of chlorinated hydrocarbons in water. Appl Spectrosc 47(9):1484–1487
8. Jakusch M, Mizaikoff B, Kellner R, Katzir A (1997) Towards a remote IR fibre-optic sensor system for the determination of chlorinated hydrocarbons in water. Sensors Actuators B Chem 38(1-3):83–87
9. Ho CK, Robinson A, Miller DR, Davis MJ (2005) Overview of sensors and needs for environmental monitoring. Sensors 5(1):4–37
10. Storey MV, van der Gaag B, Burns BP (2011) Advances in on-line drinking water quality monitoring and early warning systems. Water Res 45(2):741–747
11. Forzani ES, Zhang H, Chen W, Tao N (2005) Detection of heavy metal ions in drinking water using a high-resolution differential surface plasmon resonance sensor. Environ Sci Technol 39(5):1257–1262
12. Pashalidis I, Tsertos H (2004) Radiometric determination of uranium in natural waters after enrichment and separation by cation-exchange and liquid-liquid extraction. J Radioanal Nucl Chem 260(3):439–442
13. Leonard P, Hearty S, Brennan J, Dunne L, Quinn J, Chakraborty T, O'Kennedy R (2003) Advances in biosensors for detection of pathogens in food and water. Enzym Microb Technol 32(1):3–13
14. McCue RP, Walsh JE, Walsh F, Regan F (2006) Modular fibre optic sensor for the detection of hydrocarbons in water. Sensors Actuators B Chem 114(1):438–444
15. Harrick NJ, Beckmann KH (1974) Internal reflection spectroscopy. Characterization of solid surfaces. Springer, Boston, pp 215–245
16. Sothivelr K, Bender F, Josse F, Ricco AJ, Yaz EE, Mohler RE, Kolhatkar R (2015) Detection and quantification of aromatic hydrocarbon compounds in water using SH-SAW sensors and estimation-theory-based signal processing. ACS Sens 1(1):63–72
17. Cooper JS, Kiiveri H, Chow E, Hubble LJ, Webster MS, Müller KH, Raguse B, Wieczorek L (2014) Quantifying mixtures of hydrocarbons dissolved in water with a partially selective sensor array using random forests analysis. Sensors Actuators B Chem 202:279–285
18. Dickert FL, Achatz P, Halikias K (2001) Double molecular imprinting–a new sensor concept for improving selectivity in the detection of polycyclic aromatic hydrocarbons (PAHs) in water. Fresenius J Anal Chem 371(1):11–15
19. Hamacher C, Brito APX, Brüning IM, Wagener A, Moreira I (2000) The determination of PAH by UV-fluorescence spectroscopy in water of Guanabara Bay, Rio de Janeiro, Brazil. Rev Bras Oceanogr 48(2):167–170
20. Martin F, Otto M (1995) Multicomponent analysis of phenols in waste waters of the coal conversion industry by means of UV-spectrometry. Fresenius J Anal Chem 352(5):451–455
21. Tedetti M, Joffre P, Goutx M (2013) Development of a field-portable fluorometer based on deep ultraviolet LEDs for the detection of phenanthrene-and tryptophan-like compounds in natural waters. Sensors Actuators B Chem 182:416–423
22. Vadim Malkov, Dietmar Sievert (2009) Oil-in-water fluorescence sensor in wastewater applications. In: Paper presented at the international water conference 70th annual meeting, Orlando, FL (USA), 4–8 October 2009

23. Malkov V, Sievert D (2010) Oil-in-water fluorescence sensor in wastewater and other industrial applications. Power Plant Chem 12(3):144–154
24. Vadim B. Malkov (2010) Applications of an oil-in-water probe built upon UV-fluorescence technology. In: Paper presented at ISA automation week 2010: technology and solutions event proceedings of international society of automation conference, Houston, TX (USA), 4–7 October 2010
25. Mario Tamburri et al. (2012) Performance verification of the hach FP360sc UV fluorometer. In: Alliance for coastal technologies ACT VS12-04 UMCES/CBL 2013-018 (www.act-us.info/evaluations)

Luminescence in Photovoltaics

José Almeida Silva, João Manuel Serra, António Manuel Vallêra,
and Killian Lobato

Contents

Abstract This chapter reviews the applications of luminescence-based techniques in the photovoltaic industry, with special focus on crystalline silicon-based devices – the dominant technology in the market.

Section 1 introduces the principles of the photovoltaic effect and describes the light capture and conversion in the device. A brief description of the state-of-the-art device manufacture is then given along with a description of how power conversion efficiency of photovoltaic devices is determined.

Section 2 describes the origin of luminescence in photovoltaic devices and also describes the luminescence-based characterization of photovoltaic cells and modules.

Section 3 describes in detail how luminescence (photo- and electroluminescence) measurements are applied in the complete value chain of the PV industry, from ingot, to wafer, to device, to module, to complete infield systems.

J. A. Silva (✉), J. M. Serra, A. M. Vallêra, and K. Lobato
Instituto Dom Luiz – Faculdade de Ciências da Universidade de Lisboa, Lisbon, Portugal
e-mail: jose.silva@fc.ul.pt

© Springer Nature Switzerland AG 2019
B. Pedras (ed.), *Fluorescence in Industry*, Springer Ser Fluoresc (2019) 18: 173–212,
https://doi.org/10.1007/4243_2018_7, Published online: 8 February 2019

Section 4 briefly describes how luminescence is also relevant for emerging thin-film photovoltaic technologies.

Section 5 describes a recently developed technique, reverse bias electroluminescence, where the photovoltaic devices are inversely polarized. The emitted photons here are a result of charge carrier acceleration and consequent scattering and/or recombination in a high electric field.

Section 6 concludes this chapter with an outlook on how luminescence imaging is expected to develop in the near future, namely, how currently under development lab techniques will likely be transferred to the industrial environment.

Keywords Electroluminescence (EL) · Manufacture · Modules · Operation and maintenance (O&M) · Photoluminescence (PL) · Photovoltaic (PV) · Reliability · Silicon (Si) · Solar cells · Systems

1   Introduction

The installed capacity of photovoltaic (PV) power has been increasing exponentially in the last decades, reaching 99 GW of new installations in 2017 [1]. This growth rate is expected to continue for the next few years making the photovoltaic industry one of the fastest growing in the world. Two major issues the industry is currently tackling are cost reduction and product reliability. To sustain a steady reduction of the production costs, PV manufacturers must maintain very high production rates, typically several thousand of solar cells per hour. On the other hand, the reliability requirement demands a permanent supervision of the quality of the materials, solar cell devices, and solar modules during production and surveillance of modules and systems during their long lifespan (ca. 30 years) [2].

The standard solar cell and module characterization technique is to obtain a one-sun current vs voltage curve (IV curve), which describes the response under standard one-sun illumination (i.e., using a light source that simulates the solar spectrum). The dark IV curve which characterizes the response in the dark is also a standard characterization technique. These determine fundamental parameters, but only provide information about the average performance of the solar cell or solar module, having no spatial resolution. One of the major benefits of luminescence-based characterization techniques is that they produce spatially resolved physical descriptions of photovoltaic materials and devices allied to significantly shorter acquisition times when compared to other standard techniques. LBIC (laser beam-induced current) or photoconductance decay-based carrier lifetime measurements are standard techniques that also result in spatially resolved data; however, acquisition times are substantially longer and as such incompatible with inline process monitoring. Because of the low acquisition times (~seconds) of luminescence-based techniques, these have become standard in the PV industry.

This chapter aims to give an overview of luminescence-based techniques used by the PV community. Due to its substantial relevance in the PV market, this review is focused on crystalline silicon technologies.

The chapter is organized as follows: first a brief presentation of photovoltaic principles is made, followed by a description of the crystalline silicon (c-Si) photovoltaic technologies and specifically the aluminum-back surface field (Al-BSF) solar cell; the principles of luminescence are then presented, followed by a description of the photoluminescence (PL) and electroluminescence (EL) techniques; a few examples of the application of EL and PL to PV materials, solar cell devices, solar modules, and PV systems are discussed; the application of luminescence techniques to other PV technologies, both alternative c-Si solar cell concepts and different material devices, is touched upon; the reverse bias electroluminescence (ReBEL) technique is also described; and the chapter ends with conclusions and future prospects for the luminescence-based characterization techniques.

1.1   Principles of Photovoltaics

Photovoltaic conversion is the process by which the energy from photons generates an electrical current. The device in which this process occurs is the solar cell. A solar cell has three main functioning mechanisms: (1) electron-hole pair generation via photon absorption, (2) electron-hole pair separation at the pn junction, and the collection of the charge carriers at the respective contacts.

1.1.1   Electron-Hole Pair Creation

The basic phenomenon of the photovoltaic energy is the absorption of photons by a semiconductor that will promote the passage of one electron from the valence band (VB) to the conduction band (CB). Only photons with energies, $h\nu$, equal or greater than the gap energy (E_{gap}) between valence and the conduction band are absorbed by exciting an electron from valence band to the conduction band and generating an electron-hole pair. This phenomenon is called the photovoltaic effect (Fig. 1).

1.1.2   Charge Separation and Collection

The active part of a conventional silicon solar cell, where radiation is absorbed, is formed by a low-doped region called base and a heavily doped region called emitter. These two regions have opposite dopant charges; if the base is p-doped, the emitter is n-doped or vice versa.

p-doped or p-type silicon is a material with a deficiency of electrons also said to have an excess of holes; conversely n-doped or n-type silicon is a material with excess of electrons. When a p-type material is brought together with an n-type

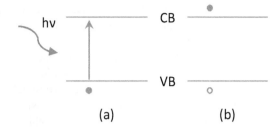

Fig. 1 Photovoltaic effect: (**a**) absorption of photon by a valence band (VB) electron; (**b**) creation of an electron-hole pair due to the promotion of the electron from to the conduction band (CB). Blue full circle denotes an electron; blue hollow circle denotes a hole

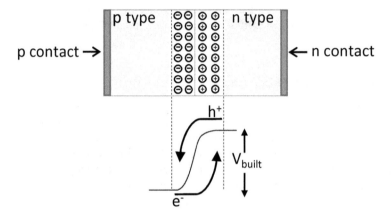

Fig. 2 The solar cell pn junction and respective built-in potential – (V_{built})

material, the excess holes in the p-type region will flow to the interface between the regions where they will recombine with the excess electrons that flow from the n-type region. The movement of the free charges (electrons and holes) leaves behind ionized atoms that form regions of fixed charges near the p-n interface, that are negative on the p-side and positive on the n-side (see Fig. 2). The creation of this region of charge inversion, named space-charge region, which is a distinct feature of the pn junction, creates a built-in potential (V_{built}) that conducts the holes to the p-side and the electrons to the n-side allowing their collection at the respective contacts [3]. Once recovered at the solar cell contacts, the free charges can be used on an external circuit.

In an ideal solar cell, all the electron-hole pairs generated due to light absorption would be recovered at the solar cell contacts, but in fact only a fraction is recovered, the remaining being lost by recombination in the semiconductor. The charge carriers (electrons and holes) recovered at the contacts have an electrical potential which is a fraction of the semiconductor gap energy and constitutes the solar cell voltage. The solar current is determined by the number of electrons and holes that are recovered in the contacts.

Fig. 3 Schematic representation of the Czochralski process to produce monocrystalline silicon ingots

Electron-hole pair recombination before reaching the solar cell contacts occurs when the electron descends from the conduction band to the valence band and, depending on the type of recombination mechanism present, one or more photons can be emitted.

1.2 Aluminum-Back Surface Field Crystalline Silicon Solar Cells and Modules

1.2.1 From Silica to the Wafer

Silicon occurs in nature mostly in the form of silica (SiO_2). To separate silicon from oxygen, SiO_2 is heated to ca. 1800°C in the presence of carbon nitride [4]. The resultant metallurgical grade silicon is further purified using the Siemens or the fluidized bed reactor processes that reduce the impurity level to a level necessary for solar photovoltaic devices[1] [4].

Following purification a crystallization step is performed to improve the crystalline quality [5]. From the perspective of the PV industry, there are two main routes for the crystallization step, the Czochralski process (see Fig. 3) and the directional solidification technique (Fig. 4) which produce respectively monocrystalline and multicrystalline silicon ingots.

In the Czochralski technique, a solid seed, made of a single crystalline silicon, is dipped into a silicon melt and is partially melted. When the seed is pulled from the melt, crystallization takes place, and the pulled crystal has the orientation of the seed crystal. By careful control, crystals with diameters up to 30 cm in diameter and 2 m in length can be obtained. The crystal quality can be very high since the crystal is not in contact with the crucible, and therefore no impurities or defects are introduced

---

[1]The quality requirements for the microelectronics industry are far more restrictive than the PV industry. In the past when the PV industry was of a far small scale, it survived on discarded microelectronic grade silicon.

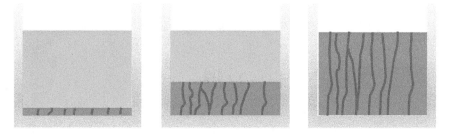

Fig. 4 Schematic representation of the directional solidification process to produce multicrystalline silicon ingots

by thermal stress during the cooling phase. However, because the quartz crucible is in direct contact with liquid silicon, it suffers partial dissolution, introducing oxygen into the melt, resulting in relatively high concentrations of oxygen in the ingot.

In multicrystalline ingot crystallization, the solidification starts at the bottom of the silicon melt ($1,410°C$). Several crystals will start growing at the bottom, resulting in a multicrystalline ingot with a columnar structure. The final part of the ingot will have lower quality because impurities coming from the crucible will diffuse into the silicon melt, and the cooling process will also introduce crystal defects such as dislocations generated by thermal stress. Ingots with dimensions of 80×80 cm^2 and 26 cm height are routinely obtained.

The electronic quality of multicrystalline materials is lower than in single crystal materials since in the former the grain boundaries (seen in Fig. 4) are in fact crystal defects that reduce the charge carrier lifetime and hence the conversion efficiency of the solar cells. Nevertheless, the ingot crystallization by the directional solidification is faster, thus having lower costs which is an advantage from the overall process cost. More details about different crystallization techniques can be found in Nakajima and Usami [5] and Luque and Hegedus [6].

After crystallization the obtained ingots are cut in bricks and then in wafers with thicknesses of ca. 200 μm. The wafering process is done by sawing the bricks with a wire. In a multiwire sawing machine, the wire passes several times through the material to increase the process throughput as shown in Fig. 5.

Before being used in the solar cell production, the wafers' surfaces have to be polished in order to remove the damaged regions introduced during the wafering step.

1.2.2 Solar Cells Production: Aluminum-Back Surface Field Solar Cell

The Al-BSF solar cell is the dominant technology in the photovoltaic industry. This technology does not retain the absolute efficiency records when one considers all types of PV technologies or only crystalline silicon technologies. Nonetheless, these have recently achieved record conversion efficiencies of 18.8% for multicrystalline [7] and 20.29% [8] monocrystalline solar cells. Furthermore, although new solar cell

Fig. 5 Multiwire sawing
machine

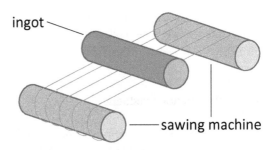

Fig. 6 Structure of an
Al-BSF silicon solar cell

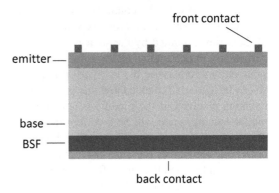

concepts (some of these new cell designs will be discussed later in this chapter) are becoming increasingly important in the market, it is expected that Al-BSF solar cells will continue to dominate the PV market in the next years [9].

Figure 6 shows the structure of the Al-BSF solar cell. As well as the base, emitter, and the front and back contacts, this cell has an aluminum-back surface field (Al-BSF) that covers all its rear part. The BSF is a region that is highly p-doped that, similar to the pn junction, creates a potential barrier that deters the flow of the minority carriers (electrons) to the rear contact, thus reducing the carrier recombination in the solar cell rear [10].

Usually the pn junction depth is between 0.2 and 0.3 μm, while the whole device thickness is around 180 μm, thus radiation absorption takes place mainly in base of the cell. In addition to the already mentioned layers, a very thin layer (~75 nm) of hydrogenated silicon nitride (SiN_x:H) is grown on top of the solar cell emitter. The presence of the SiN_x:H layer reduces carrier recombination both on the surface and bulk of the sample [11]. It also serves as an antireflective coating, thus increasing photon harvesting and hence improving the cell's photocurrent.

As previously mentioned, the radiation absorption creates electron-hole pairs that are separated in the pn junction and subsequently collected at the respective contacts. To maximize carrier collection, recombination must be suppressed as much as possible. This can be achieved firstly by ensuring good electronic quality of the substrate and secondly by using surface passivation strategies like the BSF or the growth of a dielectric layer such as SiN$_x$:H. The reader can find much more detailed explanations of the solar cell device physics in several well-established textbooks [3, 12].

1.2.3 PV Module Production

Typically, c-Si solar cells deliver a voltage in the range 0.5–0.6 V which is too low for most practical applications. To provide a higher voltage and power, a PV module is usually constructed by connecting several solar cells in series. The voltage of the resulting PV module is therefore the sum of all connected cells' voltages. The current of the series is limited by the solar cell with the lowest current. This fact makes it particularly important that all the solar cells in the module are identical and have the same current; otherwise the low-performing cells will become current sinks. Also, since the current produced by the solar cells in each moment depends on the radiation received, if the module is anisotropically irradiated, for instance, due to partial shading, there will be a current mismatch between the different solar cells. The less irradiated cells will produce lower currents and can become inversely polarized, dissipating the current coming from the neighboring cells and creating hotspots on the module. Such effect, if persistent, can create irreversible damage on the device and degrade PV module energy generation [13]. To avoid this, PV modules are equipped with bypass diodes connected in parallel with series of solar cells usually called strings. When one of the strings is producing less current, the bypass diode conducts the current produced into the remaining strings via an external circuit. This isolates the lower performing solar cells and thus avoids damaging the module [6]. Figure 7 depicts the structure of a typical PV module composed of three strings each connected to a bypass diode.

Due to their very low thicknesses, solar cells are particularly vulnerable to cracking when mechanically stressed. Also, when exposed to outdoor conditions, the solar cells are prone to corrosion. To preserve their performance during the

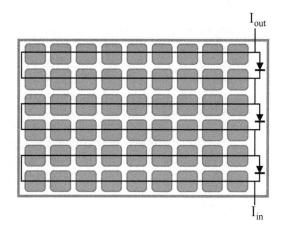

Fig. 7 PV module solar cell structure and bypass diodes. The current direction is also represented

Fig. 8 Typical layer
structure of a silicon PV
module

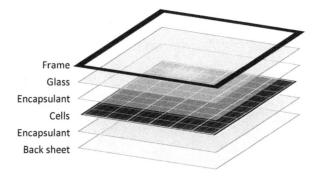

Frame
Glass
Encapsulant
Cells
Encapsulant
Back sheet

Fig. 9 One-sun and dark IV
curves

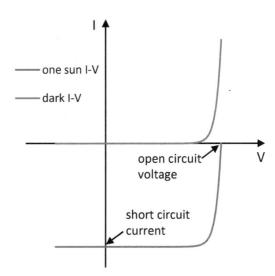

module's lifespan, solar cells are assembled in a structure that provides suitable
protection from both mechanical maltreatment and environmental variables such as
rain and moisture. Figure 8 outlines the layers that make a typical PV module. One of
the most important steps in module manufacturing is the lamination process that
encapsulates the cells, isolating them from the outside environment.

1.3   Solar Cell and Module Characterization

As previously mentioned, the most common technique to characterize PV devices is
the measurement of IV curves both with under one-sun illumination and in the dark.
In Fig. 9 the two IV curves are represented. The dark IV curve is simply the response
curve of a diode and represents the normal solar cell dark current. The one-sun IV
curve is the superposition of the dark IV curve with the current generated by light.

Fig. 10 Forward and
reverse bias current flow in a
p-type silicon solar cell

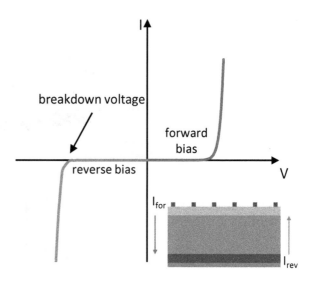

Since in a p-type solar cell the light-generated current flows in the negative direction, the one-sun IV curve is shifted to the fourth quadrant.

A solar cell or module is said to be polarized with a positive bias (or forward bias) if it is submitted to a voltage that induces a current in the same direction as the photogenerated current, and it is said to be inversely polarized (or reverse biased) if the current flows in the opposite direction. In Fig. 10 the direction followed by the electrical current in a p-type solar cell under forward and reverse bias is represented. If a PV device is functioning correctly, the voltage required to conduct current under reverse bias is high. The minimum voltage required to reach this stage is called breakdown voltage.

2   Principles of Luminescence

The recombination of electron-hole pairs in a solar cell can occur by three different mechanisms: radiative recombination, recombination through defects usually called Shockley-Read-Hall (SRH) recombination, and Auger recombination [14].

Radiative (or band-to-band) recombination is the reciprocal effect of the electron-pair generation, as such the annihilation of the electron-hole results in the emission of a photon with an energy equal to E_{gap}.

SRH recombination is a mechanism that is associated with material defects such as impurities, dislocations, grain boundaries, or surface defects. The presence of these in the material creates defect states within the band gap (i.e., between VB and CB). The defect states serve as trap states for electrons or holes, thus favoring the recombination of the charge pairs. The SRH recombination involves the emission of two or more photons with an energy that is lower than E_{gap}.

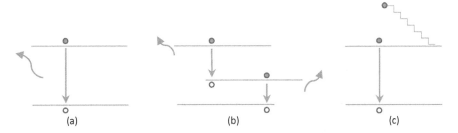

Fig. 11 Recombination mechanisms: (**a**) Radiative, (**b**) Shockley-Read-Hall, (**c**) Auger

Auger recombination is a mechanism that is significant at high levels of electron-hole pair concentration. In Auger recombination, an electron-hole pair recombines, but differently from the radiative recombination. Here the resulting energy is transmitted to a third carrier (electron or hole) that later loses its excess energy by thermalization. A representation of the three recombination mechanisms is shown in Fig. 11.

The photons emitted during electron-hole recombination correspond to the luminescence of the solar cells. The intensity of the luminescence signal depends mainly on the density of charge carriers injected, while the emitted photon energy depends on the type of recombination mechanism.

Furthermore, for semiconductors that have direct band gaps, such as gallium arsenide (GaAs), it is very likely that carrier recombination occurs via the radiative mechanism. Radiative recombination is much less likely in indirect gap semiconductors; thus, for the low-medium injection levels when Auger recombination is unimportant, recombination occurs preferentially through the SRH mechanism [3].

Using a light source and directly injecting a current are two different ways to inject charge carriers and thus stimulate luminescence in solar cells. These two different ways to induce luminescence correspond to the two techniques commonly used to characterize the quality of photovoltaic materials: photoluminescence and electroluminescence.

Photoluminescence (PL) characterization is based on the use of a light source, usually a laser, to create electron-hole pairs in the material. The photons emitted during charge carrier recombination correspond to the PL signal. One of the major advantages of PL characterization is that it is a contactless technique and as such is applicable on non-contacted materials (i.e., metallization has not been performed). Thus PL can be applied at all the stages of the photovoltaic production, from the wafer to module level.

In electroluminescence (EL) characterization, the luminescence of the material is induced by an electrical current. EL characterization requires the existence of contacts; hence, it can only be applied to finished devices (i.e., solar cells and modules).

For several decades the usefulness of luminescence measurements for the characterization of semiconductor materials has been established, becoming a standard characterization tool for both direct and indirect gap semiconductors [15, 16]. Luminescence measurements have also become a normal tool for characterizing PV materials and devices [17, 18]. However, the innovation that made PL and EL techniques of choice to characterize photovoltaic materials was the possibility of obtaining almost instant complete area luminescence images of the photovoltaic material. Livescu et al. [19] used a photoluminescence imaging system to perform real-time mapping of several direct band gap semiconductor materials such as GaAs, InP, or InP:S. A few years later, Fuyuki et al. [20] proposed a technique to photograph the electroluminescent image of a silicon solar cell based on the use of a silicon CCD camera.

2.1 Generalized Planck Equation for Spontaneous Emission

The luminescence intensity corresponds to the rate of spontaneous emission r_{sp} in the semiconductor. The generalized Planck equation describes the rate of spontaneous emission $r_{sp}(cm^{-3}s^{-1})$ for a certain energy interval $\hbar\omega_1$, $\hbar\omega_2$ [21]:

$$r_{sp} = \int_{\hbar\omega_1}^{\hbar\omega_2} \alpha_{BB}c_\gamma D_\gamma \Omega \exp\left(-\frac{\hbar\omega}{kT}\right)\exp\left(-\frac{\Delta\varepsilon_F}{kT}\right)d(\hbar\omega) \qquad (1)$$

or

$$r_{sp} = \exp\left(-\frac{\Delta\varepsilon_F}{kT}\right)\int \alpha_{BB}c_\gamma D_\gamma \Omega \exp\left(-\frac{\hbar\omega}{kT}\right)d(\hbar\omega) \qquad (2)$$

where α_{BB} is the absorption coefficient for band-band transitions, c_γ is speed of light in the medium, D_γ is the photon density of states, Ω is the solid angle into which emission occurs, T is the sample temperature, k is Boltzmann's constant, and $\Delta\varepsilon_F$ is the separation of the quasi Fermi levels within the semiconductor.

It is important to highlight the dependence of the rate of spontaneous emission on the absorption coefficient α_{BB} which allows the use of luminescence imaging to characterize the material and device properties. The dependence on $\exp\left(-\frac{\Delta\varepsilon_F}{kT}\right)$ also has important implications on the implementation of the luminescent-based characterization techniques as will be later discussed. Based on the generalized Planck's law for luminescence emission, the absorptivity of silicon solar cells' band-to-band transitions can be obtained from their luminescence spectra both for non-textured and textured surfaces [18].

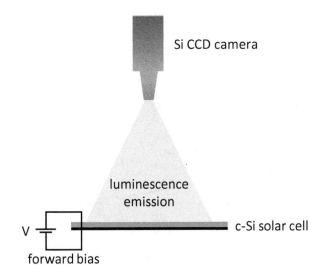

Fig. 12 Electroluminescence setup used by Fuyuki et al. [20]

2.2 Luminescence Mapping and the Use of Si CCD Camera for c-Si Solar Cells

The introduction of silicon charge-coupled device (CCD) cameras was a crucial step for the generalization of EL and PL PV characterization. The first demonstration of the capability of CCD cameras to perform luminescence mappings was reported by Fuyuki et al. [2]. Using an experimental setup like the one depicted in Fig. 12, Fuyuki et al. applied a forward bias current to a polycrystalline silicon solar cell and obtained its EL image using a cooled CCD camera.

A feature of the silicon CCD cameras is that the quantum efficiency of the Si detectors is naturally aligned with the crystalline silicon band-to-band emission peak (Fig. 13), making them particularly interesting for the characterization of crystalline silicon solar cells.

As can be observed from Fig. 13, the sensitivity of Si CCD camera and the Si emission spectrum intercept in one narrow region between 900 and 1,200 nm. Crystalline silicon materials do not emit radiation with a wavelength below that range, and Si CCD cameras do not detect photons above it. Therefore, it is easy to understand that while radiative recombination photons are detected by the Si CCD cameras, the lower energy (higher wavelength) photons due to the SRH recombination cannot be detected by the camera's sensor. Hence, the interpretation of luminescence mappings obtained with such cameras is quite straightforward.

Shown in Fig. 14 is an example of an EL image. The darker areas in the luminescence image correspond to the regions in the solar cell with a high defect density, such as grain boundaries or dislocations clusters. For such regions SRH

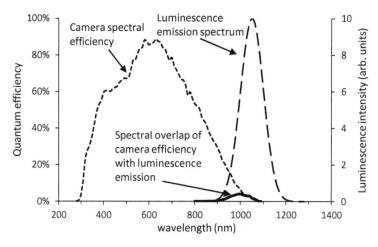

Fig. 13 Example Si CCD camera sensor quantum efficiency spectrum, silicon band-to-band emission spectrum, and camera-luminescence spectral overlap

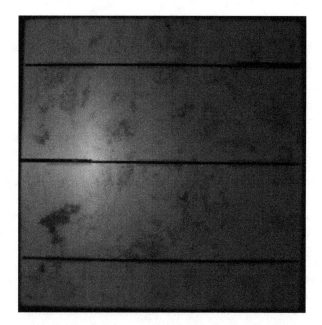

Fig. 14 Electroluminescence mapping of a multicrystalline silicon solar cell

recombination is very strong, and the emission of low energy photons $\lambda \gg 1{,}150$ nm is not detected by the camera. On the other hand, regions without defects appear bright in the image because in the absence of defect states, recombination occurs preferentially by band-to-band recombination, thus resulting in the emission of photons with $\lambda \sim 1{,}150$ nm that are detected by the camera.

The rather direct interpretation of the selective absorption of photons by the Si CCD cameras is a feature that makes such cameras particularly convenient for the characterization of silicon solar cells, reason why this equipment is frequently used for photoluminescence and electroluminescence mapping of silicon photovoltaic materials and devices.

3 Luminescence of Crystalline Silicon Materials and Solar Cells

3.1 Photoluminescence

3.1.1 Theoretical and Experimental Principles

The permanent pursuit of lowering the production costs increases the importance of fully characterizing the PV materials to benefit as much as possible from the produced materials. Mapping of the material properties such as the minority carrier lifetime is particularly useful for that purpose. Microwave photoconductance decay (μWPCD) is a powerful technique that allows carrier lifetime mapping of silicon materials, and it is very popular in research labs [22]. However, the full mapping of industrial size materials (wafers, bricks, and ingots) is rather time consuming (of the order of minutes per cm^2) as the scanning is performed point by point. PL imaging has the important advantage of performing fast and detailed mappings of the material's properties, being able to characterize a 156×156 mm^2 silicon wafer in less than 1 s [23], making it particularly suited for the use in industrial environments.

As mentioned beforehand, one of the major advantages of PL imaging when compared to EL is that it is contactless, permitting the characterization of finished photovoltaic devices without touching them. As such, this permits fast inline monitoring of the different manufacturing processing steps of photovoltaic materials – namely, ingot production, diffusion, or surface passivation.

Photoluminescence imaging was developed by researchers at the University of New South Wales, Australia [23], and quickly became a standard characterization technique in PV research and industry.

Typically, the PL experimental imaging setup includes near-infrared laser diodes coupled to a fiber to illuminate the sample and stimulate the luminescence signal, while the CCD camera captures the infrared luminescence image [24]. To prevent laser radiation scattering back to the camera, long-pass or band-pass filters are used.

The basic principle of PL imaging is that the laser irradiates continuously the surface of a semiconductor with light of a certain wavelength. Due to the radiation absorption, electron-hole pairs are generated. The subsequent electron-hole pair recombination emits the luminescence signal that is captured by the CCD camera. To separate the luminescent radiation from the much stronger reflected laser radiation

Fig. 15 Photoluminescence
setup

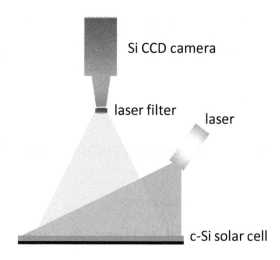

a delicate filtering process is required, which is the most challenging step of the PL measurement (see Fig. 15).

The bulk minority carrier lifetime τ_{bulk} is the most relevant parameter to characterize the quality of photovoltaic materials. It measures the semiconductors carrier recombination rate and largely determines the maximum voltage and current that can be extracted from a solar cell. Obtaining the τ_{bulk} from a sample can be complex, as usually the measured quantity is τ_{eff}, which measures the combined effect of the recombination in the bulk and at the sample's surfaces. For a crystalline silicon sample with an average surface recombination velocity, the relation between the τ_{eff} and τ_{bulk} is described by [25]:

$$\frac{1}{\tau_{\text{eff}}} = \frac{1}{\tau_{\text{bulk}}} + \frac{2S}{W} \tag{3}$$

where S is the surface recombination velocity (cm/s) admitted to be the same at the two surfaces and W is the sample thickness. From the previous equation, it can be seen that the value of τ_{eff} is largely determined by the higher of the two fractions on the right. Thus, to accurately determine τ_{bulk}, the contribution of the bulk recombination must be much higher than that from the surface (i.e., $1/\tau_{\text{bulk}} \gg 2S/W$). This condition can be met by introducing a surface coating to reduce recombination (i.e., surface passivation) or by ensuring that the sample's thickness and light penetration depth are sufficiently large so as to minimize the contribution of the recombination on the surfaces.

Obtaining quantitative measurements of carrier lifetime or solar cell minority carrier diffusion lengths with PL requires the knowledge of several parameters, such as the CCD camera's quantum efficiency or its detection solid angle. Hence the obtained signal must either be calibrated against other characterization tools or several PL images of different portions of the PL spectrum are required.

3.1.2 Wafer Characterization

The characterization of crystalline silicon wafers by PL mapping is a powerful technique to detect material defects, rank the quality of the wafers, and, when used during solar processing, assess the quality of solar cell processing. For instance, the appearance of microcracks during the production process can be detected early by PL imaging, thus avoiding critical failures during solar cell production [26].

Industrial silicon wafers typically have thicknesses around 200 μm. As such, in order to accurately determine the bulk carrier lifetime, wafer surface recombination has to be suppressed by performing a surface treatment either through a chemical bath [27] or by deposition of a dielectric film such as silicon nitride [28]. Since the implementation of these procedures is complex, they cannot be used in a production line, so different approaches must be followed. Since for as-cut silicon wafers surface recombination is too strong to allow an accurate determination of the bulk carrier lifetime and infer the effective quality of the material, the use PL in an industrial environment can only support a comparative evaluation of the wafers' quality. Haunschild et al. [29] proposed to detect material deficiencies by analyzing the contrasts and features in PL images. For this purpose, the lifetime distribution of wafers from different regions of the ingot was inspected. It was observed that PL images of multicrystalline silicon wafers show darker regions (i.e., regions of reduced lifetime) in wafer areas near the crystallization edges, probably due to iron contamination from the crucible. A similar effect was observed for PL images of wafers from the bottom and top regions of the ingot which showed, on average, lower intensities than the middle ingot wafers. Also, areas of the wafers with lower carrier lifetimes, such as grain boundaries or dislocation regions, can be easily detected in the PL images. Overall it was shown that although PL imaging cannot provide an effective evaluation of the quality of non-passivated wafer quality, the relative variation of the PL signal between wafers from different ingot regions and along the wafers is a powerful tool to detect process deficiencies [29].

3.1.3 Ingot and Brick Characterization

The advantage of using PL imaging to characterize the quality of crystalline Si ingots or bricks cut from the crystalline ingots is that for such thicknesses ($\gg$1 mm) recombination occurs mainly in the bulk (i.e., small contribution of surface recombination), avoiding the need to perform surface passivation before the measurements to determine the bulk carrier lifetime. However, the use of PL to characterize these materials and obtain carrier lifetime profiles has some hindrances. One of the main difficulties is that the doping profile along an ingot's height is not constant, and PL emission intensity is proportional to doping density. However, since such dependence is linear for carrier injection rates similar to those experienced under low injection conditions, the effect of doping variation on the PL signal can be separated from the effect of lifetime variation.

Due to dopant segregation in the liquid [30], usually the doping concentration is not constant along the ingot (brick) height[2] and considering that:

$$n_e n_h = \Delta n (\Delta n + N) \tag{4}$$

where $N = N_A, N_D$ the background doping concentration that can be either n-type-n or p-type. Using the known relation between n_e, n_h and $\Delta \varepsilon_F$ [3]:

$$n_e n_h = n_i^2 \exp\left(\frac{\Delta \varepsilon_F}{kT}\right) = \Delta n (\Delta n + N) \tag{5}$$

and for the low carrier injections typically used in PL imaging ($\Delta n \ll N$):

$$n_i^2 \exp\left(\frac{\Delta \varepsilon_F}{kT}\right) = \Delta n N \Rightarrow r_{sp} \propto \Delta n N \tag{6}$$

A simple technique to remove the doping variation effect is the photoluminescence intensity ratio (PLIR) which was initially introduced by Würfel et al. [31, 32]. By applying this technique, it is possible to directly obtain the spatially resolved diffusion lengths and carrier lifetimes. The technique is based on performing two PL measurements at two different wavelength ranges. First, a short-pass filter is introduced in front of the camera, and the first PL image is obtained; then a second PL is obtained using a long-pass filter. The quotient of the two signals cancel the exp ($\Delta \varepsilon_F/kT$) term, thus resulting in a signal with no dependence on dopant concentration, allowing the determination of the carrier lifetime distribution on ingots and bricks, regardless of the doping variations on these materials.

3.2 Electroluminescence and Photoluminescence of Finished Devices

As already mentioned, EL can only be applied on finished devices (i.e., solar PV cells and modules), while PL can be applied at any stage of the manufacturing process. When characterizing solar PV cells and modules, it might be useful to combine both EL and PL.

---

[2]Although the dopants (boron and phosphorous) usually used for silicon PV materials have segregation coefficients close to one, there is always a dopant variation along the ingot height.

3.2.1 Luminescence of Silicon Solar Cells

Luminescence mapping can be used to determine the distribution of the most important solar cell parameters and identify loss mechanisms.

The spectral external quantum efficiency (EQE) gives detailed information about the irradiation dependence of the photocurrent efficiency of the solar cell [3]. The rate at which different radiation wavelengths are absorbed at distinct depths varies according to photon energy (i.e., shorter wavelength (high energy) photons are absorbed more strongly near the front surface, and longer wavelength (low energy) photons are absorbed more uniformly throughout). Because of this, EQE spectral characterization is a valuable technique for detecting deficiencies along the device thickness. Based on the reciprocity between the EL spectral distribution and quantum efficiency of a solar cell [33], the EQE can also be determined by measuring the EL spectrum [34]. As previously mentioned, Würfel et al. [31] introduced the PLIR technique that can determine the minority carrier diffusion length on crystalline silicon materials with variable doping concentrations. The same technique can also be applied to images resultant from EL.

Since the EL intensity is highly dependent on the local voltages, localized short circuits (also named shunts), and local series resistance anomalies, these are easily detectable with luminescence imaging. The origin of shunts can be attributed to very different physical phenomena [35]. Shunts can be due to intrinsic material defects, such as grain boundaries decorated with metal impurities or dislocations; they can also be due to process-induced defects, namely, cracks, scratches, or metal contact piercing through the emitter. Localized shunts drain charge carriers from the neighboring regions of the cell and generate voltage dips that can be identified by a dark region in the EL image. Breitenstein et al. [36] used EL and PL to detect and classify shunts on crystalline silicon solar cells. They observed that the depth and lateral spread of the voltage dip depended on the shunt level and the neighboring series resistance. Hence, to accurately determine the shunt resistance using EL, it is essential to make a precise voltage calibration of the luminescence images on shunted areas.

Furthermore, the position of the shunts relatively to the solar cell metallization contacts might hinder their detection and evaluation. For instance, if the shunts are placed under metallization fingers, unless the shunts are very strong (i.e., very low shunt resistance), they are very difficult to detect by luminescence imaging. Conversely, the detection of shunts in regions surrounded by high lateral series resistance is particularly accurate. This feature can be very useful for quality control of the junction isolation process (usually performed by laser scribing). Breitenstein et al. [36] also determined the PL detection limit for point-like shunts and showed that because monocrystalline silicon cells do not have grain boundaries, PL can detect weaker shunts than in multicrystalline silicon solar cells.

Series resistance R_s, has a significant impact on the performance of a solar cell, affecting primarily the fill factor and short-circuit current both impacting on the solar cell efficiency. The series resistance of an industrial solar cell should be uniform across the device, especially in the case of monocrystalline silicon solar cells.

The local variations of the series resistance can be due to inhomogeneities of the base material, such as in the case of multicrystalline silicon substrates, or process failures. Hence monitoring R_s variations is of crucial importance for the quality control of solar cell production.

EL imaging can be used to map the solar cell series resistance, in fact given that EL intensity of a given pixel i is proportional to local voltage in that position V_i [37]:

$$\Phi_i = C_i \exp\left(\frac{qV_i}{kT}\right) \tag{7}$$

where C_i is a normalization constant that includes the solar cell optical and recombination properties and the characteristics of the experimental setup, k is the Boltzmann constant, q is the electron charge, and T is the absolute temperature.

By obtaining the EL of the solar cell for two different applied voltages V_{appl}, the value of C_i can be determined. The distribution of V_i can then be obtained by modeling the solar cell as a two-dimensional network of parallel nodes, each consisting a local resistance $R_{s,\,i}$ connected in series with a diode [37]. The local series resistance can be given by:

$$R_{s,i} = \frac{\Delta V_{R_{s,i}}}{I_i} = \frac{V_{appl} - V_i}{I_i} \tag{8}$$

where $\Delta V_{R_{s,i}} = V_{appl} - V_i$ is the local voltage drop over the series resistance and I_i is the local current intensity. As previously mentioned, the local voltages can be obtained from the EL intensity mapping and under certain operating conditions, I_i can be determined with reasonable accuracy both from the solar cell's base [38] and emitter [39], and the solar cell local series resistances can also be determined [37, 40].

3.2.2 Intrinsic and Extrinsic Defects

The defects present in a solar cell can be categorized as two different types: intrinsic and extrinsic. Intrinsic defects are related to material deficiencies such as crystallographic defects or grain boundaries. Extrinsic defects are process related and usually appear during solar cell processing; examples of extrinsic defects are wafer cracks or breakages.

The electronic trap levels associated with extrinsic defects are much deeper than the ones associated with intrinsic defects; hence, the carrier lifetime associated with extrinsic defects is less sensitive to temperature variations. On the other hand, the intrinsic defects' lifetime is more sensitive because the trap levels vary significantly with temperature [41]. Based on this principle, a clear distinction between the solar cells' intrinsic and extrinsic defects can be made calculating the difference of intensities of two images taken at two different temperatures (range 25–100°C) [41, 42].

3.3 Characterization of High-Efficiency Crystalline Solar Cells

Currently the photovoltaic industry is seeing the emergence of higher efficiency solar cell concepts based on local diffusion of emitters and BSF such as PERL (passivated emitter rear locally diffused) [43], POLO (polysilicon on oxide junction) [44], or IBC (interdigitated back-contact) [45] solar cells. Although these new solar cell architectures require a more complex manufacture, the use of the luminescence techniques to characterize these high-efficiency solar cells is rather straightforward, and there are already a few examples of the use of EL and PL to detect processing faults in these [46–48].

3.4 Luminescence of Modules and PV Systems

High-quality inspection at the system level is vital for the successful business exploitation of PV systems. Damage and degradation of cells at the module and system level are efficiently detected by luminescence techniques. Examples and case studies for damage and degradation due to extrinsic mechanical stress and intrinsic material aging are presented. Finally in this subsection, examples of technical developments are also discussed which will allow for significant increases in the throughput rate of luminescence inspection of PV systems.

3.4.1 The Case for Infield Luminescence Inspection

With the ever-decreasing cost of PV systems, the operation and maintenance of these is now subject of close inspection by those involved in PV project development. Careful attention must be given to quality control of the PV modules as they leave the factory to the point of installation. For quality assurance there are several service providers now focused on tracking module quality from factory to installation site. Shown in Fig. 16 is an example of a mobile testing station which permits the characterization of modules in this way.

Upon system sign-off, preventative maintenance and attention to module performance must be also considered so as to minimize energy production downtime. The tendency is for PV systems to consist of up to 300 modules connected in series (string in industry jargon) to operate at 1,500 V. The consequence is that string performance will be limited by, to a first approximation, the worst module in the string.

Fig. 16 Example of a mobile testing lab for PV modules. A typical system will perform visual inspections, maximum power determination at STC (standard test condition), electrical insulation tests, infrared thermography inspection, and electroluminescence (EL) testing. Image provided by H. Silva, Enertis Spain

PV system luminescence imaging is proving to be a powerful technique which allows for the fast yet detailed onsite module inspection. The operative word onsite also includes the inspection without the need for dismounting with the risk of incurring in further module damage.

Depending on system size (e.g., domestic <5 kW, commercial 5–500 kW, and utility >500 kW), the annual operation and maintenance (O&M) cost for every unit of power installed can range from 19 €/kW/year for domestic systems to 11 €/kW/year [50]. The result is that for typical values at which the electricity is sold (150 €/MWh domestic scale to 50 €/MWh utility scale), the annual cost of O&M weighs in heavily at ca 10–15% of the annual revenue. This is expected to increase because of the downward pressure on the revenue generated from the sale of electricity. As such, and apart from reducing manufacturing and installation costs of PV, the industry is also concerned in reducing the O&M costs.

Although PV modules are designed to last 25 years (or more) under rigorous outdoor conditions, these are still mechanically fragile. Also, once onsite, the PV modules can still suffer mechanical damage when installed, dropped, and stepped on, or suffer strain with inappropriate mounting. The problem herein is that much of the possible damage suffered by the modules is not apparent by visual inspection or even by measuring power output efficiency. Microcracks at the cell level are the type of damage which is not detectable by visual inspection nor is easily detectable by testing module power output. However, these microcracks will grow because of the thermal cycles PV modules will typically undergo (e.g., $-10°C$ to $80°C$ on any given day is possible), with the result being accelerated module performance degradation and hence a decrease in revenue.

Upon system installation and sign-off, there is still the requirement for undertaking O&M of PV systems. PV modules may suffer damage due to aspects that were not detected during (a) fabrication and (b) system installation and sign-off, (c) external factors, or (d) degradation mechanisms that were not detected or known and accounted for. The last point is especially important because the understanding of the material science governing module reliability is still at an embryonic stage [49]. Apart from the solar modules, inspection of the rest of the PV system is required, e.g., the mechanical fixings and electrical cabling, DC to AC inverters [50].

3.4.2 Extrinsic Module Damage

Figure 17 shows examples of damage incurred due to mishandling. A module that has been dropped may not show any physical damage upon visual inspection. However, the EL image shows that most of the cells have large dark portions, indicating severe cracking. Hailstone damage, as does stepping on modules, is not always detected by visible inspection; however, the deflection of the glass can be such that the cell underneath suffers breakages. Also shown is an example where the connection between cells has been compromised resulting in half the area of the cells being dark because of no current injection. Finally, washing modules with high-pressure water systems can also induce cracking of the modules, again because of the deflection suffered by the glass.

The microcracks and broken bus bars and fingers can be a result of a defective manufacturing process or a result of inadequate substandard module transport and storage. The damage resultant from stepping on PV modules is more likely to occur in system installations where module positioning is compact and of difficult access, such as on rooftops. Damage due to alien objects has also been reported, e.g., hailstones or stones dropped by birds. Stoicescu et al. [51] reported an example where visual inspection in a PV power plant showed damage to 1/3 of modules. However, upon luminescence imaging, it was found that essentially all the modules had suffered damage due to the hailstones.

3.4.3 Intrinsic Module Damage

Outside the O&M scope, but in the R&D scope, luminescence imaging is now also being used to further understand the infield degradation mechanisms.

Potentially induced degradation (PID) is an important degradation phenomenon in medium- to large-scale PV systems. This is a result of having several modules in

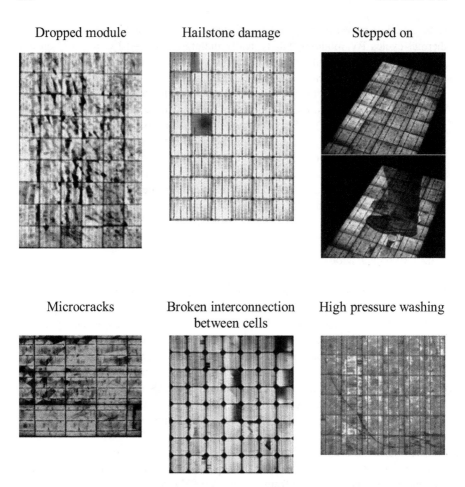

Fig. 17 Examples of electroluminescence images identifying damage suffered from mishandling by PV modules. All images, except for the stepped-on image, were obtained outdoors for functioning PV systems. The image depicting a stepped-on module shows the module before and after being stepped on. Superimposed is also the image of a person stepping on the module. All images provided by Solarzentrum Stuttgart GmbH

series to maximize string voltage, and as such those at the end of a string will have a significant voltage bias with respect to the grounded (for safety reasons) module glass and frame. The detailed mechanisms governing PID are still under examination [52], although it is generally accepted that Al and Na ion diffusion from the frame into the cells due to high electric fields is one of the main causes. The main implication is that individual cells will become inoperative because of shunting. Figure 18 shows an example of EL imaging of six modules (from a string consisting of 24) and demonstrates the increased occurrence of darker individual cells for modules closer to the negative end of the string.

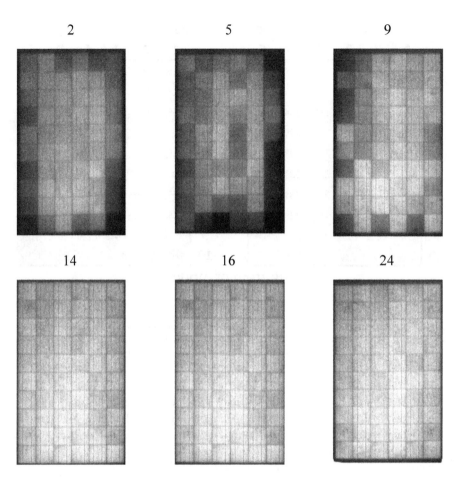

Fig. 18 Electroluminescence images of 6 modules connected in string of 24 modules: String position of the module is shown below each image. Modules closer to the negative pole (module number 1) show a checkered pattern, with dark and bright cells (evidence of PID); the cell brightness distribution for the modules closer to the positive pole (module number 24) is more uniform, thus free of PID. To note, module 1 is not shown as it was substituted before the EL measurements because it had already failed due to PID. Figure and legend adapted with permission from Martínez-Moreno et al. [53] and WIP Renewable Energies

3.4.4 A Large-Scale Case Study

Figure 19 shows data produced from a recent study using EL imaging to evaluate two types of common faults, namely, mechanical and PID, in a set of 12,500 PV modules comprising a PV power station [54]. The data shows that mechanical failures occurred preferentially at the center of the modules, i.e., where mechanical pressure will result in greater bending of the modules and hence induce cracking of the cells. The PID data is interesting as it shows that PID is more likely to occur at the

Top of modules

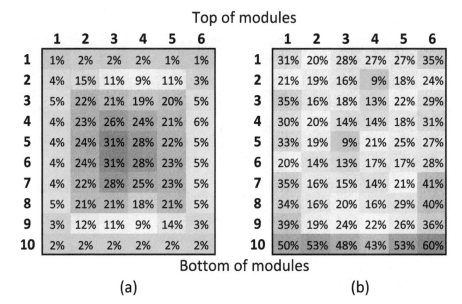

Bottom of modules

(a)   (b)

Fig. 19 Cell position failure distribution in ca 12,500 modules from a power plant analyzed by electroluminescence showing (**a**) cell mechanical damage and (**b**) potentially induced degradation. Figure and legend were adapted with permission from Koch et al. [54] and WIP Renewable Energies

border of the module where the metal frame is. However, it also shows that PID is more likely to occur for the cells at the bottom of the module. It is thought that this is due to the increased humidity in the bottom of the modules [55], either from ground humidity or due to water on the surface of the module being retained for longer on the bottom side. The data also indicates how surprisingly high failure rates can be, with a significant percentage of cells suffering mechanical damage or PID.

The examples presented above illustrate that wide-scale luminescence canvassing of PV modules is a necessity for keeping track of system performance. Köntges et al. [56] estimated from a field study that approximately 2% of modules will fail in an 11–12-year timespan, hence not meeting manufacturers' warranty claims. The reader is also encouraged to read the review by Köntges et al. [56] as it includes an extensive list of EL images of modules and cells demonstrating a comprehensive range of causes of failure.

3.4.5 Industry Standard Luminescence Inspection Systems

Table 1 shows data provided by a company that develops PV luminescence inspection systems. The system is a fixed mounting type requiring the positioning of the camera with recourse to a technician and elevators as shown in Fig. 20. The throughput rate of inspection is dependent on the type of defect which is to be

Table 1 Daily rate of inspection as a function of defect-type identification for the DaySy system developed by Solarzentrum Stuttgart GmbH

Luminescence technique	Defect type	Modules day^{-1}	Area (typical module area 2×1 m^2)		Power (typical density 150 W m^{-2})
			m^2 day^{-1}	Ha day^{-1}	KW day^{-1}
EL	PID	3,000	6,000	0.6	900
	Inactive area > 5%	2,000	4,000	0.4	600
	Microcracks	1,500	3,000	0.3	450
PL	PID	2,500	5,000	0.5	750
	Inactive area > 5%	1,750	3,500	0.35	525
	Microcracks	1,250	2,500	0.25	375

Fig. 20 Photo exemplifying the acquisition of infield luminescence stationary images of PV modules. Provided by Solarzentrum Stuttgart GmbH

mapped out, ranging from 450 kW/day for microcracks to 900 kW/day for PID. If an elevator lift is required or there is significant wind, inspection throughput can be significantly hampered.

One of the technical challenges with acquiring outdoor luminescence images is that the intensity of the module luminescence is several orders of magnitude lower than the background daylight. Techniques employed to circumvent this challenge are to use (a) indium gallium arsenide (InGaAs) sensor cameras which have a high spectral sensitivity in the ca. 1,100 nm range but suffer from sub-megapixel resolutions, (b) high-pass or band-pass optical filters to remove the visible light, and/or (c) bipolar image acquisition where the difference between the luminescing and non-luminescing images creates an image consisting of only luminescence light from the modules.

3.4.6 Future Tendencies in Luminescence Inspection Systems

To further increase the speed of inspections, mobile platform luminescence imaging is already being proposed using drones with rates of inspection up to 1 MW/h, significantly faster than a fixed system. Shown in Fig. 21 is an example of a proposed system under development. It consists of a quadcopter equipped with shortwave infrared (SWIR) camera for luminescence imaging and infrared (IR) cameras for hotspot imaging. For photoluminescence imaging, the authors also propose to equip the quadcopter with a high-powered laser diode whose beam is focused into a line.

For electroluminescence imaging, a strategy proposed is to modulate the EL emission of the modules during image acquisition [58]. EL emission can also be synchronized with image capture. This approach results in a large dataset of images which then need to be processed, namely, (1) identification of images containing the complete module, (2) module edges and corner recognition, (3) module segmentation and cropping to region, (4) parsing of EL and background images, (5) motion compensation, (6) and finally averaging for denoising.

Shown in Fig. 22 are the first published results demonstrating EL image acquisition on a moving test-bed [59]. For comparison, the indoor images taken under ideal dark room conditions are also shown. The subsequent constructed outdoor image results from a series of images acquired at a high framerate which are then processed to form the image shown. From the images it is possible to conclude that the important information pertaining to the state of the modules is discernible from the processed outdoor image. For example, modules 1 and 2 present significant problems with several cells with no luminescence. Module 3 shows some cracks and some cells with a comparatively lower luminescence intensity and one significantly darker (top right).

Outdoor electroluminescence imaging still requires that power be injected into modules by forward biasing these, which can be a complicated and time-consuming technical undertaking. This is where photoluminescence may be advantageous, apart from being more sensitive to detecting low light performance degradation and potential-induced degradation. However, PL emission tends to be of lower intensity and also requires significant illumination intensities of the modules so as to have detectable luminescence.

Shown in Fig. 23 are the photo- and electroluminescence image of a PV module. A checkered pattern of some darker cells is visible, indicating that these cells are

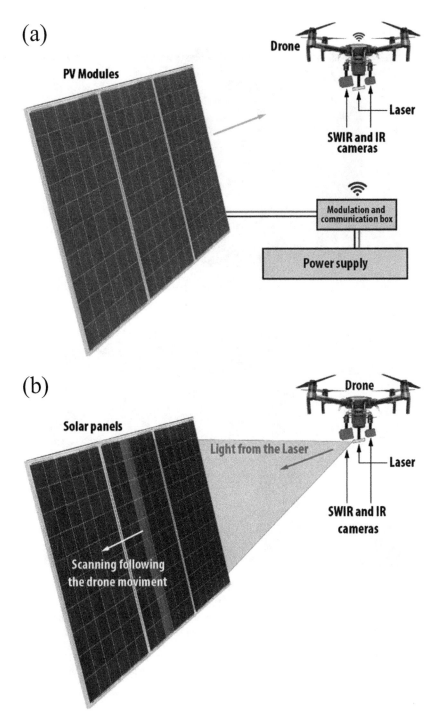

Fig. 21 Proposed (**a**) electroluminescence and (**b**) photoluminescence imaging using a quadcopter. Figure adapted with permission from Benatto et al. [57] and WIP Renewable Energies

(a)

(b)

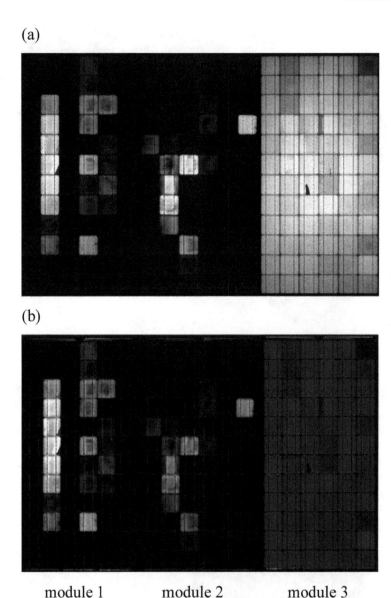

module 1   module 2   module 3

Fig. 22 EL images of three monocrystalline silicon modules acquired (**a**) indoors and stationary using a high-resolution digital single-lens (DSLR) camera with a Si-based CCD, with 30 s exposition, and (**b**) outdoors daylight at 1 m/s lateral movement speed, with an InGaAs camera OWN 640 from Raptor Photonics, with exposure time of 1 ms, 120 Hz framerate, and 25 Hz modulation frequency resulting in 162 images per panel. Images reproduced with permission from Benatto et al. [59] and DTU – Technical University of Denmark

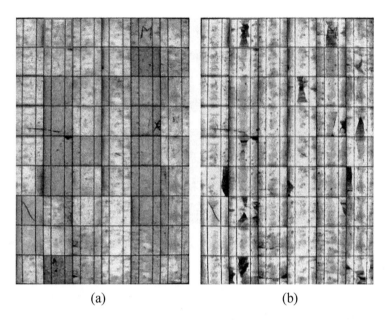

(a)   (b)

Fig. 23 (**a**) Photoluminescence and (**b**) electroluminescence of a PV module composed of multicrystalline solar cells. Images provided by Solarzentrum Stuttgart GmbH

suffering from generalized shunting, reducing free carrier densities. Cracks (dark continuous lines) are also visible. From the EL image, the checkered pattern is not discernible. However, areas which are electrically isolated and hence no charge injection occurs are visible. A clear example is the highlighted cell. The PL image looks normal, however the EL image clearly shows that the cells' central portion have lower luminescence signals.

There are different approaches for illuminating the PV modules. Diode lasers have been proposed, with an interesting solution being concentrating the laser illumination into a line which then scans over the PV module surface. The advantage of this approach is that the laser power requirement is significantly lower allowing for a less bulky integration into a quadcopter drone [57]. Shown in Fig. 24 is an example of a photoluminescence image acquired using a laser line and compared to a standard electroluminescence image. Again, in general terms, the dark patches from the EL images are bright patches in the PL image, indicating that these areas are electrically isolated and not actually suffering from significantly higher recombination. Areas which are dark in both the EL and PL images are those where shunting resistance is lower and hence recombination is higher.

It is also possible to use the Sun as the light source for photoluminescence image acquisition [60, 61]. However, because of big differences in magnitudes between background light intensity and luminescence intensity, sophisticated lock-in or bipolar techniques are still required to filter out the unwanted background signal.

Fig. 24 Electroluminescence (top) image and photoluminescence (bottom) image using a laser line scan of three multicrystalline silicon solar cells. Images reproduced with permission from Benatto et al. [57] and WIP Renewable Energies

4 Luminescence of Thin-Film Materials and Solar Cells

The use of luminescence-based techniques to characterize thin-film PV materials and devices is becoming more frequent, and recently an interesting review on the subject can be found in Abou-Ras et al. [62].

In recent years several articles describe EL- and PL-based techniques to characterize thin-film PV of technologies such as copper indium gallium selenide (CIGS) [63], cadmium telluride (CdTe) [64], and gallium arsenide (GaAs) [65]. And although the characterization of amorphous and microcrystalline material devices is cumbersome due to the prominence of tail to tail transitions which introduce important deviations to the reciprocity between quantum efficiency and EL signals [66], Gerber et al. [67] were able to characterize a-Si PV modules using EL and identify defects on the device. Finally, even if the characterization of multijunction solar cells has some complications as it is difficult to characterize the different sub-cells independently. GaInP/GaInAs/Ge multijunction solar cells have been characterized by EL, and an accurate determination of the saturation current densities of all sub-cells was demonstrated [68].

5 Reverse Bias Electroluminescence

Another characterization technique based on solar cell luminescence has recently gained importance in the characterization of solar cells, named reverse bias electro-luminescence (ReBEL) [69, 70]. Similar to EL, to perform ReBEL a current is injected on the device, but in this case the current in the injected solar cells flows in the opposite direction, i.e., under reverse bias (see Fig. 10). When performing ReBEL measurements, the solar cells are inversely polarized with voltages between 2 and 16 V, inducing significant currents in these. Differently from EL, ReBEL characterization is based on the acceleration and consequent scattering or recombination of the charge carriers in a high electric field [71]. The main purpose of ReBEL imaging is the detection and identification of solar cell voltage breakdowns.

Voltage breakdowns usually occur on localized spots and can have different origins such as metal contaminations, dislocations, or other lattice defects [70]. The presence of these defects on a solar cell locally reduces the minimum voltage at which current will start to flow when reverse biased. In a situation where a PV module is partially shaded and one solar cell is producing a lower current than the neighboring ones, it can become inversely polarized. If the voltage at which current starts to flow when inversely polarized is low at localized spots of the cell, these will start to conduct current and hence dissipate energy-generating local hotspots,[3] placing the integrity of the complete module at risk.

Recent studies were able to correlate the different type of voltage breakdowns with the voltages at which they become active. To learn more the reader can refer to the works of Breitenstein [71] and Lausch et al. [72].

In Fig. 25 an EL and ReBEL characterization of a defect region of a multicrystalline solar cell is presented. The good agreement between the two characterization techniques can be observed (see encircled region). Also, the fact that when inversely polarized the defected region becomes luminescence at a voltage of -12 V points to the existence of recombination active crystal defects probably due to the bulk contamination with metal impurities [71].

6   Conclusions and Outlook

Due to its high spatial resolution allied to the swift acquisition times, the use of PL and EL is today routinely implemented for the characterization of crystalline silicon materials and devices at all the stages of the production chain.

In terms of PV system characterization, services already exist that allow for the EL and PL imaging of complete strings and small systems. However, these are still fixed systems and may not be in the future cost competitive with foreseeable systems on autonomous mobile platforms. These are still under development and if cost effective, their widespread implementation will result in more effective preventative maintenance of PV systems but also the creation of a large dataset allowing for a detailed analysis of module failure modes. This knowledge can be feedback into the whole wafer-cell-module-system value chain so as to further improve processes to further increase PV reliability. The potential importance and cost effectiveness PV

---

[3]Hotspots on modules are known cause of solar modules to eventually ignite causing severe damage to the module, system, and location where modules are installed.

(a)

(b)

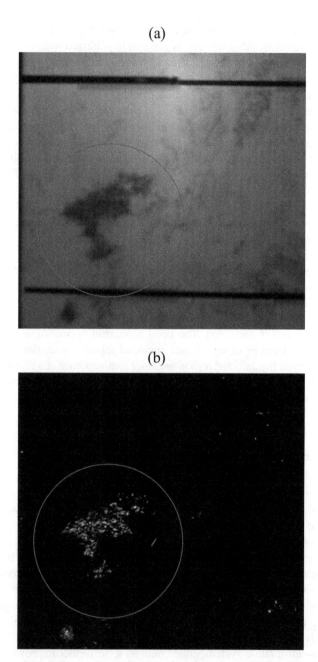

Fig. 25 Electroluminescence image of multicrystalline in forward and reversed bias; (**a**) EL, $V = 0.55$ V; (**b**) ReBEL, $V = -12$ V

luminescence imaging is such that systems are being actively developed and used in university degree courses [42, 73].

The use of EL and PL is also gaining importance in the characterization of thin-film PV devices for a plethora of technologies such as CdTe, GaAs, III-V multijunction, and perovskite solar cells.

Finally, reverse bias electroluminescence has proven to be a very useful technique that can complement the EL characterization of solar cells, allowing the detection of defects that cannot be seen with direct bias EL.

Acknowledgments The authors wish to thank Hugo Silva at Enertis Madrid, Dr. Michael Reuter and Liviu Stoicescu at Solarzentrum Stuttgart GmbH, Dr. Francisco Martínez-Moreno at the Instituto de Energía Solar – Universidad Politécnica de Madrid, Dr. Simon Koch at the Photovoltaic Institute Berlin, and Dr. Gisele A. dos Reis Benatto at the Department of Photonics Engineering, Technical University of Denmark. This chapter is significantly richer because of their willingness to openly discuss their work and to permit the use of their data. We also wish to thank WIP Renewable Energies EU PVSEC for permitting the reproduction of copyrighted material.

References

1. European Photovoltaic Industry Association (2014) Global market outlook for PV 2014–2018. European Photovoltaic Industry Association, Brussels
2. Fthenakis VM, Hyung CK, Alsema E (2008) Emissions from photovoltaic life cycles. Environ Sci Technol 42(6):2168–2174. https://doi.org/10.1021/es071763q
3. Würfel P (2009) Physics of solar cells: from basic principles to advanced concepts, 2nd edn. Wiley, New York
4. Pizzini S (2012) Advanced silicon materials for photovoltaic applications. Wiley, Chichester
5. Nakajima K, Usami N (2009) Crystal growth of silicon for solar cells. Springer, Berlin
6. Luque A, Hegedus S (2010) Handbook of photovoltaic science and engineering, 2nd edn. Wiley, Chichester
7. Zhang Y, Tao J, Chen Y et al (2016) A large-volume manufacturing of multi-crystalline silicon solar cells with 18.8% efficiency incorporating practical advanced technologies. RSC Adv 6 (63):58046–58054. https://doi.org/10.1039/c6ra05765a
8. Kim KH, Park CS, Lee JD et al (2017) Record high efficiency of screen-printed silicon aluminum back surface field solar cell: 20.29%. Jpn J Appl Phys 56:08MB25
9. International Technology Roadmap for Photovoltaic (2018) ITRPV Ninth Edition 2018 including maturity report
10. Fossum JG (1977) Physical operation of back-surface-field silicon solar cells. IEEE Trans Electron Devices 24:322–325. https://doi.org/10.1109/T-ED.1977.18735
11. Aberle AG (2000) Surface passivation of crystalline silicon solar cells: a review. Prog Photovolt Res Appl 8:473–487. https://doi.org/10.1002/1099-159X(200009/10)8:5<473::AID-PIP337>3. 0.CO;2-D
12. Sze SM, Lee KM (2012) Semiconductor devices: physics and technology, 3rd edn. Wiley, New York
13. Munoz MA, Alonso-García MC, Vela N, Chenlo F (2011) Early degradation of silicon PV modules and guaranty conditions. Sol Energy 85:2264–2274. https://doi.org/10.1016/j.solener. 2011.06.011
14. Goetzberger A, Knobloch J, Voß B (2014) Crystalline silicon solar cells. Wiley, Chichester
15. Sauer R, Weber J, Stolz J et al (1985) Dislocation-related photoluminescence in silicon. Appl Phys A Solids Surfaces 36:1–13. https://doi.org/10.1007/BF00616453

16. Yang MJ, Yamaguchi M, Takamoto T et al (1997) Photoluminescence analysis of InGaP top cells for high-efficiency multi-junction solar cells. Sol Energy Mater Sol Cells 45:331–339. https://doi.org/10.1016/S0927-0248(96)00079-7

17. Schick K, Daub E, Finkbeiner S, Wurfel P (1992) Verification of a generalized Planck law for luminescence radiation from silicon solar cells. Appl Phys A Solids Surfaces 54:109–114. https://doi.org/10.1007/BF00323895

18. Trupke T, Daub E, Wurfel P (1998) Absorptivity of silicon solar cells obtained from luminescence. Sol Energy Mater Sol Cells 53:103–114. https://doi.org/10.1016/S0927-0248(98)00016-6

19. Livescu G, Angell M, Filipe J, Knox WH (1990) A real-time photoluminescence imaging system. J Electron Mater 19:937–942. https://doi.org/10.1007/BF02652919

20. Fuyuki T, Kondo H, Yamazaki T et al (2005) Photographic surveying of minority carrier diffusion length in polycrystalline silicon solar cells by electroluminescence. Appl Phys Lett 86:1–3. https://doi.org/10.1063/1.1978979

21. Trupke T (2017) Photoluminescence and electroluminescence characterization in silicon photovoltaics. In: Reinders A, Verlinden P, van Sark W, Freundlich A (eds) Photovoltaic solar energy: from fundamentals to applications. Wiley, Chichester, pp 322–338

22. Sinton RA, Cuevas A, Stuckings M (1996) Quasi-steady-state photoconductance, a new method for solar cell material and device characterization. In: Conference record of the twenty fifth IEEE Photovoltaic specialists conference – 1996. IEEE, Washington, DC, pp 457–460

23. Trupke T, Bardos RA, Schubert MC, Warta W (2006) Photoluminescence imaging of silicon wafers. Appl Phys Lett 89:044107. https://doi.org/10.1063/1.2234747

24. Hinken D, Schinke C, Herlufsen S et al (2011) Experimental setup for camera-based measurements of electrically and optically stimulated luminescence of silicon solar cells and wafers. Rev Sci Instrum 82:033706. https://doi.org/10.1063/1.3541766

25. MacDonald DH (2001) Recombination and trapping in multicrystalline silicon solar cells. The Australian National University, Canberra

26. Demant M, Welschehold T, Oswald M et al (2016) Microcracks in silicon wafers I: inline detection and implications of crack morphology on wafer strength. IEEE J Photovolt 6:126–135. https://doi.org/10.1109/JPHOTOV.2015.2494692

27. Stephens A (1997) Effectiveness of 0.08 molar iodine in ethanol solution as a means of chemical surface passivation for photoconductance decay measurements. Sol Energy Mater Sol Cells 45:255–265. https://doi.org/10.1016/S0927-0248(96)00061-X

28. Schmidt J, Aberle AG (1997) Accurate method for the determination of bulk minority-carrier lifetimes of mono- and multicrystalline silicon wafers. J Appl Phys 81:6186–6199. https://doi.org/10.1063/1.364403

29. Haunschild J, Glatthaar M, Demant M et al (2010) Quality control of as-cut multicrystalline silicon wafers using photoluminescence imaging for solar cell production. Sol Energy Mater Sol Cells 94:2007–2012. https://doi.org/10.1016/j.solmat.2010.06.003

30. Ghandhi SK (1994) VLSI fabrication principles: silicon and gallium arsenide, 2nd edn. Wiley, New York

31. Würfel P, Trupke T, Puzzer T et al (2007) Diffusion lengths of silicon solar cells from luminescence images. J Appl Phys 101:123110. https://doi.org/10.1063/1.2749201

32. Mitchell B, Trupke T, Weber JW, Nyhus J (2011) Bulk minority carrier lifetimes and doping of silicon bricks from photoluminescence intensity ratios. J Appl Phys 109:083111. https://doi.org/10.1063/1.3575171

33. Rau U (2007) Reciprocity relation between photovoltaic quantum efficiency and electroluminescent emission of solar cells. Phys Rev B Condens Matter Mater Phys 76:1–8. https://doi.org/10.1103/PhysRevB.76.085303

34. Kirchartz T, Helbig A, Reetz W et al (2009) Reciprocity between electroluminescence and quantum efficiency used for the characterization of silicon solar cells. Prog Photovolt Res Appl 17:394–402. https://doi.org/10.1002/pip.895

35. Breitenstein O, Rakotoniaina JP, Al Rifai MH, Werner M (2004) Shunt types in crystalline silicon solar cells. Prog Photovolt Res Appl 12:529–538. https://doi.org/10.1002/pip.544
36. Breitenstein O, Bauer J, Trupke T, Bardos RA (2008) On the detection of shunts in silicon solar cells by photo- and electroluminescence imaging. Prog Photovolt Res Appl 16:325–330. https://doi.org/10.1002/pip.803
37. Trupke T, Pink E, Bardos RA, Abbott MD (2007) Spatially resolved series resistance of silicon solar cells obtained from luminescence imaging. Appl Phys Lett 90:093506. https://doi.org/10.1063/1.2709630
38. Glatthaar M, Haunschild J, Zeidler R et al (2010) Evaluating luminescence based voltage images of silicon solar cells. J Appl Phys 108:014501. https://doi.org/10.1063/1.3443438
39. Müller J, Bothe K, Herlufsen S et al (2012) Reverse saturation current density imaging of highly doped regions in silicon employing photoluminescence measurements. IEEE J Photovolt 2:473–478. https://doi.org/10.1109/JPHOTOV.2012.2201916
40. Haunschild J, Glatthaar M, Kasemann M et al (2009) Fast series resistance imaging for silicon solar cells using electroluminescence. Phys Status Solidi Rapid Res Lett 3:227–229. https://doi.org/10.1002/pssr.200903175
41. Fuyuki T, Kitiyanan A (2009) Photographic diagnosis of crystalline silicon solar cells utilizing electroluminescence. Appl Phys A Mater Sci Process 96:189–196. https://doi.org/10.1007/s00339-008-4986-0
42. Frazão M, Silva JA, Lobato K, Serra JM (2017) Electroluminescence of silicon solar cells using a consumer grade digital camera. Meas J Int Meas Confed 99:7–12. https://doi.org/10.1016/j.measurement.2016.12.017
43. Zhao J, Wang A, Altermatt P et al (1996) 24% efficient perl silicon solar cell: recent improvements in high efficiency silicon cell research. Sol Energy Mater Sol Cells 41–42:87–99. https://doi.org/10.1016/0927-0248(95)00117-4
44. Haase F, Hollemann C, Schäfer S et al (2018) Laser contact openings for local poly-Si-metal contacts enabling 26.1%-efficient POLO-IBC solar cells. Sol Energy Mater Sol Cells 186:184–193. https://doi.org/10.1016/j.solmat.2018.06.020
45. Franklin E, Fong K, McIntosh K et al (2016) Design, fabrication and characterisation of a 24.4% efficient interdigitated back contact solar cell. Prog Photovolt Res Appl 24:411–427. https://doi.org/10.1002/pip.2556
46. Moors M, Baert K, Caremans T et al (2012) Industrial PERL-type solar cells exceeding 19% with screen-printed contacts and homogeneous emitter. Sol Energy Mater Sol Cells 106:84–88. https://doi.org/10.1016/j.solmat.2012.05.006
47. Horbelt R, Hahn G, Job R, Terheiden B (2015) Void formation on PERC solar cells and their impact on the electrical cell parameters verified by luminescence and scanning acoustic microscope measurements. Energy Procedia 84:47–55. https://doi.org/10.1016/j.egypro.2015.12.294
48. Padilla M, Höffler H, Reichel C et al (2014) Surface recombination parameters of interdigitated-back-contact silicon solar cells obtained by modeling luminescence images. Sol Energy Mater Sol Cells 120:363–375. https://doi.org/10.1016/j.solmat.2013.05.050
49. Johnston S, Al-Jassim M, Hacke P et al (2016) Module degradation mechanisms studied by a multi-scale approach. In: IEEE photovoltaic specialists conference (PVSC). IEEE, New York, pp 0889–0893
50. Packard CE, Wohlgemuth JH, Kurtz SR (2012) Development of a visual inspection data collection tool for evaluation of fielded pv module condition – NREL technical report. Golden Colorado, USA
51. Stoicescu L, Reuter M, Werner JH (2014) DaySy: luminescence imaging of PV modules in daylight. In: 29th European photovoltaics solar energy conference and exhibition, Amsterdam, Netherlands, Amsterdam, Holland, pp 2553–2554
52. Luo W, Khoo YSS, Hacke P et al (2017) Potential-induced degradation in photovoltaic modules: a critical review. Energy Environ Sci 10:43–68. https://doi.org/10.1039/C6EE02271E

53. Martínez-Moreno F, Pigueiras EL, Cano JM et al (2013) On-site tests for the detection of potential induced degradation in modules. In: 28th European photovoltaic solar energy conference and exhibition, p 3313
54. Koch S, Weber T, Sobottka C et al (2016) Outdoor electroluminescence imaging of crystalline photovoltaic modules: comparative study between manual ground-level inspections and drone-based aerial surveys. In: 32nd European photovoltaic solar energy conference and exhibition energy conference and exhibition (EU PVSEC), p 1736
55. Koch S, Berghold J, Hinz C et al (2015) Improvement of a prediction model for potential induced degradation by better understanding the regeneration mechanism. In: 31st European photovoltaic solar energy conference and exhibition (EU PVSEC), Munich, Germany, pp 1813–1820
56. Köntges M, Kurtz S, Packard CE et al (2014) Review of failures of photovoltaic modules. International Energy Agency, St. Ursen
57. dos Reis Benatto GA, Riedel N, Mantel C et al (2017) Luminescence imaging strategies for drone-based PV array inspection. In: 33rd European photovoltaic solar energy conference and exhibition (EU PVSEC), Amsterdam, Holland, p 2016
58. dos Reis Benatto GA, Mantel C, Riedel N et al (2018) Outdoor electroluminescence acquisition using a movable testbed. In: NREL PV Reliability Workshop. NREL, Boulder, p 6154
59. Kurtz S (2017) 2017 NREL photovoltaic module reliability workshop. In: Kurtz S (ed) NREL photovoltaic module reliability workshop
60. dos Reis Benatto GA, Riedel N, Thorsteinsson S et al (2017) Development of outdoor luminescence imaging for drone-based PV array inspection. In: IEEE photovoltaic specialists conference
61. Bhoopathy R, Kunz O, Juhl M et al (2018) Outdoor photoluminescence imaging of photovoltaic modules with sunlight excitation. Prog Photovolt Res Appl 26:69–73. https://doi.org/10.1002/pip.2946
62. Abou-Ras D, Kirchartz T, Rau U (2016) Advanced characterization techniques for thin film solar cells, 2nd edn. Wiley, Weinheim
63. Tran TMH, Pieters BE, Ulbrich C et al (2013) Transient phenomena in Cu(In,Ga)Se2 solar modules investigated by electroluminescence imaging. Thin Solid Films 535:307–310. https://doi.org/10.1016/j.tsf.2012.10.039
64. Raguse J, McGoffin JT, Sites JR (2012) Electroluminescence system for analysis of defects in CdTe cells and modules. In: Photovoltaic specialists conference (PVSC), 2012 38th IEEE, pp 448–451
65. Hu X, Chen T, Xue J et al (2017) Absolute electroluminescence imaging diagnosis of GaAs thin-film solar cells. IEEE Photon J 9:1–9. https://doi.org/10.1109/JPHOT.2017.2731800
66. Müller TCM, Pieters BE, Kirchartz T et al (2014) Effect of localized states on the reciprocity between quantum efficiency and electroluminescence in Cu(In,Ga)Se2 and Si thin-film solar cells. Sol Energy Mater Sol Cells 129:95–103. https://doi.org/10.1016/j.solmat.2014.04.018
67. Gerber A, Huhn V, Tran TMH et al (2015) Advanced large area characterization of thin-film solar modules by electroluminescence and thermography imaging techniques. Sol Energy Mater Sol Cells 135:35–42. https://doi.org/10.1016/j.solmat.2014.09.020
68. Hoheisel R, Dimroth F, Bett AW et al (2013) Electroluminescence analysis of irradiated GaInP/GaInAs/Ge space solar cells. Sol Energy Mater Sol Cells 108:235–240. https://doi.org/10.1016/j.solmat.2012.06.015
69. Lausch D, Petter K, Von Wenckstern H, Grundmann M (2009) Correlation of pre-breakdown sites and bulk defects in multicrystalline silicon solar cells. Phys Status Solidi Rapid Res Lett 3:70–72. https://doi.org/10.1002/pssr.200802264
70. Bothe K, Ramspeck K, Hinken D et al (2009) Luminescence emission from forward- and reverse-biased multicrystalline silicon solar cells. J Appl Phys 106:104510. https://doi.org/10.1063/1.3256199

71. Breitenstein O, Bauer J, Bothe K et al (2011) Understanding junction breakdown in multicrystalline solar cells. J Appl Phys 109:071101. https://doi.org/10.1063/1.3562200
72. Lausch D, Petter K, Bakowskie R et al (2010) Identification of pre-breakdown mechanism of silicon solar cells at low reverse voltages. Appl Phys Lett 97:073506. https://doi.org/10.1063/1.3480415
73. Eissa MA, Silva J, Serra JM et al (2018) Low-cost electroluminescence system for infield PV modules. In: 35th European photovoltaic solar energy conference and exhibition (EU PVSEC), Brussels, Belgium

Time-Gated Luminescence Acquisition for Biochemical Sensing: miRNA Detection

Emilio Garcia-Fernandez, Salvatore Pernagallo, Juan A. González-Vera, María J. Ruedas-Rama, Juan J. Díaz-Mochón, and Angel Orte

Contents

Abstract Luminescence emission is a multidimensional phenomenon comprising a time-domain layer defined by its excited-state kinetics and corresponding lifetime, which is specific to each luminophore and depends on environmental conditions.

E. Garcia-Fernandez (✉), J. A. González-Vera, M. J. Ruedas-Rama, and A. Orte
Department of Physical Chemistry, Faculty of Pharmacy, University of Granada, Granada, Spain
e-mail: emiliogf@ugr.es

S. Pernagallo
DestiNA Genómica S.L., Granada, Spain

DestiNA Genomics Ltd., Edinburgh, UK

J. J. Díaz-Mochón
DestiNA Genómica S.L., Granada, Spain

Pfizer-Universidad de Granada-Junta de Andalucía Centre for Genomics and Oncological Research (GENYO), Parque Tecnológico de Ciencias de la Salud (PTS), Granada, Spain

Faculty of Pharmacy, University of Granada, Granada, Spain

© Springer Nature Switzerland AG 2019   213
B. Pedras (ed.), *Fluorescence in Industry*, Springer Ser Fluoresc (2019) 18: 213–268,
https://doi.org/10.1007/4243_2018_4, Published online: 7 March 2019

This feature allows for the discrimination of luminescence signals from species with a similar spectral profile but different lifetimes by time-gating (TG) the acquisition of luminescence. This approach represents an efficient tool for removing unwanted, usually short-lived, signals from scattered light and fluorescence interferents using luminophores with a long lifetime. Due to the emergence of time-resolved techniques using rapid excitation and acquisition methods (i.e. pulsed lasers and single-photon timing acquisition) and new long-lifetime luminophores (i.e. acridones, lanthanide complexes, nanoparticles, etc.), TG analyses can be easily applied to relevant chemical and biochemical issues. The successful application of TG to important biomedical topics has attracted the attention of the R&D industry due to its potential in the development and patenting of new probes, methods and techniques for drug discovery, immunoassays, biomarker discovery and biomolecular interactions, etc. Here, we review the technological efforts of innovative companies in the application of TG-based techniques.

Among the many currently available biomarkers, circulating microRNAs (miRNAs) have received attention since they are highly specific and sensitive to different pathological stages of numerous diseases and easily accessible from biological fluids. qPCR is a powerful and routine technique used for the detection and quantification of miRNAs, but qPCR may introduce numerous artefacts and low reproducibility during the amplification process, particularly using complex samples. Thus, due to the efficiency of TG in separating short-lived sources of fluorescence common in biological fluids, TG is an ideal approach for the direct detection of miRNAs in liquid biopsies. Recently, great efforts in the use of TG have been achieved. Our contribution is the proposal of a direct detection approach using TG-imaging with single-nucleobase resolution.

Keywords FLIM · Fluorescence · Lanthanides · Lifetime · Luminescence · miRNA · Time-gated fluorescence · Time-gating · Time-resolved fluorescence

1 Introduction

1.1 Time-Resolved Luminescence

Photoluminescence refers to the emission of UV-Vis light after the photoexcitation of a luminophore. Depending on the emitting excited state, photoluminescence is called fluorescence (singlet) or phosphorescence (triplet). The characteristic features of the emission of light depend on the sample and its chemical environment, including the emission spectra, which usually show a specular image of the absorption spectra, Stokes shift, quantum yield, anisotropy, lifetime, etc. [1, 2]. Most of these features are obtained in a steady-state acquisition mode and have been exploited for many decades in numerous applications [3]. However, in the time-resolved mode, luminescence implies an emissive relaxation or de-excitation to the ground state that results in other interesting features. In response to light,

HOMO-electrons become excited to higher energy states; then, they relax to the ground state, followed by the eventual emission of photons, which is clearly depicted in the well-known *Jablonski* energy diagram. According to the kinetics of this relaxation process, the average time spent by these electrons in the excited state is a characteristic of each luminophore and is called the luminescence lifetime, τ. This lifetime depends on several factors, particularly the chemical environment, and generally follows an exponential kinetic decay according to Eq. (1), where $[P^*]$ is the concentration of the sample in the excited state and k_r and k_{nr} are the radiative and non-radiative decay rate constants, respectively [4].

$$\frac{-d[P^*]}{dt} = (k_r + k_{nr})[P^*] \tag{1}$$

The inverse of the sum of the rate constants is called the luminescence lifetime, τ:

$$\tau = \frac{1}{k_r + k_{nr}} \tag{2}$$

The integrated Eq. (1) is an exponential law:

$$[P^*] = [P^*]_0 \cdot e^{\frac{-t}{\tau}} \tag{3}$$

The emitted intensity depends on the concentration of molecules in the excited state. After a single-pulse excitation, the luminescence intensity decays according to the lifetime of the species. Eventually, different species that are de-excited may simultaneously emit light contributing to the total fluorescence intensity:

$$I(t) = \sum A_i e^{-t/\tau_i} \tag{4}$$

$I(t)$ is the time-dependent intensity of the fluorescence, A_i is the relative amplitude of the emission of each species i and τ_i is its lifetime.

Notably, for any particular emission wavelength, we cannot distinguish among different emitters with the same steady-state features, but we can differentiate among emitters in time-resolved mode according to their different lifetime of emission.

The lifetimes of the fluorescence emitted by organic molecules are usually on the nanosecond time scale, while the phosphorescence lifetimes are longer, usually reaching the millisecond and occasionally the second time scale. Other types of luminescent emitters, such as the electronic atom orbitals in lanthanide ions or solid materials, may exhibit other luminescence emission time scales.

The precise acquisition of emission decay traces requires working in time-resolved mode with very fast excitation and acquisition setup. Single-photon timing (SPT) is an efficient technique used for accurate time-resolved acquisition [5]. SPT reconstructs the fluorescence decay of a dye based on the detection of individual photons after several excitation cycles with a very short pulse of light. This technique is usually known as time-resolved mode in the time domain.

The emission lifetimes can also be obtained in the frequency domain in which the excitation light is sent with a certain time modulation. The phase delay between the excitation radiation and the emitted light can also be related to the emission lifetime [1, 2]. However, measurements in the frequency domain are performed less frequently in time-gated analyses, which is the focus of this chapter. Hence, we focus on time-resolved measurements of emitted light after pulsed excitation.

1.2   Time-Gated Luminescence

The luminescence decay trace of a population of excited molecules represents the statistical probability of a photon to be emitted within a certain amount of time. Luminescence emission is a random phenomenon; thus, τ is the average time during which approximately half of the excited molecules have emitted a photon, but τ does not represent the precise time at which a single molecule deactivates via emission. Time-gated (TG) luminescence analyses rely on the fact that not all excited molecules spend the same amount of time in the excited state before emitting a photon; thus, if only photons arriving at the detector within a certain time-window (TW) are analysed, the signal can be filtered, and the contributions of different species can be assessed as shown in Fig. 1.

Thus, according to Eq. (3), when $t = \tau$, ~37% of the population is at the excited state, while ~63% of the population is relaxed to the ground state. Generally, for a time longer than 5τ, most excited molecules are de-excited (>99%), and at 10τ, the excited molecules are completely relaxed (100%). In addition, due to the exponential decay, the abundance of the emitted photons differs over time. While most photons

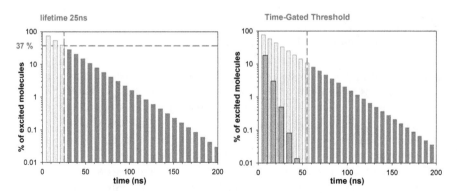

Fig. 1 (Left) Percentage of luminophores in the excited state for a molecule with a 25 ns luminescence lifetime. When $t = \tau$, ~37% of the molecules are in the excited state (dark blue), and ~63% of the fluorophores are already relaxed (light grey). (Right) Application of a 55 ns TG threshold in a simulated mixture of two molecules with 5 (grey) and 25 (blue) ns lifetimes

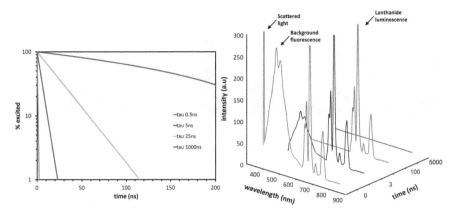

Fig. 2 (Left) Theoretical relaxation time traces of four different lifetimes: 0.5 ns (IRF), 5 ns, 25 ns and 1000 ns. (Right) Emission spectra as a function of time in a mixture of scattered light ($\tau = 0.5$ ns), background fluorescence ($\tau = 5$ ns) and long-lifetime emitting lanthanide complex ($\tau = 1000$ ns)

(63%) are emitted within τ, the remaining photons (37%) are emitted within 4τ; thus, early photons are less spread over time than late photons.

As shown in Fig. 2, we compare single exponential decay traces with different lifetimes. The shortest trace (0.5 ns) represents a rapid relaxation that is close to an instrument response function (IRF); the 5 ns and 25 ns lifetimes may correspond to a common organic fluorophore and a long-lifetime dye, respectively; finally, the longest trace (1000 ns) could represent the lifetime of a metal-ligand complex. For example, 50 ns after the excitation pulse, the only molecules with populations in the excited state have lifetimes of 25 ns and 1000 ns (approximately 13% and 95%, respectively), whereas no molecules with shorter lifetimes of 0.5 and 5 ns are in the excited state.

According to these simulations, differentiating the emission of different species is possible based on the time the photons are acquired. Identifying the optimal time threshold for removing undesired fluorescence and applying the correct time-gated detection are necessary. TG analyses only consider photons arriving at certain TW and discard the other photons. This feature is very interesting for the separation and discrimination of short- and long-lived components of fluorescence that may overlap in complex media, such as biological media, and strongly hinder the sensing mechanism of certain probes. Therefore, the use of TG to remove unwanted fluorescence requires a luminescent probe with a long lifetime. As illustrated in Fig. 2, during this process, the emission spectra collected at different detection TWs of a mixture of emissive molecules are analysed. The spectral contribution of scattered light rapidly disappears; then, the emission profile of the background fluorescence decays, and after 50 ns, only the emission of the species of interest in this example remains.

Regarding the development of the luminescence-based assays performed in complex media, there are several sources of undesired photons as follows: the

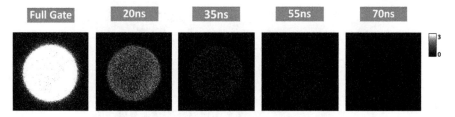

Fig. 3 Fluorescence microscopy images of a Q Sepharose bead dispersed in phosphate buffer 10 mM at pH 6 using different detection TG. From left to right, TG: 0 ns (full gate), 20 ns, 35 ns, 55 ns and 70 ns. Excitation source: 5 MHz pulsed, 440 nm laser; emission collected using a 550/50 nm bandpass filter. The size of the images is $80 \times 80 \ \mu m^2$

scattering of excitation light, autofluorescence, impurities, etc.; however, most undesired photons have short-lived lifetimes (<10 ns) and are attenuated following excitation at long wavelengths (>500 nm). Autofluorescence [6] refers to background fluorescence due to structural fluorescent groups in proteins (tryptophan <5 ns, tyrosine), cell membranes (lipofuscin), the cytosol (NADH, <1 ns), blood (porphyrins), serum (albumin, <8 ns), chlorophyll (<1 ns), B vitamins, etc. Other sources of background fluorescence from non-biological origins include scattering, impurities and nanotechnological platforms (i.e. surfaces, beads, scaffolds, etc.). Using the TG approach along with long-lived emissive probes may overcome this unwanted fluorescence. As shown in Fig. 3, the autofluorescence (550 nm) of an agarose bead (Q Sepharose, average lifetime of ca. 4 ns) vanishes after applying different acquisition TWs.

Advanced applications of fluorescence sensing require the direct detection of analytes in the working media in which unwanted background fluorescence may interfere with the desired fluorescence of the probe. As previously described, TG detection allows the separation of these two sources of fluorescence based on their differing lifetimes. This separation is particularly interesting for biomedical industries, which use chip-based platforms to perform in situ and rapid diagnoses of several target biomarkers in complex biological media (i.e. blood, serum, saliva and urine) in which autofluorescence has a strong contribution. Hence, in this chapter, we discuss the different uses of TG analyses and particularly focus on the industrial uses of such approaches in the fields of drug discovery and biomedicine. The discovery of novel biomarkers is a hot topic in the biomedical field; in particular, recently, circulating microRNAs (miRNAs) have been proposed as promising biomarkers for several diseases, such as cancer, and physiological dysfunctions. Therefore, we devoted a specific section to the application of TG analyses in the miRNA field.

2   Time-Gated Luminescence

2.1   Initial Studies

Fluorescence time-resolved techniques have achieved great development and growth due to the emergence of new fast excitation and acquisition methods, such as pulsed laser and single-photon timing (SPT) [5]; however, these techniques were applied to image microscopy only after the introduction of rapid image acquisition, such as streak or charge-coupled device (CCD) cameras, in the 1980–1990s [7, 8].

Parallel to these technical advances, recognition chemistry, including lanthanide cryptates, was seminally introduced by Prof. Jean-Marie Lehn [9], who shared the Nobel Prize for this achievement in 1987 [10]. In particular, Prof. Lehn identified the possible use of such chemistry to design new molecular devices given the distinct properties that were achieved. Subsequently, lanthanide cryptates were considered interesting photochemical supramolecular devices since they exhibited very characteristic luminescence properties [11]. Luminescence emission is characterized by a line-like spectrum and extremely long values of the luminescence lifetime, τ, both of which are caused by the forbidden electronic transition involving the f atom orbitals (see Sect. 2.2). The antenna effect of the organic moiety forming the cryptates enhances the luminescence of the metal, which remains protected within the cryptate cavity. Along with the parallel developments in time-resolved fluorimetry techniques [12, 13], the luminescent properties of lanthanide ions established the basis for new fluoroimmunoassays. The advantages of fluorescence-based techniques over radioimmunoassays, including the simplicity of the former and drawbacks related to radioactivity of the latter, were clear. However, the application of fluorescence techniques in complex matrices, such as serum, was challenging due to the intrinsic fluorescent properties of the matrix and the increased scattered excitation light. Hence, the possibility of using a set of stable luminescent molecules with very long τ values paved the way for the development of time-resolved fluorimetric immunoassays (TR-FIA), which are recognition assays based on a TG filtering analysis. Commercial kits involving Eu(III) chelates and stable antennas became available. The need for better and improved luminophores was evident, which resulted in a very active field seeking new supramolecular luminescent compounds [14, 15]. A few early examples of TR-FIA assays include the use of labelled monoclonal antibodies for the hepatitis B surface antigen [16], human chorionic gonadotrophin [17], human α-fetoprotein [18], steroid hormones [19], testosterone [20], etc.

While the drug discovery industry was highly active, these concepts were rapidly applied to imaging techniques. During the development of time-resolved microscopy, several research groups applied fluorescence time-gated acquisition to microscopy images using probes with long lifetimes to remove scattering and autofluorescence from biological media. Beverloo [21, 22] and collaborators proposed the acquisition of the delayed emission (700 μs) of phosphor crystals as a model of immunocytochemical staining, which obtained luminescence microscopy images of phosphors adsorbed over functionalized latex beads. Importantly, Marriot and collaborators developed optical microscopes capable of acquiring weak long-lived phosphorescence and delayed fluorescence of acridine orange using CCD cameras for stains in biological media [23, 24]. The luminescence of acridine orange in 3T3 cells was processed in each pixel of the image to calculate useful parameters, such as the lifetimes and phosphorescence/fluorescence ratio. Cubeddu and collaborators [25–27] developed an innovative TG fluorescence imaging technique based on a CCD video camera and a subnanosecond pulsed UV/blue excitation and applied this technique to tumour tissues. These authors emphasized the advantages of moving from the "spectral domain" to the "time domain" of fluorescence, particularly in fluorophores with similar emission spectra but different fluorescence lifetimes. For example, hematoporphyrin derivative (HPD), which is used in photodynamic therapy, has a lifetime of 14 ns, while the autofluorescence of the investigated tissues was 2–3 ns. These authors applied a delay of 5–15 ns to the acquisition of HDP fluorescence, which resulted in an increased signal-to-noise ratio in the images due to the elimination of the natural fluorescence of the tissues. Seveus used a modified time-resolved epifluorescence microscope to study the delayed luminescence of europium chelates and localize antigen C242 in malignant mucosa from the human colon [28]. Schneckenburger studied the time-resolved fluorescence of porphyrin derivative photosensitizers for medical diagnosis and biology [29–31]. These TG luminescence microscopy studies used Photosan 3 in RR1022 epithelial cells, protoporphyrin ALA in human skin and autofluorescence of teeth to detect caries.

The application of time-resolved capabilities to luminescence microscopy rapidly permitted the development of fluorescence lifetime imaging microscopy (FLIM), in which the time-resolved information is layered within the image pixels. For instance, in early studies, Kohl and co-workers proposed the delayed acquisition of fluorescence images to study ovarian carcinoma in rats using the photosensitizers Photofrin II and modified porphyrin and obtained the pseudo-colour fluorescence images shown in Fig. 4 [32]. The Lakowicz and Gadella research groups developed frequency-domain luminescence techniques based on phase-sensitive images, gain-modulated image intensifiers and CCD cameras to acquire FLIM images, thus avoiding the use of raster-scanned images. These authors applied these techniques to fluorescence lifetime images of NADH [33] and known fluorophores [34, 35]. Currently, FLIM microscopy in the time-domain and multidimensional SPT [36] are applied with high accuracy to determine the lifetimes at each pixel, although scanned acquisition is mostly required. Using this technique, accurate decay traces of emitted radiation can be obtained at each pixel, and thus, time-gating can be applied post-acquisition.

In addition, the current state-of-the-art instrumentation has paved the way for the development of new luminophores and new assays, including nanotechnological platforms. In the following sections, available materials used for efficient TG analyses and current applications are reviewed with particular emphasis on the role of different companies in the field.

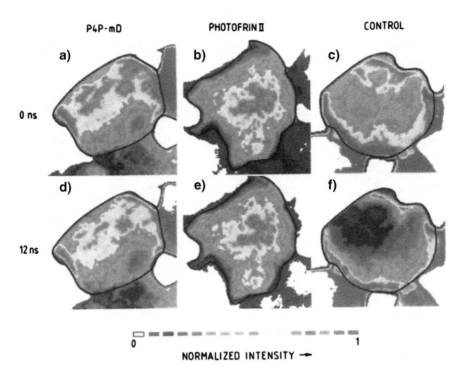

Fig. 4 Delayed fluorescence images of a tumour (ovarian carcinoma) marked with 0.1 mg of P4P-mD per kg of body weight (**a** and **d**), 5 mg of Photofrin II per kg of body weight (**b** and **e**) and control (**c** and **f**). Images were recorded with delays of 0 ns (**a**, **b** and **c**) and 12 ns (**d**, **e** and **f**). All shown images refer to the same normalized intensity scale given in false colours. Reproduced and modified with permissions from [32]

2.2 Probes for Time-Gating

For an efficient TG-based experiment, the emissive species must have a long luminescence lifetime, τ. This feature is not always trivial from the perspective of the rational design of luminophores. Here, we classify and describe the different luminescent molecules and materials that can be employed in TG analyses, which are also presented in Table 1 and Fig. 5.

2.2.1 Organic Dyes

The natural product quinine was the first widely studied fluorescent organic molecule. Subsequently, the library of organic dyes has expanded, and currently, a broad arsenal of fluorophores with a wide range of chemical and photophysical properties

Table 1 Long-lifetime luminophores

Luminophore	Emission range (nm)	Luminescence lifetime	Reference
Organic dyes	400–600	10–825 ns	[6]
Pyrene	450–550	90 ns	[37]
Acridone	420–500	14.2 ns	[38]
PureTime®14	426	15 ns	[39]
9-Aminoacridine	429–454	17 ns	[40]
Quinacridone	530–630	22.8 ns	[38]
Lucigenin	550	19.8 ns	[41]
Azaoxatriangulenium dyes	550–600	25 ns	[42, 43]
DBD dyes[a]	490–610	10–20 ns	[44]
DBO[b]	410–425	13–825 ns	[45]
EDANS	493	12.7 ns	[46]
SeTau-380	480	32.5 ns	[47, 48]
SeTau-425	545	26.2 ns	[47, 48]
Transition metal complexes	Visible to NIR	100–1000 ns	[49, 50]
Ru(bpy)$_3$[PF$_6$]$_2$	605	600 ns	[51]
Ru(bpy)$_2$(dcpby)[PF$_6$]$_2$	650	375 ns	[52]
Lanthanides	312–1530	From μs (Yb, Nd) to ms (Eu, Tb)	[53, 54]
Quantum dots	300 nm–5 μm	5–100 ns	[55, 56]
Qdot (CdSe/ZnS)	582	28.4 ns	[55, 56]
Carbon nanoparticles	400–700	6–15 ns	[57, 58]
Silicon nanoparticles	400–800	10–30 μs	[59]

[a]DBD = [1,3]dioxolo[4,5-f]-[1,3]benzodioxole
[b]DBO = 2,3-diazabicyclo[2.2.2]oct-2-ene

is available [60]. These fluorophores are well documented and specific due to the ease of derivatization, e.g. cell-penetrating agents localized in given organelles or bioconjugated chromophores used for antigen detection. Despite their great versatility, the use of organic dyes as luminescent probes is occasionally limited by the low signal-to-noise ratios observed in fluorescence microscopy due to their short emissive lifetimes (in the range 1–5 ns) and the interference from cellular autofluorescence. Although the usual range of τ values in organic fluorophores is only within a few nanoseconds, certain organic dyes exhibit unusually long-lifetime values. Dyes from the family of polyaromatic hydrocarbons, such as pyrene, usually exhibit fluorescence lifetimes greater than 5 ns and, occasionally, even longer than 100 ns. However, these molecules are highly hydrophobic and tend to have poor solubility. Moreover, these molecules can only be excited in the UV and near-UV region of the spectrum, where the autofluorescence and cellular material absorption are the strongest, which limits their potential application in investigations of complex biological systems [37].

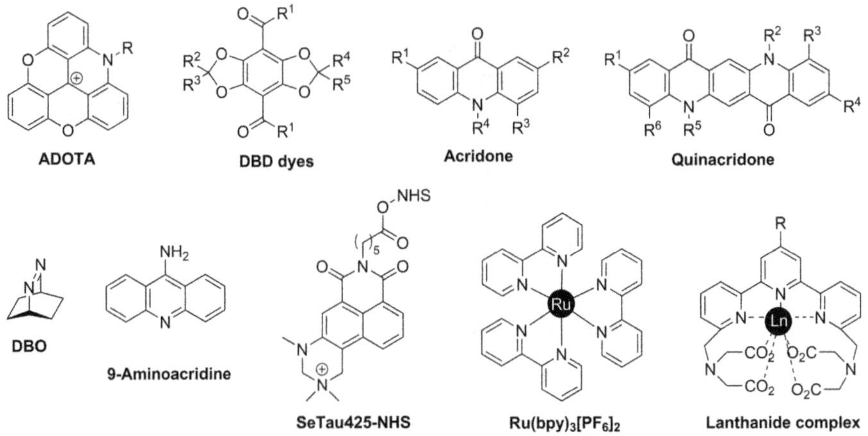

Fig. 5 Examples of representative long-lifetime luminophores

Several of the best known long-lifetime dyes are based on the acridine and acridone moieties. Dyes, such as acridone [38], 9-aminoacridine [40], lucigenin (a condensed di-acridine) [41] and 6-(9-oxo-9H-acridin-10-yl)-hexanoate (commercialized as PureTime®14) [39], exhibit fluorescence lifetime values larger than 15 ns. AssayMetrics' PureTime fluorescent dyes cover the UV-NIR spectral range. PureTime UV dyes have a fluorescence lifetime of 325 ns and are available as NHS esters for protein and peptide labelling. PureTime visible dyes have lifetimes ranging from 14 to 22 ns and are also used for labelling. Amersham Biosciences developed a family of acridones and quinacridones with long lifetimes that were specifically designed for FLIM microscopy [38]. 2,3-Diazabicyclo[2.2.2]oct-2-ene (DBO) also belongs to the long-lifetime family of dyes (13–825 ns) and can be used to perform time-resolved luminescence assays. However, DBO shows very low molar absorption coefficients, which reduces its overall sensitivity. Moreover, the absorption and emission wavelengths of DBO are very short, limiting its application in in vitro assays [45]. For instance, this dye has been extensively used as a fluorescent probe to study its inter- and intramolecular fluorescence quenching by a variety of biomolecules.

A new family of dyes based on 1,3-benzodioxole and [1,3]-dioxolo[4.5-f] benzodioxole, which are used as highly sensitive fluorescence lifetime probes, have also been used to describe their microenvironmental polarity. These dyes also exhibit long fluorescence lifetimes and have been employed to develop an assay for the discovery of inhibitors of enzymes belonging to the histone deacetylase family [44]. Similarly, the easily accessible and environment-sensitive EDANS fluorophore possesses a long fluorescence lifetime and has been widely employed in time-resolved luminescence assays [46]. For instance, azadioxatriangulenium dyes (commercialized by KU dyes®) are highly photostable and highly emissive and possess a fluorescence lifetime of approximately 15 ns [42, 43]. Azadioxatriangulenium (ADOTA) and diazaoxatriangulenium (DAOTA) are members of the triangulenium dye family and have been successfully employed in time-resolved luminescent experiments investigating complex biological systems.

Similarly, the patented SeTau dye family has attracted increasing interest in lifetime-based applications. Commercialized by SETA BioMedicals® (https://www.setabiomedicals.com), these fluorophores have a naphthalimide heterocyclic core that features a long fluorescence lifetime (in the range of 9 and 32 ns); a large Stokes shift (>100 nm); excellent chemical, thermal and photophysical stability; and water solubility [47, 48].

2.2.2 Transition Metal Complexes

During recent decades, luminescent transition metal complexes have emerged as an alternative to organic dyes in creating long-lifetime luminescence probes. These complexes are primarily based on the following d block metal ions: (a) Ru(II), Re(I), Ir(III), Rh(III) and Os(II) complexes with d^6 electronic structures; (b) Pt(II) complexes with d^8 electronic structures, such as metalloporphyrins [61]; and (c) Au(I) complexes with d^{10} electronic structures. The broad luminescence bands mostly arise from metal-to-ligand charge-transfer or metal-to-metal charge-transfer states. These dyes have attracted increasing research interest in the biosensing and bioimaging fields due to their advantageous properties, including high luminescence efficiency, tunable luminescence colours, significant Stokes shift, high photostability, relatively long emission lifetimes (100 ns to 10 µs), good water solubility and lack of dye-dye interactions [49, 50].

Metalloporphyrins are typically complexes of platinum or palladium. In particular, platinum is used due to its higher quantum yield in aqueous solutions and room temperature. The metalloporphyrins have reached a rather high level of development. However, the use of metalloporphyrins in time-resolved experiments is not as widespread as that of other dyes. The metalloporphyrins are commonly used as highly sensitive, selective and versatile labels or probes in biosensing applications [62, 63].

2.2.3 Lanthanide Complexes

The lanthanides include 14 elements from ^{58}Ce to ^{71}Lu. The spectroscopic properties of lanthanide complexes result from transitions between the states in the 4f sub-shell, which is shielded from the influence of the environment by the higher energy $6s^2$ and $5p^6$ orbitals.

Due to this shield, the 4f orbitals do not directly participate in chemical bonding, and lanthanide complexes display similar spectroscopic properties (characteristic of each metal ion) regardless of their chemical environment. Thus, the emission of lanthanides is minimally perturbed by the surrounding environment, resulting in sharp, line-like emission bands with the same fingerprint wavelengths and narrow peak widths of the corresponding free Ln(III) salts. Moreover, the f–f transitions are formally forbidden by the spin and Laporte rule and, thus, feature exceptionally long luminescence lifetimes—in the order of hundreds of microseconds to milliseconds—compared to those of classic organic dyes. Due to these outstanding photophysical properties, the lanthanides have attracted exceptional attention over the prior 30 years, and they continue to be the focus of further developments.

Since lanthanides are poor absorbers (f–f transitions are Laporte-forbidden), sensitizing chromophores (antenna) are required in their vicinity to achieve luminescence through an energy transfer process named the antenna effect [53, 54, 64]. The lanthanides Tb(III) and Eu(III) are commonly used in time-resolved luminescence assays since they can be excited by energy transfer using a variety of organic chromophores and emit relatively efficiently in the visible. The most common luminescent lanthanide probes covalently link the antenna to a chelate group containing the lanthanide (such as EDTA and DOTA) through a linker or spacer named a pendant antenna. An alternative approach involves the use of a chelating group as antennas, named chelating antennas or cryptates, for the excitation of the lanthanide ion. New and improved lanthanide complexes and cryptates are frequently reported in the literature, and researchers continue to search for large brightness and improved stability. By adding aromatic groups acting as antennas to the coordinating cage, the absorptivity and, thus, the luminescent properties of the group are largely enhanced [65]. Many different chelators can act as caging agents of Eu(III) and Tb(III) ions; thus, organic chemists have proposed the use of novel ligands to improve TG-imaging probes [66]. An in-depth understanding of the photophysical processes that govern the luminescence emission of lanthanide cryptates definitively helps in the rational design of new probes [67]. Usually, these new compounds are tested in validated TG fluoroimmunoassays. For example, new Eu(III) cryptates are tested in human C-reactive protein (hCRP) [68] or cardiac troponin I [69] TG immunoassays, among many others. The solubility and stability properties of the cryptates can be tuned to achieve specific aims. For instance, by deliberately decreasing the cell permeability of the probe [70], it can be optimized to study outer cell membrane receptors and their interaction with specific ligands using TG-imaging. The new probes are often patented for TG immunoassays or TG-imaging, such as the so-called EuroTracker dyes [71] and one of the most popular Tb(III)-based probes, i.e. the Lumi4-Tb, which is a 2-hydroxyisophthalamide-based cryptate commercialized by Lumiphore Inc. [65].

2.2.4   Luminescent Nanoparticles

The fabrication of materials on the nanometre scale paved the way for nanotechnological applications. Nanomaterials exhibit a variety of interesting features, some of them improved or unusual when compared to macroscopic materials or molecular species. For example, luminescent nanoparticles show the ability to emit radiation due to electronic transitions, with brighter luminescence, higher photostability and higher biocompatibility, compared to classical fluorescent organic dyes. Moreover, these nanoparticles can act as multivalent scaffolds for the realization of supramolecular assemblies since their high surface-to-volume ratio allows for distinct spatial domains to be functionalized, which can provide a versatile synthetic platform for the application of different sensing schemes. Due to their excellent properties, nanomaterials are among the most useful tools in biomedical research because they enable the intracellular monitoring of many different species for medical and biological purposes.

Semiconductor Quantum Dots (QDs)

Semiconductor quantum dots (QDs) and their bioconjugates consist of nanocrystals with diameters of 2–10 nm and generally contain elements from groups II and VI or groups III and V (e.g. ZnS, CdS, CdSe, PbSe, InAs or InP). If the size of the nanoparticles decreases below a critical value known as the exciton Bohr radius (typically 10 nm), the 3D confinement of charge carriers occurs, which restricts the energy states in the valence and conduction bands and results in optical properties that can be tuned by varying the particle size or internal chemical composition [55, 56]. Their emission wavelength can be finely tuned from 300 nm to 5 μm. QDs feature high quantum yields, resistance to photobleaching (even better than that of organic chromophores), broad absorption spectra and narrow and size-tunable photoluminescence emission spectra with fairly sharp emission bands covering the visible and NIR spectral ranges. Moreover, QDs present unique photoluminescence lifetime properties. QDs show multiexponential luminescence decays and comparatively long lifetimes (typically from five to hundreds of nanoseconds) that are longer than the autofluorescence decay of cells and the fluorescence lifetime of most conventional dyes [72]. Due to these optical properties, QDs are established photoluminescent platforms used for biological and sensing applications. The intracellular delivery of relatively large particles is a challenge in the use of QDs and other luminescent nanoparticles. Moreover, QDs have certain drawbacks, such as blinking of the emission if only a small number of QDs are present in the target material, nanocolloidal behaviour and controversial long-term toxicity issues [56].

Carbon Nanoparticles

Due to the possible toxicity of semiconductor QDs, the potential of carbon nanodots (CND or C-dots [57]) and nanotubes (CNT [73]) has been investigated. Carbon particles present characteristics and preferred benefits, such as no photobleaching, high thermal stability, extraordinary biocompatibility, low toxicity, easy derivatizability and no optical blinking, and their luminescence properties can be finely tuned by varying their size and/or surface. Furthermore, CNDs present large two photon excitation cross sections, paving the way for their use in photodynamic therapy and cell imaging [74]. The C-dots show multiexponential photoluminescence decays with average lifetime values of approximately 6–8 ns [57, 58, 75].

More recently, two emerging luminescent carbon nanomaterials, i.e. graphene oxide (GO) [76] and graphene quantum dots (GQDs) [77], have attracted increasing attention. GO is an atomically thin sheet of graphite that is covalently decorated with oxygen-containing functional groups either on the basal plane or its edges. GQDs are graphene sheets smaller than 100 nm that present unique optical and electronic properties due to their quantum confinement and edge effects. These materials have characteristics and advantages that are similar to those of C-dots, such as their ease of preparation and non-toxicity. Typically, graphene nanoparticles exhibit blue photoluminescence, and the luminescence lifetimes range from 5 to 7 ns [78, 79].

A new addition to the nanocarbon family is the photoluminescent nanoparticles of diamond. Photoluminescent nanodiamonds (PNDs) containing nitrogen-vacancy colour centres are a promising alternative to the family of inorganic photoluminescent probes due to the presence of embedded, perfectly photostable colour centres. These PNDs present a rather long radiative lifetime of nitrogen-vacancy colour centres, most of which is greater than 15 ns [80], rendering the PNDs appropriate for long-term imaging applications in vivo [81, 82].

Silicon Nanoparticles

Silicon nanomaterials constitute an important class of new materials with great application potential as intracellular probes. Photoluminescent porous silicon nanoparticles (SiNPs) have no toxicity due to the favourable biocompatibility of silicon and display interesting photoluminescence properties (e.g. strong luminescence and robust photostability). The luminescence lifetime of nanocrystalline silicon is on the order of microseconds (normally 10–30 μs), which is significantly longer than the nanosecond lifetimes exhibited by organic dyes, QDs and C-dots, allowing for improved visualization of biological systems [59, 83].

Other Nanoparticles

Doping is a broadly employed process that involves incorporating atoms or ions of suitable elements into host lattices to produce materials with tailored functions and properties [84]. The lifetime of dopant emission from lanthanide ions or transition metal ion-doped QDs is normally longer (from μs to ms) than that of the host, offering abundant opportunities to avoid background fluorescence in bioimaging and biosensing. Doped QDs maintain their inherent advantages while avoiding the self-quenching problem due to their considerable large Stokes shift. Two clear advantages of doped QDs, particularly doped ZnS QDs, over classical CdSe@ZnS and CdTe QDs are their longer lifetime and potentially lower cytotoxicity. In bioimaging applications, fluorescent dopants may avoid toxicity problems by generating visible or infrared emission in nanocrystals created with less-harmful elements than those currently used [85].

Neodymium-doped nanoparticles constitute another type of luminescent nanoparticles with fluorescence lifetimes of circa 100 ms that can be employed in fluorescence imaging in vivo [86]. Silica nanoparticles with encapsulated dyes, such as lanthanides or ruthenium complexes, have shown great potential as biolabels in various time-gated luminescence biodetection studies since they exhibit a long luminescence lifetime of approximately 350 μs [87, 88].

2.3 Applications

In Sect. 2.1, we described several of the earlier realizations of TG analyses, mainly in fluoroimmunoassays and TG-imaging. In the current and the following section, we focus on state-of-the-art applications using several of the luminophores described above to achieve ultrasensitive assays with multiplexing capabilities usually using nanotechnological approaches to address biologically, biochemically and biomedically relevant problems.

2.3.1 Immunoassays

As previously described, TR-FIA assays are among the most important applications of TG analyses that successfully provide physiological information. A major problem in early TR-FIA experiments was that the sensitivity of the assays was occasionally not sufficient for studying systems in which the target was expressed at low levels. Therefore, several modifications of the base analysis have been proposed to enhance the luminescence signal. The dissociation-enhanced lanthanide fluoroimmunoassay (DELFIA) method, in which the Eu(III) metal ion is released from the antibody once the reaction with the analyte is completed and protected in a micellar solution that enhances the luminescence emission, is among the most extended approaches used in clinical applications of TR-FIA [89]. The DELFIA assay has been employed to compare the performance of TR-FIA assays with that of conventional ELISAs in the detection of human tetanus antitoxin in serum [90]; the larger dynamic range in the TG analysis and the advantages of the TG approach in analyses of sera were demonstrated.

Similarly, multiplexing was another challenge in TG immunoassays. The initial realization of multiplexing in TR-FIA assays was reported by authors employing specific antibodies labelled with different luminophores. Examples of multiplexing approaches include the simultaneous detection of pregnancy-associated plasma protein A (PAPP-A) and the free β-subunit of human chorionic gonadotrophin [91]; the simultaneous detection of recombinant CP4 EPSPS and Cry3A proteins in plant sample extracts of genetically modified potatoes [92], using Eu(III) and Sm(III) cryptate-labelled antibodies; the simultaneous analysis of free and total prostate-specific antigen (PSA) [93], which is an important biomarker of prostate cancer; and the multiplex measurement of PSA along with the PSA-α1-antichymotrypsin complex [94].

Further advances in TR-FIA assays enhanced the signal and possibility of dual-recognition tests by exploiting Förster resonance energy transfer (FRET) from a donor luminophore to a nearby acceptor in a distance-dependent manner. For instance, a TG analysis of the energy transfer (TG-FRET) from a lanthanide cryptate (Eu(III) or Tb(III)) acting as a donor to a fluorescent acceptor has been commercialized by Cisbio Bioassays and is known as homogeneous time-resolved fluorescence (HTRF) [95]. The TG-FRET enhances the sensitivity and selectivity of the detection method due to the double-band emission and specific distance-dependence acceptor emission [96]. Similarly, the measurement of the luminescence signal at two wavelengths (donor and acceptor) involves an internal reference, which is used as a normalization method to avoid differences between samples or reagents [97]. Several formats have been proposed for the design of TG-FRET immunoassays (Fig. 6), including competitive assays and dual-recognition assays.

Furthermore, since the first realizations of a TG-FRET assay based on QDs acting as FRET acceptors [98, 99], many applications have involved the use of QDs as a central nanoplatform and coassembled with peptides or oligonucleotides that were labelled with either a long-lifetime luminescent lanthanide complex or a fluorescent

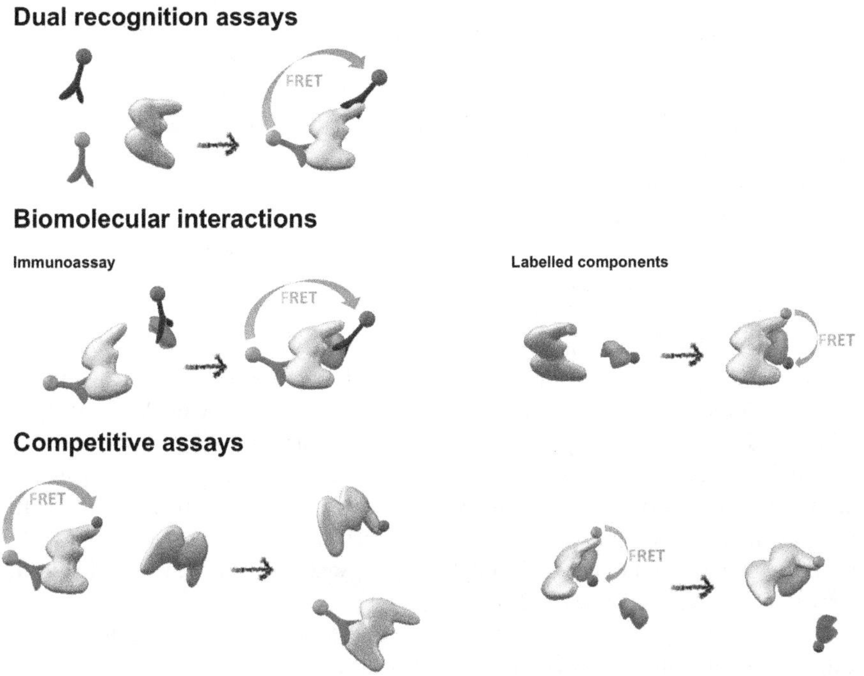

Dual recognition assays

Biomolecular interactions

Immunoassay   Labelled components

Competitive assays

Fig. 6 Different designs of TG-FRET assays used to study recognition, interactions or relative stability (competitive) using both immunoassays with labelled antibodies or directly labelled components

dye. The emission lifetime of the QD acceptors is drastically enhanced by the excitation of the energy transfer from the lanthanide complex, which permits TG-FRET signal processing for quantification and imaging. These systems also allow multiplexed biosensing based on the spectrotemporal resolution of the FRET process without requiring QDs with multiple colours [100].

2.3.2 Drug Discovery

The field of fluoroimmunoassays with clinical implications has been very active, and several companies have developed specific products, reagents and instruments for these experiments (see Sect. 3). However, TG analyses have also been widely applied in the field of drug discovery and high-throughput screening experiments. In particular, Cisbio's patented HTRF technology [95] was specifically designed for the drug discovery community.

One of the widest applications of TG-FRET and HTRF in drug discovery is related to sensing enzymatic activity, such as proteases or kinases. Kinases have attracted broad attention as validated drug targets; thus, studies investigating their activity in the presence of potential inhibitors are highly active in the field of drug discovery. Among many examples, this technique has been applied to COT kinase [101]; Rho-kinase II [102]; tyrosine kinases [103], such as Tie2 kinase [104]; and extracellular signal-regulated kinases (ERKs) [105]. A serendipitous discovery in the search for protein kinase inhibitors led to the development of kinase binding with turn-on sensors with long luminescence lifetimes without the presence of metals. The conjugates of adenosine analogues and arginine-rich peptides that include a thiophene or selenophene on one end and an energy acceptor on the other end show a striking enhancement in the emission lifetime to the hundreds of microseconds range upon binding to basophilic protein kinases. These probes, which are the so-called ACR-Lum probes, have been employed to detect kinase activity in cell extracts and live cells using TG spectroscopy and imaging [106, 107]. Many other enzymes can be studied using this methodology. For instance, the activity of heparanase, which is an enzyme with hyperactivity in malignant tumours, has been studied using HTRF assays [108]. Other recent examples of TG-FRET-based assays include the discovery of inhibitors of viral cap-methyltransferases as potential antiviral drugs [109] and inhibitors of aromatase enzyme for breast cancer treatment [110]. In an interesting multiplexing approach, the simultaneous activity of trypsin and chymotrypsin was probed using a two-step FRET relay design with FRET from a donor Tb(III) complex to a QD acceptor sensing one of the proteases, and the subsequent FRET from the QD to a red emitting fluorophore, i.e. Alexa Fluor 647, responded to the activity of the second protease [111].

Studies investigating G protein-coupled receptors (GPCRs) constitute another extensive branch that is clearly important for drug development. GPCRs represent the main mechanism of cell signalling and the recognition of extracellular molecules and, thus, have been identified as specific targets for drug development. Many different TR-FIA assays have been reported, including mechanistic and functional studies of GPCRs and specific inhibitors [95]; however, a thorough review of such reports is beyond the scope of this chapter. As an interesting example of the wide variety of studies, a TR-FIA was developed to understand the mechanistic interactions between specific GPCRs and components of traditional Chinese medicines employed as weight-loss drugs [112], which could establish a basis for metabolic disorder treatments.

An important advance in the study of GPCRs using TG-FRET experiments involves the possibility of directly tagging the studied receptor with an emissive label to avoid the use of antibody-antigen interactions. In particular, the SNAP-tag technology [113] allows for the covalent linkage of a long-lifetime luminescent probe to a functional protein. The SNAP-Lumi4-Tb reagent adds an optimized Tb-cryptate to the N-terminus of a GPCR usually without altering its function [114]. This technology has been used in TG-FRET assays to study ligand binding and competitive experiments in ApelinR [114], which is an important GPCR involved in body fluid homeostasis and cardiovascular functions; chemokine, opioid and cholecystokinin receptors [115]; the arginine-vasopressin V2 receptor, which is a drug target for the potential treatment of hypertension and several renal pathologies; and the growth hormone secretagogue receptor type 1a, which is a target for treating disorders of growth hormone secretion [116], among many other examples. A recent protocol compares the use of TG-FRET assays to study ligand binding in the parathyroid hormone receptor, which is a class B GPCR, using labelled antibodies to the use of the approach of directly using a SNAP-tag-labelled receptor [117].

2.3.3 Biomarker Discovery and Analysis

Luminescence spectroscopy offers a sensitive, easy-to-use, versatile and widespread platform for the quantitative determination of any potential analytes and biomarkers of biomedical relevance. Luminescence spectroscopy in the steady-state mode entails the use of a continuous illumination source. Under these conditions, several luminophores are promoted to the excited state and begin the emission process, whereas other molecules are subsequently excited. This approach rapidly reaches a steady situation in which the number of molecules in the excited and the ground states is in equilibrium. Under low absorbance conditions, the emitted intensity follows Kavanagh's law; thus, the total number of photons emitted is proportional to the concentration of the luminophore. The optimization of luminescent probes for specific targets is a very extensive field with the following common grounds: the emission features of the probes are modified by the presence of the analyte by any physical means in a concentration-dependent fashion. Due to their high sensitivity and versatility and relatively low cost, these techniques are among the first choices in developing sensors. Luminescent sensors can be designed by rendering the properties of such luminescent probes tunable based on the concentration of a secondary species (analyte). If long-lifetime probes are employed, TG analyses can be applied in time-delayed luminescence spectroscopy. This technique can be performed using most commercial fluorimetry spectrometers and a phosphorescence measurement accessory. The luminescence emission spectra can be obtained with a delay after an excitation pulse that is usually controlled by a mechanical chopper in the TG mode. Thus, sensors based on lanthanide complexes are generally excited by the transfer of intramolecular energy from aromatic residues that are close to the metal. Using this intramolecular transfer strategy, it has been possible to develop numerous sensors to probe a variety of analytes and biomolecules. Multiple sensing strategies have been proposed to design luminescent signals with direct responses to specific analytes. (1) The analyte can be incorporated into the coordination network, causing either an enhancement or quenching of the luminescence emission. (2) Similarly, the analyte may interact with the antenna to enhance the lanthanide excitation either covalently or noncovalently, or (3) the analyte may sequester a quenching group from the coordination network. (4) Moreover, lanthanide complexes whose conformations or structures change in the presence of an analyte can be rendered responsive by judicially positioning the antenna on the ligand. Using these approaches, many different sensors have been proposed in the literature, including sensors for metals; anions, such as fluoride, sulphide, bicarbonate, nitrate or phosphate; other species, such as thiols, ADP, ATP, NADH and hydroxyl radicals; and pH sensors. For excellent recent reviews on lanthanide-based sensors, see references [118, 119].

Sensed analytes that are related to specific biomedical applications may serve as biomarkers. The broad definition of biomarkers entails substances, structures or processes that can be directly measured in the body or its products (fluids or tissue) and used to predict normal biological functions or the outcomes of disease, effects of treatments or environmental exposure to chemicals or nutrients [120]. The search for novel, reliable and robust biomarkers of different pathological states has become the cornerstone of early diagnostics and personalized medicine. Examples of recent studies investigating novel biomarkers can be extensively found in the fields of cancer [121, 122], neurodegenerative disorders [123], cardiovascular disease [124], diabetes [125], renal function and kidney injury [126], etc.

The analysis of such biomarkers requires rapid, optimal point-of-care methodologies to provide timely diagnostics and healthcare. An early example is the use of a TG-based analysis to explore the potential of the pregnancy-associated plasma protein A (PAPP-A) in first-trimester pregnant women as a biomarker of Down syndrome [127]. In addition, a multiplexing method for the joint detection of the PAPP-A protein and free β-subunit of human chorionic gonadotrophin [91] was also proposed as a diagnostic test for Down syndrome. Importantly, these biomarkers were amply validated with several clinical tests in different countries [128, 129]. Interesting clinical applications have been feasible; for instance, plasma procalcitonin was proposed as a marker of postoperative sepsis in transplanted patients [130]. TG-FRET assays have also been employed to directly detect and quantify biomarkers, such as insulin and cortisol, which are directly related to drug discovery. For instance, the promotion of insulin production by potential antidiabetic drugs has been tested using an HTRF assay of insulin [131]. TG-FRET assays of cortisol have been developed to test the activity of the enzyme 11beta-hydroxysteroid dehydrogenase type 1 and inhibitors of that enzyme with therapeutic activity against type 2 diabetes [132]. Prostaglandin E2 is another biomarker that has been directly detected using a TG-FRET assay aiming to follow the activity of the corresponding synthase enzyme and potential inhibitors [133]. Regarding neurodegenerative diseases, a TG-FRET assay was used to detect amyloid-β peptides in brain tissues using an Eu cryptate as the donor and FRET towards an acceptor fluorophore, and this study confirmed the use of the levels of amyloid-β peptides as a biomarker of vascular dementia [134]. These examples illustrate the potential use of TG analyses for achieving rapid biomarker tests and improving the prospects of personalized medicine.

2.3.4   Biomolecular Interactions

TG analyses constitute an invaluable tool for more fundamental physiological and biochemical studies, such as studies involving biomolecular interactions. A variety of lanthanide-based biosensors reporting protein-protein and protein-nucleic acid interactions with increased intramolecular energy transfer and, therefore, luminescence upon binding to a target have been described. For instance, a highly sensitive terbium-based peptide sensor selective to RNA hairpin has been developed. Upon binding to its target, the peptide folds into an α-helical conformation that results in a large increase in luminescence [135]. A similar strategy was applied to the specific sensing of the oncogenic c-Jun transcription factor [136]. The modulation of the antenna effect by a Trp residue donor has also been successfully employed to sense post-translational modifications, such as phosphorylation. A lanthanide peptide-based biosensor has been developed to probe CDK4 kinase activity in complex media, such as melanoma cell extracts, by sensitizing a terbium complex with a unique tryptophan residue in an adjacent phosphoaminoacid binding moiety [137]. Furthermore, the intermolecular sensitization of lanthanide ions has been applied to the development of a terbium-chelating peptide sensor targeting cyclin A. Upon the interaction, the Tb^{3+} ion is placed close to a well-conserved Trp residue of cyclin A, resulting in efficient intermolecular terbium sensitization and, thus, an increase in luminescence [138].

TG-FRET approaches have been effectively exploited to study biomolecular interactions in many reports. The following basic designs used in these studies are shown in Fig. 6: either using labelled monoclonal antibodies or directly labelled components, i.e. the biomolecular interaction is probed by placing the two units, i.e. the FRET donor and acceptor, close to each other, which leads to an efficient energy transfer. TG analyses enhance the sensitivity and selectivity by discarding potential fluorescent artefacts. Since TG analyses are possible *in cellulo*, several tests have been developed to study the interaction and oligomerization of cell membrane proteins; in particular, many studies investigating GPCR interactions have been reported [95, 139]. The homo- and hetero-dimerization of GPCRs are important mechanisms for cell signalling by which the dimerized state may be the event that triggers the signalling cascade [140]. For example, protein-protein interaction studies have provided information regarding how certain proteins in the Epstein-Barr virus capsid interact with chemokine receptors in human B lymphocytes to alter the immunological response [141].

DNA recognition and hybridization assays entail a broad set of experiments involving biomolecular interactions. The TG detection of specific DNA sequences by hybridization using a capturing probe that is labelled with long-lifetime metal complexes was patented by J. R. Lakowicz in 2001 [142] due to the foreseen importance and potential of this type of analysis. Following these concepts, TG-based DNA detection through molecular interactions has rapidly grown.

Although the study of biomolecular interactions using TG spectrometry and TG plate readers has provided a substantial amount of important physiological information, the direct visualization of these interactions using imaging techniques represents an even more appealing approach, which is discussed in the following section.

2.3.5 TG-Imaging

In studies using luminescence microscopy, TG filtering has become an important advantage. TG analyses provide an extra layer of specificity to cellular imaging, enabling very high contrast imaging. TG-imaging can be easily performed using commercial microscopy systems, particularly if the luminescent probes are lanthanide-based, since the millisecond-lifetime values provide sufficient collection time because extremely rapid electronics are not required. The early realizations of TG-imaging occurred in the early 1990s [21, 28]. A pulsed light source in kHz repetition rates, a synchronized chopper and a streak camera are sufficient to modify commercial equipment [143, 144] at relatively reduced costs. Even multi-colour capabilities can be implemented with very few alterations [145, 146]. Without any additional instruments, considering the delay produced in raster scanning and mathematically accounting for the blurring effect of long-lifetime lanthanide probes, luminescence lifetimes in the millisecond time range can be obtained under a conventional confocal microscope [147]. However, in addition to the available simple approaches, more sophisticated methodologies have been developed, such as superresolution nanoscopy using TG detection in time-gated stimulated emission depletion (TG-STED) [148]. The characteristic fingerprint of the millisecond luminescence lifetimes of lanthanide complexes has also been exploited for the design of an automated microscope for the fast scanning of large areas to identify spots of interest that show luminescence that is detected at a specific TW. The so-called time-gated orthogonal scanning automated microscopy (TG-OSAM) is capable of automatic, unsupervised microsphere and cell counting [149].

It is important to highlight the differences between TG-imaging and FLIM microscopy because although both techniques use the time-resolved information of the emission decay, their main focus and aim substantially differ. In TG-imaging, the signal analysis focuses on applying the optimized detection TW, discarding the short-lived photons and obtaining the overall number of emitted luminescent photons at a given TW. Hence, the signal is the luminescence intensity but at a specific detection TW. In contrast, FLIM microscopy focuses on the determination of the luminescence lifetime, τ, at each pixel of the image. The main analytical signal is the concentration-independent τ value. However, FLIM microscopy is inherently a multidimensional technique because the total intensity emitted is the full area (integral) of the luminescence decay trace, and it also allows TG-imaging by reconstructing the intensity image with photons collected at a specific detection TW. Therefore, FLIM microscopy is a more advanced technique, because it contains the capabilities of TG-imaging. In contrast, the FLIM instrumentation is more intricate and expensive, and its use requires in-depth expertise. The advantages of TG-imaging over FLIM include its simplicity and ease of application using conventional, commercial equipment.

TG-imaging is mainly applied for the imaging and analytical sensing in the cellular interior without the problems of intrinsic cellular autofluorescence. Using emissive probes with long luminescence lifetimes, TG image filtering is particularly interesting in intracellular sensing [150]. The potential of semiconductor QD nanoparticles in TG biological imaging was first demonstrated in 2001 [151]. Subsequently, biocompatible QD nanoparticles have been used as time-gated bioimaging probes in different cancer cells [152, 153] to suppress cell autofluorescence and improve the signal. For instance, using specific nanosensors built with QDs, an FLIM imaging methodology was developed to probe the intracellular pH values [72]. Although this approach was a FLIM-based sensing method, the cellular autofluorescence was discarded by performing a TG analysis; thus, the luminescence lifetime values of the sensor were not altered by autofluorescence photons. Other pH sensors designed using mixtures of Eu(III) and Tb(III) cryptates have been applied intracellularly to probe pH changes in the cytoplasm [154] and lysosomes [155] using TG-imaging.

A widely employed method used for the specific sensing of analytes is the design of fluorogenic probes, i.e. probes prepared in a non-luminescent off state that turn into an emissive on state upon reacting with the analyte or a secondary directly related species. Many examples of fluorogenic probes specifically designed to enhance the sensitivity of TG analyses are available in the literature. For example, fluorogenic lanthanide-based probes have been reported for the intracellular detection of hypochlorous acid [156, 157], vitamin C [158], H_2S [159], biothiols [160] and singlet oxygen generation [161].

In addition to the intracellular sensing of small molecules, immunostaining techniques and TG-imaging permit the direct visualization and study of biomolecular interactions and functioning in living cells. TG-FRET imaging using lanthanide-QD-fluorophore multistep FRET relays has been used to image epidermal growth factor receptors (EFGR) via immunostaining and endosome imaging intracellularly [162]. Photoluminescent nanodiamonds have also been used to image HeLa cancer cells [81] and as probes for the intercellular transport of proteins in vivo [82]. Similarly, a multiplex TG-imaging approach was performed with the participation of researchers from PerkinElmer for the simultaneous imaging of oestrogen receptors and human epidermal growth factor receptors in human breast cancer tissue sections. Furthermore, with the participation of the company LumiSands Inc., TG-imaging was performed to discriminate SKOV3 cancer cells from A431 cancer cells

[163]. SiNPs exhibit long luminescence lifetimes in the microsecond time scale; thus, SiNPs have been conjugated to anti-HER2 antibodies for TG-imaging of SKOV3 cells, which overexpress HER2 membrane receptors. These cells were easily differentiated from A431 cancer cells, which overexpress EGFR receptors, that were immunostained with fluorescein, which is a fluorophore with a short fluorescence τ. Using a short and long detection TW, the different cells were identified in a mixed culture.

Although a comprehensive review is beyond the scope of this chapter, the abovementioned examples are representative of the vast potential of TG-imaging. TG analyses have attracted many researchers' attention due to their filtering capabilities of undesired interferences. In the following section, this aspect is discussed more in depth.

2.4 The Challenge of Luminescence Detection in Complex Media

As demonstrated by the applications described in the previous section, TG analyses performed in all solutions and heterogeneous media and the imaging of live cells have provided an enormous batch of information for biomedical studies that is particularly relevant for drug discovery, biomarker analyses and the understanding of physiological events. A common feature among these studies is that biological samples are particularly challenging due to complex matrix effects and the presence of potential interferences. Many factors can cause the matrix effect, and most causes are due to the high concentrations of multiple species in the solution. One important factor regarding the use of luminescence techniques in complex media is the large contribution of scattered light. Rayleigh and Raman scatter of excitation light inevitably occur in any fluorescence-based experiment. Importantly, if high concentrations of biomolecules and other disperse systems are present, the amount of scattered light is extensive. This extensive scattered light has two consequences. First, the scattered light can be detected by the detection device and mask the real luminescence emission or even saturate the detectors. Second, the matrix can produce scatter of the emitted luminescent photons, hence causing a lower signal than expected. Finally, the presence of species capable of absorbing light at the excitation wavelength should also be considered. If large concentrations of absorbers are present, the number of photons available to excite the probe of interest dramatically decreases, which concomitantly decreases the luminescence emission of such species. This decrease is the so-called inner filter effect.

Chemical interferences, which are species that may absorb the excitation light and emit luminescence radiation, may become a potential interference if spectral overlap occurs with the species of interest, which is another issue in such complex matrices. These emissive species are a constant factor in cellular imaging, overall gathered as the cell autofluorescence [6]. The main species causing cellular autofluorescence include NADPH, flavins, the emissive amino acids in all proteins (tryptophan and tyrosine residues), collagen, elastin, porphyrins in blood, B vitamins and chlorophyll in plant cells [164]. Interestingly, in-depth studies investigating these autofluorescence patterns in cells have demonstrated their usefulness in identifying different metabolic or pathological cellular states [165]. However, in other cellular imaging studies, these autofluorescence patterns may seriously hinder the application of luminescent sensors or physiological specific studies in live cells.

The selection of the appropriate detection TW in a TG analysis can filter these potential interferents. This time filtering is particularly powerful using lanthanide luminescence as the analytical signal because a long delay usually of several tens of microseconds between the excitation and detection window can be set to ensure that the detected photons are exclusively arising from the probe luminescence. A major advantage of TG analyses is that these analyses allow for homogeneous assays in complex media, such as cellular extracts. Many lanthanide-based sensors of small analytes have been tested in complex matrices, such as solutions containing high concentrations of disperse molecules, cell lysates or real aqueous environmental samples. For instance, lanthanide-based pH sensors have been successfully employed in solutions containing 0.4 mM of human serum albumin and in cell lysates [155]. Other cryptate sensors have been employed for the quantification of hydrogen sulphide in industrial waters and crude oil [166], and the results obtained were better than those obtained using conventional analysis methods, further demonstrating the value of TG analyses in industrial samples.

Another interesting advantage is that sensing luminophores can be included to support active materials, such as hydrogels or paper, in the design of sensing materials. In these applications, the TG approach avoids any interference from the supporting material, providing clean signals. Lanthanide-based sensors embedded in hydrogels have been reported for pH [167] and glutathione [168] TG-sensing. Interestingly, a pH-responsive hydrogel has been employed for the indirect detection of urease activity in biomedical applications [167]. Similarly, Tb-cholate hydrogels containing enzyme substrates have been embedded in paper discs for paper-based quantification of β-galactosidases and lipases [169]. Furthermore, a highly sensitive reactive sensing paper has been developed for the analysis of exhaled hydrogen sulphide using TG detection [170]. Other solid supports, such as quartz, are conventionally used in DNA hybridization and single-nucleotide polymorphism TG assays [171]. Regarding fluoroimmunoassays, several proposed TR-FIA assays could not be directly applied to plasma or whole blood samples. A proposed solution was to employ dry reagents immobilized in the wells [172], ready for the addition of the samples. Using the simple protocol of overnight drying and vacuum sealing, the TR-FIA assay of PAPP-A was successfully applied in serum, plasma and whole blood [127]. The sensing luminophores can also be embedded in polymer films. For instance, a very unique approach was proposed for the preparation of multiparametric sensing polymer films reactive simultaneously to oxygen and temperature. A temperature-sensitive film of poly(vinyl methyl ketone) containing an Eu(III) dye was combined with layers of oxygen-sensitive polystyrene film using a Pt-porphyrin dye. The use of multiple gated windows to discriminate luminescence decays from two different probes allowed for the multiplexing measurement of oxygen and temperature [173].

Furthermore, TG analyses provide a way to probe biomolecular interactions in solution in a context that is much closer to their real, complex intracellular environment. Even if a molecular interaction can be detected in vitro, the actual functional state may or may not be related to such interaction. Directly probing molecular interactions in a native tissular environment is a challenging task. However, the specificity provided by TG analyses allows these studies to be performed. For instance, a TG-FRET-based study reported the interactions between the protein p53 and different protein partners in cellular extracts [174]. Furthermore, the oligomerization state of oxytocin receptors in mammary glands of lactating rats was described using TG-FRET assays of full tissue [175]. To overcome the large autofluorescence in the full tissues, the assay was performed using a Tb(III) cryptate as the donor in the FRET process due to its very long luminescence lifetime, brightness and quantum yield.

TG-imaging has been shown to be useful for identifying specific targets in very complex matrices; for instance, this technique has been used for the identification of specific microorganisms in water and food safety inspections. In such applications, water dirt or food debris constitute examples of extremely challenging matrices that contain a large excess of particles or microorganisms that may not be the sought targets. In these samples, many microorganisms and interferent substances emit luminescent light across the whole spectral range which could mask the targets using conventional UV excitation. TG-imaging was clearly shown to overcome this problem in the direct identification of the very rare waterborne pathogens *Giardia lamblia* cysts in environmental water dirt and *Cryptosporidium parvum* oocysts in fruit juice concentrates; both microorganisms were labelled with an Eu(III) cryptate [144]. In fact, this procedure was automatized for the human-free detection of these microorganisms [176], which improved the efficiency of the current official EPA protocol [177]. Silica-encapsulated Eu(III) nanoparticles have been successfully used for the TG luminescence imaging detection of two environmental pathogens in highly fluorescent grape juice samples [87]. The results demonstrated the practical utility of the new nanoparticles as visible light-excited biolabels in TG luminescence bioassay applications.

Tissular and in vivo imaging directly benefit from the TG approach. Using nanosecond TG-imaging with triangulenium derivatives, Na,K-ATPase channels were imaged within highly autofluorescent retinal tissue sections from brown Norway rats [178]. Even the most challenging task of live organism imaging has been recently accomplished using TG-imaging and different luminophores, such as Eu(III) complexes in *Caenorhabditis elegans* [179]. An example of live organism sensing is the vitamin C burst detection in live, small planktonic crustacean, *Daphnia magna* [158], which is shown in Fig. 7. TG-imaging of tumour xenografts in live mice has also been accomplished due to the long luminescence lifetime of silicon nanoparticles [59, 83]. The combination of neodymium-doped nanoparticles and long ($\approx$100 μs) luminescence lifetimes and the incorporation of a pulse delayer into conventional infrared small animal imaging systems have allowed the

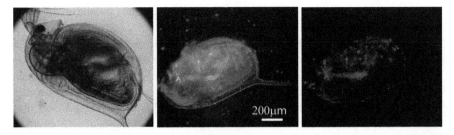

Fig. 7 Bright-field (left), full window (centre) and TG (right) luminescence images of *Daphnia magna* loaded with 1 mM vitamin C and incubated with 5.0 µM of a vitamin C, Eu(III)-cryptate sensor for 1 h. Modified from Song et al. [158] under the terms of the Creative Commons Attribution License. Copyright © 2015 Song et al

acquisition of autofluorescence-free live mice in vivo images [86]. The autofluorescence is discarded synergically by the TG-imaging approach, and the background contribution in the near-infrared (NIR) spectral region is negligible. Thus, this type of NIR-emitting nanoparticles with long luminescence lifetimes is a very promising multifunctional optical contrast agent with potential applications in numerous fields, such as in vivo three-dimensional fluorescence tomography, in vivo deep tissue therapies and real-time monitoring of thermal events in animal models.

These examples illustrate the excellent performance of TG analyses in homogeneous and heterogeneous assays and TG-imaging in complex media.

3   Time-Gated Detection of Luminescence in Industry

A great proportion of the applications described in the previous section were performed in an academic environment. However, as evidenced by the important challenges of such experiments, many research studies have rapidly transferred to industry. Several companies have been funded to develop novel luminophores, and these compounds are usually protected by exploitation patents [14, 15, 48, 71]. The development of improved luminophores for antibodies and enzyme substrates has introduced several business lines in different companies. For instance, Almac Sciences developed 9-aminoacridine peptide derivatives for a fluorescence lifetime-based assay of the activity of caspase-3, which is a cysteine protease that plays an essential role in apoptosis, using FRET from the fluorophore to a tryptophan sidechain in the peptide substrate [40]. Innotrac Diagnostics Oy developed improved Eu(III) chelates for quantitative TR-FIA assays [180].

Similarly, large worldwide companies dedicated to instrumentation development have focused on expanding the possibilities of fluorescence-based equipment (fluorescence spectrometers, plate readers, etc.) by incorporating time-resolved and TG capabilities. The interest in these techniques in the drug discovery field due to their pharmaceutical implications has increased the awareness of the possibility of entrepreneurship and business.

Several companies offer instruments based on time-resolved fluorescence for plate readers and spectrometers. PerkinElmer is among the most active companies in the field and has acquired technologies initially developed by Wallac Oy, Finland. PerkinElmer manufactures the VICTOR™ microplate reader series, which includes the benchtop multimode plate reader VICTOR Nivo™ system. PerkinElmer also manufactures the EnVision® 2105 Multimode Plate Reader, which is an ideal instrument for high-throughput screening due to its maximum sensitivity across all detection technologies. Previously, the AIO immunoanalyzer manufactured by Innotrac Diagnostics Oy measured time-resolved europium fluorescence in dried wells. Currently, Innotrac instruments are commercialized by Radiometer Medical Aps, which is located in Denmark. Other relevant instruments used in TG analyses with plate readers are listed in Table 2. Other companies offer time-resolved fluorescence spectroscopy instrumentation, such as Edinburgh Instruments (United Kingdom) with their LifeSpec II, Mini-tau and FLS100 and FS5 spectrofluorometer; Horiba Scientific (Japan), with a wide range of readers and spectrofluorometers; and PicoQuant GmbH (Germany), which is devoted to single-photon timing instrumentation, with the FluoTime series for lifetime measurements and MicroTime instruments for FLIM microscopy. Stanford Computer Optics (United States), Photonic Research Systems Ltd. (United Kingdom), LOT-QuantumDesign (Germany) and Photon Force Ltd. (United Kingdom) specialize in cameras capable of measuring time-resolved fluorescence.

In addition to these instruments, several assays based on TG detection are commercially available. The French company Cisbio commercializes homogeneous time-resolved fluorescence (HTRF®) assays [95, 181] as described above. These assays are broadly used to study kinase and protease activities based on TG-FRET. The so-called KinEASE assay consists of three biotinylated substrates, a monoclonal antibody labelled with an Eu(III) cryptate and FRET acceptor-labelled streptavidin [182]. This general assay has been validated in studies of hundreds of different kinases [181, 183]. For instance, in a joint application note by Cisbio Bioassays, BioTek Instruments Inc. and Enzo Life Sciences, the phosphorylation levels of ERK and the cAMP response element-binding protein (CREB) have been shown to dose-dependently decrease in SH-SY5Y cells treated with amyloid-β peptides [184], suggesting a physiological mechanism of cognitive impairment in Alzheimer's disease.

PerkinElmer commercializes the DELFIA® (dissociation-enhanced lanthanide fluorescence immunoassay), which is a TG intensity technology [185]. These assays were designed to detect the presence of a compound or biomolecule using lanthanide chelate-labelled reagents [186]. PerkinElmer has also developed two other different assays, i.e. the lanthanide chelate energy transfer (LANCE®) and LANCE Ultra TR-FRET assays. These two assays are simple, highly sensitive and highly reproducible immunoassays used to study cell cytotoxicity and cell proliferation. The LANCE assay is primary applied to measure caspase-3 activity using a high-throughput screening approach [187].

Table 2 Plate readers used for time-resolved luminescence measurements

Entry	Instrument	Manufacturer
1	Victor 1420 multilabel plate counter	PerkinElmer Life Sciences, Wallac OY, Turku, Finland
2	1230 Arcus fluorometer for time-resolved measurement of DELFIA enhanced fluorescence in tube format	LKB Wallac, Turku, Finland
3	1234 DELFIA fluorometer for time-resolved measurement of enhanced fluorescence in plate format	LKB Wallac, Turku, Finland
4	DELFIA platewash	PerkinElmer Life Sciences, Wallac OY, Turku, Finland
5	AIO® immunoanalyzer	Innotrac Diagnostics Oy, Turku, Finland
6	CLARIOstar® spectrofluorimeter	BMG Labtech, Ortenberg, Germany
7	PHERAstar® FSX spectrofluorimeter	BMG Labtech, Ortenberg, Germany
8	FLUOstar® Omega	BMG Labtech, Ortenberg, Germany
9	POLARstar® Omega	BMG Labtech, Ortenberg, Germany
10	Nanotaurus®	Edinburgh Instruments, United Kingdom
11	LF502 NanoScan® FLT-TRF	IOM, Germany
12	Synergy Neo2 Multi-Mode Reader	BioTek, Vermont, USA
13	Synergy H1 Multi-Mode Reader	BioTek, Vermont, USA
14	Synergy 2 Multi-Mode Microplate Reader	BioTek, Vermont, USA
15	Cytation 5 Cell Imaging Multi-Mode Reader	BioTek, Vermont, USA
17	EnVision 2105 multimode plate reader	PerkinElmer Life Sciences, Wallac OY, Turku, Finland
18	EnSpire™ multimode plate reader	PerkinElmer Life Sciences
19	EnSight™ Multimode Plate Reader	PerkinElmer Life Sciences
20	EnVision 2105 multimode plate reader	PerkinElmer Life Sciences
21	FlexStation 3 Multi-Mode Microplate Reader	Molecular Devices LL, Sunnyvale, California, Estados Unidos
22	TUNE-SpectraMax® Paradigm® Multi-Mode Microplate Detection Platform	Molecular Devices LL, Sunnyvale, California, Estados Unidos
23	Infinite® 200 PRO	TECAN, Zürich, Switzerland
24	Spark® Multi-Mode Microplate Reader	TECAN, Zürich, Switzerland
25	Varioskan Lux reader	Thermo Fisher, USA
26	SENSE multimodal plate readers	HIDEX (Finland)
27	Tristar^2 S LB 942 Multimode Microplate Reader	Berthold Technologies GmbH & Co. KG, Bad Wildbad, Germany
28	Mithras Multimode Microplate Reader LB 940	Berthold Technologies GmbH & Co. KG, Bad Wildbad, Germany

PerkinElmer has also developed a homogeneous, PCR-based assay with TG detection to simultaneously amplify and quantify DNA alleles using lanthanide probes and quenchers. This methodology, i.e. the so-called TruPoint-PCR and competitive TruPoint-PCR assays, was commercially released in 2004 [188]. Subsequently, this assay was improved by the company using nonoverlapping FRET acceptors in an anti-Stokes energy transfer [189]. Other companies have also contributed with new assays. For instance, researchers from GlaxoSmithKline have employed an HTRF screening platform to identify inhibitors of the NOD1, which is a receptor involved in several inflammatory disorders, signalling pathway [190]. Edinburgh Instruments proposed an efficient method for data treatment to avoid false positives in time-resolved FRET experiments investigating kinase activity [191]. Both the absolute values of activity and the kinetics of the interactions are important for the pharmaceutical industry in drug discovery [192]. Therefore, the company Bayer HealthCare proposed a TG-FRET-based method to probe drug-target association and dissociation rates [193] and applied this method in proof-of-concept experiments to investigate enzymatic activity, protein-protein interactions and G protein-coupled receptors (GPCRs), which are the three main current paradigms in drug discovery.

The company KinaSense LLC developed a family of time-resolved luminescence biosensors of protein kinases based on peptide substrates that enhance lanthanide ion luminescence upon phosphorylation, enabling the rapid, sensitive screening of kinase activity. These substrates chelate lanthanide ions directly upon phosphorylation, eliminating the need for chemical labelling with a separate lanthanide chelate and resulting in a higher lanthanide luminescence intensity and longer luminescence lifetime. These researchers used curated proteomic data from endogenous kinase substrates and known Tb(III)-binding sequences to build a generalizable in silico pipeline using tools that generate, screen, align and select potential phosphorylation-dependent Tb(III)-sensitizing substrates that are most likely to be kinase specific. This approach was used to develop several substrates that are selective to specific kinase families and amenable to high-throughput screening applications. Overall, this strategy represents a pipeline for developing efficient and specific assays for virtually any tyrosine kinase using high-throughput screening-compatible lanthanide-based detection. The tools provided in the pipeline also have the potential to be adapted to identify peptides for other purposes, including other enzyme assays or protein-binding ligands [194–197]. Recently, a more flexible strategy for a multiplexed, antibody-free kinase assay was reported using TG-FRET between QD nanoparticles and phosphorylation-dependent lanthanide-sensitizing peptide biosensors [195].

Finally, collaborative studies among companies have enriched the field through the design of novel assays based on TG analyses. For example, collaborations among the companies Biosyntan GmbH, Novartis Pharma AG and AssayMetrics Ltd. created a fluorescence lifetime-FRET-based assay for studying the activity of tyrosine kinases [198] and serine proteases [199] and the identification of protease inhibitors [39]. Amersham Biosciences, and subsequently, GE Healthcare collaborated to create a patented method for measuring enzymatic activity based on time-resolved fluorescence via acridone and quinacridone dyes [200, 201].

4 Detection of miRNA by Luminescence

4.1 miRNAs as Important Biomarkers

Cancer is a devastating family of diseases involving numerous interconnected biochemical processes and constitutes a medical challenge in our current society. Interestingly, the progress of the disease starts much early than its first clinical symptoms, highlighting the importance of early diagnosis using relevant, accessible and specific biomarkers. Rapid, non-invasive, multi-analyte tests represent a paradigm for early cancer diagnostics [202].

Recently, circulating microRNAs (miRNAs) have attracted much interest as valuable biomarkers due to the following important features: miRNAs are very specific to numerous diseases, their concentration is sensitive to the different pathological stages and they are easily accessible in biological fluids (plasma, serum, urine, etc.) obtained in liquid biopsies. miRNAs are small non-coding RNA fragments (18–22 nucleotides) whose main function is to regulate the expression of certain genes, normally by silencing them, when hybridized with the correspondent region of messenger RNA, resulting in the degradation or repression of the gene [203, 204]. In vertebrate organisms, up to 1000 different miRNAs regulate at least 30% of the genes. Currently, several pathological processes are known to cause miRNA deregulation, and significant differences exist between regular and pathological conditions. miRNA expression levels have been recently studied to identify specific biomarkers for breast [205], lung [206] and prostate cancer [207], diabetes [208] and leukaemia [209].

4.2 Steady-State Luminescence Detection of miRNA

Overall, quantitative PCR (qPCR) is the most popular technique used for miRNA profiling mainly due to its high sensitivity. qPCR is used in combination with the retro-transcription reaction (RT) to quantify the expression levels of specific miRNAs of interest. This technique comprises the following two steps: first, a reverse transcription from miRNA to cDNA and second, a qPCR. The following are the two most common strategies used for reverse transcription: polyadenylation of the miRNAs and the use of oligo (dT) primers and miRNA-specific stem-loop primers [210].

qPCR allows for the quantification of the synthesis of the PCR product at each amplification cycle in real time. This process allows for a quantitative analysis of the quantity of the initial product of reverse transcription (cDNA of target miRNA). The signal that is generated and quantified is represented by the fluorescence emitted by fluorescent dyes that bind the DNA molecules produced at each amplification cycle. qPCR signals can be generated using the following two different technologies:

1. Molecules intercalate into the synthesized DNA helix and show visible fluorescence staining only after they are incorporated into the neo-synthesized DNA strands. The emitted fluorescence of these molecules increases proportionally to the number of DNA strands produced. Therefore, the quantity of amplified product can be determined at each amplification cycle, and at the end of the phase of extension, the emitted radiation of the fluorophore can be detected. The most commonly used intercalator is SYBR® Green, which is an asymmetrical cyanine dye that is intercalated into the double strand of the DNA during the amplification reaction. The resulting DNA-SYBR® Green complex absorbs blue light at 488 nm and emits green light at 522 nm. During the denaturing phase, SYBR green is free in the reaction mixture; then, during the annealing phase, SYBR green is positioned in a nonspecific manner in the minor grooves of the DNA. During the elongation phase, the dye intercalates into the DNA molecule and, following excitation, emits fluorescence proportional to the number of copies of DNA produced during the amplification [211].

2. Fluorescent synthetic oligonucleotide constructs can selectively provide fluorescence to the amplified segments (probes). In this approach, the fluorescent signal is detected only as a result of the probe's hybridization with the DNA target under interrogation. The following two types of probes are typically used: (1) TaqMan probes and (2) hybridization probes. (1) TaqMan probes comprise dual-labelled oligomers with fluorophores at each end, a reporter and a fluorophore quencher. If the probe hybridized with the target DNA, the quencher is close to the fluorophore and blocks the emission of the fluorescent signal. During the elongation phase, in each amplification cycle, the $5'–3'$ exonuclease activity of Taq polymerase cleaves the dual-labelled probe. Thus, the reporter is released into the reaction mixture and moves away from the quencher, resulting in a fluorescence signal. In qPCR experiments using TaqMan probes, the fluorescent signal depends on the exonuclease activity of the Taq DNA polymerase. (2) Hybridization probes allow for the detection of the signal by hybridizing to the target sequence [212]. There are different models of hybridization probes, including a probe that exploits the energy transfer process. FRET probes are formed using two differently labelled probes, i.e. an oligonucleotide labelled at the $3'$-end with the donor dye and an oligonucleotide labelled at the $5'$-end with a FRET acceptor. The two probes are designed to hybridize with the target DNA. The detected signal is proportional to the quantity of hybrid probe; thus, the resulting fluorescence signal enables quantitative measurements of the accumulation of the product during PCR [213]. Other hybridization probe variants consist of beacon and scorpion probes [214, 215].

More recently, techniques that address the challenges of existing miRNA assays and maintain the high analytical sensitivity of qPCR have been reported. A major challenge in the detection of biomedically relevant amounts of miRNAs is the very low concentrations at which they are expressed and detected in bodily fluids. To solve this problem and achieve multiplexing analysis capabilities, Krylov and his team proposed a capillary electrophoresis separation using hybridization DNA probes labelled with a fluorescent dye and *drag tags* that vary the mobility of different miRNA targets and allow for the separation [216]. These authors further improved the method and boosted the sensitivity of the fluorimetric-based capillary electrophoresis assay using an isotachophoresis preconcentration step [217].

David M. Rissin and colleagues recently described a PCR-free method for the detection of miRNAs that integrates the dynamic chemical approach developed by DestiNA Genomics Ltd. [218–220] with the Simoa™ (Single Molecule Array) technology (Quanterix). A peptide nucleic acid (PNA) probe complementary to the miRNA sequence of interest was conjugated to superparamagnetic beads. These beads were incubated with the miRNA sample, and a biotinylated reactive nucleobase was added. When a target molecule with an exact match in sequence hybridized to the capture probe, the reactive nucleobase is covalently attached to the backbone of the probe by a dynamic covalent chemical reaction. Then, the single molecules of the biotin-labelled probe were labelled with streptavidin-β-galactosidase, and the beads were resuspended in a fluorogenic enzyme substrate, loaded into an array of femtolitre wells and imaged using zero-mode waveguides for single-molecule fluorescence imaging [221]. This dynamic chemical approach has also been applied to another fluorescent platform (Luminex xMAP®) for the successful detection of miRNAs [222].

Although sufficiently powerful, the examples described above address the problems derived from complex matrices using different approaches. TG analyses offer an additional layer of filtering that has also shown to be useful in miRNA detection, which is discussed in detail in the following section.

4.3   Time-Gated Luminescence Detection of miRNA

Routine and conventional analyses of circulating miRNAs are based on qPCR, which may introduce numerous artefacts and low reproducibility during the amplification process [223]. Therefore, identifying a reliable, robust and sensitive method for the quantification of miRNAs in biological fluids is warranted. An extremely sensitive technique, such the fluorimetry, in conjunction with a tool that removes the complex background signal of biological fluids, such as the TG detection method, may contribute to the use of miRNAs as successful biomarkers.

The requirement for simple, homogeneous assays capable of detecting low copy numbers of miRNAs in complex matrices led to the application of the TG approach for enhanced selectivity. A patent placed by J. R. Lakowicz for the detection of specific DNA sequences in complex matrices claimed several approaches under different configurations, including the TG detection of a capturing probe labelled with long-lifetime metal complexes, particularly of Ru, Os, Re, Rh, Ir, W or Pt [142]. However, very few variations of the methods proposed by Lakowicz are required for the application of such technology to miRNA detection.

A direct proof-of-concept experiment investigating the application of TG analyses to miRNAs was reported in 2012 by L. Jiang and colleagues [224]. In this study, a capturing probe was bound to magnetic beads, and an additional probe containing biotin was added. When the target miRNAs were detected, both the miRNA and the additional tagging probe were captured. Then, Eu(III) complex-labelled streptavidin was added for the TG analysis, which was performed using a time-resolved plate reader. Although not explicitly mentioned, the authors used the DELFIA protocol; thus, the Eu(III) signal was enhanced by the micellar solution [185, 225]. This perhaps could be the main reason for the very low limit of detection reported of 20 fM using clean, optimum solutions. Although the authors of this study mentioned the possibility of differentiating single point mutations, the validity of this claim must be considered carefully. Because a single common capturing probe was used, the multiplexing capability was not implemented. The results showed that at the same concentration, single or triple nucleotide mutants of let-7f miRNA exhibited approximately 20% of the signal of the target sequence. These results were used to support the specificity of the assay. However, because the assay was based on a single dye, this study was only feasible because the concentrations of the target and mutant sequences were previously known. In a real case scenario, using this approach, differentiating a certain concentration of target miRNA from a concentration five times larger in the mutant sequence would be impossible. In fact, the coexistence of different mutants could contribute to the total signal, and discriminating one signal from the others is impossible. Hence, although this study was a preliminary report, the low limit of detection achieved by these authors demonstrated the high potential of the TG analysis in miRNA analysis [224].

An important further step is the possibility of multiplexing several sequences in parallel and identifying single point mutations on the miRNA sequence. In a very powerful approach capable of the multiplex detection of different miRNA sequences in a single step, TG-FRET from a Tb-complex donor to either dye-fluorophore or QD acceptors of different colours has been proposed.

Niko Hildebrandt and his group demonstrated that QDs can act as efficient FRET acceptors from the Tb complex (Lumi4) and that using QDs of different sizes, low amounts of target biomolecules could be detected in parallel [226]. These authors applied this concept to the multiplex detection of three different miRNA sequences, i.e. miRNA-20a, miRNA-20b and miRNA-21 [227]. QDs of three different colours which carried a capturing DNA strand could hybridize to the Tb-complex-labelled reporter strand. Once the target miRNA sequence was also present, the full structure hybridized, and FRET was detected from the Tb complex to the QD (Fig. 8a). Each reporter was specific to an miRNA target, which facilitated the linkage to a specifically coloured QD. Using a parallel approach, the authors employed a similar concept in which the acceptors were cyanine dyes (Cy3.5, Cy5 and Cy5.5) tagging additional probes that could hybridize adaptor probes, connecting the Tb complex donor probe and the target miRNA. A T4 DNA ligase-catalysed reaction could fix the whole complex and achieve stable FRET from the Tb donor to the cyanine acceptor (Fig. 8b) [228]. In both cases, the authors used the FRET ratio, which is defined as the TG-filtered emission of the acceptor over the TG-filtered emission of the Tb complex donor, as the analytical parameter in a similar manner to the use of this parameter in the HTRF approach. The multiplexing determination required a

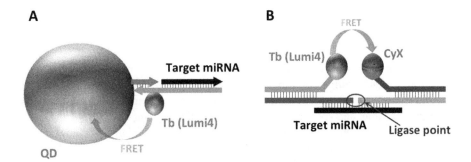

Fig. 8 TG-FRET approaches used for the detection of miRNAs using probes containing a Tb complex as donor and QD (**a**) [227] or cyanine dye (**b**) [228] acceptors

careful accounting of the spectral crosstalk between the different channels [229]. The limits of the detection of the three miRNAs were below 1.5 nM in a multiplexing detection of the three targets in 5% [228] or 10% [227] serum samples. This study confirmed the promising approach for the development of point-of-care, rapid, PCR-free homogeneous assays with potential clinical applications.

One problem in the homogeneous detection and quantification of miRNAs in biological fluids is their low copy numbers (concentration). Hence, the amplification of the luminescent signal is a valid technique to achieve a sensitive analysis that is free of enzymatic amplification reactions. Considering this approach, X. Y. Chen and colleagues reported a method for amplifying the signal of the target sequence using lanthanide-doped nanoparticles. Once the target sequence is detected, the lanthanide ions are released from the nanoparticles using mild acidic conditions, and the signal enhanced in micellar solution containing appropriate antenna moieties [230]. The authors applied this concept to the development of a highly sensitive assay for miRNA21 in which a molecular beacon opened upon binding the miRNA target, and a sensing sequence carrying biotin is also hybridized. Then, avidin and biotinylated Eu(III)-loaded nanoparticles were added to the system. After washing, the acidification and micellar enhancing solution were added, and the signal was read using TG luminescence spectrometry. Because each nanoparticle carries several Eu(III) ions, a single miRNA21 molecule provides a high lanthanide emission signal, allowing for the desired amplification for an ultrasensitive assay [231]. In this study, the authors reported an extraordinary limit of detection in the femtomolar range. However, these results must be carefully considered given the lack of linearity in the semi-log plots of the 10 fM–100 pM region of the miRNA concentrations. The authors also tested the specificity of the assay towards single- and triple-mismatched sequences at 1 nM concentration of all miRNA strands based on a decrease in the fluorescence signal, but differentiating between 0.5 nM of the correct target and 1 nM of the single-nucleotide polymorph is impossible, and the situation could be even more difficult if mixtures of the correct and mismatched sequence coexist. Therefore, further improvements in state-of-the-art methods for the detection and quantification of low concentrations of miRNAs are necessary for actual clinical application.

4.4   Time-Gated Detection of miRNAs Using FLIM

In this context, we sought to exploit commercial products with the strengths discussed in the previous sections: the well-known sensitivity of fluorescence techniques along with the ability of TG detection to easily remove background interferences in complex media and the increasing interest in miRNAs as biomarkers for early, accurate and specific diagnosis of several important diseases.

A simple approach to these ideas is to develop a *lab-on-a-chip*-based application to directly detect miRNA and avoid the use of amplification by polymerases. This application should fulfil the following industrial requirements for final commercialization: detection of relevant biomarkers; affordable materials, technology and distribution; easy application in final research centres, hospitals, etc.; non-expertise in the final user; and optimized usefulness/cost ratio. We suggested a simple design as shown in Fig. 9. The chip used as a platform supports a sample well in which the capturing beads are fixed. After the in situ reaction with hybridization probes (reporters) and subsequent washing steps, the chip may be ready for TG luminescence detection.

To achieve this goal, we collaborated with important companies in the fields of biotechnology (DestiNA Genomics) [232] and optoelectronics detection (Optoi Microelectronics) [233] to further implement these concepts through a collaborative consortium involving universities and companies from Germany, Italy, Spain and Brazil called *miRNA-DisEASY* [234].

As reported elsewhere, we considered two different approaches for the direct detection of miRNAs by fluorescence lifetime imaging microscopy (FLIM).

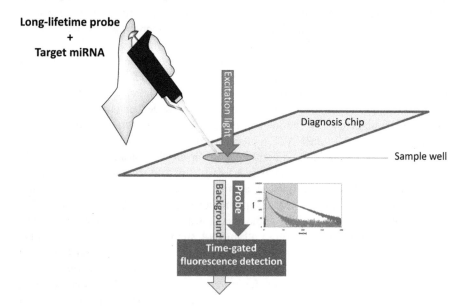

Fig. 9 Chip-based design for the direct detection of miRNA using a TG analysis

Both approaches employ a chemical reporter (capture probes) of the target miRNA and the use of peptide nucleic acids (PNA) that easily bind to the complementary oligonucleotide strand due to the absence of negative charges in their skeleton. For the fluorescent probe, we used SeTau425 from SETA BioMedicals, which is a chemically and photophysically stable fluorophore with a lifetime of 25 ns. Using the differences in the charges between the negative target miRNA and neutral reporter PNA, we use beads with a positively charged surface, Q Sepharose (QSph, GE Healthcare), to separate the unwanted and unreacted material from the target duplex. Fluorescence was detected by placing the sample over a coverslip in a FLIM confocal microscope (MicroTime 200, PicoQuant GmbH, Germany) using an excitation source of a 5 MHz pulsed, 440 nm laser, and the emission was collected using a 550/50 nm bandpass filter. After the acquisition, TG filtering is performed to eliminate all sources of undesired fluorescence mainly originating from the beads. We focused our attention on miRNA122, which is a miRNA used in vitro to assess the cellular toxicity of new drugs and as a biomarker for the diagnosis of severe liver failure.

In the first approach (A), a PNA probe (Rep1*) labelled with the fluorescent probe SETau425 is used to bind the target miRNA122 by hybridization. After the subsequent washing steps and separation of unreacted materials, the hybridized duplex (D1*: Rep1*-miRNA122) remains bound to the QSph beads. Target duplexes D1* were detected and analysed in the confocal microscope. Figure 10 shows FLIM images of QSph beads incubated in the presence and absence of D1* with different intensities and lifetimes. Expectedly, D1* has an average decay of

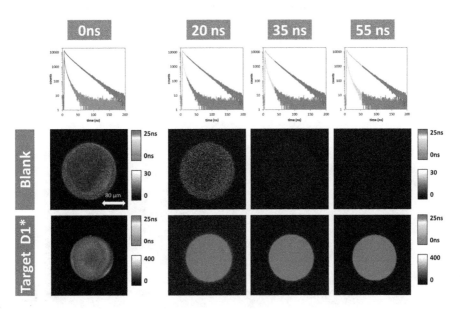

Fig. 10 Average decays of fluorescence and FLIM images at different TG values for a D1* 1 μM solution and a blank solution (QSph beads in phosphate buffer pH 6). The size of the images is 80×80 μm^2

fluorescence that is similar to that of the fluorescent probe, i.e. approximately 25 ns, and the blank solution of QSph autofluorescence. The loading of D1* onto the QSph beads is demonstrated by applying the principles of TG to remove the background fluorescence. Based on the different decay of the fluorescence, we can obtain the photons at a certain TW at which only SeTau425 fluoresces. Notably, the narrower the TW, the lower the number of detected photons. Hence, finding a compromise based on the background levels is recommended. The repetition rate of 5 MHz provides a total TW of 200 ns, and the application of TG from 55 ns completely eliminates the autofluorescence from QSph as shown in Fig. 10, while the D1* FLIM images preserve emission to a certain degree.

We sought to determine the limit of detection using our current FLIM setup in the described approach. The output signal strongly depends on the following parameters: the selected analysis TW as shown in Fig. 10, the time of acquisition per pixel, the loading of duplexes per bead, the time of incubation, etc. We optimized these parameters to achieve a better performance as follows: selection of TG of 55 ns and greater, the use of a single bead, increasing the time of acquisition per pixel and optimizing the time of incubation. Figure 11 shows the incubation of a 10 nM D1* solution without stirring using a single bead for capturing. The changes in the intensity and lifetime indicate the loading of D1* onto QSph beads over time. The detection is clearly visualized by applying different analysis TWs to the acquired photons.

We prepared different solutions of D1* (100 pM–1 μM) and performed an analysis using the described optimized approach with a detection TW of 55 ns onwards. We detected and distinguished up to subnanomolar concentrations of D1* (Fig. 12).

DestiNA Genomics is a company [235] that specializes in nucleic acid testing with proprietary technology. This company has developed methodologies using PNA capture probes combined with aldehyde modified nucleobases, i.e. the so-called SMART bases, to detect oligonucleotides and single-nucleotide

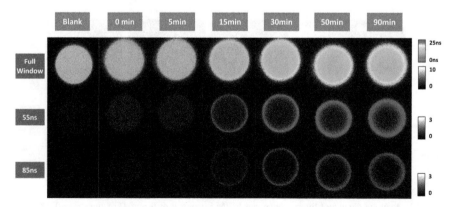

Fig. 11 FLIM images of a 10 nM D1* solution (PB pH 6) directly incubated with a QSph bead at different times using two different analysis TWs. The size of the images is 80×80 μm^2

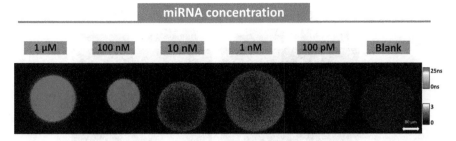

Fig. 12 TG-FLIM images (detection TW 55 ns onwards) of different D1* solutions compared with a blank solution only containing beads in phosphate buffer at a pH of 6. The size of the images is $80 \times 80\ \mu m^2$

polymorphism. DestiNA Genomics' patented dynamic chemistry is capable of not only directly detecting miRNAs but also providing single-base resolution [221, 236], adding an extra source of specificity to analyses of miRNAs. This approach is a differential and unique feature of our method compared to other approaches used for TG-based detection of miRNAs.

Using this approach, we studied the recognition of two different miRNA122 sequences. One sequence, i.e. miRNA122-18G, has a guanidine nucleobase at a certain position of the sequence, while the other, i.e. miRNA122-18A, bears an adenine at the same position. Notably, approach A cannot differentiate among the single point mutations on the miRNA as demonstrated in Fig. 13. However, using the DestiNA advanced methodology, we suggested a different approach (B) in which an unlabelled PNA probe (Rep2) complementary to miRNA122 is modified at a certain position with an absent nucleobase that is substituted for a reactive amine, thus creating a *pocket* in probe Rep2. In addition, a SMART cytosine nucleobase (SMBC) is labelled with SeTau425 (SMBC*). Rep2 and SMBC* are added in a single step to the solution containing the target miRNA and a reducing agent, i.e. NaBH$_3$CN. In either miRNA122 sequence, Rep2 and miRNA122 hybridize (D2), but only if the target miRNA contains a guanidine position in front of the pocket of Rep2, SMBC* dynamically enters the pocket and covalently is fixed with NaBH$_3$CN to D2*. Hence, the labelled duplex D2* only forms with miRNA122-18G but not with miRNA122-18A. Capturing and separating the unreacted material are performed similarly using QSph beads that are subsequently imaged using confocal FLIM microscopy.

In contrast to approach A, our results using approach B demonstrate the ability of SMBC* along with Rep2 to distinguish miRNA122-18G from miRNA122-18A and highlight the challenge of detecting single point mutations using conventional hybridization techniques. As shown in Fig. 13, the TG-FLIM images demonstrate the detection of miRNA122-18G using both methodologies. Similarly, miRNA122-18A binds Rep1*, yielding a positive result in approach A, while SMBC* hinders the dynamic incorporation into Rep2, resulting in a negative result in approach B.

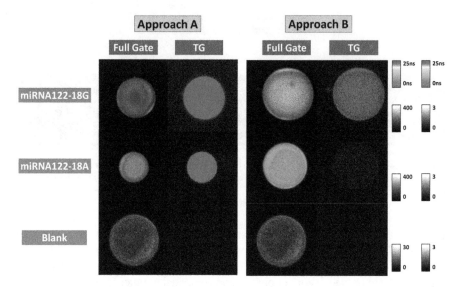

Fig. 13 Comparison of approaches A and B differentiating single point mutations in miRNA122. Target oligonucleotides, i.e. miRNA122-18G and miRNA122-18A, were reacted with Rep1* (approach A) and Rep2 + SMBC* (approach B) for final D1* and D2 concentrations of 1 μM. FLIM images of the captured duplexes in QSph beads and a control blank solution are shown. FLIM images at the full detection window are compared with TG-FLIM images (detection TW of 55 ns onwards). The size of the images is $80 \times 80 \ \mu m^2$

The combination of advanced chemical recognition and TG fluorescence detection represents a potential and promising methodology for directly detecting low concentrations of miRNAs at a single-nucleobase resolution, which is, indeed, a future pathway for in situ detection kits for liquid biopsies and other sources of low amounts of miRNAs, including complex matrices. In addition, this methodology has the potential to simultaneously interrogate certain positions in oligonucleotides using four different SMART nucleobases functionalized with four different fluorophores, allowing for multiplexing determinations. Thus, four fluorophores with different lifetimes could be required.

5 Outlook

The emission of luminescent photons is a multidimensional phenomenon involving several aspects, including emission kinetics. Time-resolved luminescence acquisition allows for the detection of photons depending on the time of emission after an excitation pulse and the discrimination of emitters with similar steady-state profiles but different lifetimes. Thus, time-gated acquisition filters short- and long-lived components of luminescence by applying certain detection TW during the detection, which is particularly useful for analysing species of interest in complex media that have a usually short-lived background intensity with undesired contribution to the total signal.

Advances in pulsed excitation and ultrafast acquisition techniques have led to the development of time-resolved fluorescence microscopy for the investigation of complex samples, including biological processes in solution, tissues and cells. The resolution and contrast of microscopy images have been improved using TG acquisition approaches due to the elimination of short-lived fluorescence background. Thus, the improvement of long-lived fluorophores is also crucial, and much progress has been achieved using organic, metal-complex, lanthanides, nanoparticles and other luminophores. This progress has attracted much interest in industrial and technological areas, which has led to the development of new instruments, assays and probes for biotechnological applications.

Due to recent advances in TG fluorescence lifetime microscopy, this instrumentation has become widely accessible, allowing for its application to resolve cutting-edge problems of biological interest, in particular, the detection of new precise miRNA biomarkers. The combination of fluorescence imaging sensitivity and the discrimination of background interferences by TG acquisition paves the path for direct and in situ detection of analytes in biological fluids obtained in liquid biopsies. In the future, these technologies could provide breakthrough diagnostic and prognostic methods for early screening of biological samples across the population, allowing for the detection of an eventual cancer and, therefore, increasing the chances of the survival of patients.

Acknowledgements The authors acknowledge funding from the European Union's Horizon 2020 research and innovation programme under the Marie Sklodowska-Curie grant agreement No 690866 (miRNA-DisEASY) and grants CTQ2017–85658-R from the Spanish Ministry of Economy and Competitiveness and the European Regional Development Fund (ERDF).

References

1. Lakowicz JR (2006) Principles of fluorescence spectroscopy3rd edn. Springer, New York. https://doi.org/10.1007/978-0-387-46312-4
2. Valeur B, Berberan-Santos MN (2012) Molecular fluorescence: principles and applications. Wiley, Weinheim. https://doi.org/10.1002/9783527650002
3. Williams RT, Bridges JW (1964) Fluorescence of solutions: a review. J Clin Pathol 17:371–394
4. Demas JN (1983) Excited state lifetime measurements. Academic, New York
5. O'Connor DV, Phillips D (1984) Time-correlated single photon counting. Academic, New York
6. Berezin MY, Achilefu S (2010) Fluorescence lifetime measurements and biological imaging. Chem Rev 110:2641–2684. https://doi.org/10.1021/cr900343z
7. Schneckenburger H (1985) Fluorescence techniques in biotechnology. Trends Biotechnol 3:257–261. https://doi.org/10.1016/0167-7799(85)90025-3
8. Schneckenburger H, Seidlitz HK, Eberz J (1988) New trends in photobiology. J Photochem Photobiol B 2:1–19. https://doi.org/10.1016/1011-1344(88)85033-4

9. Lehn JM (1978) Cryptates: the chemistry of macropolycyclic inclusion complexes. Acc Chem Res 11:49–57. https://doi.org/10.1021/ar50122a001
10. Nobelprize.org. Nobel Media AB (2014) Jean-Marie Lehn - facts. http://www.nobelprize.org/nobel_prizes/chemistry/laureates/1987/lehn-facts.html. Accessed 28 Dec 2017
11. Sabbatini N, Guardigli M, Lehn J-M (1993) Luminescent lanthanide complexes as photochemical supramolecular devices. Coord Chem Rev 123:201–228. https://doi.org/10.1016/0010-8545(93)85056-A
12. Dickson EF, Pollak A, Diamandis EP (1995) Ultrasensitive bioanalytical assays using time-resolved fluorescence detection. Pharmacol Ther 66:207–235. https://doi.org/10.1016/0163-7258(94)00078-H
13. Gudgin Dickson EF, Pollak A, Diamandis EP (1995) Time-resolved detection of lanthanide luminescence for ultrasensitive bioanalytical assays. J Photochem Photobiol B 27:3–19. https://doi.org/10.1016/1011-1344(94)07086-4
14. Mathis G, Lehn JM (1986) Macropolycyclic complexes of rare earth metals and application as fluorescent labels. Patent number: EP0180492 A1
15. Lehn JM, Mathis G, Alpha B, Deschenaux R, Jolu E (1996) Rare earth cryptates, processes for their preparation, synthesis intermediates and application as fluorescent tracers. Patent number: US5534622 A
16. Siitari H, Hemmilä I, Soini E, Lövgren T, Koistinen V (1983) Detection of hepatitis B surface antigen using time-resolved fluoroimmunoassay. Nature 301:258–260. https://doi.org/10.1038/301258a0
17. Pettersson K, Siitari H, Hemmilä I, Soini E, Lövgren T, Hänninen V, Tanner P, Stenman UH (1983) Time-resolved fluoroimmunoassay of human choriogonadotropin. Clin Chem 29:60–64
18. Suonpää MU, Lavi JT, Hemmilä IA, Lövgren TN (1985) A new sensitive assay of human alpha-fetoprotein using time-resolved fluorescence and monoclonal antibodies. Clin Chim Acta 145:341–348. https://doi.org/10.1016/0009-8981(85)90044-0
19. Lövgren TNE (1987) Time-resolved fluoroimmunoassay of steroid hormones. J Steroid Biochem 27:47–51. https://doi.org/10.1016/0022-4731(87)90293-7
20. Bertoft E, Eskola JU, Näntö V, Lövgren T (1984) Competitive solid-phase immunoassay of testosterone using time-resolved fluorescence. FEBS Lett 173:213–216. https://doi.org/10.1016/0014-5793(84)81049-2
21. Beverloo HB, van Schadewijk A, van Gelderen-Boele S, Tanke HJ (1990) Inorganic phosphors as new luminescent labels for immunocytochemistry and time-resolved microscopy. Cytometry 11:784–792. https://doi.org/10.1002/cyto.990110704
22. Beverloo HB, van Schadewijk A, Bonnet J, van der Geest R, Runia R, Verwoerd NP, Vrolijk J, Ploem JS, Tanke HJ (1992) Preparation and microscopic visualization of multicolor luminescent immunophosphors. Cytometry 13:561–570. https://doi.org/10.1002/cyto.990130603
23. Marriott G, Clegg RM, Arndt-Jovin DJ, Jovin TM (1991) Time resolved imaging microscopy. Phosphorescence and delayed fluorescence imaging. Biophys J 60:1374–1387. https://doi.org/10.1016/S0006-3495(91)82175-0
24. Marriott G, Heidecker M, Diamandis EP, Yan-Marriott Y (1994) Time-resolved delayed luminescence image microscopy using an europium ion chelate complex. Biophys J 67:957–965. https://doi.org/10.1016/S0006-3495(94)80597-1
25. Cubeddu R, Taroni P, Valentini G, Canti G (1992) Use of time-gated fluorescence imaging for diagnosis in biomedicine. J Photochem Photobiol B 12:109–113. https://doi.org/10.1016/1011-1344(92)85023-N
26. Cubeddu R (1993) Time-gated imaging system for tumor diagnosis. Opt Eng 32:320. https://doi.org/10.1117/12.60754
27. Cubeddu R, Taroni P, Valentini G, Ghetti F, Lenci F (1993) Time-gated fluorescence imaging of Blepharisma red and blue cells. Biochim Biophys Acta 1143:327–331. https://doi.org/10.1016/0005-2728(93)90204-S

28. Seveus L, Väisälä M, Syrjänen S, Sandberg M, Kuusisto A, Harju R, Salo J, Hemmilä I, Kojola H, Soini E (1992) Time-resolved fluorescence imaging of europium chelate label in immunohistochemistry and *in situ* hybridization. Cytometry 13:329–338. https://doi.org/10.1002/cyto.990130402

29. Schneckenburger H, Feyh J, Götz A, Jocham D, Unsöld E (1986) Time-resolved fluorescence of hematoporphyrin derivative in tumor cells and animal tissues. In: Waidelich W, Kiefhaber P (eds) Laser/optoelectronics in medicine/laser/optoelektronik in der medizin. Springer, Berlin, pp 70–73

30. Schneckenburger H, Koenig K, Dienersberger T, Hahn R (1994) Time-gated microscopic imaging and spectroscopy. In: Fercher AF, Lewis A, Podbielska H, Schneckenburger H, Wilson T (eds) Microscopy, holography, and interferometry in biomedicine. SPIE, Budapest, p 124

31. Koenig K (1994) Time-gated microscopic imaging and spectroscopy in medical diagnosis and photobiology [also Erratum 33(11)3828(Nov1994)]. Opt Eng 33:2600. https://doi.org/10.1117/12.177101

32. Kohl M, Neukammer J, Sukowski U, Rinneberg H, Wöhrle D, Sinn HJ, Friedrich EA (1993) Delayed observation of laser-induced fluorescence for imaging of tumors. Appl Phys B Lasers Opt 56:131–138. https://doi.org/10.1007/BF00332192

33. Lakowicz JR, Szmacinski H, Nowaczyk K, Johnson ML (1992) Fluorescence lifetime imaging of free and protein-bound NADH. Proc Natl Acad Sci U S A 89:1271–1275. https://doi.org/10.1073/pnas.89.4.1271

34. Lakowicz JR, Szmacinski H, Nowaczyk K, Berndt KW, Johnson M (1992) Fluorescence lifetime imaging. Anal Biochem 202:316–330. https://doi.org/10.1016/0003-2697(92)90112-K

35. Gadella TWJ, Jovin TM, Clegg RM (1993) Fluorescence lifetime imaging microscopy (FLIM): spatial resolution of microstructures on the nanosecond time scale. Biophys Chem 48:221–239. https://doi.org/10.1016/0301-4622(93)85012-7

36. Becker W (2012) Fluorescence lifetime imaging--techniques and applications. J Microsc 247:119–136. https://doi.org/10.1111/j.1365-2818.2012.03618.x

37. Birks JB, Dyson DJ, Munro IH (1963) 'Excimer' fluorescence. II. Lifetime studies of pyrene solutions. Proc R Soc A 275:575–588. https://doi.org/10.1098/rspa.1963.0187

38. Smith JA, West RM, Allen M (2004) Acridones and quinacridones: novel fluorophores for fluorescence lifetime studies. J Fluoresc 14:151–171

39. Boettcher A, Gradoux N, Lorthiois E, Brandl T, Orain D, Schiering N, Cumin F, Woelcke J, Hassiepen U (2014) Fluorescence lifetime-based competitive binding assays for measuring the binding potency of protease inhibitors in vitro. J Biomol Screen 19:870–877. https://doi.org/10.1177/1087057114521295

40. Maltman BA, Dunsmore CJ, Couturier SCM, Tirnaveanu AE, Delbederi Z, McMordie RAS, Naredo G, Ramage R, Cotton G (2010) 9-Aminoacridine peptide derivatives as versatile reporter systems for use in fluorescence lifetime assays. Chem Commun 46:6929–6931. https://doi.org/10.1039/c0cc01901a

41. Ruedas-Rama MJ, Orte A, Hall EAH, Alvarez-Pez JM, Talavera EM (2012) A chloride ion nanosensor for time-resolved fluorimetry and fluorescence lifetime imaging. Analyst 137:1500–1508. https://doi.org/10.1039/c2an15851e

42. Bora I, Bogh SA, Rosenberg M, Santella M, Sørensen TJ, Laursen BW (2016) Diazaoxatriangulenium: synthesis of reactive derivatives and conjugation to bovine serum albumin. Org Biomol Chem 14:1091–1101. https://doi.org/10.1039/c5ob02293b

43. Sørensen TJ, Thyrhaug E, Szabelski M, Luchowski R, Gryczynski I, Gryczynski Z, Laursen BW (2013) Azadioxatriangulenium: a long fluorescence lifetime fluorophore for large bio-molecule binding assay. Methods Appl Fluoresc 1:025001. https://doi.org/10.1088/2050-6120/1/2/025001

44. Wawrzinek R, Ziomkowska J, Heuveling J, Mertens M, Herrmann A, Schneider E, Wessig P (2013) DBD dyes as fluorescence lifetime probes to study conformational changes in proteins. Chemistry 19:17349–17357. https://doi.org/10.1002/chem.201302368

45. Nau WM, Greiner G, Rau H, Wall J, Olivucci M, Scaiano JC (1999) Fluorescence of 2,3-diazabicyclo[2.2.2]oct-2-ene revisited: solvent-induced quenching of the n,π*-excited state by an aborted hydrogen atom transfer. J Phys Chem A 103:1579–1584. https://doi.org/10.1021/jp984303f

46. Petersen KJ, Peterson KC, Muretta JM, Higgins SE, Gillispie GD, Thomas DD (2014) Fluorescence lifetime plate reader: resolution and precision meet high-throughput. Rev Sci Instrum 85:113101. https://doi.org/10.1063/1.4900727

47. Patsenker LD, Tatarets AL, Povrozin YA, Terpetschnig EA (2011) Long-wavelength fluorescence lifetime labels. Bioanal Rev 3:115–137. https://doi.org/10.1007/s12566-011-0025-2

48. Patsenker LD, Yermolenko IG, Fedyunyaeva IA, Obukhova YN, Semenova ON, Terpetschnig EA (2011) Highly water-soluble, cationic luminescent labels. US 2011/0143387A1. https://patents.google.com/patent/US20110143387A1/en?oq=US+2011%2f0143387A1

49. Zhao Q, Huang C, Li F (2011) Phosphorescent heavy-metal complexes for bioimaging. Chem Soc Rev 40:2508–2524. https://doi.org/10.1039/c0cs00114g

50. Ma D-L, He H-Z, Leung K-H, Chan DS-H, Leung C-H (2013) Bioactive luminescent transition-metal complexes for biomedical applications. Angew Chem Int Ed Engl 52:7666–7682. https://doi.org/10.1002/anie.201208414

51. Browne WR, Coates CG, Brady C, Matousek P, Towrie M, Botchway SW, Parker AW, Vos JG, McGarvey JJ (2003) Isotope effects on the picosecond time-resolved emission spectroscopy of tris(2,2′-bipyridine)ruthenium (II). J Am Chem Soc 125:1706–1707. https://doi.org/10.1021/ja0289346

52. Terpetschnig E, Dattelbaum JD, Szmacinski H, Lakowicz JR (1997) Synthesis and spectral characterization of a thiol-reactive long-lifetime Ru(II) complex. Anal Biochem 251:241–245. https://doi.org/10.1006/abio.1997.2253

53. Heffern MC, Matosziuk LM, Meade TJ (2014) Lanthanide probes for bioresponsive imaging. Chem Rev 114:4496–4539. https://doi.org/10.1021/cr400477t

54. Eliseeva SV, Bünzli J-CG (2010) Lanthanide luminescence for functional materials and bio-sciences. Chem Soc Rev 39:189–227. https://doi.org/10.1039/b905604c

55. Silvi S, Credi A (2015) Luminescent sensors based on quantum dot-molecule conjugates. Chem Soc Rev 44:4275–4289. https://doi.org/10.1039/c4cs00400k

56. Wegner KD, Hildebrandt N (2015) Quantum dots: bright and versatile in vitro and in vivo fluorescence imaging biosensors. Chem Soc Rev 44:4792–4834. https://doi.org/10.1039/c4cs00532e

57. Baker SN, Baker GA (2010) Luminescent carbon nanodots: emergent nanolights. Angew Chem Int Ed Engl 49:6726–6744. https://doi.org/10.1002/anie.200906623

58. Dekaliuk MO, Viagin O, Malyukin YV, Demchenko AP (2014) Fluorescent carbon nanomaterials: "quantum dots" or nanoclusters? Phys Chem Chem Phys 16:16075–16084. https://doi.org/10.1039/c4cp00138a

59. Joo J, Liu X, Kotamraju VR, Ruoslahti E, Nam Y, Sailor MJ (2015) Gated luminescence imaging of silicon nanoparticles. ACS Nano 9:6233–6241. https://doi.org/10.1021/acsnano.5b01594

60. Lavis LD, Raines RT (2008) Bright ideas for chemical biology. ACS Chem Biol 3:142–155. https://doi.org/10.1021/cb700248m

61. Soini AE, Seveus L, Meltola NJ, Papkovsky DB, Soini E (2002) Phosphorescent metalloporphyrins as labels in time-resolved luminescence microscopy: effect of mounting on emission intensity. Microsc Res Tech 58:125–131. https://doi.org/10.1002/jemt.10129

62. Finikova OS, Cheprakov AV, Vinogradov SA (2005) Synthesis and luminescence of soluble meso-unsubstituted tetrabenzo- and tetranaphtho[2,3]porphyrins. J Org Chem 70:9562–9572. https://doi.org/10.1021/jo051580r

63. Papkovsky DB, O'Riordan TC (2005) Emerging applications of phosphorescent metalloporphyrins. J Fluoresc 15:569–584. https://doi.org/10.1007/s10895-005-2830-x
64. Bünzli J-CG (2010) Lanthanide luminescence for biomedical analyses and imaging. Chem Rev 110:2729–2755. https://doi.org/10.1021/cr900362e
65. Moore EG, Samuel APS, Raymond KN (2009) From antenna to assay: lessons learned in lanthanide luminescence. Acc Chem Res 42:542–552. https://doi.org/10.1021/ar800211j
66. Mohamadi A, Miller LW (2016) Brightly luminescent and kinetically inert lanthanide bioprobes based on linear and preorganized chelators. Bioconjug Chem 27(10):2540–2548. https://doi.org/10.1021/acs.bioconjchem.6b00473
67. Bünzli J-CG (2015) On the design of highly luminescent lanthanide complexes. Coord Chem Rev 293–294:19–47. https://doi.org/10.1016/j.ccr.2014.10.013
68. Wang Q, Nchimi Nono K, Syrjänpää M, Charbonnière LJ, Hovinen J, Härmä H (2013) Stable and highly fluorescent europium(III) chelates for time-resolved immunoassays. Inorg Chem 52:8461–8466. https://doi.org/10.1021/ic400384f
69. Sund H, Blomberg K, Meltola N, Takalo H (2017) Design of novel, water soluble and highly luminescent europium labels with potential to enhance immunoassay sensitivities. Molecules 22(10):E1807. https://doi.org/10.3390/molecules22101807
70. Delbianco M, Sadovnikova V, Bourrier E, Mathis G, Lamarque L, Zwier JM, Parker D (2014) Bright, highly water-soluble triazacyclononane europium complexes to detect ligand binding with time-resolved FRET microscopy. Angew Chem Int Ed Engl 53:10718–10722. https://doi.org/10.1002/anie.201406632
71. Butler SJ, Delbianco M, Lamarque L, McMahon BK, Neil ER, Pal R, Parker D, Walton JW, Zwier JM (2015) EuroTracker® dyes: design, synthesis, structure and photophysical properties of very bright europium complexes and their use in bioassays and cellular optical imaging. Dalton Trans 44:4791–4803. https://doi.org/10.1039/c4dt02785j
72. Orte A, Alvarez-Pez JM, Ruedas-Rama MJ (2013) Fluorescence lifetime imaging microscopy for the detection of intracellular pH with quantum dot nanosensors. ACS Nano 7:6387–6395. https://doi.org/10.1021/nn402581q
73. Kruss S, Hilmer AJ, Zhang J, Reuel NF, Mu B, Strano MS (2013) Carbon nanotubes as optical biomedical sensors. Adv Drug Deliv Rev 65:1933–1950. https://doi.org/10.1016/j.addr.2013.07.015
74. Liu J-H, Cao L, LeCroy GE, Wang P, Meziani MJ, Dong Y, Liu Y, Luo PG, Sun Y-P (2015) Carbon "quantum" dots for fluorescence labeling of cells. ACS Appl Mater Interfaces 7:19439–19445. https://doi.org/10.1021/acsami.5b05665
75. Ortega-Liebana MC, Encabo-Berzosa MM, Ruedas-Rama MJ, Hueso JL (2017) Nitrogen-induced transformation of vitamin C into multifunctional up-converting carbon nanodots in the visible-NIR range. Chemistry 23:3067–3073. https://doi.org/10.1002/chem.201604216
76. Eda G, Lin Y-Y, Mattevi C, Yamaguchi H, Chen H-A, Chen I-S, Chen C-W, Chhowalla M (2010) Blue photoluminescence from chemically derived graphene oxide. Adv Mater 22:505–509. https://doi.org/10.1002/adma.200901996
77. Loh KP, Bao Q, Eda G, Chhowalla M (2010) Graphene oxide as a chemically tunable platform for optical applications. Nat Chem 2:1015–1024. https://doi.org/10.1038/nchem.907
78. Wang F, Gu Z, Lei W, Wang W, Xia X, Hao Q (2014) Graphene quantum dots as a fluorescent sensing platform for highly efficient detection of copper(II) ions. Sensors Actuators B Chem 190:516–522. https://doi.org/10.1016/j.snb.2013.09.009
79. Röding M, Bradley SJ, Nydén M, Nann T (2014) Fluorescence lifetime analysis of graphene quantum dots. J Phys Chem C 118:30282–30290. https://doi.org/10.1021/jp510436r
80. Vaijayanthimala V, Cheng P-Y, Yeh S-H, Liu K-K, Hsiao C-H, Chao J-I, Chang H-C (2012) The long-term stability and biocompatibility of fluorescent nanodiamond as an in vivo contrast agent. Biomaterials 33:7794–7802. https://doi.org/10.1016/j.biomaterials.2012.06.084
81. Faklaris O, Garrot D, Joshi V, Druon F, Boudou J-P, Sauvage T, Georges P, Curmi PA, Treussart F (2008) Detection of single photoluminescent diamond nanoparticles in cells and

study of the internalization pathway. Small 4:2236–2239. https://doi.org/10.1002/smll.200800655

82. Kuo Y, Hsu T-Y, Wu Y-C, Chang H-C (2013) Fluorescent nanodiamond as a probe for the intercellular transport of proteins in vivo. Biomaterials 34:8352–8360. https://doi.org/10.1016/j.biomaterials.2013.07.043

83. Gu L, Hall DJ, Qin Z, Anglin E, Joo J, Mooney DJ, Howell SB, Sailor MJ (2013) In vivo time-gated fluorescence imaging with biodegradable luminescent porous silicon nanoparticles. Nat Commun 4:2326. https://doi.org/10.1038/ncomms3326

84. Bryan JD (2005) Doped semiconductor nanocrystals: synthesis, characterization, physical properties, and applications. In: Karlin KD (ed) Progress in inorganic chemistry. Wiley, Hoboken, pp 47–126

85. Wu P, Yan X-P (2013) Doped quantum dots for chemo/biosensing and bioimaging. Chem Soc Rev 42:5489–5521. https://doi.org/10.1039/c3cs60017c

86. Del Rosal B, Ortgies DH, Fernández N, Sanz-Rodríguez F, Jaque D, Rodríguez EM (2016) Overcoming autofluorescence: long-lifetime infrared nanoparticles for time-gated in vivo imaging. Adv Mater 28:10188–10193. https://doi.org/10.1002/adma.201603583

87. Tian L, Dai Z, Zhang L, Zhang R, Ye Z, Wu J, Jin D, Yuan J (2012) Preparation and time-gated luminescence bioimaging applications of long wavelength-excited silica-encapsulated europium nanoparticles. Nanoscale 4:3551–3557. https://doi.org/10.1039/c2nr30233k

88. Song C, Ye Z, Wang G, Jin D, Yuan J, Guan Y, Piper J (2009) Preparation and time-gated luminescence bioimaging application of ruthenium complex covalently bound silica nanoparticles. Talanta 79:103–108. https://doi.org/10.1016/j.talanta.2009.03.018

89. Wessel D, Flügge UI (1984) A method for the quantitative recovery of protein in dilute solution in the presence of detergents and lipids. Anal Biochem 138:141–143. https://doi.org/10.1016/0003-2697(84)90782-6

90. Maple PA, Jones CS, Andrews NJ (2001) Time resolved fluorometric immunoassay, using europium labelled antihuman IgG, for the detection of human tetanus antitoxin in serum. J Clin Pathol 54:812–815

91. Qin Q, Christiansen M, Lövgren T, Nørgaard-Pedersen B, Pettersson K (1997) Dual-label time-resolved immunofluorometric assay for simultaneous determination of pregnancy-associated plasma protein A and free beta-subunit of human chorionic gonadotrophin. J Immunol Methods 205:169–175

92. Bookout JT, Joaquim TR, Magin KM, Rogan GJ, Lirette RP (2000) Development of a dual-label time-resolved fluorometric immunoassay for the simultaneous detection of two recombinant proteins in potato. J Agric Food Chem 48:5868–5873. https://doi.org/10.1021/jf000841p

93. Mitrunen K, Pettersson K, Piironen T, Björk T, Lilja H, Lövgren T (1995) Dual-label one-step immunoassay for simultaneous measurement of free and total prostate-specific antigen concentrations and ratios in serum. Clin Chem 41:1115–1120

94. Zhu L, Leinonen J, Zhang W-M, Finne P, Stenman U-H (2003) Dual-label immunoassay for simultaneous measurement of prostate-specific antigen (PSA)-alpha1-antichymotrypsin complex together with free or total PSA. Clin Chem 49:97–103. https://doi.org/10.1373/49.1.97

95. Degorce F, Card A, Soh S, Trinquet E, Knapik GP, Xie B (2009) HTRF: a technology tailored for drug discovery - a review of theoretical aspects and recent applications. Curr Chem Genomics 3:22–32. https://doi.org/10.2174/1875397300903010022

96. Bazin H, Trinquet E, Mathis G (2002) Time resolved amplification of cryptate emission: a versatile technology to trace biomolecular interactions. J Biotechnol 82:233–250

97. Mabile M, Mathis G, Jolu EJP, Pouyat D, Dumont C (1996) Method of measuring the luminescence emitted in a luminescent assay. Patent number: US5527684 A

98. Hildebrandt N, Charbonnière LJ, Beck M, Ziessel RF, Löhmannsröben H-G (2005) Quantum dots as efficient energy acceptors in a time-resolved fluoroimmunoassay. Angew Chem Int Ed Engl 44:7612–7615. https://doi.org/10.1002/anie.200501552

99. Charbonnière LJ, Hildebrandt N, Ziessel RF, Löhmannsröben H-G (2006) Lanthanides to quantum dots resonance energy transfer in time-resolved fluoro-immunoassays and luminescence microscopy. J Am Chem Soc 128:12800–12809. https://doi.org/10.1021/ja062693a

100. Algar WR, Wegner D, Huston AL, Blanco-Canosa JB, Stewart MH, Armstrong A, Dawson PE, Hildebrandt N, Medintz IL (2012) Quantum dots as simultaneous acceptors and donors in time-gated Förster resonance energy transfer relays: characterization and biosensing. J Am Chem Soc 134:1876–1891. https://doi.org/10.1021/ja210162f

101. Jia Y, Quinn CM, Clabbers A, Talanian R, Xu Y, Wishart N, Allen H (2006) Comparative analysis of various in vitro COT kinase assay formats and their applications in inhibitor identification and characterization. Anal Biochem 350:268–276. https://doi.org/10.1016/j.ab. 2005.11.010

102. Schröter T, Minond D, Weiser A, Dao C, Habel J, Spicer T, Chase P, Baillargeon P, Scampavia L, Schürer S, Chung C, Mader C, Southern M, Tsinoremas N, LoGrasso P, Hodder P (2008) Comparison of miniaturized time-resolved fluorescence resonance energy transfer and enzyme-coupled luciferase high-throughput screening assays to discover inhibitors of Rho-kinase II (ROCK-II). J Biomol Screen 13:17–28. https://doi.org/10.1177/1087057107310806

103. Gracias V, Ji Z, Akritopoulou-Zanze I, Abad-Zapatero C, Huth JR, Song D, Hajduk PJ, Johnson EF, Glaser KB, Marcotte PA, Pease L, Soni NB, Stewart KD, Davidsen SK, Michaelides MR, Djuric SW (2008) Scaffold oriented synthesis. Part 2: design, synthesis and biological evaluation of pyrimido-diazepines as receptor tyrosine kinase inhibitors. Bioorg Med Chem Lett 18:2691–2695. https://doi.org/10.1016/j.bmcl.2008.03.021

104. Liu J, Lin TH, Cole AG, Wen R, Zhao L, Brescia M-R, Jacob B, Hussain Z, Appell KC, Henderson I, Webb ML (2008) Identification and characterization of small-molecule inhibitors of Tie2 kinase. FEBS Lett 582:785–791. https://doi.org/10.1016/j.febslet.2008.02.003

105. Ayoub MA, Trebaux J, Vallaghe J, Charrier-Savournin F, Al-Hosaini K, Gonzalez Moya A, Pin J-P, Pfleger KDG, Trinquet E (2014) Homogeneous time-resolved fluorescence-based assay to monitor extracellular signal-regulated kinase signaling in a high-throughput format. Front Endocrinol (Lausanne) 5:94. https://doi.org/10.3389/fendo.2014.00094

106. Vaasa A, Ligi K, Mohandessi S, Enkvist E, Uri A, Miller LW (2012) Time-gated luminescence microscopy with responsive nonmetal probes for mapping activity of protein kinases in living cells. Chem Commun (Camb) 48:8595–8597. https://doi.org/10.1039/c2cc33565d

107. Enkvist E, Vaasa A, Kasari M, Kriisa M, Ivan T, Ligi K, Raidaru G, Uri A (2011) Protein-induced long lifetime luminescence of nonmetal probes. ACS Chem Biol 6:1052–1062. https://doi.org/10.1021/cb200120v

108. Enomoto K, Okamoto H, Numata Y, Takemoto H (2006) A simple and rapid assay for heparanase activity using homogeneous time-resolved fluorescence. J Pharm Biomed Anal 41:912–917. https://doi.org/10.1016/j.jpba.2006.01.032

109. Aouadi W, Eydoux C, Coutard B, Martin B, Debart F, Vasseur JJ, Contreras JM, Morice C, Quérat G, Jung M-L, Canard B, Guillemot J-C, Decroly E (2017) Toward the identification of viral cap-methyltransferase inhibitors by fluorescence screening assay. Antivir Res 144:330–339. https://doi.org/10.1016/j.antiviral.2017.06.021

110. Ji J, Lao K, Hu J, Pang T, Jiang Z, Yuan H, Miao J, Chen X, Ning S, Xiang H, Guo Y, Yan M, Zhang L (2014) Discovery of novel aromatase inhibitors using a homogeneous time-resolved fluorescence assay. Acta Pharmacol Sin 35:1082–1092. https://doi.org/10.1038/aps.2014.53

111. Algar WR, Malanoski AP, Susumu K, Stewart MH, Hildebrandt N, Medintz IL (2012) Multiplexed tracking of protease activity using a single color of quantum dot vector and a time-gated Förster resonance energy transfer relay. Anal Chem 84:10136–10146. https://doi.org/10.1021/ac3028068

112. Zhang S, Ma Y, Li J, Ma J, Yu B, Xie X (2014) Molecular matchmaking between the popular weight-loss herb Hoodia gordonii and GPR119, a potential drug target for metabolic disorder. Proc Natl Acad Sci U S A 111:14571–14576. https://doi.org/10.1073/pnas.1324130111

113. Keppler A, Gendreizig S, Gronemeyer T, Pick H, Vogel H, Johnsson K (2003) A general method for the covalent labeling of fusion proteins with small molecules in vivo. Nat Biotechnol 21:86–89. https://doi.org/10.1038/nbt765

114. Valencia C, Dujet C, Margathe J-F, Iturrioz X, Roux T, Trinquet E, Villa P, Hibert M, Dupuis E, Llorens-Cortes C, Bonnet D (2017) A time-resolved FRET cell-based binding

assay for the apelin receptor. ChemMedChem 12:925–931. https://doi.org/10.1002/cmdc.201700106

115. Zwier JM, Roux T, Cottet M, Durroux T, Douzon S, Bdioui S, Gregor N, Bourrier E, Oueslati N, Nicolas L, Tinel N, Boisseau C, Yverneau P, Charrier-Savournin F, Fink M, Trinquet E (2010) A fluorescent ligand-binding alternative using Tag-lite® technology. J Biomol Screen 15:1248–1259. https://doi.org/10.1177/1087057110384611

116. Leyris J-P, Roux T, Trinquet E, Verdié P, Fehrentz J-A, Oueslati N, Douzon S, Bourrier E, Lamarque L, Gagne D, Galleyrand J-C, M'kadmi C, Martinez J, Mary S, Banères J-L, Marie J (2011) Homogeneous time-resolved fluorescence-based assay to screen for ligands targeting the growth hormone secretagogue receptor type 1a. Anal Biochem 408:253–262. https://doi.org/10.1016/j.ab.2010.09.030

117. Emami-Nemini A, Roux T, Leblay M, Bourrier E, Lamarque L, Trinquet E, Lohse MJ (2013) Time-resolved fluorescence ligand binding for G protein-coupled receptors. Nat Protoc 8:1307–1320. https://doi.org/10.1038/nprot.2013.073

118. Thibon A, Pierre VC (2009) Principles of responsive lanthanide-based luminescent probes for cellular imaging. Anal Bioanal Chem 394:107–120. https://doi.org/10.1007/s00216-009-2683-2

119. Aulsebrook ML, Graham B, Grace MR, Tuck KL (2017) Lanthanide complexes for luminescence-based sensing of low molecular weight analytes. Coord Chem Rev 375:191–220. https://doi.org/10.1016/j.ccr.2017.11.018

120. Strimbu K, Tavel JA (2010) What are biomarkers? Curr Opin HIV AIDS 5:463–466. https://doi.org/10.1097/COH.0b013e32833ed177

121. Detassis S, Grasso M, Del Vescovo V, Denti MA (2017) microRNAs make the call in cancer personalized medicine. Front Cell Dev Biol 5:86. https://doi.org/10.3389/fcell.2017.00086

122. Borrebaeck CAK (2017) Precision diagnostics: moving towards protein biomarker signatures of clinical utility in cancer. Nat Rev Cancer 17:199–204. https://doi.org/10.1038/nrc.2016.153

123. Jeromin A, Bowser R (2017) Biomarkers in neurodegenerative diseases. Adv Neurobiol 15:491–528. https://doi.org/10.1007/978-3-319-57193-5_20

124. Wang J, Tan G-J, Han L-N, Bai Y-Y, He M, Liu H-B (2017) Novel biomarkers for cardiovascular risk prediction. J Geriatr Cardiol 14:135–150. https://doi.org/10.11909/j.issn.1671-5411.2017.02.008

125. Dorcely B, Katz K, Jagannathan R, Chiang SS, Oluwadare B, Goldberg IJ, Bergman M (2017) Novel biomarkers for prediabetes, diabetes, and associated complications. Diabetes Metab Syndr Obes 10:345–361. https://doi.org/10.2147/DMSO.S100074

126. Kashani K, Cheungpasitporn W, Ronco C (2017) Biomarkers of acute kidney injury: the pathway from discovery to clinical adoption. Clin Chem Lab Med 55:1074–1089. https://doi.org/10.1515/cclm-2016-0973

127. Qin Q-P, Christiansen M, Pettersson K (2002) Point-of-care time-resolved immuno-fluorometric assay for human pregnancy-associated plasma protein A: use in first-trimester screening for Down syndrome. Clin Chem 48:473–483

128. Tsukerman GL, Gusina NB, Cuckle HS (1999) Maternal serum screening for Down syndrome in the first trimester: experience from Belarus. Prenat Diagn 19:499–504. https://doi.org/10.1002/(SICI)1097-0223(199906)19:6<499::AID-PD555>3.0.CO;2-6

129. Niemimaa M, Suonpää M, Perheentupa A, Seppälä M, Heinonen S, Laitinen P, Ruokonen A, Ryynänen M (2001) Evaluation of first trimester maternal serum and ultrasound screening for Down's syndrome in Eastern and Northern Finland. Eur J Hum Genet 9:404–408. https://doi.org/10.1038/sj.ejhg.5200655

130. Prieto B, Llorente E, González-Pinto I, Alvarez FV (2008) Plasma procalcitonin measured by time-resolved amplified cryptate emission (TRACE) in liver transplant patients. A prognosis marker of early infectious and non-infectious postoperative complications. Clin Chem Lab Med 46:660–666

131. Tousch D, Lajoix A-D, Hosy E, Azay-Milhau J, Ferrare K, Jahannault C, Cros G, Petit P (2008) Chicoric acid, a new compound able to enhance insulin release and glucose uptake. Biochem Biophys Res Commun 377:131–135. https://doi.org/10.1016/j.bbrc.2008.09.088
132. Wang D-Y, Lu Q, Walsh SL, Payne L, Modha SS, Scott MJ, Sweitzer TD, Ames RS, Krosky DJ, Li H (2008) Development of a high-throughput cell-based assay for 11beta-hydroxysteroid dehydrogenase type 1 using BacMam technology. Mol Biotechnol 39:127–134. https://doi.org/10.1007/s12033-008-9050-y
133. Goedken ER, Gagnon AI, Overmeyer GT, Liu J, Petrillo RA, Burchat AF, Tomlinson MJ (2008) HTRF-based assay for microsomal prostaglandin E2 synthase-1 activity. J Biomol Screen 13:619–625. https://doi.org/10.1177/1087057108321145
134. Lewis H, Beher D, Cookson N, Oakley A, Piggott M, Morris CM, Jaros E, Perry R, Ince P, Kenny RA, Ballard CG, Shearman MS, Kalaria RN (2006) Quantification of Alzheimer pathology in ageing and dementia: age-related accumulation of amyloid-beta(42) peptide in vascular dementia. Neuropathol Appl Neurobiol 32:103–118. https://doi.org/10.1111/j.1365-2990.2006.00696.x
135. Penas C, Pazos E, Mascareñas JL, Vázquez ME (2013) A folding-based approach for the luminescent detection of a short RNA hairpin. J Am Chem Soc 135:3812–3814. https://doi.org/10.1021/ja400270a
136. Pazos E, Jiménez-Balsa A, Mascareñas JL, Vázquez ME (2011) Sensing coiled-coil proteins through conformational modulation of energy transfer processes—selective detection of the oncogenic transcription factor c-Jun. Chem Sci 2:1984. https://doi.org/10.1039/c1sc00108f
137. González-Vera JA, Bouzada D, Bouclier C, Eugenio Vázquez M, Morris MC (2017) Lanthanide-based peptide biosensor to monitor CDK4/cyclin D kinase activity. Chem Commun (Camb) 53:6109–6112. https://doi.org/10.1039/c6cc09948c
138. Pazos E, Torrecilla D, Vázquez López M, Castedo L, Mascareñas JL, Vidal A, Vázquez ME (2008) Cyclin A probes by means of intermolecular sensitization of terbium-chelating peptides. J Am Chem Soc 130:9652–9653. https://doi.org/10.1021/ja803520q
139. Newton P, Harrison P, Clulow S (2008) A novel method for determination of the affinity of protein: protein interactions in homogeneous assays. J Biomol Screen 13:674–682. https://doi.org/10.1177/1087057108321086
140. Maurel D, Comps-Agrar L, Brock C, Rives M-L, Bourrier E, Ayoub MA, Bazin H, Tinel N, Durroux T, Prézeau L, Trinquet E, Pin J-P (2008) Cell-surface protein-protein interaction analysis with time-resolved FRET and snap-tag technologies: application to GPCR oligomerization. Nat Methods 5:561–567. https://doi.org/10.1038/nmeth.1213
141. Vischer HF, Nijmeijer S, Smit MJ, Leurs R (2008) Viral hijacking of human receptors through heterodimerization. Biochem Biophys Res Commun 377:93–97. https://doi.org/10.1016/j.bbrc.2008.09.082
142. Lakowicz JR (2001) Method and composition for detecting the presence of a nucleic acid sequence in a sample. US6200752B1. https://patents.google.com/patent/US6200752B1/
143. Rajendran M, Miller LW (2015) Evaluating the performance of time-gated live-cell microscopy with lanthanide probes. Biophys J 109:240–248. https://doi.org/10.1016/j.bpj.2015.06.028
144. Jin D, Piper JA (2011) Time-gated luminescence microscopy allowing direct visual inspection of lanthanide-stained microorganisms in background-free condition. Anal Chem 83:2294–2300. https://doi.org/10.1021/ac103207r
145. Zhang L, Zheng X, Deng W, Lu Y, Lechevallier S, Ye Z, Goldys EM, Dawes JM, Piper JA, Yuan J, Verelst M, Jin D (2014) Practical implementation, characterization and applications of a multi-colour time-gated luminescence microscope. Sci Rep 4:6597. https://doi.org/10.1038/srep06597
146. Soini AE, Kuusisto A, Meltola NJ, Soini E, Seveus L (2003) A new technique for multiparameter imaging microscopy: use of long decay time photoluminescent labels enables multiple color immunocytochemistry with low channel-to-channel crosstalk. Microsc Res Tech 62:396–407. https://doi.org/10.1002/jemt.10389
147. Grichine A, Haefele A, Pascal S, Duperray A, Michel R, Andraud C, Maury O (2014) Millisecond lifetime imaging with a europium complex using a commercial confocal

microscope under one or two-photon excitation. Chem Sci 5:3475–3485. https://doi.org/10.1039/C4SC00473F

148. Vicidomini G, Moneron G, Han KY, Westphal V, Ta H, Reuss M, Engelhardt J, Eggeling C, Hell SW (2011) Sharper low-power STED nanoscopy by time gating. Nat Methods 8:571–573. https://doi.org/10.1038/nmeth.1624

149. Lu Y, Xi P, Piper JA, Huo Y, Jin D (2012) Time-gated orthogonal scanning automated microscopy (OSAM) for high-speed cell detection and analysis. Sci Rep 2:837. https://doi.org/10.1038/srep00837

150. Ruedas-Rama MJ, Alvarez-Pez JM, Crovetto L, Paredes JM, Orte A (2015) FLIM strategies for intracellular sensing. In: Kapusta P, Wahl M, Erdmann R (eds) Advanced photon counting. Springer, Cham, pp 191–223

151. Dahan M, Laurence T, Pinaud F, Chemla DS, Alivisatos AP, Sauer M, Weiss S (2001) Time-gated biological imaging by use of colloidal quantum dots. Opt Lett 26:825–827

152. Mandal G, Darragh M, Wang YA, Heyes CD (2013) Cadmium-free quantum dots as time-gated bioimaging probes in highly-autofluorescent human breast cancer cells. Chem Commun (Camb) 49:624–626. https://doi.org/10.1039/c2cc37529j

153. Bouccara S, Fragola A, Giovanelli E, Sitbon G, Lequeux N, Pons T, Loriette V (2014) Time-gated cell imaging using long lifetime near-infrared-emitting quantum dots for autofluorescence rejection. J Biomed Opt 19:051208. https://doi.org/10.1117/1.JBO.19.5.051208

154. Liu M, Ye Z, Xin C, Yuan J (2013) Development of a ratiometric time-resolved luminescence sensor for pH based on lanthanide complexes. Anal Chim Acta 761:149–156. https://doi.org/10.1016/j.aca.2012.11.025

155. Smith DG, McMahon BK, Pal R, Parker D (2012) Live cell imaging of lysosomal pH changes with pH responsive ratiometric lanthanide probes. Chem Commun (Camb) 48:8520–8522. https://doi.org/10.1039/c2cc34267g

156. Liu X, Guo L, Song B, Tang Z, Yuan J (2017) Development of a novel europium complex-based luminescent probe for time-gated luminescence imaging of hypochlorous acid in living samples. Methods Appl Fluoresc 5:014009. https://doi.org/10.1088/2050-6120/aa61af

157. Liu X, Tang Z, Song B, Ma H, Yuan J (2017) A mitochondria-targeting time-gated luminescence probe for hypochlorous acid based on a europium complex. J Mater Chem B 5:2849–2855. https://doi.org/10.1039/C6TB02991D

158. Song B, Ye Z, Yang Y, Ma H, Zheng X, Jin D, Yuan J (2015) Background-free in-vivo imaging of vitamin C using time-gateable responsive probe. Sci Rep 5:14194. https://doi.org/10.1038/srep14194

159. Dai Z, Tian L, Song B, Ye Z, Liu X, Yuan J (2014) Ratiometric time-gated luminescence probe for hydrogen sulfide based on lanthanide complexes. Anal Chem 86:11883–11889. https://doi.org/10.1021/ac503611f

160. Dai Z, Tian L, Ye Z, Song B, Zhang R, Yuan J (2013) A lanthanide complex-based ratiometric luminescence probe for time-gated luminescence detection of intracellular thiols. Anal Chem 85:11658–11664. https://doi.org/10.1021/ac403370g

161. Sun J, Song B, Ye Z, Yuan J (2015) Mitochondria targetable time-gated luminescence probe for singlet oxygen based on a β-diketonate-europium complex. Inorg Chem 54:11660–11668. https://doi.org/10.1021/acs.inorgchem.5b02458

162. Afsari HS, Cardoso Dos Santos M, Lindén S, Chen T, Qiu X, van Bergen En Henegouwen PMP, Jennings TL, Susumu K, Medintz IL, Hildebrandt N, Miller LW (2016) Time-gated FRET nanoassemblies for rapid and sensitive intra- and extracellular fluorescence imaging. Sci Adv 2:e1600265. https://doi.org/10.1126/sciadv.1600265

163. Tu C-C, Awasthi K, Chen K-P, Lin C-H, Hamada M, Ohta N, Li Y-K (2017) Time-gated imaging on live cancer cells using silicon quantum dot nanoparticles with long-lived fluorescence. ACS Photonics 4:1306–1315. https://doi.org/10.1021/acsphotonics.7b00188

164. Monici M (2005) Cell and tissue autofluorescence research and diagnostic applications. Biotechnol Annu Rev 11:227–256. https://doi.org/10.1016/S1387-2656(05)11007-2

165. Zipfel WR, Williams RM, Christie R, Nikitin AY, Hyman BT, Webb WW (2003) Live tissue intrinsic emission microscopy using multiphoton-excited native fluorescence and second

harmonic generation. Proc Natl Acad Sci U S A 100:7075–7080. https://doi.org/10.1073/pnas. 0832308100

166. Thorson MK, Ung P, Leaver FM, Corbin TS, Tuck KL, Graham B, Barrios AM (2015) Lanthanide complexes as luminogenic probes to measure sulfide levels in industrial samples. Anal Chim Acta 896:160–165. https://doi.org/10.1016/j.aca.2015.09.024

167. Surender EM, Bradberry SJ, Bright SA, McCoy CP, Williams DC, Gunnlaugsson T (2017) Luminescent lanthanide cyclen-based enzymatic assay capable of diagnosing the onset of catheter-associated urinary tract infections both in solution and within polymeric hydrogels. J Am Chem Soc 139:381–388. https://doi.org/10.1021/jacs.6b11077

168. Chen X, Wang Y, Chai R, Xu Y, Li H, Liu B (2017) Luminescent lanthanide-based organic/ inorganic hybrid materials for discrimination of glutathione in solution and within hydrogels. ACS Appl Mater Interfaces 9:13554–13563. https://doi.org/10.1021/acsami.7b02679

169. Gorai T, Maitra U (2016) Supramolecular approach to enzyme sensing on paper discs using lanthanide photoluminescence. ACS Sens 1:934–940. https://doi.org/10.1021/acssensors. 6b00341

170. Zhang R, Liu S, Wang J, Han G, Yang L, Liu B, Guan G, Zhang Z (2016) Visualization of exhaled hydrogen sulphide on test paper with an ultrasensitive and time-gated luminescent probe. Analyst 141:4919–4925. https://doi.org/10.1039/c6an00830e

171. Hashino K, Ikawa K, Ito M, Hosoya C, Nishioka T, Makiuchi M, Matsumoto K (2007) Application of a fluorescent lanthanide chelate label on a solid support device for detecting DNA variation with ligation-based assay. Anal Biochem 364:89–91. https://doi.org/10.1016/j. ab.2007.02.004

172. Lövgren T, Meriö L, Mitrunen K, Mäkinen ML, Mäkelä M, Blomberg K, Palenius T, Pettersson K (1996) One-step all-in-one dry reagent immunoassays with fluorescent europium chelate label and time-resolved fluorometry. Clin Chem 42:1196–1201

173. Nagl S, Stich MIJ, Schäferling M, Wolfbeis OS (2009) Method for simultaneous luminescence sensing of two species using optical probes of different decay time, and its application to an enzymatic reaction at varying temperature. Anal Bioanal Chem 393:1199–1207. https://doi. org/10.1007/s00216-008-2467-0

174. Leblanc V, Delaunay V, Claude Lelong J, Gas F, Mathis G, Grassi J, May E (2002) Homogeneous time-resolved fluorescence assay for identifying p53 interactions with its protein partners, directly in a cellular extract. Anal Biochem 308:247–254

175. Albizu L, Cottet M, Kralikova M, Stoev S, Seyer R, Brabet I, Roux T, Bazin H, Bourrier E, Lamarque L, Breton C, Rives M-L, Newman A, Javitch J, Trinquet E, Manning M, Pin J-P, Mouillac B, Durroux T (2010) Time-resolved FRET between GPCR ligands reveals oligomers in native tissues. Nat Chem Biol 6:587–594. https://doi.org/10.1038/nchembio.396

176. Lu Y, Jin D, Leif RC, Deng W, Piper JA, Yuan J, Duan Y, Huo Y (2011) Automated detection of rare-event pathogens through time-gated luminescence scanning microscopy. Cytometry A 79:349–355. https://doi.org/10.1002/cyto.a.21045

177. U.S. EPA (2005) Method 1623: Cryptosporidium and Giardia in Water by Filtration/IMS/FA. EPA 815-R-05-002

178. Rich RM, Stankowska DL, Maliwal BP, Sørensen TJ, Laursen BW, Krishnamoorthy RR, Gryczynski Z, Borejdo J, Gryczynski I, Fudala R (2013) Elimination of autofluorescence background from fluorescence tissue images by use of time-gated detection and the AzaDiOxaTriAngulenium (ADOTA) fluorophore. Anal Bioanal Chem 405:2065–2075. https://doi.org/10.1007/s00216-012-6623-1

179. Zhu Z, Song B, Yuan J, Yang C (2016) Enabling the triplet of tetraphenylethene to sensitize the excited state of Europium(III) for protein detection and time-resolved luminescence imaging. Adv Sci 3:1600146. https://doi.org/10.1002/advs.201600146

180. Von Lode P, Rosenberg J, Pettersson K, Takalo H (2003) A europium chelate for quantitative point-of-care immunoassays using direct surface measurement. Anal Chem 75:3193–3201

181. Jia Y (2008) Current status of HTRF($^{®}$) technology in kinase assays. Expert Opin Drug Discov 3:1461–1474. https://doi.org/10.1517/17460440802518171
182. Cisbio Bioassays, Codolet, France Application Note: HTRF KinEASE: a universal expanded platform to address Serine/Threonine & Tyrosine kinases. https://www.cisbio.com/drug-discov ery/htrf-kinease-universal-expanded-platform-address-serinethreonine-tyrosine-kinases. Accessed 29 Jan 2018
183. Harbert C, Marshall J, Soh S, Steger K (2008) Development of a HTRF kinase assay for determination of Syk activity. Curr Chem Genomics 1:20–26. https://doi.org/10.2174/1875397300801010020
184. Larson B, Gonzalez-Moya A, Wolff A, Luty W. Application Note: analysis of the effect of aggregated β-amyloid on cellular signaling pathways critical for memory in Alzheimer's disease. http://www.enzolifesciences.com/about-us/collaborations-at-work/neuroscience/analy sis-of-the-effect-of-aggregated-b-amyloid-on-cellular-signaling-pathways-critical-for-memory-in-alzheimer%27s-disease/. Accessed 29 Jan 2018
185. PerkinElmer I (2012) DELFIA assays: flexible and sensitive tools for monoclonal antibody development. Technical note. PerkinElmer, Hopkinton
186. Allicotti G, Borras E, Pinilla C (2003) A time-resolved fluorescence immunoassay (DELFIA) increases the sensitivity of antigen-driven cytokine detection. J Immunoassay Immunochem 24:345–358. https://doi.org/10.1081/IAS-120025772
187. Karvinen J, Hurskainen P, Gopalakrishnan S, Burns D, Warrior U, Hemmilä I (2002) Homogeneous time-resolved fluorescence quenching assay (LANCE) for caspase-3. J Biomol Screen 7:223–231. https://doi.org/10.1177/108705710200700306
188. Ylikoski A, Elomaa A, Ollikka P, Hakala H, Mukkala V-M, Hovinen J, Hemmilä I (2004) Homogeneous time-resolved fluorescence quenching assay (TruPoint) for nucleic acid detection. Clin Chem 50:1943–1947. https://doi.org/10.1373/clinchem.2004.036616
189. Laitala V, Hemmilä I (2005) Homogeneous assay based on anti-Stokes' shift time-resolved fluorescence resonance energy-transfer measurement. Anal Chem 77:1483–1487. https://doi.org/10.1021/ac048414o
190. Rickard DJ, Sehon CA, Kasparcova V, Kallal LA, Haile PA, Zeng X, Montoute MN, Poore DD, Li H, Wu Z, Eidam PM, Emery JG, Marquis RW, Gough PJ, Bertin J (2014) Identification of selective small molecule inhibitors of the nucleotide-binding oligomerization domain 1 (NOD1) signaling pathway. PLoS One 9:e96737. https://doi.org/10.1371/journal.pone.0096737
191. Gakamsky DM, Dennis RB, Smith SD (2011) Use of fluorescence lifetime technology to provide efficient protection from false hits in screening applications. Anal Biochem 409:89–97. https://doi.org/10.1016/j.ab.2010.10.017
192. De Witte WEA, Wong YC, Nederpelt I, Heitman LH, Danhof M, van der Graaf PH, Gilissen RAHJ, de Lange ECM (2016) Mechanistic models enable the rational use of in vitro drug-target binding kinetics for better drug effects in patients. Expert Opin Drug Discov 11:45–63. https://doi.org/10.1517/17460441.2016.1100163
193. Schiele F, Ayaz P, Fernández-Montalván A (2015) A universal homogeneous assay for high-throughput determination of binding kinetics. Anal Biochem 468:42–49. https://doi.org/10.1016/j.ab.2014.09.007
194. Lipchik AM, Perez M, Bolton S, Dumrongprechachan V, Ouellette SB, Cui W, Parker LL (2015) KINATEST-ID: a pipeline to develop phosphorylation-dependent terbium sensitizing kinase assays. J Am Chem Soc 137:2484–2494. https://doi.org/10.1021/ja507164a
195. Cui W, Parker LL (2016) Modular, antibody-free time-resolved LRET kinase assay enabled by quantum dots and Tb(3+)-sensitizing peptides. Sci Rep 6:28971. https://doi.org/10.1038/srep28971
196. Lipchik AM, Perez M, Cui W, Parker LL (2015) Multicolored, Tb^{3+}-based antibody-free detection of multiple tyrosine kinase activities. Anal Chem 87:7555–7558. https://doi.org/10.1021/acs.analchem.5b02233

197. Cui W, Parker LL (2015) A time-resolved luminescence biosensor assay for anaplastic lymphoma kinase (ALK) activity. Chem Commun (Camb) 51:362–365. https://doi.org/10.1039/c4cc07453j

198. Pritz S, Meder G, Doering K, Drueckes P, Woelcke J, Mayr LM, Hassiepen U (2011) A fluorescence lifetime-based assay for abelson kinase. J Biomol Screen 16:65–72. https://doi.org/10.1177/1087057110385817

199. Doering K, Meder G, Hinnenberger M, Woelcke J, Mayr LM, Hassiepen U (2009) A fluorescence lifetime-based assay for protease inhibitor profiling on human kallikrein 7. J Biomol Screen 14:1–9. https://doi.org/10.1177/1087057108327328

200. Whateley JG (2003) Fluorescence-based methods for measuring enzyme activity. US20030228646A1

201. Whateley JG (2010) Methods for measuring enzyme activity. US7727739B2. https://patents.google.com/patent/US7727739

202. Cohen JD, Li L, Wang Y, Thoburn C, Afsari B, Danilova L, Douville C, Javed AA, Wong F, Mattox A, Hruban RH, Wolfgang CL, Goggins MG, Dal Molin M, Wang T-L, Roden R, Klein AP, Ptak J, Dobbyn L, Schaefer J, Silliman N, Popoli M, Vogelstein JT, Browne JD, Schoen RE, Brand RE, Tie J, Gibbs P, Wong H-L, Mansfield AS, Jen J, Hanash SM, Falconi M, Allen PJ, Zhou S, Bettegowda C, Diaz L, Tomasetti C, Kinzler KW, Vogelstein B, Lennon AM, Papadopoulos N (2018) Detection and localization of surgically resectable cancers with a multi-analyte blood test. Science 359(6378):926–930. https://doi.org/10.1126/science.aar3247

203. Bartel DP (2004) MicroRNAs: genomics, biogenesis, mechanism, and function. Cell 116:281–297

204. Kosaka N, Iguchi H, Ochiya T (2010) Circulating microRNA in body fluid: a new potential biomarker for cancer diagnosis and prognosis. Cancer Sci 101:2087–2092. https://doi.org/10.1111/j.1349-7006.2010.01650.x

205. Zhao H, Shen J, Medico L, Wang D, Ambrosone CB, Liu S (2010) A pilot study of circulating miRNAs as potential biomarkers of early stage breast cancer. PLoS One 5:e13735. https://doi.org/10.1371/journal.pone.0013735

206. Hu Z, Chen X, Zhao Y, Tian T, Jin G, Shu Y, Chen Y, Xu L, Zen K, Zhang C, Shen H (2010) Serum microRNA signatures identified in a genome-wide serum microRNA expression profiling predict survival of non-small-cell lung cancer. J Clin Oncol 28:1721–1726. https://doi.org/10.1200/JCO.2009.24.9342

207. Sita-Lumsden A, Dart DA, Waxman J, Bevan CL (2013) Circulating microRNAs as potential new biomarkers for prostate cancer. Br J Cancer 108:1925–1930. https://doi.org/10.1038/bjc.2013.192

208. Guay C, Regazzi R (2013) Circulating microRNAs as novel biomarkers for diabetes mellitus. Nat Rev Endocrinol 9:513–521. https://doi.org/10.1038/nrendo.2013.86

209. Cattaneo M, Pelosi E, Castelli G, Cerio AM, D'Angiò A, Porretti L, Rebulla P, Pavesi L, Russo G, Giordano A, Turri J, Cicconi L, Lo-Coco F, Testa U, Biunno I (2015) A miRNA signature in human cord blood stem and progenitor cells as potential biomarker of specific acute myeloid leukemia subtypes. J Cell Physiol 230:1770–1780. https://doi.org/10.1002/jcp.24876

210. Benes V, Castoldi M (2010) Expression profiling of microRNA using real-time quantitative PCR, how to use it and what is available. Methods 50:244–249. https://doi.org/10.1016/j.ymeth.2010.01.026

211. Zipper H, Brunner H, Bernhagen J, Vitzthum F (2004) Investigations on DNA intercalation and surface binding by SYBR Green I, its structure determination and methodological implications. Nucleic Acids Res 32:e103. https://doi.org/10.1093/nar/gnh101

212. Bustin SA (2000) Absolute quantification of mRNA using real-time reverse transcription polymerase chain reaction assays. J Mol Endocrinol 25:169–193. https://doi.org/10.1677/jme.0.0250169

213. Okamura Y, Kondo S, Sase I, Suga T, Mise K, Furusawa I, Kawakami S, Watanabe Y (2000) Double-labeled donor probe can enhance the signal of fluorescence resonance energy transfer (FRET) in detection of nucleic acid hybridization. Nucleic Acids Res 28:E107

214. Tyagi S, Kramer FR (1996) Molecular beacons: probes that fluoresce upon hybridization. Nat Biotechnol 14:303–308. https://doi.org/10.1038/nbt0396-303

215. Ng CT, Gilchrist CA, Lane A, Roy S, Haque R, Houpt ER (2005) Multiplex real-time PCR assay using Scorpion probes and DNA capture for genotype-specific detection of *Giardia lamblia* on fecal samples. J Clin Microbiol 43:1256–1260. https://doi.org/10.1128/JCM.43.3.1256-1260.2005

216. Wegman DW, Krylov SN (2011) Direct quantitative analysis of multiple miRNAs (DQAMmiR). Angew Chem Int Ed Engl 50:10335–10339. https://doi.org/10.1002/anie.201104693

217. Wegman DW, Ghasemi F, Khorshidi A, Yang BB, Liu SK, Yousef GM, Krylov SN (2015) Highly-sensitive amplification-free analysis of multiple miRNAs by capillary electrophoresis. Anal Chem 87:1404–1410. https://doi.org/10.1021/ac504406s

218. Pernagallo S, Ventimiglia G, Cavalluzzo C, Alessi E, Ilyine H, Bradley M, Diaz-Mochon JJ (2012) Novel biochip platform for nucleic acid analysis. Sensors (Basel) 12:8100–8111. https://doi.org/10.3390/s120608100

219. Bowler FR, Reid PA, Boyd AC, Diaz-Mochon JJ, Bradley M (2011) Dynamic chemistry for enzyme-free allele discrimination in genotyping by MALDI-TOF mass spectrometry. Anal Methods 3:1656. https://doi.org/10.1039/c1ay05176h

220. Bowler FR, Diaz-Mochon JJ, Swift MD, Bradley M (2010) DNA analysis by dynamic chemistry. Angew Chem Int Ed Engl 49:1809–1812. https://doi.org/10.1002/anie.200905699

221. Rissin DM, López-Longarela B, Pernagallo S, Ilyine H, Vliegenthart ADB, Dear JW, Díaz-Mochón JJ, Duffy DC (2017) Polymerase-free measurement of microRNA-122 with single base specificity using single molecule arrays: detection of drug-induced liver injury. PLoS One 12:e0179669. https://doi.org/10.1371/journal.pone.0179669

222. Venkateswaran S, Luque-González MA, Tabraue-Chávez M, Fara MA, López-Longarela B, Cano-Cortes V, López-Delgado FJ, Sánchez-Martín RM, Ilyine H, Bradley M, Pernagallo S, Díaz-Mochón JJ (2016) Novel bead-based platform for direct detection of unlabelled nucleic acids through Single Nucleobase Labelling. Talanta 161:489–496. https://doi.org/10.1016/j.talanta.2016.08.072

223. Del Vescovo V, Meier T, Inga A, Denti MA, Borlak J (2013) A cross-platform comparison of affymetrix and Agilent microarrays reveals discordant miRNA expression in lung tumors of c-Raf transgenic mice. PLoS One 8:e78870. https://doi.org/10.1371/journal.pone.0078870

224. Jiang L, Duan D, Shen Y, Li J (2012) Direct microRNA detection with universal tagged probe and time-resolved fluorescence technology. Biosens Bioelectron 34:291–295. https://doi.org/10.1016/j.bios.2012.01.035

225. Hemmilä I, Dakubu S, Mukkala VM, Siitari H, Lövgren T (1984) Europium as a label in time-resolved immunofluorometric assays. Anal Biochem 137:335–343. https://doi.org/10.1016/0003-2697(84)90095-2

226. Geissler D, Charbonnière LJ, Ziessel RF, Butlin NG, Löhmannsröben H-G, Hildebrandt N (2010) Quantum dot biosensors for ultrasensitive multiplexed diagnostics. Angew Chem Int Ed Engl 49:1396–1401. https://doi.org/10.1002/anie.200906399

227. Qiu X, Hildebrandt N (2015) Rapid and multiplexed microRNA diagnostic assay using quantum dot-based Förster resonance energy transfer. ACS Nano 9:8449–8457. https://doi.org/10.1021/acsnano.5b03364

228. Jin Z, Geißler D, Qiu X, Wegner KD, Hildebrandt N (2015) A rapid, amplification-free, and sensitive diagnostic assay for single-step multiplexed fluorescence detection of microRNA. Angew Chem Int Ed Engl 54:10024–10029. https://doi.org/10.1002/anie.201504887

229. Geißler D, Stufler S, Löhmannsröben H-G, Hildebrandt N (2013) Six-color time-resolved Förster resonance energy transfer for ultrasensitive multiplexed biosensing. J Am Chem Soc 135:1102–1109. https://doi.org/10.1021/ja310317n

230. Zhou S, Zheng W, Chen Z, Tu D, Liu Y, Ma E, Li R, Zhu H, Huang M, Chen X (2014) Dissolution-enhanced luminescent bioassay based on inorganic lanthanide nanoparticles. Angew Chem Int Ed Engl 53:12498–12502. https://doi.org/10.1002/anie.201405937

231. Lu L, Tu D, Liu Y, Zhou S, Zheng W, Chen X (2018) Ultrasensitive detection of cancer biomarker microRNA by amplification of fluorescence of lanthanide nanoprobes. Nano Res 11:1–10. https://doi.org/10.1007/s12274-017-1629-9
232. DestiNA Genomics Ltd. http://www.destinagenomics.com. Accessed 24 Jan 2018
233. Optoi Microelectronics. http://www.optoi.com/. Accessed 27 Jan 2018
234. miRNA-DisEASY: microRNA biomarkers in an innovative biophotonic sensor kit for high-specific diagnosis. https://optoi.com/en/applications/research-and-development/projects/mirna-diseasy-home-page. Accessed 24 Jan 2018
235. Destina Genomics. In: www.destinagenomics.com; http://www.destinagenomics.com. Accessed 27 Jan 2018
236. Bradley M, Diaz-Mochon JJ (2009) Nucleobase characterisation. WO2009037473A2. https://patents.google.com/patent/WO2009037473A2/

Thermally Activated Delayed Fluorescence Emitters for Light-Emitting Diodes and Sensing Applications

João Avó, Tiago Palmeira, and Fernando B. Dias

Contents

Abstract Thermally activated delayed fluorescence (TADF) has revamped the scientific and technological interest in metal-free organic fluorescent compounds in recent years. The application of TADF emitters in organic light-emitting diodes (OLEDs) resulted in highly energy-efficient devices that promise to replace metal-complex systems based on iridium(III) and platinum(II) in a near future.

Three quarters of the excitons that are created by the electrical current driving an OLED are non-emissive triplet states, therefore unable to generate electroluminescence. The maximum device efficiency is thus limited to 25%. OLED emitters based on metal complexes respond to this problem by promoting emission directly from the triplet state, which is induced by the presence of the heavy metal that enhances spin-orbit coupling interactions. Remarkably, OLEDs with internal quantum efficiency of nearly 100% have been fabricated with metal complexes, owning to the fast intersystem crossing (ISC) and room-temperature phosphorescent properties of these materials. However, while the heavy-metal complexes have many advantages,

J. Avó and T. Palmeira
CQFM-IN and iBB-Institute for Bioengineering and Biosciences, Instituto Superior Técnico, Universidade de Lisboa, Lisboa, Portugal

F. B. Dias (✉)
Department of Physics, Durham University, Durham, UK
e-mail: f.m.b.dias@durham.ac.uk

© Springer Nature Switzerland AG 2019
B. Pedras (ed.), *Fluorescence in Industry*, Springer Ser Fluoresc (2019) 18: 269–292,
https://doi.org/10.1007/4243_2019_8, Published online: 14 February 2019

they also show significant problems when applied in light-emitting diodes. These are scarce and expensive materials that create environmental challenges and are affected by strong degradation in the blue spectral region. These issues, therefore, may create difficulties for the utilization of metal complexes in areas that require high-volume manufacturing, such as in lighting and display technologies, and alternative materials free of heavy metals are needed. TADF molecules allow for efficient triplet harvesting with no use of heavy-metal atoms and appear to improve device stability in the blue region. In addition, they display interesting properties that grant sensitivity to several parameters of the surrounding media, making it an ideal tool for optical sensing applications. TADF research toward application in lighting devices started in 2012 and had not yet entered in commercial applications, as of mid-2018. This chapter covers the principles governing the mechanism behind the TADF process, the recent developments on TADF emitter design, and their planned applications in commercial devices.

Keywords Electroluminescence · Lighting devices · OLEDs · Organic emitters · Thermally activated delayed fluorescence

Abbreviations

ΔE_{ST}	Singlet-triplet energy gap
A	Acceptor moiety
CT	Charge transfer
Cz	Carbazole
D	Donor moiety
$DBTO_2$	Dibenzothiophene-S,S-dioxide
DF	Delayed fluorescence
$DPSO_2$	Diphenylsulfoxide
EQE	External quantum efficiency
ESIPT	Excited-state intramolecular proton transfer
FRET	Förster resonance energy transfer
HOMO	Highest occupied molecular orbital
IQE	Internal quantum efficiency
ISC	Intersystem crossing
L	Ligand
LUMO	Lowest unoccupied molecular orbital
M	Metal center
MLCT	Metal-to-ligand charge transfer
OLED	Organic light-emitting diode
PF	Prompt fluorescence
PH	Phosphorescence
PhOLED	Phosphorescence-based organic light-emitting diode

PLQY	Photoluminescence quantum yield
PN	Phthalonitrile
RISC	Reverse intersystem crossing
S_1	Lowest excited singlet state
SOC	Spin-orbit coupling
T_1	Lowest triplet excited state
TADF	Thermally activated delayed fluorescence
TTA	Triplet-triplet annihilation

1 Principles of TADF

Molecular fluorescence (Fig. 1) can take place by two different unimolecular mechanisms, upon photoexcitation: prompt fluorescence (PF) or thermally activated delayed fluorescence (TADF). In the PF mechanism, emission occurs following $S_n \leftarrow S_0$ photon absorption and excited-state relaxation to the lowest excited singlet state (S_1). The TADF mechanism, on the other hand, occurs via the triplet manifold: once S_1 is attained, the molecule undergoes intersystem crossing (ISC) to the triplet state, where it relaxes to the lowest triplet excited state (T_1), and, once in T_1, may undergo a reverse ISC (RISC) process back to S_1, due to thermal activation [1, 2]. This is possible because TADF molecules are designed with a very small energy difference between the two lowest excited states S_1 and T_1 (ΔE_{ST}) which is often smaller than 100 meV. As the T_1 lifetime is long enough (typically in the order of microseconds or milliseconds), the probability of triplets to be up-converted to the singlet state increases significantly [3]. Depending on the ΔE_{ST} value and the ISC probability, the RISC mechanism may involve a number of cycles ($\bar{n}$), between singlet and triplet states in the form $S_1 \rightarrow T_1 \rightarrow S_1$, according to Eq. 1, where Φ_T and

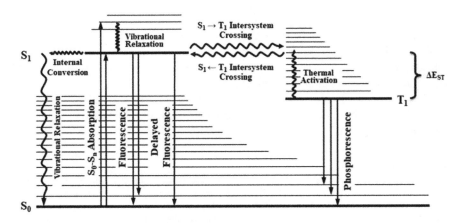

Fig. 1 Perrin-Jablonski diagram. Adapted from [2]

Φ_S represent, respectively, the triplet formation yield and the yield of triplet states that are up-converted to the singlet manifold [4].

$$\bar{n} = \frac{I_{DF}}{I_{PF}} = \frac{1}{\left(\frac{1}{\Phi_T \Phi_S}\right) - 1} \tag{1}$$

TADF emission occurs in the same spectral region as PF, but with significant fundamental differences. While PF occurs in the nanosecond time regime, TADF displays a much longer lifetime, in the micro- to millisecond domain. Basically, the TADF decays with the lifetime of the triplet state. In addition, this type of emission is thermally activated, and, thus, its intensity is significantly temperature dependent, increasing in intensity with temperature increase [5, 6].

Owing to its mechanism, TADF is a relatively rare process to be observed, due to the singular conditions that are required: small ΔE_{ST}, high ISC and RISC probabilities, and long T_1 lifetime. Most molecules used as fluorophores have relatively large ΔE_{ST}, compared to the thermal energy at room temperature, or have slow ISC rates. This is because they have been designed to promote strong radiative decay from S_1 with short decay time. In addition, even when a molecule displays low ΔE_{ST} and high ISC probability, the RISC rate constant must still be fast enough to compete with other deactivation channels of T_1; this is also unusual. For these reasons, TADF was only observed in a few xanthene dyes, aromatic ketones, porphyrins, polycyclic aromatic hydrocarbons, and fullerenes, until recently [6–10].

In 2009, scientists realized that, due to the unimolecular nature of this process, TADF could allow harvesting triplet states with a theoretical efficiency of 100%. This discovery renovated the interest in TADF molecules for their potential application in OLED devices, and, since then, the number of reported molecules displaying this mechanism increased dramatically, with several new molecules with emission ranging from blue to red being reported every year [11–13].

The sudden attention toward TADF is due to the mechanism of electrical excitation in OLEDs, since spin statistics determines that upon electron-hole recombination, singlet and triplet states are formed in a one-to-three ratio [14]. TADF emitters, being able to harvest all triplet excitons, allow the internal quantum efficiency (IQE) of lighting devices to reach 100% without resorting to the use of heavy-metal-based phosphors. This symbolized a breakthrough in lighting industry, representing the advent of a third generation of OLEDs[1] [15].

As referred previously, one of the most important aspects to achieve efficient TADF is the minimization of the ΔE_{ST}, since it will result in a fast RISC process according to Eq. 2 [16].

---

[1]The first generation of OLEDs emerged in 1996 and was based on emitter with prompt fluorescence, displaying a maximum IQE of 25%. In the second generation, fluorescent emitters were replaced with heavy-metal, mainly iridium and platinum, phosphorescent complexes, which allowed $T_1 \rightarrow S_0$ transition due to increased spin-orbit coupling, ultimately allowing an IQE of 100%.

$$k_{\text{RISC}} = A\, e^{\left(-\frac{\Delta E_{ST}}{RT}\right)} \tag{2}$$

The energy of the lowest singlet (S_1) and triplet (T_1) excited states is determined by the orbital energy (E_{orb}), the electron repulsion energy (K), and the exchange energy (J) terms, i.e., the first-order quantum-mechanical correction involving electron-electron repulsion which affects the two unpaired electrons in the excited state (one electron in the HOMO and the other in the LUMO) [17]. For states with the same electronic configuration, the three components contribute equally, and the ΔE_{ST} will depend only on the exchange energy (Eqs. 3–5).

$$E_{S1} = E_{\text{orb}} + K + J \tag{3}$$

$$E_{T1} = E_{\text{orb}} + K - J \tag{4}$$

$$\Delta E_{ST} = E_{S1} - E_{T1} = 2J \tag{5}$$

Since J is given by Eq. 6, where φ and ψ represent the HOMO and LUMO wave functions, respectively, and e is the electron charge, it becomes evident that ΔE_{ST} can be minimized by reducing the overlap between the HOMO and LUMO orbitals.

$$J = \iint \phi(r_1)\psi(r_2)\left(\frac{e^2}{r_1 - r_2}\right)\phi(r_2)\psi(r_1)dr_1 dr_2 \tag{6}$$

One strategy to minimize the HOMO-LUMO overlap uses electron-rich donor (D) and electron-deficient acceptor (A) moieties arranged in D-A or D-A-D architectures that exhibit transitions with strong charge transfer (CT) character. To reduce HOMO and LUMO overlap even further, it is common to twist D and A moieties around the D-A axis by promoting steric hindrance, until near-orthogonality is achieved (Fig. 2). These features contribute significantly to reduce the ΔE_{ST} value, enabling efficient RISC, but, at the same time, lead to low radiative efficiencies, as suggested by the Franck-Condon principle [18]. Therefore, the need to minimize ΔE_{ST} and still achieve significant radiative decay from S_1 need to be effectively balanced in the molecular design of D-A systems for TADF emission.

Fig. 2 Schematic representation of HOMO-LUMO overlap and resulting singlet-triplet energy gap. Small singlet-triplet energy splitting is normally achieved when strong donor and acceptor units are used

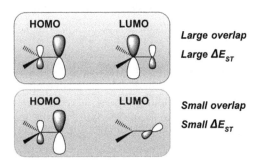

2   History of TADF

TADF is currently one of the most interesting topics in luminescence-related research, and the number of TADF emitters with great efficiency increased exponentially in the last 6 years. However, since its first publication, the understanding of the TADF mechanism has evolved throughout the years, and there are still many open questions, which could have impact on the design of more efficient TADF emitters in the different spectral regions. There are reports that pinpoint the first observation of TADF before 1941. However, it was Gilbert Lewis, David Lipkin, and Theodore Magel who published for the first time a study entirely dedicated to TADF [19]. Using fluorescein dissolved in boric acid, the authors measured the absorption, luminescence spectra (at different temperatures), and phosphorescence (PH) and TADF lifetimes. They determined the ΔE_{ST} for fluorescein using two different methods: the energy difference between TADF and PH emission maxima – spectroscopic value – and through the variation of the TADF decay rate constant with temperature [20]. At this point, the nature of the metastable state was still unknown. The assignment of the triplet as the metastable state was suggested by A. Terenin (1943) and confirmed by Gilbert Lewis and Michael Kasha (1944), who presented evidence that PH emission was originated from the triplet state while studying the energy of PH peaks of several organic molecules [21].

After the identification of the triplet as the metastable state, studies on TADF progressed to a better understanding of the mechanism and to the development of experimental methods, complementary to those introduced by Lewis, Lipkin, and Magel. In particular, Rosenberg and Shombert presented, in 1960, a method perfecting that introduced by Jablonski, in 1935, based on the ratio of TADF and PH intensities, at different temperatures [22]. This same method was later presented by Parker and Hatchard, in 1961, when studying TADF from eosin in glycerol and ethanol [7]. In this work, Parker and Hatchard introduced also terminology, used by some authors, to distinguish between delayed fluorescence created by TADF and bimolecular triplet-triplet annihilation (TTA). For TADF, the designation was e-type (from eosin), whereas for TTA, the designation was p-type (from pyrene). Five years later, Leach and Migirdicyan (1966) discussed the properties and different mechanisms of TADF and TTA [23]. One year later, Kropp and Dawson (1967) presented, for the first time, a study of the deuteration effect on the photophysical parameters, principally TADF, of coronene [8]. In 1968, Wilkinson and Horrocks introduced, for the first time, the expression "thermally activated delayed fluorescence," which gave origin to the widely used acronym TADF [24]. During the 1970s and 1980s, the photophysical properties of several molecules with TADF, essentially in ketones and porphyrins, were studied by several groups [9, 10, 25]. This has led to the development of new applications for this kind of molecules. Two of the most noteworthy applications that were developed during this time were its use as a luminescent probe, to characterize the membrane fluidity through lifetime of the rotational diffusion, and as an optical thermometer [25, 26].

The discovery of fullerenes, in 1985, opened a new area of investigation. Owing to its spherical-shaped structure and photophysical properties (high triplet formation quantum yield and small spectroscopic ΔE_{ST}), fullerenes were demonstrated to display TADF, first by Berberan-Santos and Garcia, in 1996, with fullerene C_{70}, and, 1 year later, by Berberan-Santos and co-workers, this time with fullerene C_{60} [5, 27]. In 2000, Weisman and co-workers published a study, based on the TADF properties, using time-resolved measurements, to elucidate the thermodynamic and kinetic properties of pristine C_{70} and derivatives [28]. In 2006, Baleizão and Berberan-Santos published the TADF characterization of C_{70}, dissolved in a solid matrix, and a year later, the application of C_{70} as temperature and oxygen sensor was reported by Berberan-Santos and co-workers [6, 29]. Four years later, in 2011, Baleizão and Berberan-Santos introduced, for the first time, the effect of carbon-13 (^{13}C) enrichment on the photophysical properties of fullerene C_{70} [30]. Finally, in 2013, Berberan-Santos and co-workers, following the work published in 2007, studied $^{13}C_{70}$ system as an oxygen and temperature sensor, reaching extraordinary detection limits [31].

The introduction of fullerenes into the TADF domain brought up a new application as dual sensor, for oxygen and temperature measurements. However, despite the striking results with the application of fullerenes as sensors, TADF was still missing a leading technological application. In 2009, Adachi and co-workers published the turning-point paper by introducing the application of the TADF as a triplet-harvesting mechanism in OLEDs, using a tin-based molecule.

3 Design and Properties of TADF Emitters

Significant scientific and technological interest was devoted to the development of novel TADF emitters since 2012, when the first TADF OLED using a pure organic emitter was published by Adachi and co-workers [32]. These luminescent compounds fall under three different categories, depending on their molecular structure: metal-organic complexes, donor-acceptor systems, and fullerenes. In each of these classes, the TADF processes occur from different photophysical processes, and, thus, their design follows distinct rules.

3.1 Metal-Organic Complexes

Prior to the discovery of TADF as a potential tool for OLED application, studies on this subject focused mainly on rare-metal phosphors (second- and third-row elements) with inherent large spin-orbit coupling (SOC) interactions that allow overcoming the spin-forbidden nature of phosphorescence ($T_1 \rightarrow S_0$). However, the tin-porphyrin complex OLED developed by Adachi and co-workers brought light to a different strategy for harvesting triplet excitons in lighting devices. By employing

d^{10} metal complexes, such as Cu(I), Ag(I), and Au(I), instead of d^6 or d^8 iridium(III) or platinum(II) complexes, the low SOC could be compensated by a small ΔE_{ST}, thus enabling RISC. This feature arises from a distinct charge transfer from the metal center (M) to a sterically hindered ligand (L), leading to a strong M-L charge separation, effectively working as a formal oxidation of M(I) to M(II) in the excited state [33]. In addition, the sterically constrained ligand and the d^{10} full d-orbitals render the low-lying triplet (T_1) high stability, by reducing the non-radiative deactivation pathway and the internal quenching of excited states, respectively [34–37]. The fundamental understanding of these properties leads to an abrupt interest in the design of novel metal-complex emitters for OLEDs. Copper(I) is the most widely investigated d^{10} metal for the preparation of TADF emitters that have low-lying metal-to-ligand charge transfer (MLCT) excited states and small ΔE_{ST}. However, the typical distorted tetrahedral geometry around the Cu(I) atom in these complexes usually leads to emitters with low luminescence quantum yield due to increased non-radiative decay of distorted excited states. This limitation can be overcome through the application of bulky chromophoric ligands that provide a firm framework around the metal center, suppressing both non-radiative internal conversion and environmental effects. These structural features were already present in a bis(triphenylphosphine)phenanthroline copper(I) complex reported in 1980, which is believed to be the first TADF-emitting Cu(I) complex (Fig. 3) [38]. The mechanistic study of the photophysical properties of this compound contributed to the development of various derivatives exhibiting TADF.

For many years, the common strategy for the design of novel emitters consisted in increasing the bulkiness of the chromophoric ligands, which greatly increased the photoluminescence quantum yield (PLQY). Devices prepared with such Cu(I) complexes yielded promising results, with OLEDs and light-emitting electrochemical cells achieving external quantum efficiencies (EQE) up to 16% [39–41]. However, these were cationic complexes that required the use of counterions, which usually lead to unexpected effects on the emissive properties under the strong electrical field used in OLEDs [42]. Therefore, neutral Cu(I) complexes were prepared using formally charged sterically hindered ligands (Fig. 4) [43–45]. These emitters displayed excellent PLQY, decent emission decay times, and good color stability in devices. In addition, the higher stability of these complexes allowed for fabrication of OLEDs via thermal sublimation, resulting in devices with high current efficiency and high EQE.

TADF-emitting copper(I) complexes can also be prepared using different structural organizations. Three-coordinate complexes (Fig. 5) have gathered significant interest, since the trigonal planar geometry suffers no distortion in the excited state as

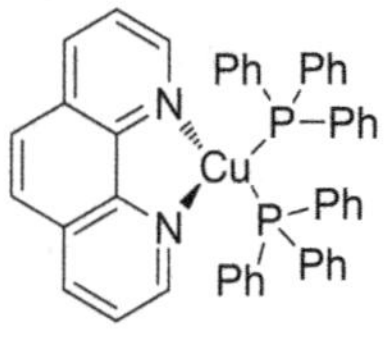

Fig. 3 Molecular structure of bis(triphenylphosphine) phenanthroline copper(I), the first Cu(I) complex exhibiting TADF

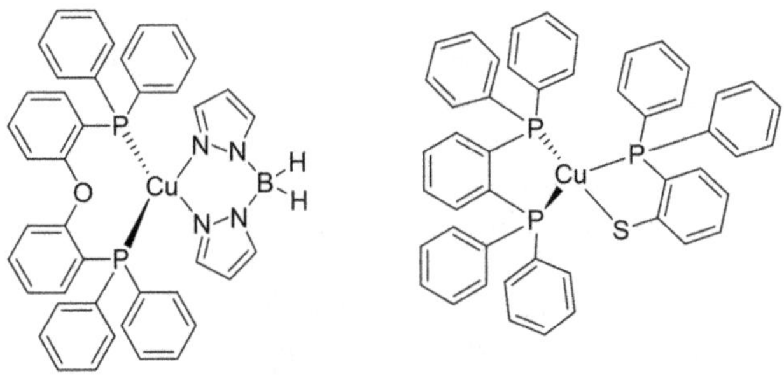

Fig. 4 Molecular structures of neutral Cu(I) complexes with TADF

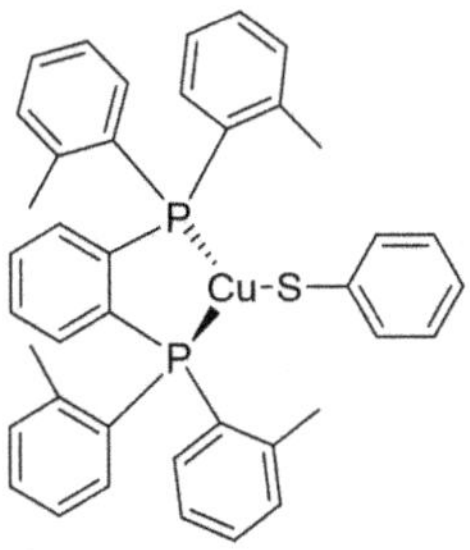

Fig. 5 Molecular structure of a tricoordinate Cu(I) complex with TADF

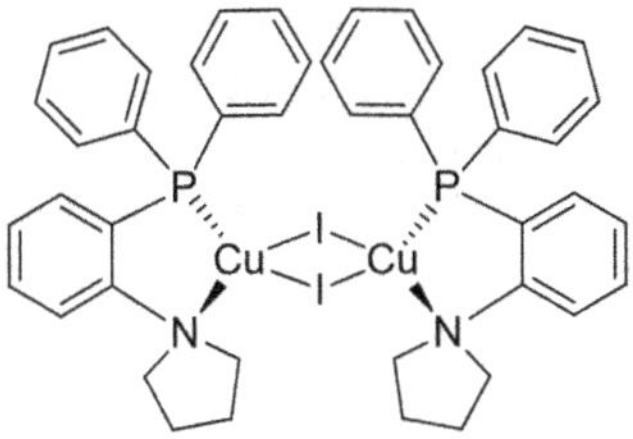

Fig. 6 Molecular structure of a binuclear Cu(I) complex with TADF

the tetrahedral geometry of four-coordinate analogs [46]. Thus, these compounds tend to exhibit high luminescence efficiency, and OLEDs prepared with such complexes have achieved remarkable EQEs, up to 21.3% [47].

The structural rearrangement between ground and emissive excited states can also be prevented in four-coordinate binuclear Cu(I) complexes (Fig. 6), due to the strong covalency of the Cu_2N_2 core. In addition, these complexes do not require counterions and are, thus very stable and can be deposited by sublimation methods for device fabrication. Four-coordinate Cu(I) complexes exhibit excellent emissive properties (PLQY up to 80%), which arise from (metal+halide)-to-ligand charge transfer transitions, and the corresponding devices show comparable performances to current rare-metal-based PhOLEDs [48, 49].

Similar to the described Cu(I) complexes, Ag(I) and Au(I) complexes also have d^{10} configuration, and their photophysical properties follow the same principles, being potentially applicable for TADF OLEDs. In fact, the occurrence of TADF in Ag(I) [50] and Au(I) [51] complexes has been reported and gathered interest in recent years. In parallel to what occurred in Cu(I) systems, four-coordinate mononuclear compounds tend to exhibit poor-to-medium PLQY [33] due to geometry distortion in the excited state and, therefore, were not used successfully in OLED fabrication. Recently, reports on different geometries, such as trigonal planar Au(I) and tetranuclear Ag(I) complexes, demonstrated that these limitations can be overcome, and decent emissive properties are obtainable. However, low compound stability and solubility problems are still a limitation in these systems, and successful OLED fabrication has not been achieved.

3.2 Purely Organic TADF Emitters

Despite the promising potential of TADF-emitting Cu(I) complexes, the majority of research studies are focused on purely organic compounds. This is because small organic luminophores offer distinct advantages over metal-organic complexes in terms of precise molecular structure, achievable high purity from recrystallization or sublimation, tailorable modification, and low structural distortion in the excited state [52]. However, to achieve efficient RISC and, ultimately, TADF in purely organic compounds, special design rules must be followed. Most donor units used for the design of TADF emitters fall into the aromatic amine group, e.g., carbazole (Cz), diphenylamine, phenoxazine, or phenothiazine (Fig. 7), owing to their strong electron-donating ability and stable, high-energy, triplet states. On the other hand, the choice of acceptor units is rather crucial, as it not only affects ΔE_{ST} but also greatly influences radiative properties such as emission color, excited-state lifetime, and TADF efficiency.

Cyano-based acceptors, e.g., phthalonitriles (PN) and tetracyanobenzenes, are widely used to build TADF emitters, due to the remarkable electron affinity of the cyano group. Such emitters usually display high PLQY, short delayed fluorescence (DF) lifetimes, and green-to-yellow emission [53–55]. Triazine and heptazine are aromatic heterocycles that also have great electron affinity and, thus, are also extensively applied in TADF emitter preparation. The stronger electron-withdrawing character of these groups increases the CT character of the excited state, which

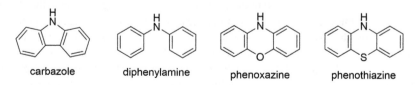

| carbazole | diphenylamine | phenoxazine | phenothiazine |

Fig. 7 Examples of commonly used donor units in organic TADF emitters

usually shifts the emission to longer wavelengths, increases DF lifetime, and reduces overall PLQY [56–58]. When efficient blue TADF emission is desired, π-conjugation length, the redox potential, and degree of conjugation of donor and acceptor moieties must be considered. To this end, diphenyl sulfoxide (DPSO$_2$) and dibenzothiophene-S,S-dioxide (DBTO$_2$) (Fig. 8) are used as acceptor moieties to construct TADF emitters, since the tetrahedral geometry of the sulfonyl group limits the conjugation of the π-system, and the hexavalent sulfur atom attached to two oxygen atoms provides the necessary electronegativity and electron-withdrawing features for charge transfer to occur in the excited state [59]. TADF emitters bearing these units tend, in fact, to exhibit their emission at shorter wavelengths (blue-to-green range), albeit with larger ΔE_{ST} values and, consequently, longer DF lifetimes [60–62].

The number of D and A units and their derivatization position also play a critical role in the properties of the final emitter. For instance, studies in isomers of Cz-DBTO$_2$ emitters showed that TADF is switched off altogether when the Cz units change from positions 2 and 8 to 3 and 7 (Fig. 9), due to an increase in ΔE_{ST}. Several works with Cz-PN derivatives with varying number of D units and substitution positions resulted in emitters with very distinct properties [61, 63]. For example, by varying the relative position of D and A units in tetracarbazolyl-dicyanobenzene isomers, the emission color shifts from sky blue to yellow, and

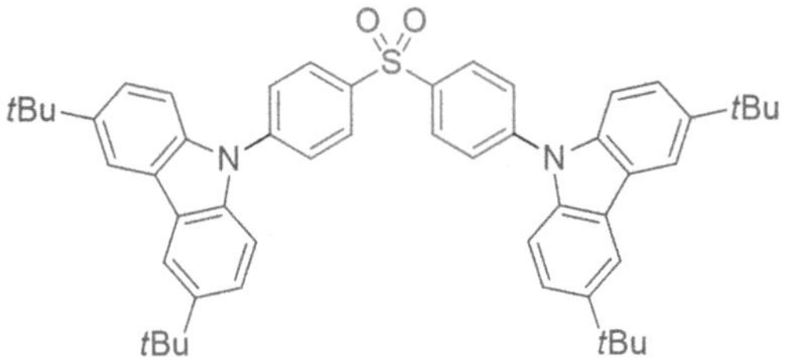

Fig. 8 Example of a TADF emitter bearing DPSO$_2$ as acceptor unit

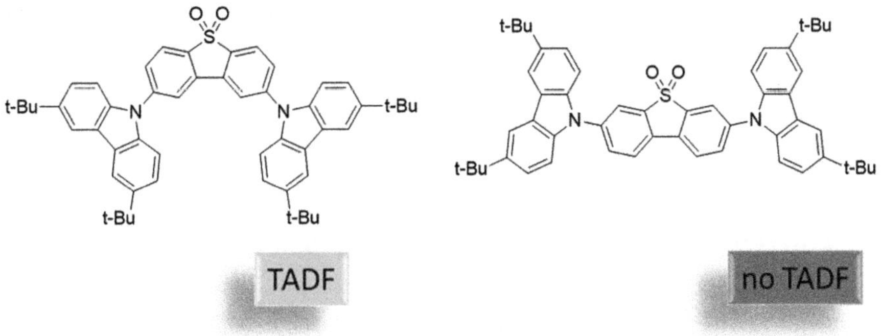

Fig. 9 Molecular structure of two regioisomers of Cz-DBTO$_2$ derivatives with distinct DF properties [62]

Fig. 10 Molecular structure
of C_{70}, a spherical fullerene
exhibiting TADF

C_{70}

the EQE of corresponding devices increases from 8.5% to 36% [15]. Using the same
acceptor units, the removal of carbazolyl donors greatly shifts the emission to shorter
wavelengths and slightly lowers the singlet-triplet splitting [64].

The tuning of TADF emission properties can also be done through additional
changes in the molecular structure. The electron density and donating character of D
units can be increased through the functionalization with alkyl or aryl moieties, as
demonstrated in studies of Cz-PN derivatives, whose emission shifts to longer
wavelengths upon addition of methyl and phenyl units. This approach can also be
used to increase bulkiness of D and A moieties to achieve near-orthogonality around
D-A axis, improving HOMO and LUMO separation and reducing ΔE_{ST}, and
enhance PLQY by suppressing vibrational quenching [65]. However, this approach
can sometimes lead to unexpected results, and room-temperature phosphorescence
may compete with the RISC process [66, 67]. Despite the advances of recent years
on the design of TADF molecules, further studies are still necessary to optimize
molecular structures with improved tuning of their emission and improve stability.

Fullerenes are a special class of organic molecules all composed of carbon atoms
that have gathered significant technological interest in recent years, especially
carbon nanotubes, due to their remarkable tensile strength, high conductivity, and
peculiar photophysical properties [68]. The fluorescence quantum yield of fullerenes, especially from C_{60} to C_{70}, which are the most common, is very low, near
10^{-3}. The triplet formation quantum yield, on the other hand, almost reaches the
unity, being 0.995 [4–6]. In addition, fullerenes have also a long triplet lifetime, a
small ΔE_{ST}, and an efficient RISC process, which arise from their large and
delocalized π-caged system and highly symmetrical curved structure (Fig. 10).
These features make fullerenes ideal candidates for TADF materials, but their low
solubility in several solvents and their low fluorescence quantum yield have hindered their application in lighting devices. However, fullerenes have been studied in
sensing applications with very promising results, which will be discussed later in this
chapter.

4   Current Challenges for Lighting Applications

Although TADF emerged as a promising approach to achieve efficient blue emission
in OLEDs and to solve some of the issues that affect metal-complex phosphorescent
OLEDs, most efficient TADF molecules exhibit emission in the green region.
Although blue and red emitters have been described recently, their performance in

lighting devices is still suboptimal, and optimized structures are needed for these spectral regions. The design of blue TADF emitters is difficult because it requires weak donors and acceptors, so the CT emission is not strongly shifted to lower energies. However, these features also result in higher ΔE_{ST} values [69]. On the other hand, red TADF emission is strongly affected by non-radiative decay, in agreement with the energy gap law, which competes with the RISC process. Moreover, the TADF mechanism itself is far from being fully understood, which limits the design of new and more efficient emitters in these two regions. Experimental evidence seems to indicate that RISC involves not only the CT singlet and triplet states but also triplet states that are localized on the D or A units of the TADF molecule, mediated by strong vibronic coupling [70]. The importance of such effects is still uncertain as it is the possibility of hyperfine coupling interactions being also involved in the RISC mechanism.

A widespread problem in the development of TADF-based devices is the decrease in EQE at high current densities, commonly known as the efficiency "roll-off" effect [71, 72]. This phenomenon is attributed to exciton quenching, promoted by singlet-triplet and triplet-triplet annihilation mechanisms, and polaron-exciton interactions; therefore, it is more pronounced in systems with long luminescence lifetimes (Fig. 11) [73]. In addition to efficiency roll-off, long-lived excited states also lead to low stability and reduced operation lifetime of the devices, due to secondary reactions that result in hole-transport and electron-transport material degradation. Thus, devices with larger ΔE_{ST} values, such as blue TADF emitters, are particularly affected by these problems. Several approaches were devised to circumvent roll-off and stability problems, most of which are focused on the choice

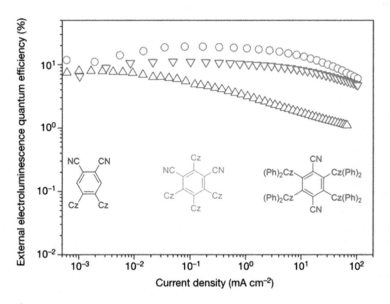

Fig. 11 Evidence of efficiency "roll-off" of OLED prepared from TADF-emitting PN-Cz derivatives. Adapted from [64]

of host materials for TADF emitters and device preparation steps in order to reduce excited-state lifetime [74–76]. However, great research efforts aim to optimize the TADF mechanism in order to promote faster RISC rates, thus decreasing luminescence lifetimes.

One more drawback affecting TADF OLEDs is the restriction of their application in small devices, due to the mandatory vacuum sublimation of the emitting layer step. In order to prepare large-area displays, methods involving solution processing of the emitters have attracted attention owing to their simple fabrication processes and relatively low cost. However, the performance of the resulting devices is usually considerably inferior, mostly due to the heterogeneity in the emitter dispersion and poor interface separation that arise upon solvent application [77].

The key to overcome these challenges lies in the fundamental understanding of the structure-property relationship of TADF emitters. However, to simultaneously achieve small ΔE_{ST}, high PLQY, and a fast rate of the reverse intersystem crossing with high stability and processability in a readily synthesizable molecular structure is still a great challenge. New models and challenging views for designing new types of TADF molecules could have a profound effect on the development of highly efficient TADF optoelectronic materials.

5   Recent Advances in TADF Materials

In recent years, new strategies to achieve efficient TADF in organic molecules that are not based on small D-A molecules were developed. Supramolecular approaches have been explored with relative success, both in polymeric and dendrimer structures and in exciplex systems [78–80]. Exciplexes are formed when a molecule in the excited state interacts with a different molecule in the ground state, forming a short-lived complex that is dissociative in the ground state. In the case of TADF-emitting exciplexes, they are essentially a charge transfer state whose emission occurs because of an electron transition from the LUMO of an acceptor to the HOMO of a donor (Fig. 12). Due to the resulting large spatial separation of these orbitals, triplet and singlet energy levels are remarkably low. This approach consists of mixing hole-transport-type donor materials with electron-transport-type acceptor materials and has been used extensively in the preparation of devices [81, 82]. However, due to the bimolecular nature of the emitting species, non-radiative decays are very significant, and the corresponding EQE values have so far not exceed 15% [81].

TADF-emitting polymers have also been recently developed, mostly to be applied in solution preparation methods, such as spin-coating, die-casting, or inkjet printing for low-cost displays. Achieving efficient TADF in polymers is, however, challenging due to the strong non-radiative internal conversion affecting the triplet state of high molecular weight compounds. In addition, the triplet population in polymers and oligomers is often efficiently quenched by operative intramolecular and intermolecular TTA. Therefore, additional design rules should be followed when designing effective TADF polymers, in order to efficiently separate the

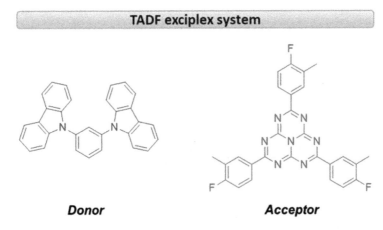

Fig. 12 Molecular structures of two molecules that compose a TADF exciplex system

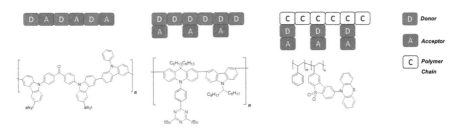

Fig. 13 Schematic representation of possible monomer arrangements and molecular structures of representative TADF polymers

HOMO and LUMO orbitals and achieve low ΔE_{ST} while confining the excitons in the luminophore to improve PLQY (Fig. 13) [83, 84]. Devices prepared from TADF polymers through wet processes usually have lower performances than their OLED counterparts, with EQE lower than 10% [85].

Dendrimer TADF emitters are a particular case of TADF polymers, in which the molecular structure is perfectly branched with exact molecular weights. Owing to their unique structure with a high steric hindrance, they are generally highly soluble and amorphous and can isolate chromophores at the core to prevent concentration quenching. Carbazole dendrimers are the most representative emitters of this class, being prepared of an electron-poor core, to which carbazole units are radially attached (Fig. 14) [79]. Devices prepared with these emitters through wet processes have displayed remarkable properties, with good EQE values and reduced efficiency roll-off at high current densities, which represents a decent alternative for low-cost display fabrication [86].

Currently, different mechanisms to enable TADF are emerging as powerful tools for the development of novel emitters. Owing to their electronic abilities with empty

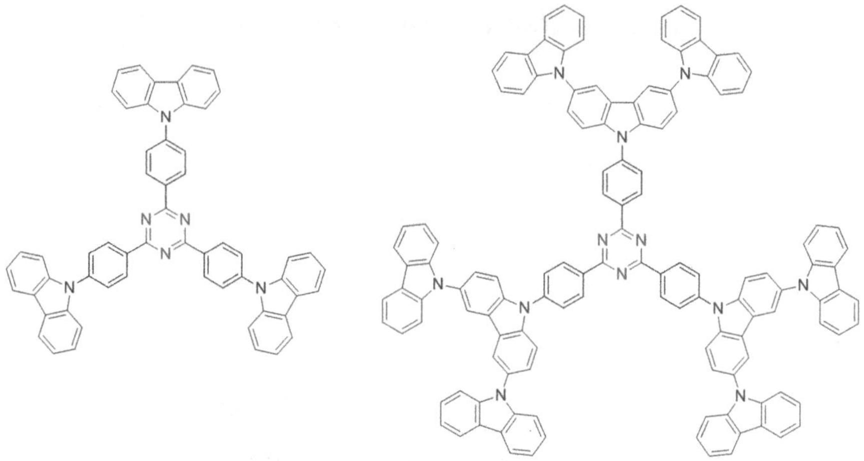

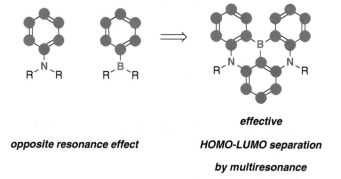

Fig. 14 Molecular structure of dendrimeric emitters exhibiting TADF

opposite resonance effect

effective

HOMO-LUMO separation

by multiresonance

Fig. 15 Schematic representation of efficient HOMO-LUMO separation by multi-resonance effect. The localized HOMO and LUMO are indicated by blue and red circles, respectively

p orbitals, boron-containing molecules have gathered considerable attention for the preparation of novel emitters. In particular, triarylboron moieties have been coupled with nitrogen-containing aromatic compounds to obtain rigid, polycyclic, aromatic frameworks that display multi-resonance effect. This phenomenon arises from the opposite resonance effects of the nitrogen and boron atoms and allows efficient separation of HOMO and LUMO without the use of donor and acceptor moieties nor sterically hindered structures (Fig. 15).

The resulting compounds exhibit high PLQY, sharp emission in the blue region, and negligible concentration quenching, leading to highly efficient blue-emitting OLEDs, with significantly lower efficiency roll-off than conventional TADF-based OLEDs [87].

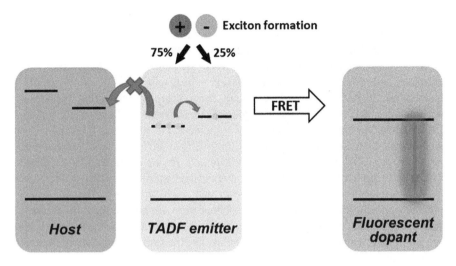

Fig. 16 Schematic illustration of proposed energy transfer mechanism in the emitter dopant/ assistant dopant/host matrix system of hyperfluorescence TADF

Similarly to multi-resonance effects, additional strategies to achieve efficient HOMO and LUMO separation were developed. Excited-state intramolecular proton transfer (ESIPT) is also emerging as a promising tool for this purpose. This is essentially a tautomerization process that leads to the formation of a keto form in the excited state of an enol molecule. The keto form commonly exhibits a better separation of its HOMO and LUMO and, therefore, promotes efficient RISC. However, this strategy is still significantly unexplored, and devices have not exceeded 14% in EQE [88].

Finally, and maybe, the most promising approach to address current TADF-based OLED limitation is the hyperfluorescence strategy. This combines TADF emitters with conventional fluorescent dyes to promote Förster resonance energy transfer (FRET) from the TADF molecule to the emissive dye (Fig. 16). This mechanism results in relative shorter emission lifetimes and, thus, improved operational stability and reduced efficiency roll-off. In addition, due to the flexibility in the selection of the fluorophore, narrow emission bands and high color purity across the visible spectrum are achievable [89]. Due to these features, hyperfluorescence-based OLEDs are the most probable TADF-based devices to reach the display market in the near future.

6 Sensing and Imaging

The search for methods to overcome the restrictions imposed by the charge recombination spin statistics that limits maximum IQE to 25% in lighting devices was the major driving force for the development of TADF-based materials. However, several researchers have taken a different view at this photophysical phenomenon and start to investigate its potential for different applications. Since TADF emission arises from triplet states, it is inherently sensitive to properties of the surrounding media, such as oxygen concentration and viscosity. In addition, the RISC process is thermally assisted, rendering TADF emission to be also sensitive to the surrounding temperature. These features make TADF emitters highly interesting for sensing applications, and several works exploring their potential have been reported [31, 90–93]. As previously mentioned, fullerenes are a special class of carbon compounds that display strong TADF emission at high temperatures but were never successfully used in lighting applications. However, fullerenes have been extensively applied in oxygen and temperature sensors, often coupled with another emitter to allow ratiometric measurements with high sensitivity. Oxygen sensors prepared from fullerene dispersions in polymer films are among the most sensitive, with limits of detection below 1 ppmv [31]. Temperature sensors prepared with oxygen-impermeable polymers also display high sensitivity and broad working ranges [90].

Cu(I) complexes have also been employed in oxygen sensors due to the luminescence quenching of the long-lived triplet excited state in TADF and phosphorescence emission processes [91].

TADF emitters, in particular, fluorescein derivatives, have been tested as luminescent probes for microscopy imaging [92, 93]. Owing to its relatively long emission lifetime, TADF represents a powerful tool for imaging techniques, enabling time-gated acquisition methods that significantly improve signal-to-noise ratio by eliminating autofluorescence and light-scattering effects. In addition, the TADF-emitting probes do not require the application of heavy-metal complexes that pose toxicity and availability problems. However, the application of TADF emitters as luminescent probes remains significantly unexplored, with only a small number of works on this topic.

7   Conclusions and Future Perspectives

The recent developments in the study of TADF emitters led to excellent electroluminescent performances exceeding those of conventional phosphorescent complex-based devices, owing to the triplet-harvesting capability in these compounds that promote device IQE close to 100%. Furthermore, TADF emitters circumvent the use of heavy metals and lanthanides, reducing the cost of starting materials and, at the same time, providing flexibility and processability of the luminescent materials. However, the TADF mechanism is still not fully understood, and further fundamental studies remain of great scientific and technological interest. While multimolecular approaches have been used to circumvent current limitations, the design of TADF emitters with optimal photophysical properties is the key to achieve efficient and durable devices. Optimization of the molecular structure aims higher PLQY values, smaller ΔE_{ST} values, shorter triplet lifetimes, and narrower emission bands, resulting in devices with better efficiency, more durability, and improved color purity. Solution processing remains the most promising approach for low-cost large-area devices, although several issues must be addressed before efficient devices are prepared, such as improving solubility of TADF emitter in common solvents while maintaining desirable emissive properties. Macromolecular TADF materials, such as polymeric emitters, represent a promising tactic to address this issue, although the photophysical mechanisms are rather complex and require further study to determine the effect of molecular aggregation or conformation changes, for instance.

Apart from their application as the integral part of the emissive layer in OLEDs, TADF materials are also being tested as host materials, hole- or electron-transport materials, and exciton blocking layers, and devices with excellent performances are expected to appear in the near future in commercial applications such as mobile phones, wearable displays, and luminophores.

Although there are no TADF-based products available in the market, several companies are developing products that will reach the market soon. Kyulux is a Japan-based company founded in 2015 to develop next-generation materials for OLED displays and lighting using exclusively licensed technology from Kyushu University. In 2018, Kyulux engaged in a joint development with Samsung and LG focusing on blue TADF and Hyperfluorescence™ technology that aims to provide the world's first commercially viable Hyperfluorescence™ AMOLED in mid-2019. With more than 15 years old of existence, Novaled is one of the most influential companies in OLED industry, specializing in high-efficiency, long-lifetime devices. In 2015, Novaled integrated the European Commission-funded PHEBE project with the device manufacturer Astron-FIAMM and research units from TU Dresden, Durham University, and Kaunas University with the goal of preparing iridium-free efficient OLED devices based on organic TADF emitters. Cynora is a Germany-based company founded in 2008 that focuses on the use of TADF technology for the development of OLEDs. Recently, after extending the Joint Development Agreement with LG Display, Cynora has presented their latest achievements in the deep-blue emitter materials for different approaches of the TADF technology, self-emitting or co-emitting, which also includes Hyperfluorescence^{TM}-type approach.

Acknowledgements Fundação para a Ciência e a Tecnologia (FCT) is acknowledged for funding fellowships SFRH/BPD/120599/2016 (JA) and SFRH/BD/118525/2016 (TP) and project PTDC/QUIQFI/32007/2017.

References

1. Parker CA (1968) Photoluminescence of solutions. Elsevier, Amsterdam
2. Valeur B, Berberan-Santos MN (2012) Molecular fluorescence. Principles and applications. Wiley-VCH, Weinheim
3. Yersin H, Mataranga-Popa L, Li SW, Czernwieniec R (2018) Design strategies for materials showing thermally activated delayed fluorescence and beyond: towards the fourth-generation OLED mechanism. J Soc Inf Disp 26:194–199. https://doi.org/10.1002/jsid.654

4. Baleizão C, Berberan-Santos MN (2007) Thermally activated delayed fluorescence as a cycling process between excited singlet and triplet states. Application to the fullerenes. J Chem Phys 126:204510. https://doi.org/10.1063/1.2734974
5. Berberan-Santos MN, Garcia JMM (1996) Unusually strong delayed fluorescence of C70. J Am Chem Soc 118:9391–9394. https://doi.org/10.1021/ja961782s
6. Baleizão C, Berberan-Santos MN (2006) A molecular thermometer based on the delayed fluorescence of C70 dispersed in a polystyrene film. J Fluoresc 16:215–219. https://doi.org/10.1007/s10895-005-0049-5
7. Parker CA, Hatchard C (1961) Triplet-singlet emission in fluid solutions. Phosphorescence of eosin. Trans Faraday Soc 57:1894–1904. https://doi.org/10.1039/TF9615701894
8. Kropp JL, Dawson WR (1967) Radiationless deactivation of triplet coronene in plastics. J Phys Chem 71:4499–4506. https://doi.org/10.1021/j100872a054
9. Callis JB, Gouterman M, Jones Y, Henderson BH (1971) Porphyrins XXII: fast fluorescence, delayed fluorescence, and quasiline structure in palladium and platinum complexes. J Mol Spectrosc 39:410–420. https://doi.org/10.1016/0022-2852(71)90212-8
10. Brown RE, Singer LA, Parks JH (1972) Prompt and delayed fluorescence from benzophenone. Chem Phys Lett 14:193–195. https://doi.org/10.1016/0009-2614(72)87176-8
11. Endo A, Ogasawara M, Takahashi A, Yokoyama D, Kato Y, Adachi C (2009) Thermally activated delayed fluorescence from Sn(4+)-porphyrin complexes and their application to organic light emitting diodes – a novel mechanism for electroluminescence. Adv Mater 20:4802–4806. https://doi.org/10.1002/adma.200900983
12. Volz D (2016) Review of organic light-emitting diodes with thermally activated delayed fluorescence emitters for energy-efficient sustainable light sources and displays. J Photon Energ 6:020901. https://doi.org/10.1117/1.JPE.6.020901
13. Cai X, Li X, Xie G, He Z et al (2016) "Rate-limited effect" of reverse intersystem crossing process: the key for tuning thermally activated delayed fluorescence lifetime and efficiency roll-off of organic light emitting diodes. Chem Sci 7:4264–4275. https://doi.org/10.1039/c6sc00542
14. Yersin H, Rausch AF, Czerwieniec R, Hofbeck T, Fischer T (2011) The triplet state of organo-transition metal compounds. Triplet harvesting and singlet harvesting for efficient OLEDs. Coord Chem Rev 255:2622–2652. https://doi.org/10.1016/j.ccr.2011.01.042
15. Uoyama H, Goushi K, Shizu K et al (2012) Highly efficient organic light-emitting diodes from delayed fluorescence. Nature 492:234–238. https://doi.org/10.1038/nature11687
16. Dias FB (2015) Kinetics of thermal-assisted delayed fluorescence in blue organic emitters with large singlet–triplet energy gap. Philos Trans R Soc A Math Phys Eng Sci. 373. pii: 20140447. doi: https://doi.org/10.1098/rsta.2014.0447
17. Turro NJ, Scaiano JC, Ramamurthy V (2010) Principles of molecular photochemistry: an introduction. University Science Books, Mill Valley
18. Rajamalli P, Senthilkumar N, Gandeepan P, Huang PY et al (2016) A new molecular design based on thermally activated delayed fluorescence for highly efficient organic light emitting diodes. J Am Chem Soc 138:628–634. https://doi.org/10.1021/jacs.5b10950
19. Lewis GN, Lipkin D, Magel T (1941) Reversible photochemical processes in rigid media. A study of the phosphorescent state. J Am Chem Soc 63:3005–3018. https://doi.org/10.1021/ja01856a043
20. Jablonski A (1935) Über den Mechanismus der Photolumineszenz von Farbstoffphosphoren. Z Physik 94:38–46. https://doi.org/10.1007/BF01330795
21. Lewis GN, Kasha M (1944) Phosphorescence and the triplet state. J Am Chem Soc 66:2100–2116. https://doi.org/10.1021/ja01240a030
22. Rosenberg JL, Shombert DJ (1960) The phosphorescence of adsorbed acriflavine. J Am Chem Soc 82:3252–3257. https://doi.org/10.1021/ja01498a006
23. Leach S, Migirdicyan E (1966) Fluorescence a longue durée de vie de composés organiques. In: Haissinsky M (ed) Actions chimiques et biologiques des radiations. 9th edn. Masson, Paris, pp 117–186
24. Wilkinson F, Horrocks AR (1968) Phosphorescence and delayed fluorescence of organic substances. In: Bowen EJ (ed) Luminescence in chemistry. Van Nostrand, London, pp 116–153

25. Garland P, Moore CH (1979) Phosphorescence of protein-bound eosin and erythrosin. A possible probe for measurements of slow rotational mobility. Biochem J 183:561–572. https://doi.org/10.1042/bj1830561
26. Fister JC, Rank D, Harris JM (1995) Delayed fluorescence optical thermometry. Anal Chem 67:4269–4275. https://doi.org/10.1021/ac00119a011
27. Salazar FA, Fedorov A, Berberan-Santos MN (1997) A study of thermally activated delayed fluorescence in C_{60}. Chem Phys Lett 271:361–366. https://doi.org/10.1016/S0009-2614(97)00469-7
28. Bachilo SM, Benedetto AF, Weisman RB et al (2000) Time-resolved thermally activated delayed fluorescence in C_{70} and 1,2-$C_{70}H_2$. J Phys Chem A 104:11265–11269. https://doi.org/10.1021/jp002742k
29. Baleizão C, Nagl S, Schäferling M et al (2008) Dual fluorescence sensor for trace oxygen and temperature with unmatched range and sensitivity. Anal Chem 80:6449–6457. https://doi.org/10.1021/ac801034p
30. Baleizão C, Berberan-Santos MN (2011) The brightest fullerene: a new isotope effect in molecular fluorescence and phosphorescence. Chemphyschem 12:1247–1250. https://doi.org/10.1002/cphc.201100156
31. Kochmann S, Baleizão C, Berberan-Santos MN, Wolfbeis OS (2013) Sensing and imaging of oxygen with parts per billion limits of detection and based on the quenching of the delayed fluorescence of (13)C_{70} fullerene in polymer hosts. Anal Chem 85:1300–1304. https://doi.org/10.1021/ac303486f
32. Uoyama H, Goushi K, Shizu K, Nomura H, Adachi C (2012) Highly efficient organic light-emitting diodes from delayed fluorescence. Nature 492:234–238. https://doi.org/10.1038/nature11687
33. Osawa M, Kawata I, Ishii R, Igawa S, Hashimoto M, Hoshino M (2013) Application of neutral d^{10} coinage metal complexes with an anionic bidentate ligand in delayed fluorescence-type organic light-emitting diodes. J Mater Chem C 1:4375–4383. https://doi.org/10.1039/C3TC30524D
34. Zhang Q, Zhou Q, Cheng Y et al (2006) Highly efficient electroluminescence from green-light-emitting electrochemical cells based on Cu^I complexes. Adv Funct Mater 16:1203–1208. https://doi.org/10.1002/adfm.200500691
35. Deaton JC, Switalski SC, Kondakov DY et al (2010) E-type delayed fluorescence of a phosphine-supported Cu2(mu-NAr2)2 diamond core: harvesting singlet and triplet excitons in OLEDs. J Am Chem Soc 132:9499–9508. https://doi.org/10.1021/ja1004575
36. Zhang Q, Komino T, Huang S et al (2012) Triplet exciton confinement in green organic light-emitting diodes containing luminescent charge-transfer Cu(I) complexes. Adv Funct Mater 22:2327–2336. https://doi.org/10.1002/adfm.201101907
37. Wallesch M, Volz D, Zink DM et al (2014) Bright coppertunities: multinuclear Cu(I) complexes with N-P ligands and their applications. Chemistry 20:6578–6590. https://doi.org/10.1002/chem.201402060
38. Blasse G, McMillin DR (1980) On the luminescence of bis (triphenylphosphine) phenanthroline copper (I). Chem Phys Lett 70:1–3. https://doi.org/10.1016/0009-2614(80)80047-9
39. Palmer C, Mcmillin DR (1987) Singlets, triplets, and exciplexes: complex, temperature-dependent emissions from (2,9-dimethyl-1,10 phenanthroline)bis (triphenylphosphine)copper (1+) and (1,10phenanthroline) (triphenylphosphine)copper(1+). Inorg Chem 26:3837–3840. https://doi.org/10.1021/ic00270a004
40. Kuang S, Cuttell DG, McMillin DR et al (2002) Synthesis and structural characterization of Cu(I) and Ni(II) complexes that contain the bis[2 (diphenylphosphino)phenyl]ether ligand. Novel emission properties for the Cu(I) species. Inorg Chem 41:3313–3322. https://doi.org/10.1021/ic0201809
41. Cuttell DG, Kuang S, Fanwick PE, McMillin DR, Walton RA (2002) Simple Cu(I) complexes with unprecedented excited-state lifetimes. J Am Chem Soc 124:6–7. https://doi.org/10.1021/ja012247h

42. Tao Y, Yuan K, Xu P et al (2014) Thermally activated delayed fluorescence materials towards the breakthrough of organoelectronics. Adv Mater 17:7931–7958. https://doi.org/10.1002/adma.201402532

43. Czerwieniec R, Yu J, Yersin H (2011) Blue-light emission of Cu(I) complexes and singlet harvesting. Inorg Chem 50:8293–8301. https://doi.org/10.1021/ic200811a

44. Czerwieniec R, Yu J, Yersin H (2012) Correction to blue-light emission of Cu(I) complexes and singlet harvesting. Inorg Chem 51:1975–1975. https://doi.org/10.1021/ic202653h

45. Igawa S, Hashimoto M, Kawata I et al (2013) Highly efficient green organic light-emitting diodes containing luminescent tetrahedral copper(I) complexes. J Mater Chem C 1:542–551. https://doi.org/10.1039/C2TC00263A

46. Hashimoto M, Igawa S, Yashima M et al (2011) Highly efficient green organic light-emitting diodes containing luminescent three-coordinate copper(I) complexes. J Am Chem Soc 133:10348–10351. https://doi.org/10.1021/ja202965y

47. Osawa M (2014) Highly efficient blue-green delayed fluorescence from copper(I) thiolate complexes: luminescence color alteration by orientation change of the aryl ring. Chem Commun 50:1801–1803. https://doi.org/10.1039/C3CC47871H

48. Zink DM, Bächle M, Baumann T et al (2013) Synthesis, structure, and characterization of dinuclear copper(I) halide complexes with P^N ligands featuring exciting photoluminescence properties. Inorg Chem 52:2292–2305. https://doi.org/10.1021/ic300979c

49. Leitl MJ, Küchle F, Mayer HA et al (2013) Brightly blue and green emitting Cu(I) dimers for singlet harvesting in OLEDs. J Phys Chem A 117:11823–11836. https://doi.org/10.1021/jp402975d

50. Gan XM, Yu R, Chen XL et al (2018) Controlled syntheses of cubic and hexagonal $ZnIn_2S_4$ nanostructures with different visible-light photocatalytic performance. Dalton Trans 40:2607–2613. https://doi.org/10.1039/C0DT01435D

51. Osawa M, Aino M, Nakagura T et al (2018) Near-unity thermally activated delayed fluorescence efficiency in three- and four-coordinate Au(I) complexes with diphosphine ligands. Dalton Trans 47:8229–8239. https://doi.org/10.1039/C8DT01097H

52. Yang Z, Mao Z, Xie Z et al (2017) Recent advances in organic thermally activated delayed fluorescence materials. Chem Soc Rev 46:915–1016. https://doi.org/10.1039/C6CS00368K

53. Cho YJ, Yook KS, Lee JY (2014) High efficiency in a solution-processed thermally activated delayed-fluorescence device using a delayed-fluorescence emitting material with improved solubility. Adv Mater 26:6642–6646. https://doi.org/10.1002/adma.201402188

54. Kretzschmar A, Patze C, Schwaebel ST, Bunz UHF (2015) Development of thermally activated delayed fluorescence materials with shortened emissive lifetimes. J Org Chem 80:9126–9131. https://doi.org/10.1021/acs.joc.5b01496

55. Park IS, Lee SY, Adachi C, Yasuda T (2016) Full-color delayed fluorescence materials based on wedge-shaped phthalonitriles and dicyanopyrazines: systematic design, tunable photophysical properties, and OLED performance. Adv Funct Mater 26:1813–1821. https://doi.org/10.1002/adfm.201505106

56. Chang CH, Kuo MC, Lin WC et al (2012) A dicarbazole–triazine hybrid bipolar host material for highly efficient green phosphorescent OLEDs. J Mater Chem 22:3832–3838. https://doi.org/10.1039/C2JM14686J

57. Marghad I, Clochard MC, Ollier N et al (2015) Thermally activated delayed fluorescence evidence in non-bonding transition electron donor-acceptor molecules. Proc SPIE 9566:956629. https://doi.org/10.1117/12.2186802

58. Takahashi T, Shizu K, Yasuda T, Togashi K, Adachi C (2014) Donor-acceptor-structured 1,4-diazatriphenylene derivatives exhibiting thermally activated delayed fluorescence: design and synthesis, photophysical properties and OLED characteristics. Sci Technol Adv Mater 15:034202. https://doi.org/10.1088/1468-6996/15/3/034202

59. Grabowski ZR, Rotkiewicz K, Rettig W (2003) Structural changes accompanying intramolecular electron transfer: focus on twisted intramolecular charge-transfer states and structures. Chem Rev 103:3899–4032. https://doi.org/10.1021/cr9407451

60. Zhang Q, Li J, Shizu K et al (2012) Design of efficient thermally activated delayed fluorescence materials for pure blue organic light emitting diodes. J Am Chem Soc 134:14706–14709. https://doi.org/10.1021/ja306538w

61. Dias FB, Bourdakos KN, Jankus V et al (2013) Triplet harvesting with 100% efficiency by way of thermally activated delayed fluorescence in charge transfer OLED emitters. Adv Mater 25:3707–3714. https://doi.org/10.1002/adma.201300753

62. Huang B, Qi Q, Jiang W et al (2014) Thermally activated delayed fluorescence materials based on 3,6-di-*tert*-butyl-9-((phenylsulfonyl)phenyl)-9*H*-carbazoles. Dyes Pigm 111:135–144. https://doi.org/10.1016/j.dyepig.2014.06.008

63. Huang R, Avó J, Northey T et al (2017) The contributions of molecular vibrations and higher triplet levels to the intersystem crossing mechanism in metal-free organic emitters. J Mater Chem C 5:6269–6280. https://doi.org/10.1039/C7TC01958K

64. Nishimoto T, Yasuda T, Lee SY, Kondo R, Adachi C (2014) A six-carbazole-decorated cyclophosphazene as a host with high triplet energy to realize efficient delayed-fluorescence OLEDs. Mater Horiz 1:264–269. https://doi.org/10.1039/C3MH00079F

65. Zhang D, Cai M, Zhang Y, Zhang D, Duan L (2016) Sterically shielded blue thermally activated delayed fluorescence emitters with improved efficiency and stability. Mater Horiz 3:145–151. https://doi.org/10.1039/C5MH00258C

66. Huang R, Ward JS, Kukhta NA, Avó JA et al (2018) The influence of molecular conformation on the photophysics of organic room temperature phosphorescent luminophores. J Mater Chem C 6:9238–9247. https://doi.org/10.1039/c8tc02987c

67. Ward JS, Nobuyasu RS, Batsanov AS et al (2016) The interplay of thermally activated delayed fluorescence (TADF) and room temperature organic phosphorescence in sterically-constrained donor–acceptor charge-transfer molecules. Chem Commun 52:2612–2615. https://doi.org/10.1039/C5CC09645F

68. Kroto HW, Heath JR, O'Brien SC, Curl RF, Smalley RE (1985) C_{60}: buckminsterfullerene. Nature 318:162–163. https://doi.org/10.1038/318162a0

69. Dias FB, Penfold T, Monkman AP (2017) Photophysics of thermally activated delayed fluorescence molecules. Methods Appl Fluoresc 9:012001. https://doi.org/10.1088/2050-6120/aa537e

70. Gibson J, Monkman AP, Penfold TJ (2016) The importance of vibronic coupling for efficient reverse intersystem crossing in thermally activated delayed fluorescence molecules. Chemphyschem 17:2956–2961. https://doi.org/10.1002/cphc.201600662

71. Masui K, Nakanotani H, Adachi C (2013) Analysis of exciton annihilation in high-efficiency sky-blue organic light-emitting diodes with thermally activated delayed fluorescence. Org Electron 14:2721–2726. https://doi.org/10.1016/j.orgel.2013.07.010

72. Murawski C, Leo K, Gather MC (2013) Efficiency roll-off in organic light-emitting diodes. Adv Mater 17:6801–6827. https://doi.org/10.1002/adma.201301603

73. Zhang Y, Zhang D, Cai M et al (2016) Towards highly efficient red thermally activated delayed fluorescence materials by the control of intra-molecular π-π stacking interactions. Nanotechnology 27:094001. https://doi.org/10.1088/0957-4484/27/9/094001

74. Nakanotani H, Masui K, Nishide J, Shibata T, Adachi C (2013) Promising operational stability of high-efficiency organic light-emitting diodes based on thermally activated delayed fluorescence. Sci Rep 3:2127. https://doi.org/10.1038/srep02127

75. Im Y, Lee JY (2014) Above 20% external quantum efficiency in thermally activated delayed fluorescence device using furodipyridine-type host materials. Chem Mater 26:1413–1419. https://doi.org/10.1021/cm403358h

76. Chatterjee T, Wong KT (2018) Perspective on host materials for thermally activated delayed fluorescence organic light emitting diodes. Adv Optical Mater. doi: https://doi.org/10.1002/adom.201800565

77. Duan L, Hou L, Lee T-W et al (2010) Solution processable small molecules for organic light-emitting diodes. J Mater Chem 20:6392–6407. https://doi.org/10.1039/B926348A

78. Nikolaenko AE, Cass M, Bourcet F, Mohamad D, Roberts M (2015) Thermally activated delayed fluorescence in polymers: a new route toward highly efficient solution processable OLEDs. Adv Mater 25:7236–7240. https://doi.org/10.1002/adma.201501090

79. Albrecht K, Matsuoka K, Fujita K, Yamamoto K (2015) Carbazole dendrimers as solution-processable thermally activated delayed-fluorescence materials. Angew Chem 54:5677–5682. https://doi.org/10.1002/anie.201500203

80. Li J, Nomura H, Miyazaki H, Adachi C (2014) Highly efficient exciplex organic light-emitting diodes incorporating a heptazine derivative as an electron acceptor. Chem Commun 50:6174–6176. https://doi.org/10.1039/C4CC01590H

81. Liu XK, Chen Z, Zheng CJ, Liu CL et al (2015) Prediction and design of efficient exciplex emitters for high-efficiency, thermally activated delayed-fluorescence organic light-emitting diodes. Adv Mater 27:2378–2383. https://doi.org/10.1002/adma.201405062

82. Hung WY, Chiang PY, Lin SW et al (2016) Balance the carrier mobility to achieve high performance exciplex OLED using a triazine-based acceptor. ACS Appl Mater 8:4811–4818. https://doi.org/10.1021/acsami.5b11895. Interfaces

83. Nobuyasu RS, Ren Z, Griffiths GC et al (2016) Rational design of TADF polymers using a donor–acceptor monomer with enhanced TADF efficiency induced by the energy alignment of charge transfer and local triplet excited states. Adv Optical Mater 4:597–607. https://doi.org/10.1002/adom.201500689

84. Zhu Y, Zhang Y, Yao B et al (2016) Synthesis and electroluminescence of a conjugated polymer with thermally activated delayed fluorescence. Macromolecules 49:4373–4377. https://doi.org/10.1021/acs.macromol.6b00430

85. Lee SY, Yasuda T, Komiyama H, Lee J, Adachi C (2016) Thermally activated delayed fluorescence polymers for efficient solution-processed organic light-emitting diodes. Adv Mater 28:4019–4024. https://doi.org/10.1002/adma.201505026

86. Luo J, Gong S, Gu Y, Chen T, Li Y, Zhong C, Xie G, Yang C (2016) Multi-carbazole encapsulation as a simple strategy for the construction of solution-processed, non-doped thermally activated delayed fluorescence emitters. J Mater Chem C 4:2442–2446. https://doi.org/10.1039/C6TC00418K

87. Shiu YJ, Cheng YC, Tsai WL, Wu CC, Chao CT, Lu CW, Chi Y, Chen YT, Liu SH, Chou T (2016) Pyridyl pyrrolide boron complexes: the facile generation of thermally activated delayed fluorescence and preparation of organic light-emitting diodes. Angew Chem 24:3017–3021. https://doi.org/10.1002/anie.201509231

88. Wu K, Zhang T, Wang Z, Wang L, Zhan L, Gong S, Zhong C, Lu ZH, Zhang S, Yang C (2018) De novo design of excited-state intramolecular proton transfer emitters via a thermally activated delayed fluorescence channel. J Am Chem Soc 140:8877–8886. https://doi.org/10.1021/jacs.8b04795

89. Nakanotani H, Higuchi T, Furukawa T, Masui K, Morimoto K, Numata M, Tanak H, Sagara Y, Yasuda T, Adachi C (2014) High-efficiency organic light-emitting diodes with fluorescent emitters. Nature Comm 5:4016. https://doi.org/10.1038/ncomms5016

90. Augusto V, Baleizao C, Berberan-Santos MN, Farinha J (2010) Oxygen-proof fluorescence temperature sensing with pristine C70 encapsulated in polymer nanoparticles. J Mater Chem 20:1192–1197. https://doi.org/10.1039/B920673F

91. Smith CS, Branham CW, Marquardt BJ, Mann KR (2010) Oxygen gas sensing by luminescence quenching in crystals of Cu(xantphos)(phen)+ complexes. J Am Chem Soc 132:14079–14085. https://doi.org/10.1021/ja103112m

92. Xiaong X, Song F, Wang J, Zhang Y, Xue Y, Sun L, Jiang N, Gao P, Tian L, Peng X (2014) Thermally activated delayed fluorescence of fluorescein derivative for time-resolved and confocal fluorescence imaging. J Am Chem Soc 136:9590–9597. https://doi.org/10.1021/ja502292p

93. Li T, Yang D, Zhai L, Wang S, Zhao B, Fu N, Wang L, Tao Y, Huang W (2017) Thermally activated delayed fluorescence organic dots (TADF Odots) for time-resolved and confocal fluorescence imaging in living cells and in vivo. Adv Sci 4:1600166. https://doi.org/10.1002/advs.201600166

Explosives Detection: From Sensing to Response

Liliana Marques Martelo, Lino Forte Marques, Hugh Douglas Burrows, and Mário Nuno Berberan-Santos

Contents

L. M. Martelo (✉)
Centro de Química-Física Molecular (CQFM), The Institute of Nanoscience and Nanotechnology (IN) and The Institute for Bioengineering and Biosciences (iBB), Instituto Superior Técnico, University of Lisbon, Lisbon, Portugal

Department of Chemistry, University of Coimbra, Coimbra, Portugal
e-mail: liliana.martelo@tecnico.ulisboa.pt

L. F. Marques
Institute of Systems and Robotics (ISR), University of Coimbra, Coimbra, Portugal

H. D. Burrows
Department of Chemistry, University of Coimbra, Coimbra, Portugal

M. N. Berberan-Santos
Centro de Química-Física Molecular (CQFM), The Institute of Nanoscience and Nanotechnology (IN) and The Institute for Bioengineering and Biosciences (iBB), Instituto Superior Técnico, University of Lisbon, Lisbon, Portugal

© Springer Nature Switzerland AG 2019
B. Pedras (ed.), *Fluorescence in Industry*, Springer Ser Fluoresc (2019) 18: 293–320,
https://doi.org/10.1007/4243_2019_9, Published online: 7 March 2019

Abstract The purpose of this chapter is to summarize the state of art of fluorescence-based explosive sensors in a simple way, focusing especially on the research progress. Importantly, the advances in this field are organized in the different strategies and improvements in the exploitation of fluorescence explosives detection. Mechanisms of fluorescence explosives detection are reviewed, not only fluorescence-quenching-based mechanisms but also some novel mechanism which are applied in explosives detection. We also focused our discussion on several fluorescent probes that can be used for the detection of explosives, in this way the discussion is organized based on their structures. Some design and data treatment requirements for obtaining an optical sensor are discussed, including statistical analysis and signal modelling. At the end of the chapter, future directions and perspectives are presented and discussed.

Keywords Explosives detection · Fluorescence sensing · Optical sensing · Terrorism and countermeasures · Trace analysis

1 Introduction: Background of Explosives Detection

Explosives continue to be a threat material because of their destruction potential, low cost, ease of manufacturing (especially from common materials) and freely available online resources on their production and use.

An explosive is defined as a material that can be made to undergo a very rapid and self-propagating chemical decomposition resulting in the formation of more stable material, accompanied by a strong pressure increase [1].

Based on structure and potential, explosives have been classified into many types (Scheme 1). Basically, explosives are classified as low and high explosives, and

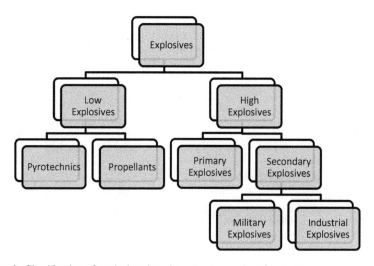

Scheme 1 Classification of explosives based on structure and performance

both types are further classified into different forms. Low explosives or propellants burn at relatively low rates (cm s^{-1}), while high explosives detonate at larger velocities (km s^{-1}). Low explosives include propellants, smokeless powder, black powder, pyrotechnics, etc. Low explosives contain oxygen for combustion which sequentially produces a gas that forms an explosion. The chemical reaction propagates with such rapidity that the rate of reaction in material exceeds the speed of sound. High explosives have again been subdivided into two groups according to their function in the explosive, i.e. primary explosives and secondary explosives.

Primary explosives, which include lead azide and lead styphnate, are highly susceptible to initiation and are often referred as "initiating explosives" because they can be used to ignite secondary explosives. Secondary explosives, which include nitroaromatics and nitroamines, are much more common at military sites. Secondary explosives are often used as main charge explosives because they are formulated to detonate only under specific circumstances. Secondary explosives can be roughly categorized into melt-pour explosives which are based on nitroaromatics, such as trinitrotoluene (TNT) and dinitrotoluene (DNT), and plastic-bonded explosives which are based on a binder and crystalline explosive, such as 3,5-trinitrroperhydro-1,3,5-triazine (RDX). Plastic explosive means an explosive material in flexible or elastic sheet form, formulated with one or more high explosives which in their pure form have a vapour pressure of less than 10^{-4} Pa at a temperature of 25°C. Such explosives are formulated with a binder material, and the so-formed mixture is malleable and flexible at room temperature [2].

Secondary explosive compounds are classified into six broad classes based on their structures [3] (Fig. 1). Typically, nitroaromatic compounds (NACs) such as 2,4,6-trinitrotoluene (TNT) and 2,4-dinitrotoluene (2,4-DNT) are the primary military explosives and the principal components in the unexploded landmines worldwide. Nitramines and nitrate ester, for example, 3,5-trinitrroperhydro-1,3,5-triazine (RDX) and pentaerythritol tetranitrate (PETN), are the main components of highly energetic plastic explosives, such as C-4 (91% RDX) and Semtex (40–76% PETN) [4]. Ammonium salts such as ammonium nitrate (AN) and ammonium phosphate (AP) are usually used for industrial applications, typically in solid rockets propellants [5]. Peroxide-based explosives like triacetone triperoxide (TATP) and hexamethylene triperoxide diamine (HTMD) have grown rapidly as homemade explosives because they can be easily synthesized from inexpensive and available materials [6]. Most highly energetic explosives are nitro-substituted (nitrated) compounds, which still have the top priority for detection.

As nitrated explosives are sensitive to shock, friction and impact, it is desirable to use detection methods that permit contact-free analysis. The demand for detecting hidden explosives in transportation, car bombs and buried explosives in warzones has also led to intense interest in the ultrasensitive detection of nitrated explosives in the vapour phase. However, explosive vapours are difficult to detect due to

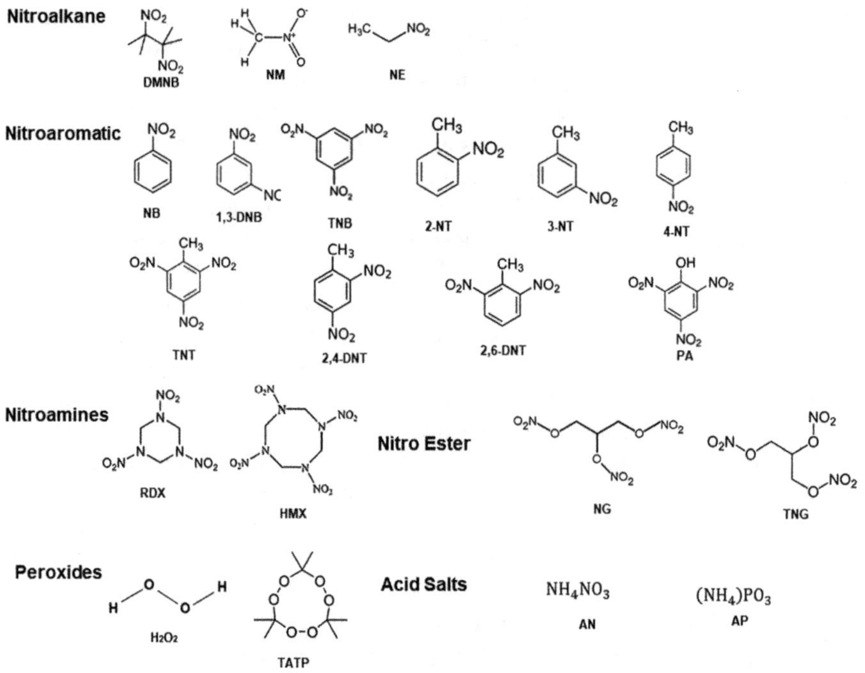

Nitroalkane NO_2 DMNB   NM   NE

Nitroaromatic   NB   1,3-DNB   NC   TNB   2-NT   3-NT   4-NT

TNT   2,4-DNT   2,6-DNT   PA

Nitroamines   RDX   HMX   **Nitro Ester**   NG   TNG

Peroxides H_2O_2 TATP   **Acid Salts** NH_4NO_3 AN $(NH_4)PO_3$ AP

Fig. 1 Six principal chemical categories of explosives

Table 1 Vapour pressure of some explosives and other chemical and physical properties [9, 10]

Explosive	Molecular weight	Melting point (°C)	Boiling point (°C)	Vapour pressure (20°C/Torr)
Nitromethane (NM)	61.04	−29	100–103	2.8×10^{-1}
2,4,6-Trinitrotoluene (TNT)	227.13	80.1–80.6	240 (explodes)	1.1×10^{-6}
Pentaerythritol tetranitrate (PETN)	316	141.3	190 (decomposes)	3.8×10^{-10}
2,4,6-Trinitrophenol (PA)	229.11	122	300	5.8×10^{-9}
Nitroglycerin (NG)	227	13.2	–	2.6×10^{-6}
Tetranitro-triazacyclohexane (RDX)	222.26	204.1	–	4.1×10^{-9}

several issues. First, most explosives have low volatility at room temperature (Table 1), especially for those with the highest priority to be detected (e.g. TNT, 1.33×10^{-4} Pa at 20°C; RDX, 5.33×10^{-7} Pa at 20°C; and PETN, 5.06×10^{-8} Pa at 20°C) [7, 8]. In addition, explosives are often wrapped in a plastic packaging which further blocks the escape of explosive vapours.

The physical properties of each class of nitrated explosives differ significantly, making the detection of broad-class explosives quite a challenge. For example, there is a demand for the direct detection of RDX and PETN, both of which lack nitroaromatic rings and have the highest lowest unoccupied molecular orbital (LUMO) energies and even lower vapour pressure, making them hard to be detected by common optical approaches used, for example, for NACs. Finally, many everyday chemicals are likely to interfere, leading to false positives. In field tests, the false positives often come from environmental contaminants, and the false negatives are related to the lack of sensitivity. For instance, bleach is a strong oxidant and is likely to generate a false-positive alarm in TATP or HMTD sensing [10]. All of these issues make it extremely difficult to achieve the sensitive and selective detection of a broad range of explosives by a single material.

Currently, the commercially available methods for the detection of explosives are trained canines [11–13], metal detectors [14] and ion mobility detectors (IMS) [15–19]. While each method has specific advantages, their use is not problem-free. Trained canines are reliable for the detection of explosives due to the powerful olfactory system. Well-trained canines have been widely used in the field test and at transportation sites such as airports or train stations, to identify and discriminate between different explosives. However, canine training is very expensive, and dogs easily get tired for continuous sensing; thus this method is not well-suited for widespread and long-time detection. Metal detectors are an indirect and very efficient technique for landmines and weapon detection packaged in metals. However, this technique is not directly sensitive to explosives and thus cannot be applied for transportation site screening. IMS is commonly used for explosives detection system in airports and has sensitivity down to nanogram or picogram levels for common explosives, but this technique lacks sensitivity for a broad range of explosives, such as PETN and RDX, which greatly limits its overall utility. Moreover, IMS requires sophisticated protocols such as calibration, which, along with its poor portability and high cost, makes it unsuitable for real-time field detection. Similar limitations apply to other detection systems, such as gas chromatography coupled with mass spectrometry (GC-MS) [20], quartz microbalance (QCM), surface plasmon resonance (SPR), electrochemical methods, immunoassay, etc. [12, 21, 22]. There thus remains an urgent need for innovative detection strategies that are not only low cost and user-friendly but also highly sensitive and selective.

One particularly attractive and promising approach involves the use of optical methods, which offer many benefits over other detection techniques, such as low cost, good portability and high sensitivity and selectivity [23]. Consequently, absorbance (colorimetric) and fluorescence responses are the focus of many recently developed optical sensors for explosives detection [24–26]. Typically, fluorescence-based detection is one to three orders of magnitude more sensitive and has wider linear ranges compared to absorbance-based methods. Moreover, the source and detector of a fluorescence method could be easily incorporated into a handheld device for the field detection of explosives. Thus, the fluorescence-based method has the great promise to be applied in the rapid, sensitive and selective detection of explosives and therefore becomes the focus of discussion in this chapter.

2   Mechanisms for Fluorescence-Based Explosives Detection

Due to the electron-withdrawing nitro groups in the benzene ring, nitroaromatics have empty π^* orbitals of low energy, which are good acceptors of electrons, so electron-rich luminescent molecules can be strongly quenched by nitroaromatics via photoinduced electron transfer (PET), fluorescence resonance energy transfer (FRET) or electron exchange energy transfer is also possible [27, 28].

The explosive, acting as a quencher, Q, reduces the intensity of light emitted by the fluorescent indicator. Emission intensity (I_0) in the absence of the analyte is high, but the intensity in the presence of analyte (I) depends on the concentration of analyte or quencher ([Q]). Static quenching depends on the binding of the analyte to indicator, as characterized by the binding association constant (K_A), whereas collisional quenching is a function of the intrinsic fluorescence lifetime (τ_0) and the rate of analyte-indicator collision rate (k_q). The Stern-Volmer equation is a function of static and collisional quenching:

$$\frac{I_0}{I} = \left(1 + K_A[Q]\right) + \left(1 + k_q\tau_0[Q]\right) \tag{1}$$

Frequently, a Stern-Volmer quenching efficiency (K_{SV}) is reported, in which K_{SV} equals either K_A or $k_q\tau_0$, as steady-state emission measurements cannot distinguish between the two quenching pathways.

Both static and collisional quenching often involve a mechanism of photoinduced electron transfer between the excited state of the luminescent molecule and the ground state of the explosive's compounds. Static quenching is more common in explosives detection due to the strong binding for explosives binding to many luminescent molecules. Static quenching involves the formation of a ground-state adduct between fluorophore and quencher, which, due to rapid electron transfer upon photoexcitation, inhibits the fluorescence. This ground-state interaction is only evident in a decrease in the steady-state emission intensity (I): unbound fluorophore still emits with an excited-state lifetime ($\tau = \tau_0$) unaffected by quencher. In contrast, collisional quenching leads to an equivalent decrease in both the excited-state lifetime and steady-state emission intensity (I_0/I).

Static and dynamic quenching can also be differentiated by their differing dependence on temperature and viscosity. Higher temperatures result in faster diffusion and hence larger amounts of collisional quenching, while for static quenching, higher temperatures will typically promote the dissociation of weakly bound complexes and hence result in lower amounts of nonfluorescent fluorophore-quencher complexes.

One additional method to distinguish static from dynamic quenching is the examination of the absorption spectra of the fluorophore. Collisional quenching only affects the excited state of the fluorophores, and consequently no change in the absorption spectra is expected. In contrast, ground-state complex formation will frequently lead to changes of the absorption spectrum of the fluorophore.

Additionally, the quenching constant k_q can also be used for discrimination between static and dynamic quenching; k_q is calculated using the ratio of Stern-Volmer quenching constant (K_{SV}) to unquenched fluorescence lifetime (τ_0). For dynamic quenching, diffusion-controlled quenching typically results in values of k_q near $10^{10}\,\mathrm{M}^{-1}\,\mathrm{s}^{-1}$, while for static quenching, the k_q value is generally several orders of magnitude larger than $10^{10}\,\mathrm{M}^{-1}\,\mathrm{s}^{-1}$ [3].

2.1 Photoinduced Electron Transfer (PET)

As referred previously, many explosives are nitrated compounds, which makes them electron-deficient, such that they could bind to electron-rich fluorophores through donor-acceptor (D-A) interaction [29]. In PET, the excited state of fluorophores (D) is likely to donate an electron to the ground state of an explosive compound (A) as shown in Fig. 2. In PET, a complex is formed between the electron donor and the acceptor. This complex can return to the ground state without emission of a photon, but in some case, an exciplex can be observed. Finally, the extra electron on the acceptor returns to the electron donor.

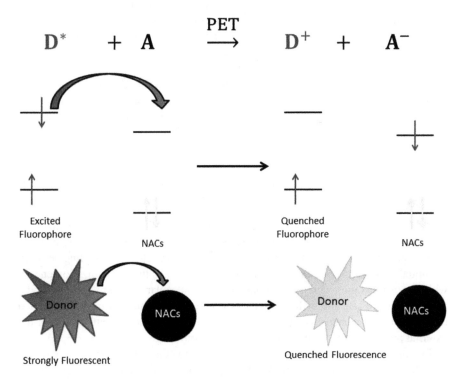

Fig. 2 Schematic of photoinduced electron transfer (PET) process

PET plays a major role in the fluorescence-quenching process and provides useful insights into the development of explosive fluorescence arrays. In a given solvent, the standard Gibbs energy $\Delta_{ET}G^0$ for the above reaction (Fig. 2) can be expressed using the redox potential E^0; the excitation energy $\Delta E_\infty = h\upsilon_\infty$, that is, the difference in energy between the lowest vibrational levels of the excited state and the ground state; and the Coulombic energy of the formed pair [30]:

$$\Delta_{ET}G^0 = F\left(E_{\underline{D}^+}^{\ 0} - E_{\underline{A}^-}^{0}\right) - N_a h\upsilon_\infty - \frac{N_a e^2}{4\pi\varepsilon R} \tag{2}$$

where F is the Faraday constant ($F = 96{,}485$ C mol^{-1}), e is the electron charge, ε is the permittivity of the solvent and R is the distance between the two ions in the pair. This equation can be obtained from a thermodynamic cycle (Born-Haber cycle).

The redox potentials can be determined by electrochemical measurements in the same solvent or estimated from potentials measured in other solvents. Estimation by theoretical calculations using the energy levels of the lowest unoccupied molecular orbital (LUMO) and the highest occupied molecular orbital (HOMO) is also possible.

Equation (2) allows evaluating the thermodynamic feasibility of photoinduced electron transfer for a given pair in each solvent. If the reaction is not diffusion-limited, the reaction rate k_R, denoted here k_{ET} for electron transfer, can be determined. Two cases are possible: if the interaction between the donor and acceptor in the encounter pair is strong, this encounter pair (DA)* is called an "exciplex" or if the interaction between the donor and acceptor in the encounter pair (D* ... A) is weak, the rate constant k_{ET} can be estimated from Marcus theory.

The quenching efficiency is directly proportional to the electron transfer. For conjugated polymers, electron transfer is based on $\pi - \pi$ interaction or stacking in a D-A system. The semiclassical limit of Marcus theory has been developed to analyse the PET process, the electron transfer rate (k_{ET}) is expressed as [31, 32]:

$$k_{ET} = A\exp\left(-\frac{\Delta G^0}{kT}\right) = 2\frac{\pi^{2/3}}{h\sqrt{\lambda kT}}V^2\exp\left[-\frac{\left(\Delta G^0 + \lambda\right)^2}{4\lambda kT}\right] \tag{3}$$

where h, k and T are the physical parameters such as Planck constant, the Boltzmann constant and the temperature in kelvin respectively. ΔG^0 is the standard Gibbs free energy difference of the electron transfer reaction, V is the electron coupling between the initial state (D*A) and the final state (D$^+$A$^-$), and λ is the reorganization energy which is a relaxation energy of adjusting the molecular structure for producing a new stabilized sate and includes two contributions. One the internal part λ_e related to the geometry changes of the D and the A, and the second contribution is the external part λ_e related to the changes in the surrounding.

Photoinduced electron transfer contributes to the fluorescence quenching by nitroaromatic explosives, which accounts for the detection for most of the fluorescence-based explosives detection. The PET mechanics also leads to fluorescence enhancement and is used for explosives detection in recent years. It is believed that sensing the enhancement of fluorescence is superior to quenching due to several reasons [33]. Firstly, the appearance of a bright signal on a completely dark background is qualitatively easier to detect than a dimming of an already bright signal and is little affected by the fluorescence background, thus leading to higher sensitivity. Secondly, turn-on signals result from a stoichiometric binding event rather than from a collisional encounter.

2.2 Resonance Energy Transfer (RET)

The energy transfer mechanism has been also used to develop several explosive sensors and can enhance the fluorescence-quenching efficiency and improve sensibility. In Förster resonance energy transfer (FRET), an initially excited molecule (donor) returns to the ground state, while simultaneously the transferred energy promotes an electron on the acceptor to the excited state as shown in Fig. 3.

According to FRET theory, the rate of energy transfer depends on the relative orientation of the donor and the acceptor dipoles, the extent overlap of the fluorescence emission spectrum of the donor (fluorophore) and the absorption spectrum of

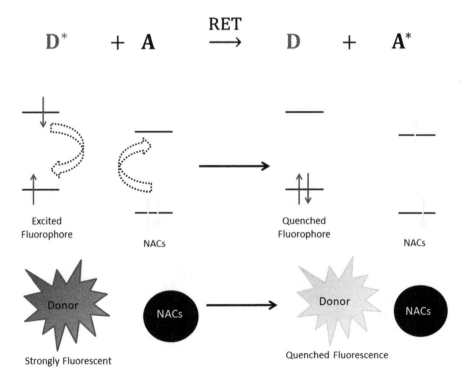

Fig. 3 Schematic of resonance energy transfer (RET) process

the acceptor (analyte) and finally the distance between the donor and the acceptor [30, 34]. The efficiency (E) of energy transfer between the donor and the acceptor can be calculated by the following equation:

$$E = 1 - \frac{F}{F_0} = \frac{R_0^6}{R_0^6 + r_0^6} \tag{4}$$

where F and F_0 are the relative fluorescence intensities of the donor in the presence and the absence of an acceptor (analyte), respectively. r_0 represents the distance between the donor and acceptor, and R_0 is the critical distance or the Förster radius at which energy transfer efficiency equals 50%, which is generally in the range of 1–8 nm. It should be noted the characteristic inverse sixth power dependence on the distance (Eq. 4). The expression obtained by Förster is:

$$R_0^6 = \frac{9\,(\ln 10)k^2 \Phi_D^0}{128\pi^5 N_a n^4}\, J \tag{5}$$

where k^2 is the orientation factor, Φ_D^0 is the fluorescence quantum yield of the donor in the absence of transfer, n is the refractive index of the medium, J is the spectral integral and can be written both in wavenumber and in wavelength scales,

$$J = \int F_V(\bar{v})\, \varepsilon(\bar{v})\, \frac{d\bar{v}}{\bar{v}^4} = \int F_\lambda(\lambda)\, \epsilon(\lambda)\, \lambda^4\, d\lambda \tag{6}$$

the quantities F_V and F_λ represent the normalized emission spectrum of the excited donor, defined with respect to wavenumber ($\int F_V(\bar{v}) d\bar{v} = 1$) or wavelength ($\int F_\lambda(\lambda)\, d\lambda = 1$).

In recent years, FRET mechanism has been widely used by several research groups for explosives detection to improve the sensitivity and selectivity. Almost all works apply this mechanism through fluorescence-quenching phenomena to detect explosives, which could fall into two broad categories. One is the direct utilization of the unique spectrum of explosives, through specifically designed fluorophore materials, in which the emission spectrum of the fluorophore efficiently overlaps with the absorption spectrum of explosives. The second method utilizes electron-deficient properties of nitro-explosives and the formation of a Meisenheimer complex. Feng et al. [35] synthetized novel amine-functionalized mesoporous silica nanoparticles containing poly(p-phenylenevinylene) (PPV) providing a facile strategy to detect TNT through FRET, allowing the quantitative detection of TNT with the detection limit of 6×10^{-7} M. Compared with the electron transfer mechanism, the FRET mechanism improves the sensitivity of TNT detection. Additionally, in the presence of TNT, TNT and amino groups form Meisenheimer complexes (TNT-amine complexes) between the electron-deficient aromatic rings and electron-rich amine ligands through a charge-transfer complexing interaction. The resultant amino groups of these novel fluorescent inorganic-organic hybrid mesoporous nanoparticles can chemically recognize and absorb TNT molecules by the formation of the Meisenheimer complex. This complex absorbs in the green region of the visible spectrum and strongly supresses the fluorescence emission of the PPV through FRET in the mesoporous channels.

Most researchers have used PET and FRET mechanism separately in the individual sensing platform, but recently some groups have used a combination of PET and FRET mechanism in dual wavelength nitro-explosives assay using a single material, which provides a higher sensitivity for TNT detection. In the work developed by Wang et al. [36], they synthetized a polyethylenimine polymer derivative with pyrene moieties, a sensing material for the detection of tetryl and TNT in aqueous systems. The combination of PET and FRET mechanisms allows an expansion of the dynamic range for TNT in aqueous solution to seven orders of magnitude, from 22 ppt up to 225 ppm.

2.3   Energy Transfer by Electron Exchange

Energy transfer by electron exchange (Dexter's mechanism) occurs between a donor and an acceptor, where the excited donor has an electron in the LUMO (Fig. 4). This excited-state electron is transferred to the acceptor, and the acceptor then transfers a ground-state electron back to the donor. The electron comes from the HOMO of the acceptor, so the acceptor is left in an excited state. Dexter interaction is a short-range phenomenon and depends on spatial overlap of donor and acceptor molecular orbitals.

In contrast to the inverse sixth power dependence on distance for the dipole-dipole mechanism, an exponential dependence applies to the exchange mechanism. The rate constant can be written [30]:

$$k_T^{ex} = \frac{2\pi}{h} KJ' \exp(-2r/L) = k_0 \exp\left[\frac{2(r - R_c)}{L}\right] \tag{7}$$

where J' is the overlap integral

$$J' = \int_0^\infty F_D(\lambda)d\lambda = \int_0^\infty \varepsilon_A(\lambda)d\lambda - 1 \tag{8}$$

with the normalization

$$\int_0^\infty F_D(\lambda)d\lambda = \int_0^\infty \varepsilon_A(\lambda)d\lambda = 1 \tag{9}$$

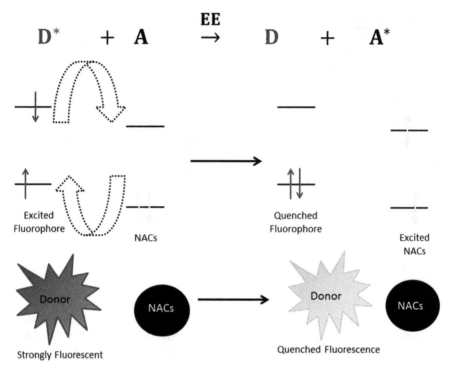

Fig. 4 Schematic of energy exchange (EE) process

and R_c is the distance of closest approach (collisional radius) and L is the average Bohr radius. Because K (Eq. 7) is a constant that is not related to any spectroscopic data, it is more difficult to fully characterize the exchange mechanism experimentally.

3 Fluorophores for Explosives Detection

3.1 Conjugated Polymers

Fluorescent-conjugated polymers have been employed extensively as sensing materials in recent years. One of the features that has led to their increasing use as sensors relates to their ability to detect low concentrations of target analytes owing to amplified quenching [37]. Quenching amplification is a direct consequence of the conjugated polymer's ability to work as a highly efficient energy transport medium via the migration of excited states (excitons) through the polymer chain [38] (Fig. 5). The first examples of these polymers were developed by the research group of *T*. Swager using poly(phenylene ethynylene) (*PPE*) with cyclophane receptors

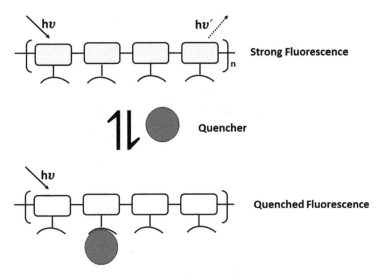

Fig. 5 Amplified fluorescent quenching mechanism of CPs by quenchers. Adapted from [42]

integrated in the polymeric backbone for coordination with paraquat (a powerful quencher) [39–41]. Since this work appeared, fluorescence quenching-amplifying polymers have been extensively employed as sensing materials to detect metal cations, anions and neutral molecules.

Poly(*p*-phenylene ethynylene) (*PPE*) has been widely employed in the specific filed of sensing explosives. Typically, it possesses a high fluorescence quantum yield in solution with a blue emission. *PPE* and its derivatives are one of the most prominent families of fluorescent CPs used for the detection of NACs. The detection mechanism involves the fluorescence quenching via electron transfer from electron-rich aryl group on polymer chains, which forms favourable π-π interactions to bind the electron-poor nitroaromatics. However, in the solid-state *PPE* tends to suffer self-quenching because of inter-chain aggregation [43]. Consequently efficient sensing requires the incorporation of bulky groups within the polymer backbone to create an appropriate distance between chains and prevent self-quenching.

J. S. Yang and co-workers prepared *PPEs* derivatives *1* and *2* (Fig. 6) and showed their accentuated fluorescence quenching in the presence of TNT and 2,4-DNT vapours [44, 45]. The intensity of emission quenched was influenced by the thickness of the thin film and the time of exposure of the explosive's vapours. The best system was obtained with 25-Å-thick films and an exposure time of 60 s (100% and 75% quenching with 2,4-DNT and TNT vapours, respectively, for the *PPE 1*). This same PPE *1* was incorporated in a portable prototype that was used for the detection of landmines, recording the air level of TNT in contaminated environment (concentration levels of femtograms of TNT per mL of air) [46]. These features of pentiptycene derivatives have been successfully utilized in a commercially available device, called FIDO® (FLIR Inc.), for real-time monitoring of buried explosives and landmines and roadside bombs and the detection of suspected bomb markers and some other homeland security-related fields.

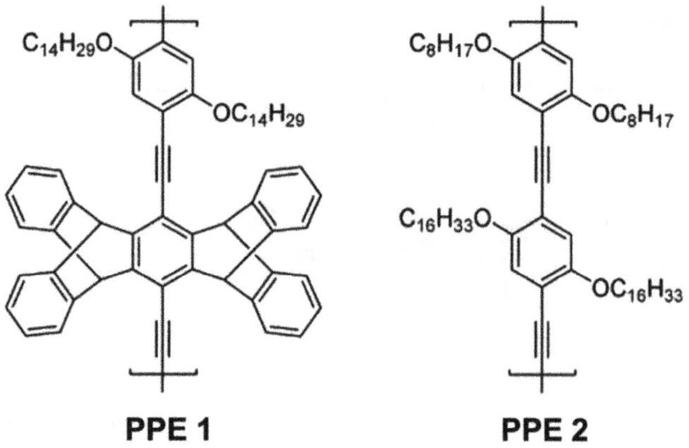

PPE 1   **PPE 2**

Fig. 6 Structures of PPE *1* and PPE *2*

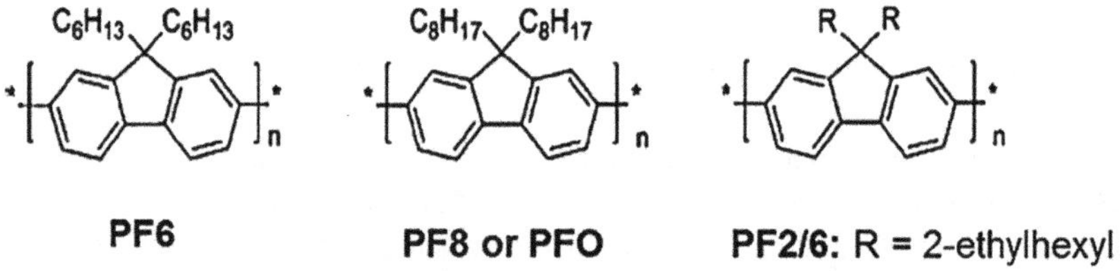

PF6   **PF8 or PFO**   **PF2/6: R = 2-ethylhexyl**

Fig. 7 Poly(9,9-dialkylfluorene) (PDAFs)

Polyfluorenes (*PFs*) have emerged as versatile semiconducting materials with applications in various polymer optoelectronic devices, such as light-emitting devices, lasers, solar cells, memories, field-effect transistors and sensors. *PFs* are regarded as the simplest regular stepladder-type poly(*para*-phenylene)s (*PPPs*), in which two phenyl rings are locked into a plane via the C-9 carbon of the fluorene units. Poly(9,9-dialkylfluorenene)s (*PDAFs*) have a good solubility owing to the introduction of alkyl chains, including poly(9,9-dihexylfluorene) (*PF6*) (Fig. 7), poly(9,9-di-*n*-octylfluorene) (*PF8* or *PFO*) and poly(9,9-di(2-ethylhexyl)fluorene) (*PF2/6*). *PDAFs* possess relatively large band gaps, which are suitable for blue-light-emitting materials (Fig. 7) due to their photoluminescence efficiency of more than 80% in solution and approximately 50% in solid film. *PDAFs* show excellent thermal stability but a low glass transition temperature (Tg) of 75°C and limited spectral stability [47]. Their highly emissive fluorescence can be turned through the entire visible range.

Recently Martelo et al. used *PFO* and its homologue poly[9,9-dioctylfluorenyl-2,7-diyl)-co-bithiophene]) (*F8T2*) in ethyl cellulose films as sensors of nitrobenzene (NB) and 1,3-dinitrobenzene (DNB) vapours at room temperature [48]. To increase the sensitivity towards detection of nitroaromatic vapours of the *P8T2* in ethyl cellulose films, they have tested the effect of adding different plasticizers [49] to the ethyl cellulose matrix. The studies indicated up to 96% quenching of light emission (Fig. 8), accompanied by a marked decrease in the fluorescence lifetime, upon exposure of the films of F8T2 in ethyl 1,3-dinitrobenzene (DNB) vapours at room temperature.

This sensor was tested in a dynamic setup composed of two Mass Flow Controllers (MFCs Dwyer GFC-2102), one controlling the flow of clean air and the other controlling the flow of saturated nitrobenzene vapour, obtained from a bubbler at constant temperature and pressure. Both MFCs are controlled from MATLAB through a microcontroller (Microchip PIC24FV16KM202) that sets the references to the MFCs, defining the composition of the nitrobenzene-clean air mixture [50]. The sensor employed a differential approach where a reference signal from the excitation was used to normalize the detection signal, providing high dynamic range and mitigating drift effects. As expected, this approach guaranteed very good sensitivity and fast response time. Figure 9 shows the sensor response after 20 s exposure to dilute NB vapours.

The sensor prototype showed very fast response (a few seconds) to the presence of small concentrations of the target analyte but also showed a long recovery time which limits its potential applications. This slow recovery may result from the design of the sampling chamber and from the time required for desorption of the nitrobenzene molecules from the polymer surface. For in-field applications, the fast response to the presence of a small concentration of nitroaromatics is a very positive characteristic. Although the recovery time is rather long (several minutes), this can be acceptable for scenarios where the detection of frequent changes in analyte level is not required.

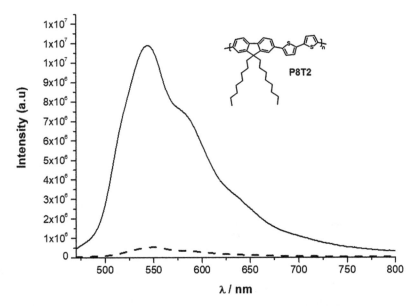

Fig. 8 Emission spectra of the F8T2 ethyl cellulose film with 1% (w/w) of PEG 3400 in the absence (solid line) and after 3 min of exposure to 2,3-dinitrobenzene (1.83×10^{-5} M) vapour (dashed lines). Excitation wavelength was 450 nm. *Inset*: The chemical structure of the P8T2

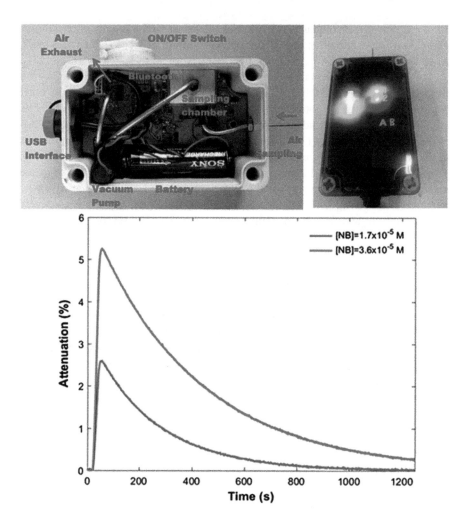

Fig. 9 *Top*: prototype sensor device. *Bottom*: attenuation of the fluorescence emission of the F8T2 in ethyl cellulose when exposed to NB (1.7×10^{-5} M) vapours for 20 s

3.2   Small-Molecule Fluorophores for Explosives Detection

Small molecules provide advantages for additional detection strategies, such as a simple synthesis, diverse pathways of fluorescence quenching and ability to detect a wide range of explosives. The principal difference between the polymeric systems described above and small-molecule-based detection is based on the physical mechanism of fluorophore quenching. Polymer-based sensors frequently detect explosives through static quenching due to the large number of weak binding interactions with analyte in contrast; small molecules typically work through collisional quenching. Generally, small molecules are divided into organic and inorganic ones. Functional molecules have been appearing showing considerable potential in sensing explosives in the form of oligofluorophores, self-assembling small fluorophores or the doping of these molecules into a matrix, which can enhance binding and/or introduce exciton migration.

Polycyclic aromatic hydrocarbons (*PAHs*) are organic molecules that consist of fused aromatic rings without the presence of heteroatoms (e.g. pyrene (*Pyr*), anthracene, perylene, naphthalene, etc.). They have a remarkable electron donor nature due to their electron-rich structure, allowing the formation of charge-transfer complexes with electron acceptor molecules. Considering the fact that they also facilitate π-π interactions with aromatic compounds, *PAHs*, these constitute an important class of small fluorophores for nitrated explosives, NACs in particular.

Over many years, *Pyr* and its derivatives have drawn substantial attention as fluorescent probes in various applications due to their interesting photophysical properties, high stability and planar aromatic surface which tend to form strong charge-transfer complexes with electron-deficient analytes [51]. The fluorescence spectra show monomer emission characteristics in the near VIS-UV range (410 nm) at low concentration. Upon increasing the concentration of *Pyr* in solution the fluorescence peak shifts bathochromically to the visible range, to give a new broad emission which is assigned to the formation of an excimer [52]. Both *Pyr* monomer and excimer emission can be quenched by NACs molecules. This simple approach has been explored by Goodpaster and McGuffin [53] who reported the separation of complex explosives mixtures (and their degradation products) with capillary liquid chromatography and the individual detection of NACs by measuring pyrene emission quenching by laser-induced fluorescence. For nitroaromatic explosives, these authors found that the Stern-Volmer quenching constants increased with the number of nitro groups (TNT > 2,4-DNT > 2-NT), whereas the effectiveness of quenching did not follow this trend for non-aromatic nitrated explosives. The method showed certain interferences from other species, such as organic amines, nitriles, halides and inorganic anions also induced a partial quenching of pyrene emission. This method was successfully applied to determine RDX, HMX, TNT, nitromethane (NM) and ammonium nitrate in various commercial explosives samples. Focsaneanu and Scaiano [54] compared the quenching profiles of *Pyr* monomers and excimers and proposed that the ratio of monomer-to-excimer emission allowed discrimination towards different explosives.

One neglected aspect in explosives detection involves picric acid (PA), Fig. 1, although its explosive power is higher of TNT and it is also widely used in the manufacture of rocket fuel and in fireworks [55]. Because of this, the development of a highly sensitive and rapid method to detect TNP is desirable. Peng et al. [56] developed a colorimetric and fluorescent chemosensor for the detection of PA-based isonicotinohydrazide and anthracene in solution. Like pyrene, anthracene is also a PAH which can emit strong fluorescence at 420 nm for the monomer and emit above 470 nm for excimer formation. When interacting with a compound containing strong electron-withdrawing groups, as PA, fluorescence quenching of anthracene occurs due to the formation of a nonfluorescent complex.

Du et al. [57] developed a fluorescent sensing film obtained by self-assembling pyrene moieties on a glass surface via a spacer containing a benzene ring. The introduction of benzene ring as spacer increases the π-π stacking between pyrene moieties on the end of each benzene spacer with the NACs in aqueous solution. With this design, they have reached a low detection limit for PA (1.0×10^{-8} mol/L) and demonstrated the sensing process is reversible.

Ding et al. [58] combined α-cyanostilbene with primary amine and secondary amine to design a small chemosensors for PA detection in solution and solid phase. The 1:1 host guest complex formed between the chemosensors and PA enhanced the fluorescence quenching; in the case of the best system, nearly 97% quenching of fluorescence emission was recorded. The best calculated limit of detection for PA was 1.85×10^{-7} M.

3.3 Nanoparticles

Nanotechnology has recently become one of the most exciting approaches in analytical chemistry. A wide variety of nanomaterials, especially nanoparticles with different properties, have found broad application in many kinds of analytical methods. Owing to their small size (normally in the range of 1–100 nm), nanoparticles exhibit unique chemical, physical and electronic properties that are different from those of bulk materials and can be used to construct novel and improved sensing devices. Matter at the nanoscale shows different chemical and physical properties, and in certain cases, enhancement occurs in physical and chemical properties. This enhancement can arise from various sources such as increases in the surface area and confinement effects [28, 38, 59]. These nanosensors are designed to overcome the many problems and limitations of traditional detection systems such as selectivity, sensitivity, size and cost. Implementing nanomaterials in these applications has advantages such as high surface area and improved surface activity, providing higher signal-to-noise ratios with unique electrical and optical properties which can be useful for highly sensitive molecular detection. The detection mechanisms of nanosensors for NACs can be separated in two main categories: electrochemical and optical. Here, we will only consider and discuss some new applications and systems for the optical nanosensors.

In a pioneering work, Goldman et al. developed QDs-based chemosensor for specific detection of TNT on the nanoscale [60]. The sensor consists of anti-TNT-specific antibody fragments, specifically immobilized on the surface of CdSe-ZnS QDs. The use of an antibody fragment instead of a full antibody provides a more compact QD conjugate and is better suited for FRET, as distances between donor and antibody-bound acceptor are substantially reduced. The hybrid sensor consists of anti-TNT-specific antibody fragments attached to a hydrophilic QD via metal-affinity coordination. A dye-labelled TNT analogue is bound in the antibody binding site and quenches the QD photoluminescence via proximity-induced FRET. The results showed that in the absence of TNT, there was FRET between dye molecules and QDs, while in the presence of TNT, which displaced the dye molecules, the FRET was eliminated and a concentration-dependent recovery of the QD photoluminescence resulted.

Enkin et al. [61] prepared luminescent Ag nanoclusters (NCs) stabilized by nucleic acids as optical labels for the detection of the explosives PA, TNT and RDX. The sensing schemes consist of two parts, a nucleic acid with the nucleic acid-stabilized Ag NCs and a nucleic acid functionalized with electron-donating units, including L-DOPA, L-tyrosine and 6-hydroxy-L-DOPA, self-assembled on a nucleic acid scaffold. The formation of donor-acceptor complexes between the nitro-substituted explosives, with the sensing molecule, concentrates the explosives in close proximity to the Ag NCs, leading to an electron transfer quenching of the luminescence of the Ag NCs by the explosive molecule in study. The sensitivities of the analytical platforms are controlled by the electron-donating properties of the donor substituents, and 6-hydroxy-L-DOPA was found to be the most sensitive donor. Picric acid, TNT and RDX are analysed with detection limits corresponding to 5.2×10^{-12} M, 1.0×10^{-12} M and 3.0×10^{-12} M, respectively, using the 6-hydroxy-L-DOPA-modified Ag NCs sensing module.

Toal et al. [62] developed suspended colloidal oligo(tetraphenyl)silole nanoparticles in tetrahydrofuran and water which offer a method to detect TNT in aqueous media. They observed luminescence quenching upon adding successive aliquots of TNT to the nanoparticles as low as 100 ppm in pH 7. The lifetime recorded in the same condition suggests a static quenching mechanism, as previously observed for TNT and oligosilole in toluene solution [63]. In the case of static quenching, the Stern-Volmer constant (K_{SV}) is an association constant between the quencher and the receptor sites.

4   Substrate Considerations

The effectiveness of any optical sensor array will be determined not only by the choice of the fluorophore but also will be influenced by the substrate and the morphology of the substrate upon which they will be placed. Sensitivity, reliability, accuracy, response time, interferences and lifetime of the array will influence the performance of the optical sensor.

4.1   Printed Arrays

The desired properties of printed substrates include unresponsiveness towards gas and liquids, high surface area, optical transparency or highly reflective and a stability over a wide range of pH [64, 65]. A simple method for array manufacture involves printing the fluorophores on the surface of reverse phase silica gel plates, acid-free paper or mesoporous polymer membranes made from material such as cellulose acetate or polyvinylidene difluoride (PVDF).

Inkjet printing technology is usually classified as either continuous inkjet printing (CIJ) or drop-on-demand (DoD) printing: the two are distinguished by the physical process by which the drops are generated [66]. CIJ printing involves the ejection of a continuous stream of liquid through an orifice (nozzle), which then breaks up under surface tension forces into a stream of drops. This natural breakup under surface tension forces is enhanced by modulating the flow through the nozzle at an appropriate frequency, often by a piezoelectric transducer behind the nozzle. In drop-on-demand printing, the liquid is ejected from the printhead only when a drop is required: the production of each drop occurs rapidly in response to a trigger signal. A DoD printhead usually contains multiple nozzles (typically 100–1,000, although special printheads may contain only a single nozzle), and instead of drop ejection resulting from external fluid pressure as in CIJ printing, the drop's kinetic energy derives from sources located within the printhead, very close to each nozzle. Typical drop diameters in DoD printing range from 10 to 50 μm, corresponding to drop volumes between 1 and 70 pL; the drop diameter is like that of the nozzle from which it is ejected. DoD printing can employ small volumes of liquid, unlike CIJ printing in which a substantial recirculating volume is required.

Although digital inkjet printing is an attractive non-contact technique for dispensing fluorophores, other printing methods, including flexographic, screen and gravure printing, have been considered. Compared to inkjet printing, all of these methods are contact techniques with nondigital pattern formation. Roll-to-roll techniques are feasible for large area patterning of fluorophores and offer a possible method for large-area fabrication surfaces.

Flexographic printing offers a high-speed reel-to-reel alternative to inkjet printing with the larger flexographic printing presses running at 50–500 m/min [67]. The viscosity of the flexographic ink is usually between 50 and 500 mPa·s and surface tension between 25 and 45 mN/m. There is also a small pressure between 50 and 150 N applied when transferring ink to the substrate.

Currently, screen printing is the most widely used printing technique in the manufacturing of disposable biosensors. Typically, screen printing uses a flat printing technique in batch processing, but it is also adaptable to a roll-to-roll process, such as rotary screen printing, which enables a higher throughput. Rotary screen printing uses a cylindrical screen that rotates in a fixed position rather than a flat screen that is raised and lowered over the same print. The screen is a negative of the required image and the print resolution and the wet film thickness (3–100 μm) depends on the density of the mesh and the ink properties, typically being in the range of 50–100 μm [68]. Generally, the screen printing fluids contain a functional material, a binding agent and a solvent. The solvent provides the suitable viscosity and volatility for thermal curing, whereas the binding agent improves the mechanical strength and substrate adhesion of the printed film. The typical viscosity of the screen printing inks is 500–50,000 mPa s. The rheological requirements and curing conditions (high temperature or UV cure) for the screen printing inks make the formulation of dyes challenging.

Gravure (rotogravure) printing is one of the highest volumes printing processes, reaching the printing speed of 20–1,000 m/min and the print resolution is comparable to flexographic printing. The print patterns are engraved as a discrete cell into a rotary cylinder. During the printing process, the engraved cells are filled with the ink, and a flexible doctor blade is used to remove excess ink. The ink is transferred when the roll is brought into contact with the substrate. The gravure rolls have a long lifetime, but are expensive to produce, which makes gravure printing a technique mainly suitable for very large print volumes. In order to achieve a good ink transfer, a relatively high print pressure (1–5 MPa) is used, and the ink should have a rather low viscosity of 50–200 mPa·s.

Porous sol-gel glasses can also provide excellent matrices for chemically responsive fluorophores. An effective sensor can be made by adding responsive dyes to ormosils prepared from suitable silane precursors in low volatility solvents [69]. The physical and chemical properties of the matrix, such as hydrophobicity and porosity, can be easily modified using organically modified sol-gel formulations and different silane precursors, depending on the solubility of the dye. Sol-gel methods are particularly useful because they permit direct fabrication of multicomponent materials in different configurations (monoliths, coatings, foams and fibres) without powder intermediates or the use of expensive processing technologies, such as vacuum methods. The diversity of materials which can be obtained has made the sol-gel method an important synthesis route in several domains of research. Sol-gel thin film processing offers a number of advantages including low-temperature processing, ease of fabrication and precise microstructural and chemical control. The sol-gel-derived film or layer had a high specific surface area and an external surface whose rich chemistry allows ease of functionalization by suitable responsive dyes [70, 71].

4.2   Fibre-Optic Arrays

A powerful method, especially for fluorescent sensors, is the use of fibre-optic arrays. Optical fibres consist of a cylindrical core and a surrounding cladding, both made of silica. The core is generally doped with germanium to make its refractive index slightly higher than that of the cladding, which results in light propagation by total internal reflection (TIR). Light propagating through an optical fibre consists of two components: the guided field in the core and the exponentially decaying evanescent field in the cladding. In a uniform-diameter fibre, the evanescent field decays to almost zero within the cladding. Thus, light propagating in uniform-diameter cladded fibres cannot interact with the fibre's surroundings [72].

For achieving simultaneous, multi-analyte, high-density and high-output sensing analysis, optical fibre-based sensor arrays were developed using bundles comprising thousands of individual single-core fibres which are individually modified with a diverse sensing chemistry using a random assembly method. In general, the fibre arrays (with a total size of 1 mm) contain from a few thousand up to a hundred thousand individual fibres with size of 2–10 μm prepared by the polished array in acid solution [73]. This structure enables individual wells to be recognized by the optical fibre defining its base, providing a high-density array of microwells that can be simultaneously and independently interrogated by light. Excitation light is introduced into the non-functionalized end of the fibre; emission signals from individual sensors return through the fibre and are magnified and projected onto a charge-coupled device (CCD) camera, leading to the simultaneous observation of all sensors [74].

More recently, Dasary et al. [75] developed a miniaturized fibre-optic fluorescence analyser for detection of picric acid explosive. Picric acid can form a fluorescent Meisenheimer complex with nucleophiles like N,N'-Diisopropylcarbodiimide (DPC). PA forms an orange fluorescent zwitterionic spirocyclic Meisenheimer complex with DPC with an emission at 575 nm. This complex is exclusively formed by adding PA and measuring the concentration of PA by just recording the emission. Based on this Meisenheimer complex, the authors developed a miniaturized fluorescence sensor equipped with a portable 532 nm DPSS green laser which could detect picric acid as low as 100 nM (25 ppb) with excellent reproducibility. The validation results for picric acid concentration were in good agreement with high-performance liquid chromatography analysis.

5 Requirements for Optical Chemical Sensors

An optical chemical sensor is a device whose signal is functionally related to the concentration of an analyte adsorbed by the sensing element of this device and causes a change in a controlled optical parameter upon the interaction of electromagnetic radiation with the sensing element. The action of optical sensors is based on fundamental optical phenomena such as fluorescence, phosphorescence luminescence, absorption, reflection, polarization, interference, etc. The following types of optical chemical sensors are used for detecting explosives vapours and microparticles: fluorescence (luminescence), chemiluminescence, surface-enhanced Raman, surface plasmon resonance, waveguide (absorption integrated optical, waveguide interferometric and ring resonator based) and colorimetric sensors [20, 22]. Most optical chemical sensor systems follow the architecture summarized in Fig. 10.

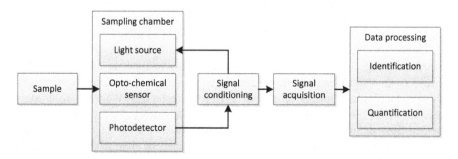

Fig. 10 Diagram for optical chemical sensors

5.1 Analytical Characteristics of Optical Chemical Sensors in Detection of Explosives Vapours

The main analytical characteristics of optical sensors important for the detection of explosives vapours are the limit of detection (LOD), selectivity, sensitivity, dynamic range, settling time (response time) and recovery time.

The sensor sensitivity is $S = {}^d X_{dC}$, where X is the sensor signal, and C is the analyte concentration. The limit of detection (LOD) is defined by the noise floor of the instrument. For an instrument with zero mean Gaussian noise and σ_n standard deviation, the limit of detection is defined as $3\sigma_n$ (i.e. a value with 99.7% probability of being above the noise signal). The sensor settling time t_s is the time interval from the onset of measurement to the achievement of a sensor signal level. In the detection of explosive traces and microparticles, the minimum measured concentration or minimum detected mass of explosives traces on the surface is used as a characteristic. The dynamic range is the measuring span, from the LOD until the concentration that saturates the sensor. The sensor may not reach a steady-state value. In this case, the concept of exposure time is used; this is a time which adsorbing surface was exposed to analyte vapours. The sensor recovery time is the interval from the termination of analyte vapour supply into the sensor to the moment of a signal drop below the value of N. Selectivity is a ratio between sensor sensitivity coefficients to analyte x and interfering substance y, i.e. S_x/S_y [76].

5.2 Statistical Analysis and Modelling Signal

Chemical sensors suffer frequently from selectivity problems, responding to multiple substances, including normal substances found in common operating environments, like water vapour or organic volatiles. These substances interfere with the sensor response degrading the whole detection ability. If the detection threshold is set too low, the system will provide a high false alarm rate. In contrast, if the detection threshold is set too high, the system will perform poorly and will not detect low concentration targets. This selectivity problem can be addressed either using materials with better selectivity or by the use of arrays of materials responding both to the target substance and to the substances expected to interfere with the system. Additionally, to provide more selective sensor systems, this multiple element approach can also support multi-target detection and classification. This classification can be accomplished through a large variety of pattern classification methods which transform a vector of measurements (the measurements from each sensor element) into a set of targets whose presence may or may not have been detected. A possible outcome of this classification process may be the probability of detection for a given substance. Complementary to this classification problem, we may be interested in quantifying the concentration of the substance. In these cases, some type of regression should be used to evaluate the measurement vector against a representative calibration database.

Before analysis, the raw data from the sensors are preprocessed. Choice of the appropriate preprocessing technique is crucial for the performance of the pattern classifier. After preprocessing the data obtained from large sensors arrays is compressed into lower-dimensional data with minimum information loss. This dimensionality reduction can be performed with several compression techniques, like principal component analysis (PCA), wavelet transform, etc. To make models from calibration sets of data, different techniques are used. For that purpose, projection to latent structures or partial least square regression (PLS) is a useful tool [77].

This processed data set is finally classified with the help of artificial neural networks (ANN), fuzzy logic or statistical techniques. Artificial neural networks attempt to mimic the functioning of human brain with mathematical models [78, 79], and, therefore, to perform a task, it needs a knowledge base. This knowledge base is developed through suitable learning processes which may be supervised or unsupervised. There are several standard supervised learning techniques like error-correction learning, memory-based learning, Hebbian learning, competitive learning, Boltzmann learning, etc. ANNs consist of a large number of parallel connected arithmetic units which are called neurons. A neuron can be mathematically described as a non-linear, parameterized, bounded function, whose inputs are the linear combination of all the inputs x_i weighted by parameters w_i, and the output is obtained through a non-linear function of the combined as $y = f\left(w_0 + \sum_{i=1}^{n} w_i x_i\right)$, where w_0 is the bias [80, 81]. Statistical methods can be either biased, in which case the evaluation algorithm is told of the class identities of individual cases, or unbiased (or model-free), where all cases are evaluated identically regardless of class identity. Unbiased methods are generally used to evaluate a data set providing a semi-quantitative idea of the quality of the data set and follow simple, straightforward algorithms. Biased methods, on the other hand, can provide significantly more power and utility information and also can be predictive, allowing for class assignment of new experimental cases by using a training set.

There are a variety of statistical methods available to deal with high-dimensional data; the three most common approaches are hierarchical cluster analysis (HCA), principal component analysis (PCA) and linear discriminant analysis (LDA) [82].

5.2.1   Hierarchical Cluster Analysis (HCA)

Hierarchical cluster analysis (HCA) is an agglomerative clustering technique where the clusters are determined from Euclidean distance between experimental data. The nearest-neighbour points are paired into a single cluster which is then paired with other nearest-neighbour points until all points and clusters are connected to each other [83, 84]. The resultant dendrogram shows connectivity and some measure of distance between each of the pairs. The criterion mostly used for clustering in HCA is Ward's minimum variance method which minimizes the total clusters variances.

There are three limitations to the HCA technique. The first involves the limitation of all unbiased methods which is not possible to predict any analysis. The second one is the dendrogram created using HCA must be re-generated with the addition of a new analyte, so comparing dendrogram is only possible for qualitative purpose. And last, there is a limitation in interpreting the noise data. Despite these limitations, dendrogram provide a straightforward method of showing cluster similarity semi-quantitatively.

5.2.2   Principal Component Analysis (PCA)

Principal component analysis is a linear transformation that transforms a n-dimensional data space into another n-dimensional data space arranged in decreasing variance of data. Since the data in the last dimensions after this transform contain minimal variance, their relevance for a classification process may be minimal and consequently they may be discarded in the classification process. A plot from PCA transformed data is generally easier to visualize than the original data set. PCA method also provides information about the variability of a sample set.

As an example, Zeng and Zhang [85] designed a colorimetric array containing 16 dye formulations to detect specifically strong oxidants and peroxide-based explosives. This array probes only a small chemical reactivity, and after PCA the first two dimensions contain 95% of the data variance.

Like HCA, PCA is an unbiased method that is best suited for evaluation of data sets rather than prediction. However, PCA can make rudimental prediction if the data set is low dimensional and has a large separation among sample classes.

5.2.3   Linear Discriminant Analysis (LDA)

Like PCA, linear discriminant analysis (LDA) can be also employed for dimensional reduction. However, rather than transforming the data from highest to lowest variance, this method provides a set of dimensions that best separates data into already known classes. LDA can be used to predict the identity of unknown samples by using a training set such as PCA, but in this case, the dimensional components are linearly transformed to maximize differentiability between the samples showing a better ability to differentiate among samples classes.

One of the limitations of LDA is related to sample size. All statistical methods require large observations in order to determine a useful data set (variance), however, as sample class covariances must also be determined to allow comparison among classes. Because of this, the covariance matrix tends to be unstable when sample size is not significantly larger than the number of sample classes analysed, and this is more problematic for high-dimensional data [83]. For these reason, LDA can give drastically fluctuating results with small samples sizes, when compared with PCA and HCA. As an improvement on LDA, tensor discrimination analysis (TDA) [84, 85] which is an array generalization of LDA and it is able to take advantages of high dimensionally. The TDA is used to classify multi-way array measurements instead than one-way vector measurements.

6 Limitation, Opportunities and Future Challenges

The fluorescence-based methods for explosives detection accomplish most of the criteria required in field for an effective sensing, such as sensitivity, selectivity, portability and low cost.

NAC explosives have been intensely investigated with recognized detection methods; however the direct detection of RDX and PETN and the increasing variety of explosive compounds (improvised explosives) are still a challenge. It has been reported that using the current sensing tools as Fido XT, IMS and mass spectrometry, it is difficult to directly detect RDX and PETN, and generally only degradation products could be detected.

For any real application, chemosensors need to satisfy several criteria such as being low cost, environmental-friendliness and stability. PPE-conjugated polymers are one class of materials which have been used in commercial devices. However, these polymers require complicated synthetic routes which is a clear limitation for large production. Other popular fluorescent materials such as PAHs and quantum dots also have some problems such as toxicity which may contaminate the environment. These issues require the research and development of novel, low-cost, environmental-friendly and stable fluorescent materials.

The fluorescence quenching mechanism still dominates in fluorescence-based explosive detection; however, fluorescence enhancement is in principle a more sensitive technique due to the weaker effect of fluorescence background. Furthermore, spectral shift and lifetime changes are also expected to be applied for explosives sensing soon.

The use of sensor arrays is another promising approach to increase the poor selectivity in explosives detection. In addition, the data from sensor arrays can be processed with artificial neural network in order to detect and differentiated a broad range of explosives.

Given the current trends on new fluorophore design and progress in simulation and data modelling, fluorescence sensing of explosives is expected to have a promising future.

Acknowledgements This work was supported by Fundação para a Ciência e Tecnologia (Portugal) within project FAPESP/20107/2014. L. Martelo was supported by Fundação para a Ciência e Tecnologia (FCT, Portugal) with a Postdoctoral Fellowship (SFRH/BPD/121728/2016). H. D. Burrows is grateful for funding from "The Coimbra Chemistry Centre" which is supported by the Fundação para a Ciência e a Tecnologia (FCT), through the programmes UID/QUI/UI0313/ 2019 and COMPETE.

References

1. Meyer R, Köhler J, Homburg A (2015) Explosives, 7th edn. Wiley, Weinheim
2. Woodfin RL (2007) Trace chemical sensing explosives. Wiley, Hoboken
3. Sun X, Wang Y, Lei Y (2015). Chem Soc Rev 44:8019–8061
4. Cooper PW (1996) Explosives engineering. Wiley, Hoboken
5. Aziz A, Mamat R, Ali W, Perang M (2015). J Eng Appl Sci 10(15):6188–6191
6. Schulte-Ladbeck R, Vogel M, Karst U (2006). Anal Bioanal Chem 386:559–565
7. Östmark H, Wallin S, Ang HG (2012). Propellants Explos Pyrotch 37:12–33
8. Harper RJ, Almirall JR, Furton KG (2005). Talanta 67:313–327
9. Schulte-Ladbeck R, Kolla P, Karst U (2002). Analyst 127:1152–1154
10. Johnen D, Heuwieser W, Fischer-Tenhagen C (2013). Appl Anim Behav Sci 148:201–208
11. Ong T, Mendum T, Guertsen G, Kelley J, Ostrinskaya A, Kunz R (2017). Anal Chem 89:6482–6490
12. Caygill JS, Davis F, Higson SPJ (2012). Talanta 88:14–29
13. Brown KE, Greenfield MT, McGrane SD, More DS (2016). Anal Bioanal Chem 408:35–47
14. Bello R (2013). Front Sci 3(1):27–42
15. Moore DS (2004). Appl Phys Lett 75(8):2499–2512
16. Steinfeld JL, Wormhoudt J (1998). Annu Rev Phys Chem 49:203–232
17. Cheng S, Dou J, Wang W, Chen C, Hua L, Zhou Q, Hou K, Li J, Li H (2013). Anal Chem 85:319–326
18. Tabrizchi M, Ilbeigi V (2010). J Hazard Mater 176:692–696
19. Ewing RG, Alkison DA, Eiceman GA, Ewing GJ (2001). Talanta 54:515–529
20. Singh S (2007). J Hazard Mater 144:15–28
21. Marshall M, Oxley J (eds) (2009) Aspects of explosives detection, 1st edn. Elsevier, Oxford
22. Mokalled L, Al-Husseini M, Kabalan KY, El-Hajj A (2014). Int J Sci Eng Res 5(6):337–350
23. Evans RC, Douglas P, Burrows HD (eds) (2013) Applied photochemistry. Springer, Dordrecht
24. Germain MG, Knapp MJ (2009). Chem Soc Rev 38:2543–2555
25. Pablos JL, Sarabia LA, Ortiz MC, Mendína A, Muñoz A, Serna F, García FC, García JM (2015). Sens Actuat B Chem 212:18–27
26. Pablos JL, Trigo-López M, Serna F, García FC, García JM (2014). RSC Adv 49:25562–25568
27. Wang S, Li N, Pan W, Tang B (2012). Trends Anal Chem 39:3–37
28. Ma Y, Wang S, Wang L (2015). Trends Anal Chem 65:13–21
29. Salinas Y, Martínez-Máñez R, Marcos MD, Sancenón F, Costero AM, Parra M, Gil S (2012). Chem Soc Rev 41:1261–1296
30. Valeur B, Berberan-Santos MN (2012) Molecular fluorescence, principles and applications, 2nd edn. Wiley, Weinheim
31. He G, Yang N, Wang H, Ding L, Yin S, Fang Y (2011). Macromolecules 44:4759–4766
32. Nie H, Lv Y, Yao L, Pan Y, Zhao Y, Li P, Sun G, Ma Y, Zhang M (2014). J Hazard Mater 264:474–480
33. Wu J, Liu W, Ge J, Zhang H, Wang P (2011). Chem Soc Rev 40:3483–3495
34. Lakowicz J (2006) Principles of fluorescence spectroscopy3th edn. Springer, New York
35. Feng L, Li H, Qu Y, Lu C (2012). Chem Commun 48:4633–4635
36. Wang Y, La A, Brückner C, Lei Y (2012). Chem Commun 48:9903–9905
37. Fan L, Zhang Y, Murphy CB, Angell SE, Parker MFL, Flynn BR, Jones Jr WE (2009). Coord Chem 253:410–422
38. Kundu S, Patra A (2017). Chem Rev 117:712–757
39. McQuade DT, Pullen AE, Swager TM (2000). Chem Rev 100:2537–2574
40. Amara JP, Swager TM (2005). Macromolecules 38(22):9091–9094
41. Wasnick JH, Mello CM, Swager TM (2005). J Am Chem Soc 127(10):3400–3405
42. Feng X, Liu L, Wang S, Zhu D (2010). Chem Soc Rev 39:2411–2419
43. Cotts PM, Swager TM, Zhou Q (1996). Macromolecules 29:7323–7338
44. Yang JS, Swager TM (1998). J Am Chem Soc 120:5321–5322

45. Yang JS, Swager TM (1998). J Am Chem Soc 120:11864–11873
46. Cumming CJ, Aker C, Fisher M, Fox M, La Grone MJ, Reust D, Rockley MG, Swager TM, Towers E, Willians V (2001). IEEE Trans Geosci Remote Sens 39:1119–1128
47. Xie L, Yin C, Lai W, Fan Q, Huang W (2012). Prog Polym Sci 37:1192–1264
48. Martelo L, Neves T, Figueiredo J, Marques L, Fedorov A, Charas A, Berberan-Santos MN, Burrows H (2017). Sensors 17(11):2532
49. Martelo L, Valente A, Fonseca S, Burrows H, Marques A, Forster M, Scherf U, Peltzer M, Jiménez A (2012). Polym Int 61:1023–1030
50. Neves T, Marques L, Martelo L, Burrows HD (2014). IEEE Sens Proc 11:1415–1418
51. Shanmugaraju S, Mukherjee PS (2015). Chem Commun 51:16014–16032
52. Martelo L, Fedorov A, Berberan-Santos MN (2015). J Phys Chem B 119(48):15023–15029
53. Goodpaster JV, McGuffin VL (2001). Anal Chem 73:2004–2011
54. Focsaneanu KS, Scaiano JC (2005). Photochem Photobiol Sci 4:817–821
55. Akhavan J (2011) The chemistry of explosives, 3rd edn. RSC, Cambridge
56. Peng Y, Zhang A, Dong M, Wang Y (2011). Chem Commun 47:4505–4507
57. Du H, He G, Liu T, Ding L, Fang Y (2011). J Photochem Photobiol A 217:356–362
58. Ding A, Yang L, Zhang Y, Zhang G, Kong L, Zhang X, Tian Y, Tao X, Yang J (2014). Chem Eur J 20:12215–12222
59. Akhgari F, Fattahi H, Oskei YM (2015). Sens Actuat B 221:867–878
60. Goldman ER, Medintz IL, Whitley JL, Hayhurst A, Clapp AR, Uyeda HT, Deschamps JR, Lassman ME, Mattoussi H (2005). J Am Chem Soc 127:6744–6751
61. Enkin N, Sharon E, Golub E, Willner I (2014). Nano Lett 14:4918–4922
62. Toal SJ, Magde D, Trogler WC (2005). Chem Commun 0:5465–5467
63. Sohn H, Sailor MJ, Magde M, Trogler WC (2003). J Am Chem Soc 125:3821–3830
64. Jerónimo PCA, Araújo AN, Conceição M, Montenegro BSM (2007). Talanta 72:13–27
65. Daly R, Harrington TS, Martin GD, Hutcchings IM (2015). Int J Pharm 494:554–567
66. Derby B (2010). Annu Rev Mater Res 40:395–414
67. Kipphan H (2001) Handbook of print meida, 1st edn. Springer, New York
68. Ihalainen P, Määttänen A, Sandler N (2015). Int J Pharm 494:585–592
69. Owens GJ, Singh RK, Foroutan F, Alqaysi M, Han CM, Mahapatra C, Kim HW, Knowles JC (2016). Prog Mater Sci 77:1–79
70. Leung A, Shankar PM, Mutharasan R (2007). Sens Actuat B 125:688–703
71. Pantano P, Walt DR (1996). Chem Mater 8:2832–2835
72. Walt DR (2006). BioTechniques 41:529–535
73. Thomsen V, Schatzlein D, Mercuro D (2003). Spectroscopy 18(12):112–114
74. Winquist F, Lundström I, Wide P (1999). Sens Actuat B 58:512–517
75. Dasary SSR, Singh AK, Lee KS, Yu H, Ray PC (2018). Sens Actuat B 255:1646–1654
76. Guiterrez-Osuna R (2002). IEEE Sensors J 2:189–202
77. Pioggia G, Ferro M, Di Francesco F, Ahluwalia A, De Rossi D (2008). Bioinspir Biomim 3:1–11
78. Goodner K, Dreher G, Rouseff R (2001). Sens Actuat B 80:261–266
79. Banerjee R, Tudu B, Bandyopadhyay R, Bhattacharyya N (2016). J Food Eng 190:110–121
80. Shannon WD (2007). Handbook Statist 27:342–366
81. Massart DL, Kaufman L (1983) The interpretation of analytical chemical data by the use of cluster analysis. Wiley, New York
82. Askim JR, Mahmoudi M, Suslick KS (2013). Chem Soc Rev 42:8649–8682
83. Lin H, Suslick KS (2010). J Am Chem Soc 132:15519–15521
84. Li H, Kim MK, Altman N (2010). Ann Stat 38:1094–1121
85. Zeng P, Zhang W (2013). Top Appl Stat 55:213–217

New Trends in Fluorescent Reporters in Biology

L. Della Ciana

Contents

Abstract Fluorescent dyes play a prominent role in the detection and quantification of components in biological samples. Analytical methods based on fluorescent dyes include fluorescence microscopy, fluorescent immunoassays, nucleic acid assays, flow cytometry, and fluorescent imaging. During the past decade, most research effort by synthetic chemistry companies focused on three different areas: (a) improvement of existing cyanine dyes, (b) development of new dyes with large Stokes-shifts (LSS), and (c) development of dye assemblies, namely, fluorescent conjugated polymers (FCP) and fluorescent dye-doped silica nanoparticles (FSNP).

Keywords 4,4-Difluoro-4-bora-3a,4a-diaza-indacene (BODIPY) · Cyanines · Flow cytometry · Fluorescent conjugated polymers (FCP) · Fluorescent dye-doped silica nanoparticles (FSNP) · Förster resonance energy transfer (FRET) · InGaN (indium gallium nitride) · InGaP (indium gallium phosphide) · Large stokes-shift (LSS) dyes ·

L. Della Ciana (✉)
Cyanagen srl, Bologna, Italy
e-mail: dellaciana.leopoldo@cyanagen.com

© Springer Nature Switzerland AG 2019
B. Pedras (ed.), *Fluorescence in Industry*, Springer Ser Fluoresc (2019) 18: 321–340,
https://doi.org/10.1007/4243_2018_3, Published online: 2 February 2019

Laser diode excitation · Major histocompatibility complex (MHC) · Multicolor
panels · Near infrared (NIR) · Plasma protein binding (PPB)

1 Introduction

Reporters based on fluorescent dyes play a prominent role in the detection and
quantification of components in biological samples. Analytical methods based on
fluorescent dyes include fluorescence microscopy, fluorescent immunoassays,
nucleic acid assays, flow cytometry, and fluorescent imaging.

During the past decade, most research effort by synthetic chemistry companies
focused on three different areas: (a) improvements of cyanine dyes, (b) development
of new dyes with large Stokes-shifts, and (c) development of dye assemblies.

Fluorescent reporter dyes based on cyanines, first introduced during the last
decade of the twentieth century [1, 2], have become one the most popular dye
types for labeling biomolecules [3]. Their success is linked to a unique combination
of properties, including extremely high molar absorptivity, often exceeding
200,000 M^{-1} cm^{-1}, medium to high fluorescence quantum yield, and extreme
chemical versatility, which allows fine-tuning for specific applications. Absorption
and emission wavelengths of cyanines span the entire visible and near-infrared
spectrum, up to about 850 nm. However, they have a tendency to form dimer and
higher aggregates on aqueous solutions due to ring stacking or hydrophobic interactions, with a reduction of performance due to self-quenching. In addition, many
cyanine dyes exhibit poor chemical and/or photochemical stability; they also often
exhibit non-specific binding behavior with respect to biomolecules, leading, in the
most severe cases, to aggregation or even precipitation of labeled proteins. For these
reasons, a considerable amount of effort was devoted to the elimination or at least
mitigation of these weaknesses.

For organic fluorophores, the difference between band maxima of the absorption
and emission spectra, or Stokes-shift, is typically rather small, in most cases comprised between 10 and 30 nm. A small Stokes-shift makes it difficult to cleanly
separate the fluorescent signal from the scattered excitation light. In addition, dye
self-quenching limits the optimum dye/protein loading to a few dyes per protein.
Finally, it is desirable in many cases or even necessary to distinguish a plurality of
labeled analytes from one another upon excitation with the same light source and
different emission wavelengths of the fluorophore labels. Another important stimu-
lus for the development of new fluorophores with large Stokes-shifts (LSS) has been
the introduction of powerful, yet relatively inexpensive and compact excitation
sources, such as laser diodes and other solid-state lasers in the UV and visible
spectral regions.

Reporter dyes organized in large assemblies of individual fluorophores represent
another area under intensive development. Two approaches have been particularly
successful, one based on fluorescent conjugated polymer dyes (FCP) and the other
on fluorescent dye-doped silica nanoparticles (FSNP). The new fluorescent reporters

obtained by either of these two methods are several orders of magnitude brighter than single-molecule fluorophores and in many cases considerably more photostable. They can be produced in tunable emission wavelength sets with a single excitation source, enabling their use in multicolor panels. For these reasons, they are expected to have a major impact in techniques requiring a very high level of sensitivity, such as flow cytometry, where the greatly increased sensitivity will enable the identification of cell populations with lower receptor density than previously possible and resolve very dim cell populations.

2 Improvement of Cyanine Dyes

Fluorescent cyanine dyes are widely used in biological applications requiring high sensitivity [3]. They generally possess very narrow and intense absorption bands and good fluorescence quantum yields. However, labeling dyes based on cyanines often have the tendency to aggregate and form dimers, especially in aqueous solution, due to the planarity of their π-system. Formation of dimers and higher aggregates may produce a shift in absorption maximum in comparison with the monomer dye molecule and, more significantly, reduce their fluorescence. In addition, compounds with insufficient hydrophilicity undergo non-specific interactions with various surfaces and biomolecules; finally, the photostability of some cyanines, especially dicarboindocyanines such as Cy5 and Cy 5.5, is mediocre. Therefore, many efforts were at reducing the undesirable properties of these dyes by suitable modifications in the dye structure. For example, cyanine compounds were synthesized by Life Technologies Corp. having one of the common methyl groups in the 3-position of the terminal indole heterocycle substituted by a ω-carboxyalkyl function (Fig. 1) [4].

Similarly, at the Institut für Diagnostikforschung GmbH, one of the common methyl groups was substituted by a ω-alkyl sulfonic acid function [5]. Cyanines where the common methyl groups in the 3-positions of the terminal indole

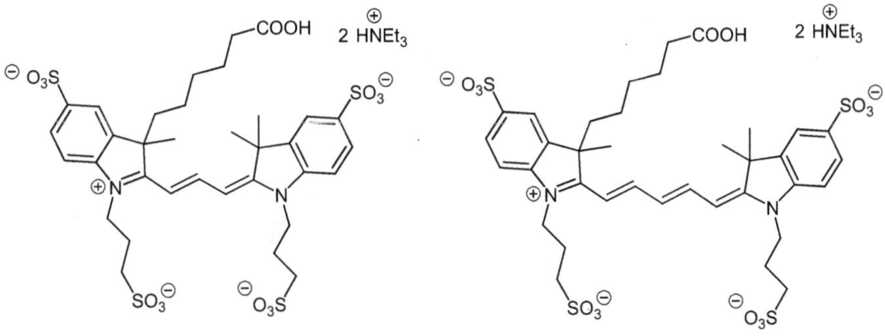

Fig. 1 Modified Cy3 (**1**) and Cy5 (**2**) labeling dyes according to [4]

heterocycles were simultaneously substituted, one with a ω-carboxyalkyl function and the other an N-ω-alkyl sulfonic acid function, were also reported by Thermo Fisher Scientific (Fig. 2) [6].

In a related development, cyanine compounds were synthesized (Pierce Biotechnology Inc./Dyomics GmbH) with additional hydrophilic substituents (Fig. 3) [7].

The introduction of N-phenylsulfonate substituents in indocyanines was reported by Illumina Cambridge Ltd. to increase fluorescence and photochemical stability with respect to dyes of the same structural class (Fig. 4) [8]. The N-phenyl substituent on the end group is not planar relative to the rest of the dye molecule, and the presence of a solvent sphere surrounding the negatively charged sulfonate group acts synergistically to decrease dye interactions and aggregation. In addition, the N-phenyl substituents increase the rigidity of the dye molecule resulting in a decrease in non-radiative energy loss from the excited state of the cyanine dyes in solution. A further benefit of N-phenyl substituents is the increase in dye photostability, possibly through a decrease in reactivity of the methine carbon next

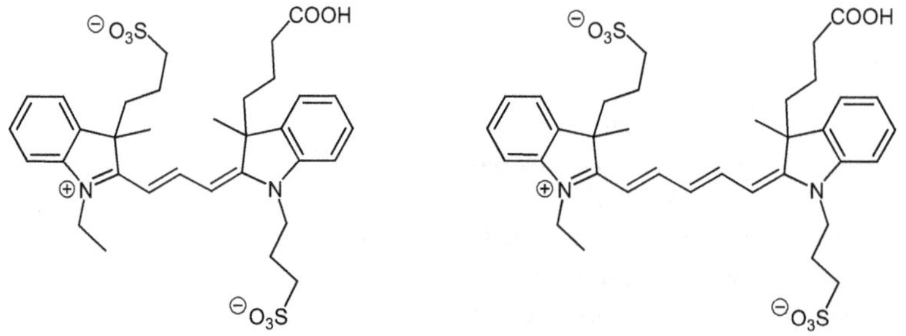

Fig. 2 Modified Cy3 (3) and Cy5 (4) labeling dyes according to [6]

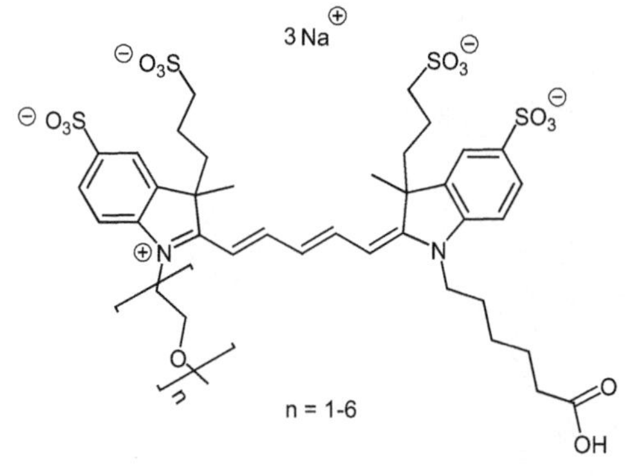

Fig. 3 Modified Cy5 (5) labeling dye according to [7]

Fig. 4 Modified Cy5 (6 and 7) and Cy 5.5 (8) labeling dyes according to [8]

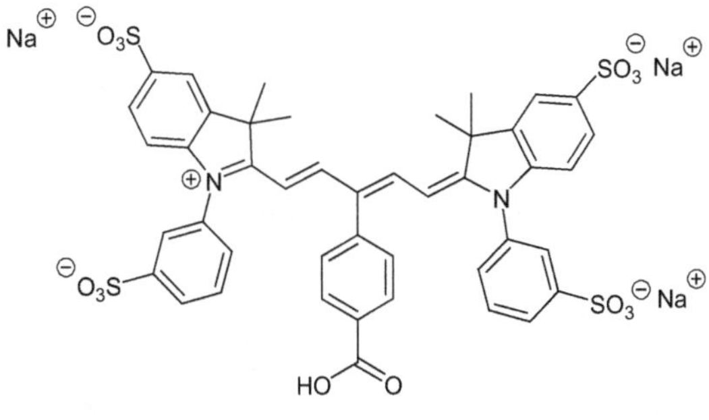

6

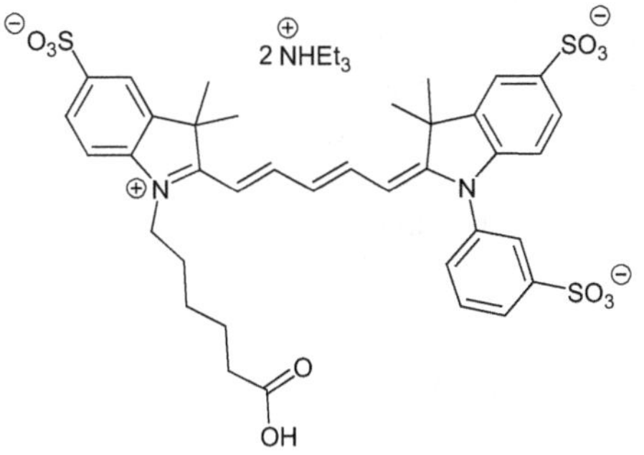

7

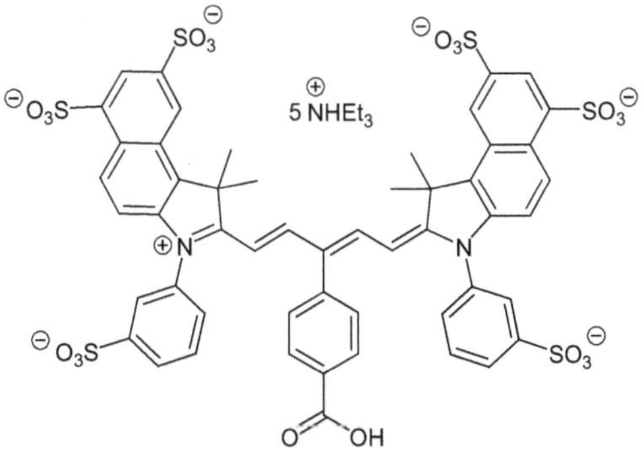

8

to the indole nitrogen toward singlet oxygen or perhydroxyl (•OOH) radicals. Dye **6** with an absorption maximum at 658 nm shows an improved stability compared to Alexa 647. After 25 days of irradiation, it retained 85% of its initial fluorescence intensity. Similarly, Dye **7** exhibits an absorption maximum at 647 nm in water and is more stable in water compared to the structurally similar dyes Cy5 and Alexa

647, while Dye **8** (Abs$_{max}$ at 686 nm) exhibits better stability and is brighter than commercially available Dy681.

3 Large Stokes-Shift Dyes

For most fluorophores, the separation between the absorption and the emission maxima, known as Stokes-shift, is quite small, with typical values in the 10–30 nm range. This characteristic imposes a significant limitation to sensitivity due to Rayleigh and Raman scattering. Moreover, in applications requiring multiple labeling, for example, conjugation to proteins such as antibodies, self-quenching becomes prevalent even at very small dye-to-protein ratio. The emerging of high-throughput screening, or multiplexing, by which multiple fluorophore signals are excited by single-wavelength light and can be combined into one single experiment has also stimulated the development of sets of dyes with tunable, large Stokes-shifts (LSS). In the following sections, LSS dyes are listed according to the excitation wavelength of available laser diodes.

3.1 UV Excitation

Currently available UV laser diodes are based on InGaN (indium gallium nitride) and emit at 375 nm. A series of fluorescent marker dyes with UV excitation (<400 nm) and large Stokes-shifts was developed by Dyomics, based on benzoxazoles [9]. In addition, these dyes possess high photostability and storage stability and high fluorescence quantum yields (Fig. 5 and Table 1).

3.2 Violet Excitation

The violet laser diode is based on InGaN (indium gallium nitride) and emits at 405 nm. Its introduction has stimulated a considerable research effort, aimed at developing dyes optimized for excitation at this wavelength and, at the same time, possessing a large Stokes-shift and a high fluorescence quantum yield. The majority of such dyes are either coumarin or pyridyloxazole derivatives.

Thus, Diwu et al. of Becton Dickinson describe a water-soluble, halogenated 7-hydroxy-coumarin with absorption maximum at 415 nm, which closely matches the 405 nm diode laser excitation, and an emission at 499 nm [10], such as Dye **14** (Fig. 6).

The incorporation of halogen atoms in the 6- and/or 8-position of the coumarin core structure lowers the pK$_a$ and allows these dyes to be useful near physiological pH. Nevertheless, a strong pH dependency is observed up to pH 6–6.5, and the

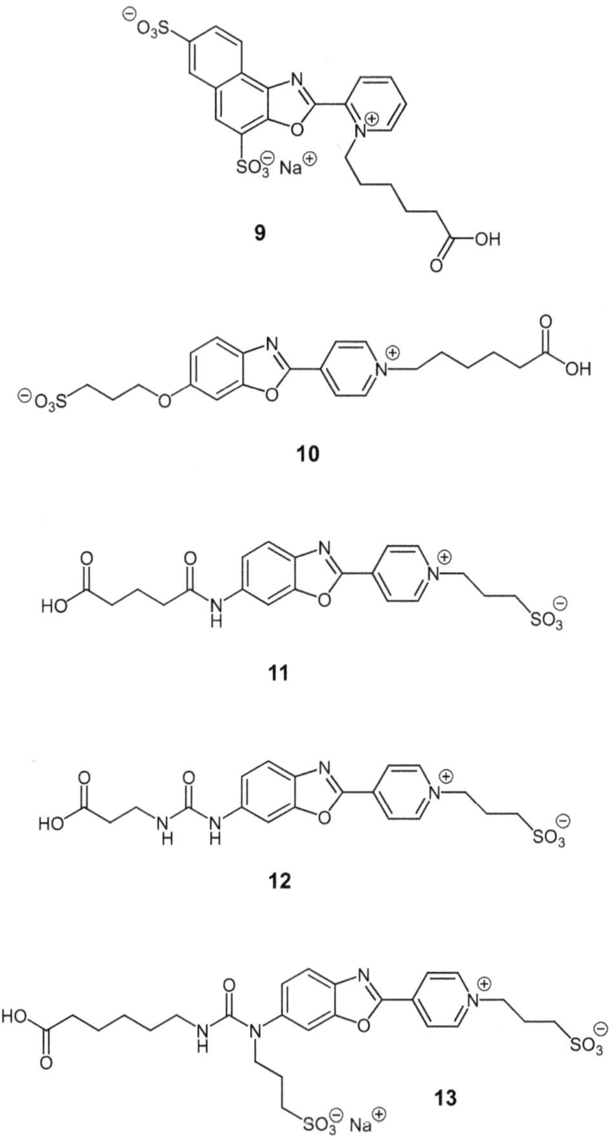

Fig. 5 Large Stokes-shift labeling dyes based on benzoxazoles [9]

chlorine atoms may impart a heavy atom effect with consequent reduction of fluorescent quantum yield as well as reduce water solubility. Furthermore, the anionic form of these dyes, required for fluorescence, may be undesirable in some applications as described by Jin et al. [11]. On the contrary, 7-amino coumarins synthesized by Molecular Probes Inc. are known to be pH independent over a much wider pH range and show excellent photostability and high fluorescence quantum

Table 1 Optical properties of LSS dyes **9–13**

Compound	Exc/Abs peak (nm)	Ext. coeff $(M^{-1} cm^{-1})$	Em peak (nm)	QY	Stokes-shift (nm)
9	368	13,200	474	0.85	106
10	383	22,000	516	0.70	133
11	375	16,000	543	0.62	168
12	397	20,000	572	0.28	175
13	353	16,000	614	0.29	261

In PBS buffer

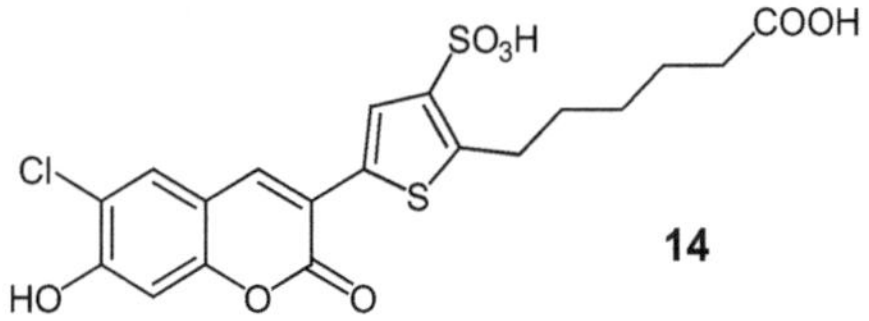

Fig. 6 A violet laser diode excitable LSS labeling dye based on 7-hydroxy-coumarin [10]

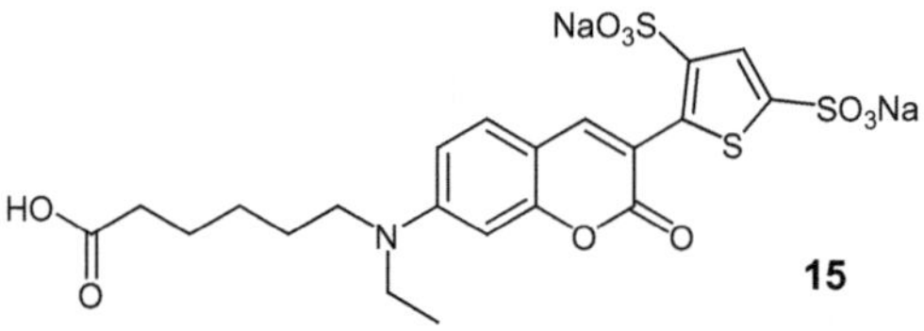

14

Fig. 7 A violet laser diode excitable LSS labeling dye based on 7-amino-3-thienyl coumarin [13]

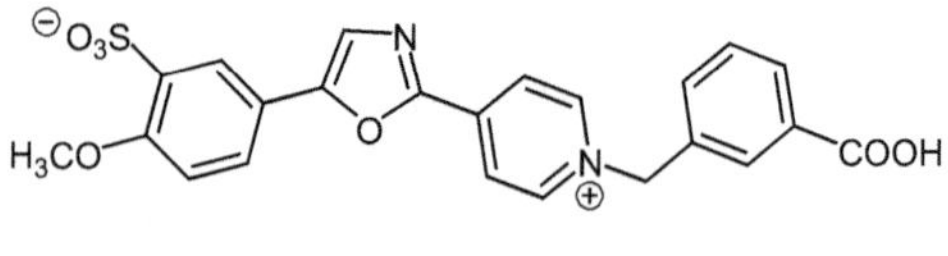

15

Fig. 8 Cascade Yellow, a pyridoxazole-based LSS labeling dye which can be excited with a violet laser diode [14]

16

yields [12]. Recently, 7-amino-3-thienyl coumarin dyes were developed at Cyanagen Srl, which show large Stokes-shifts of at least 80 nm with efficient excitability by a 405 nm violet excitation source and high fluorescence intensities in the green [13]. In particular, Dye **15** (Fig. 7) has an absorption maximum at 412 nm, while the emission maximum is at 507 nm, resulting in a Stokes-shift of 95 nm. This, combined with its excellent fluorescence quantum yield of 0.7 makes it quite suitable for applications such as multicolor flow cytometry.

Another class of dyes which combine efficient excitation with the 405 nm diode laser, good fluorescence quantum yield, and unusually high Stokes-shift is based on pyridoxazole derivatives. An important example is Dye **16** (Fig. 8), known as Cascade Yellow, which exhibits an absorption maximum at 400 nm, while its emission shifts to 550 nm [14].

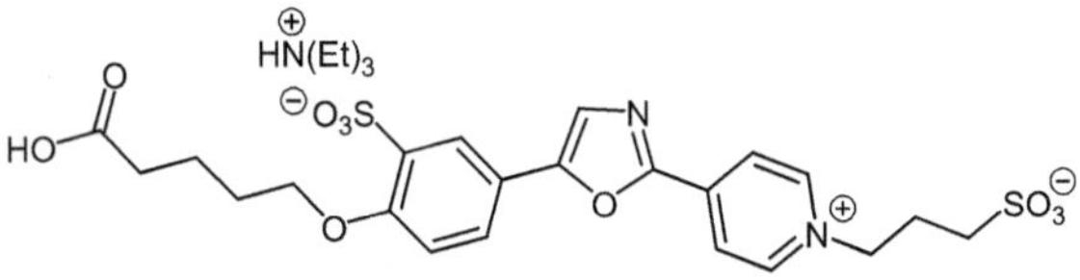

Fig. 9 A violet laser diode excitable LSS labeling dye based on pyridoxazole [15]

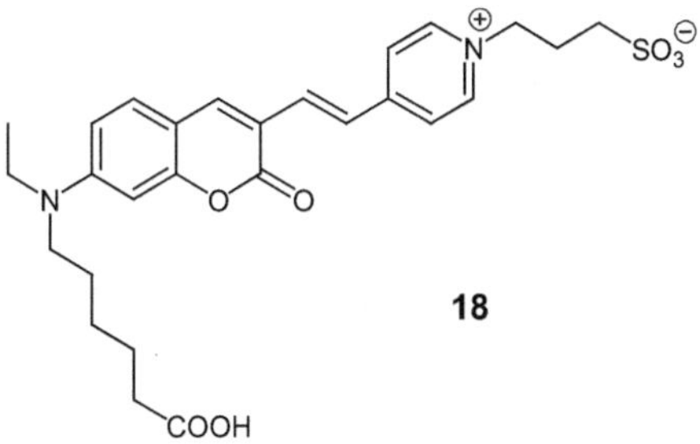

17

Fig. 10 Blue laser diode excitable LSS labeling dyes based on derivatives of coumarin polymethines [16]

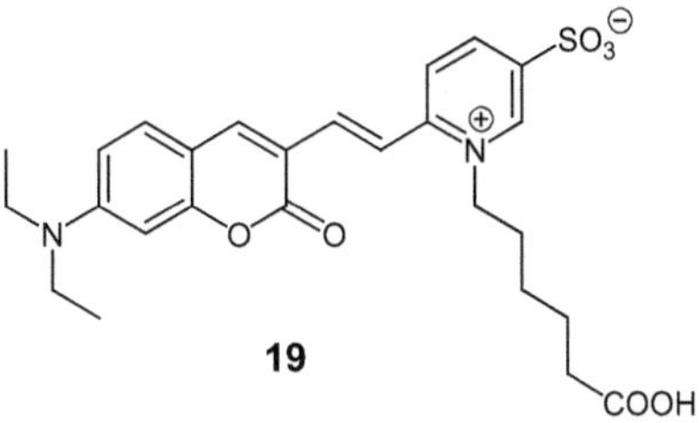

18

19

An improved version of Cascade Yellow is Dye **17** (Fig. 9), developed by Life Technologies Corp., which exhibits similar absorption and emission wavelength, but is brighter, has no emission at 488 nm and has "very little emission in the Pacific Blue filter off the violet" [15].

3.3 Blue Excitation

The replacement of the 488 nm Ar-ion laser with 473 and 488 nm InGaN diode lasers has generated great interest in dyes that absorb at this wavelength and exhibit high Stokes-shifts. For example, MegaStokes Dyes **18** (DY-485XL) and **19** (DY-480XL) developed by Dyomics are coumarin-based polymethine dyes tailored to an excitation wavelength between 470 and 500 nm (Fig. 10 and Table 2) [16].

Another group of LSS dyes which can be efficiently excited by the 488 nm diode laser are derivatives of amine-substituted tricyclic dyes, such as Dyes **20** (AZ 285) and **21** (AZ 312) developed by ATTO-TEC GmbH (Fig. 11 and Table 3) [17].

Table 2 Optical properties of LSS Dyes **18** and **19**

Compound	Exc/Abs peak (nm)	Ext. coeff (M^{-1} cm^{-1})	Em peak (nm)	QY	Stokes-shift (nm)
18	485	50,000	560	–	70
19	500	50,000	630	–	130

In ethanol

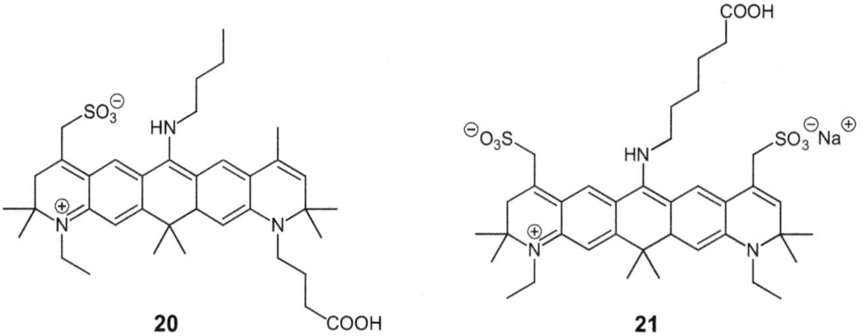

Fig. 11 Blue laser diode excitable LSS labeling dyes based on derivatives of amine-substituted tricyclic dyes [17]

Table 3 Optical properties of LSS Dyes **20** and **21**

Compound	Exc/Abs peak (nm)	Ext. coeff (M^{-1} cm^{-1})	Em peak (nm)	QY	Stokes-shift (nm)
20	484	–	653	32	169
21	490	–	670	26	180

In PBS buffer

3.4 Green Laser Diode Excitation

Recently introduced, InGaN-based 514 and 532 nm *true* green laser diodes, are efficient excitation sources for MegaStokes Dyes **22** (DY-520XL) and **23** (DY-521XL) developed by Dyomics GmbH (Fig. 12 and Table 4) [16].

Older, 552 and 561 nm InGaP (indium gallium phosphide)-based green laser diodes are more suitable excitation sources for 4,4-difluoro-4-bora-3a,4a-diaza-indacene (BODIPY) push-pull compounds incorporating a naphthyridine acceptor site and dimethylamino aryl linked BODIPY donor [18]. Thus, compound **24** (Fig. 13) absorbs at 554 nm while emitting at 740 nm, with a very large Stokes-shift of 185 nm.

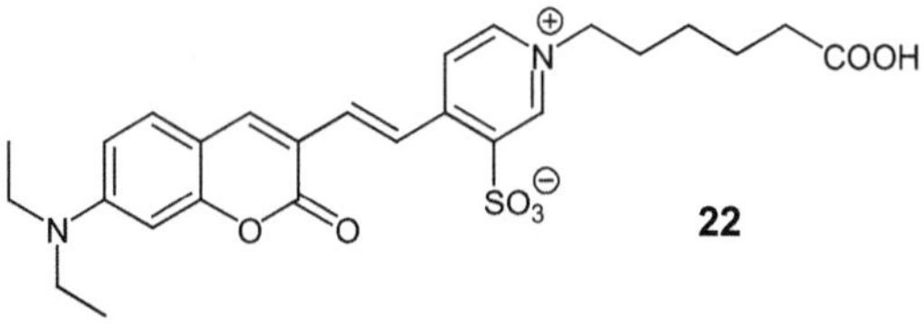

Fig. 12 Green laser diode excitable LSS labeling dyes based on derivatives of coumarin polymethines [16]

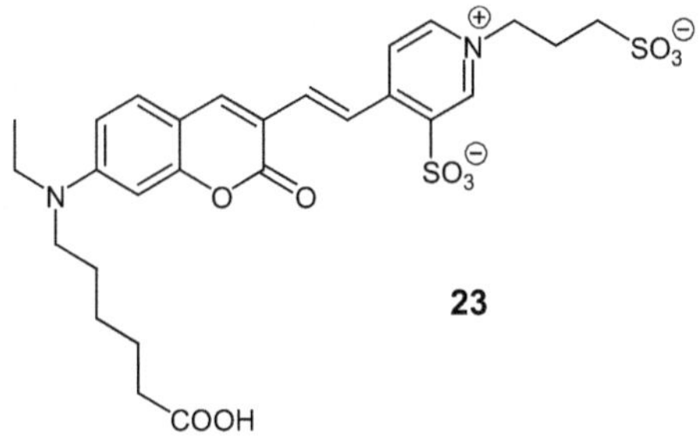

Table 4 Optical properties of LSS Dyes **22** and **23**

Compound	Exc/Abs peak (nm)	Ext. coeff (M^{-1} cm^{-1})	Em peak (nm)	QY	Stokes-shift (nm)
22	520	50,000	664	–	114
23	523	50,000	668	–	145

In ethanol

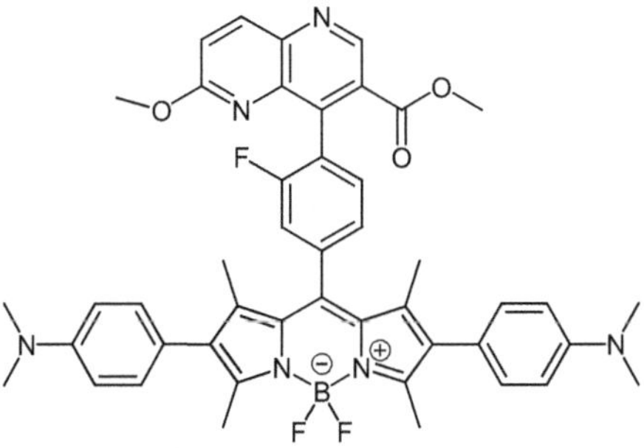

24

Fig. 13 Green laser diode excitable LSS labeling dyes based on *meso*-amine substituted push-pull BODIPY derivative [18]

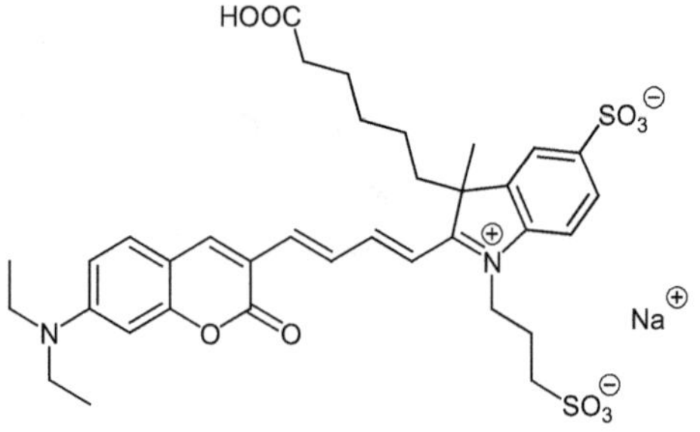

Fig. 14 Orange laser diode
excitable LSS labeling dye
based on coumarin
polymethines [16]

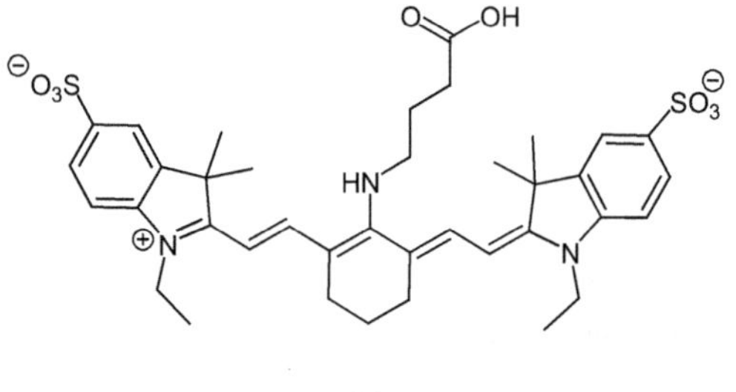

25

Fig. 15 Red laser diode
excitable LSS labeling dye
based on *meso*-amine-
substituted heptamethine
cyanine [19]

26

3.5 Orange Diode-Pumped Solid-State (DPSS) Laser Excitation

MegaStokes Dye **25** (DY-601XL) by Dyomics (Fig. 14), with an absorption maximum at 606 nm (ethanol; molar absorbance 85,000 M^{-1} cm^{-1}), can be efficiently excited with the recently introduced diode-pumped solid-state (DPSS) 594 nm laser. Its emission occurs at 663 nm (ethanol), with a Stokes-shift of 57 nm [16].

3.6 Red Excitation

The majority of near-infrared (NIR) emitting dyes is based on the tricarbocyanine core structure. These compounds have exceedingly high extinction coefficients (200,000 to 300,000 M^{-1} cm^{-1}), good quantum yields of fluorescence (>10%), and high chemical flexibility. Their Stokes-shift is rather small, typically around 20–25 nm. However, when a nitrogen atom is introduced in the *meso* (γ-methine) position of the polymethine chain, a band broadening is observed, accompanied by a large increase in Stokes-shift. For example, Dye **26** (Fig. 15) exhibits a broad absorption band

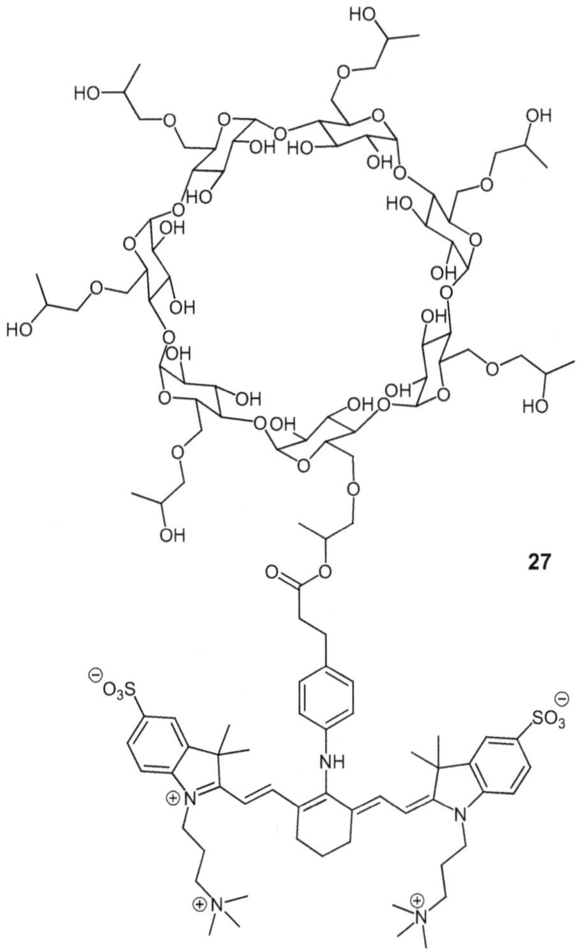

Fig. 16 2-Hydroxypropyl-β-cyclodextrin/*meso*-amino heptamethine indocyanine conjugate [20, 21]

centered at 625 nm, with an emission maximum at 755 nm, and thus a Stokes-shift of 116 nm [19].

IgG labeling with Dye **26** shows a linear increase in fluorescence up to a 30-fold dye/protein ratio. These spectroscopic characteristics are well suited for multitasking NIR analysis, IgG labeling, in vivo imaging, fluorescence microscopy, and FRET applications.

Recently, *meso*-amino heptamethine indocyanine dyes were developed by Cyanagen for use in the noninvasive real-time transcutaneous assessment of kidney function [20]. A 2-Hydroxypropyl-β-cyclodextrin dye conjugate **27** (Fig. 16) was injected intravenously into conscious rats, and its elimination was measured transcutaneously by means of a miniaturized plaster device. This consisted of two

light emitting diodes (LEDs) with an excitation wavelength at 700 nm and a
photodiode for emission wavelength detection at 790 nm, very closely matching
the spectral properties of the dye ($\lambda_{abs} = 704$ nm; $\lambda_{em} = 790$ nm). This large Stokes-
shift, NIR agent outperforms existing NIR commercial dyes such as IRDye800CW
regarding skin accumulation and plasma protein binding (PPB). It is excreted
efficiently through the kidneys into urine without significant reabsorption and
secretion in the kidney tubules [21].

4 Dye Assemblies

4.1 Fluorescent Conjugated Polymers

Water-soluble, fluorescent conjugated polymers (FCP) are efficient light-gathering
macromolecules, which are finding increasing use for biological detection in a
variety of applications including immunoassays and nucleic acid assays. Their
advantage over non-interacting small molecule reporters is related to their collective
response. This improves important optical characteristics, in particular Förster resonance energy transfer (FRET) and fluorescence efficiency. Properly designed FCP
can provide ultrasensitive detection methods for protein/nucleic acid detection. BD
Horizon Brilliant Violet polymer dyes were developed from a polymer dye technology pioneered by Sirigen [22, 23]. The core structure of these conjugated polymers
is represented by general formula **28** (Fig. 17), where E is an end group, A is an aryl
or heteroaryl group, and WSG is a water-solubilizing group. Pendent reactive groups
can be attached to the E or A groups in order to provide chain extension, conjugation,
or cross-linking.

The first Brilliant Violet polymer dye to be developed was BV421 [24]. Its
absorption maximum at 407 nm closely matches the 405 nm laser diode, while its
emission occurs at 421 nm. BV421 is dramatically brighter than Pacific Blue, a
popular fluorophore with similar excitation and emission spectra. In fact, BV421
ranks among the brightest of all fluorophores typically used in multicolor flow
cytometry. Subsequently Brilliant Violet polymers were produced in an array of
other colors [24]. This goal was achieved in two ways. The first involved tuning the

Fig. 17 General structure
of Brilliant Violet
fluorescent polymer dyes

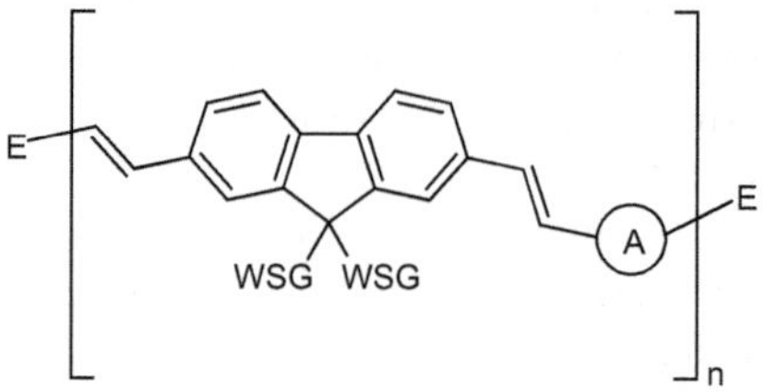

28

Table 5 Optical properties of Brilliant Violet polymer dyes

Name	Type of dye	Exc/Abs peak (nm)	Em peak (nm)	Stokes-shift (nm)
BV421	Base polymer	407	421	106
BV510	Base polymer	405	510	133
BV570	Tandem BV421 + Cy3	407	570	168
BV605	Tandem BV421 + Cy3.5	407	605	175
BV650	Tandem BV421 + Dy610	407	650	261
BV711	Tandem BV421 + Dy682	407	711	304
BV786	Tandem BV421 + Dy752	407	786	379

polymer backbone to emit light at a longer wavelength, as in the case of BV 510, with absorption and emission maximum at 405 and 510 nm, respectively. The second option involved the production of energy tandem reporters, which comprise a donor structure (BV421) and an acceptor dye (for fluorescent emission, like a rhodamine, cyanine, or Alexa dye). Brilliant Violet polymer dyes can all be excited with the 405 nm violet diode laser and emit at a range of wavelengths as summarized in Table 5.

Brilliant Violet reporters are dramatically brighter than other UV-violet excitable dyes and are comparable in brightness to polymeric fluorophores of natural origin such as phycoerythrin (PE) and allophycocyanin (APC). Thus, they are especially suited for cytometric assays requiring high sensitivity, such as MHC multimer staining or detection of intracellular antigens. Furthermore, these reporters are an excellent option for fluorescence imaging, two-photon microscopy, and imaging cytometry. They are expected to have a dramatic impact on the design and implementation of multicolor panels for high-sensitivity immunofluorescence assays.

4.2 Dye-Doped Fluorescent Silica Nanoparticles (FSNP)

In recent years, dye-doped fluorescent silica nanoparticles (FSNP) have been gaining increased interest as a way to overcome the limitations of individual dyes. Properly designed FSNP can be 2–3 orders of magnitude brighter than single fluorophores and are significantly more photostable [25–27].

The preparation of FSNP is based on simple, yet versatile and inexpensive sol-gel techniques [28]. Three main synthetic strategies are employed: (a) the Stöber-Van Blaaderen method [29, 30], (b) the reverse (water-in-oil) microemulsion method [31, 32], and the direct micelle-assisted method [32–34]. In all cases, the main silica precursor is tetraethoxysilane (TEOS), while fluorescent dye molecules are preferably incorporated in the silica matrix as trialkoxysilane derivatives [35, 36]. The condensation of the fluorophores within the silica matrix prevents leaching of the dye from the particle.

Silica NPs are chemically inert and physically stable, but surface silanol groups provide a means of molecular functionalization [25–27]. This can allow "targeting" of functionality or particular cell types. In addition, these NPs can be biofunctionalized with a wide array of different molecular recognition probes, such as antibodies or aptamers [25–27].

A large number of molecular dyes have been used for doping silica nanoparticles, including dansyl, coumarins, fluorescein, rhodamines, cyanines, BODIPYs, ruthenium, and iridium complexes [25–27]. Each nanoparticle contains a large number of fluorophores and reaches a molar extinction coefficient well above 10^6 M^{-1} cm^{-1}. The silica matrix protects the dyes segregated inside the nanoparticle from the adverse effects of oxygen, thus increasing their photostability and, in some cases, the average quantum yield of fluorescence.

As in the case of single fluorophores, a large Stokes-shift is desirable, since it helps reducing scattering interferences. The most straightforward approach is based on the incorporation of commercial fluorophores with an intrinsically large Stokes-shift into the silica matrix [37]. The relatively small number of dyes with these properties, however, limits this approach. An alternative, more versatile approach is to exploit energy transfer processes between one or more fluorophores locked inside the nanoparticle [38, 39]. Sets of nanoparticles with emissions of different colors, but excitation at the same wavelength, can also be obtained [40, 41].

Recently, pluronic-silica nanoparticles doped with various dyes with optimized photophysics were prepared, exhibiting an almost complete energy transfer among them [42]. Since the only analytical signal produced by these FSNP is the emission of the lowest energy fluorophore, multiple analyses can be performed with a single excitation and without separation steps. In flow cytometry, this feature allows the use of only one emission channel per kind of nanoparticle without the typical interferences observed in the case of a multiband emission. Furthermore, a significant gain in the signal-to-noise ratio is obtained because of a high pseudo-Stokes-shift.

5 Conclusions

Recent developments in the field of fluorescent reporters in biology include the optimization of cyanine dyes, synthesis of new dyes with large Stokes-shifts, and dye assemblies.

Cyanine dyes, introduced as fluorescent reporters in the late twentieth century, have enjoyed widespread popularity, due to their high brightness and versatility. However, they suffer from several drawbacks, especially their tendency to aggregate, their often poor chemical/photochemical stability, and their undesirable interactions with proteins. These shortcomings were successfully addressed, resulting in considerably improved dyes.

The availability of powerful InGaN solid-state lasers, emitting in the green to near-UV spectral region, has greatly stimulated the synthesis of fluorescent reporter dyes, which can be excited at these wavelengths. Many such dyes were designed as members of a set, all with the same excitation wavelength and tunable, large Stokes-shifts. Such sets of dyes are especially valuable in emerging applications, i.e., high-throughput screening and multiplexing. In the NIR region, the development of large Stokes-shift dyes has enabled the noninvasive measurement of glomerular filtration rate in the kidneys.

Finally, the last decade has witnessed the rapid growth of dye assemblies, as a way to overcome the limitations of individual fluorophores. Two main approaches were followed, water-soluble, fluorescent conjugated polymers (FCP) and dye-doped fluorescent silica nanoparticles (FSNP). Both technologies offer extreme brightness, sensitivity, increased stability of the single dyes, and versatility. Fluorescent reporters based on FCP and FSNP are expected to have an increasing impact on cytometry, fluorescence imaging, two-photon microscopy, and the design and implementation of multicolor panels for high-sensitivity immunofluorescence assays.

References

1. Strekowski L, Lipowska M, Patonay G (1992) Substitution reactions of a nucleofugal group in heptamethine cyanine dyes. Synthesis of an isothiocyanato derivative for labeling of proteins with a near-infrared chromophore. J Org Chem 57:4578–4580
2. Mujumdar RB, Ernst LA, Mujumdar SR, Lewis CJ, Waggoner AS (1993) Cyanine dye labeling reagents: sulfoindocyanine succinimidyl esters. Bioconjug Chem 4:105–111
3. Hermanson GT (2013) Bioconjugate techniques. Academic Press, London
4. Leung WY, Cheung CY, Yue S (2010) Modified carbocyanine dyes and their conjugates. US Pat 7820824
5. Licha K, Riefke B, Semmler W, Speck U, Hilger CS (2010) Near infrared imaging agent. US Pat 7655217
6. Brush CK, He K, Czerney PT, Wenzel M (2011) Hydrophilic labels for biomolecules. US Pat 7951959
7. Hermanson G, Czerney PT, Desai S, Wenzel MS, Dworecki BR, Lehmann FG, Nlend MC (2016) Cyanine compounds. US Pat Appl 15/042328
8. Romanov N, Anastasi C, Liu X (2015) Dyes for labelling molecular ligands. US Patent 9085698
9. Schweder BG, Wenzel MS, Frank WG, Lehmann FG, Czerney PT (2016) Marker dyes for UV and short wave excitation with high stokes shift based on benzoxazoles. US Pat 9453010
10. Diwu Z, Dubrovsky T, Abrams B, Liao J, Meng Q (2013) Reactive heterocycle-substituted 7-hydroxycoumarins and their conjugates. US Pat 8431416
11. Jin X, Uttamapinant C, Ting AY (2011) Synthesis of 7-aminocoumarin by Buchwald–Hartwig cross coupling for specific protein labeling in living cells. Chembiochem 12:65–70
12. Wang HY, Leung WY, Mao F (1997) Sulfonated derivatives of 7-aminocoumarin. US Pat 5696157
13. Jansen TP, Rodeghiero G, Foglietta D, Perciaccante R, Della Ciana L (2017) Novel coumarin dyes and conjugates thereof. US Pat Appl 15/458801
14. Haugland RP (2013) The handbook: a guide to fluorescent probes and labeling technologies, 11th edn. Invitrogen Corp., Eugene
15. Buller G, Liu J, Yue S, Bradford JA (2017) Violet laser excitable dyes and their method of use. US Pat 9644099
16. Czerney P, Wenzel M, Schweder B, Lehmann F (2009) Compounds based on polymethines. US Pat 7563907

17. Zilles A, Drexhage KH, Kemnitzer NU, Arden-Jacob J, Hamers-Schneider M (2017) Amine-substituted tricyclic fluorescent dyes. US Pat 9651490
18. Moriarty RD, Martin A, Adamson K, O'Reilly E, Mollard P, Forster RJ, Keyes TE (2014) The application of water soluble, mega-Stokes-shifted BODIPY fluorophores to cell and tissue imaging. J Microsc 253:204–218
19. Bertolino CA, Caputo G, Barolo C, Viscardi G, Coluccia S (2006) Novel heptamethine cyanine dyes with large stokes' shift for biological applications in the near infrared. J Fluoresc 16:221–225
20. Della Ciana L, Gretz N, Perciaccante R, Rodeghiero F, Geraci S, Huang J, Herrera Pérez Z, Pill J, Weinfurter S (2015) US Pat Appl 14/713,459
21. Huang J, Weinfurter S, Daniele C, Perciaccante R, Federica R, Della Ciana L, Pill J, Gretz N (2017) Zwitterionic near infrared fluorescent agents for noninvasive real-time transcutaneous assessment of kidney function. Chem Sci 8:2652–2660
22. Gaylord BS, Hong JW, Fu TJ, Sun, CJ, Baldocchi R (2013) Fluorescent methods and materials for directed biomarker signal amplification. US Pat 8354239
23. Liu B, Wang C, Zhan R, Bartholomew GP, Hong JW, Wheeler JM, Gaylord BS (2015) Signal amplified biological detection with conjugated polymers US Pat 8969509
24. Chattopadhyay PK, Gaylord B, Palmer A, Jiang N, Raven MA, Lewis G, Roederer M (2012) Brilliant violet fluorophores: a new class of ultrabright fluorescent compounds for immunofluorescence experiments. Cytometry A 81:456–466
25. Bonacchi S, Genovese D, Juri R, Montalti M, Prodi L, Rampazzo E, Zaccheroni N (2011) Luminescent silica nanoparticles: extending the frontiers of brightness. Angew Chem Int Ed 50:4056–4066
26. Bae SW, Tan W, Hong JI (2012) Fluorescent dye-doped silica nanoparticles: new tools for bioapplications. Chem Commun 48:2270–2282
27. Montalti M, Prodi L, Rampazzo E, Zaccheroni N (2014) Dye-doped silica nanoparticles as luminescent organized systems for nanomedicine. Chem Soc Rev 43:4243–4268
28. Hench LL, West JK (1990) The sol-gel process. Chem Rev 90:33–72
29. Stöber W, Fink A, Bohn E (1968) Controlled growth of monodisperse silica spheres in the micron size range. J Colloid Interface Sci 26:62–69
30. Van Blaaderen A, Vrij A (1992) Synthesis and characterization of colloidal dispersions of fluorescent, monodisperse silica spheres. Langmuir 8:2921–2931
31. Zanarini S, Rampazzo E, Della Ciana L, Marcaccio M, Marzocchi E, Montalti M, Paolucci F, Prodi L (2009) Ru(bpy)$_3$ covalently doped silica nanoparticles as multicenter tunable structures for electrochemiluminescence amplification. J Am Chem Soc 131:2260–2267
32. Bagwe RP, Yang C, Hilliard R, Tan W (2004) Optimization of dye-doped silica nanoparticles prepared using a reverse microemulsion method. Langmuir 20:8336–8342
33. Zanarini S, Rampazzo E, Bonacchi S, Juris R, Marcaccio M, Montalti M, Paolucci F, Prodi L (2009) Iridium doped silica-PEG nanoparticles: enabling electrochemiluminescence of neutral complexes in aqueous media. J Am Chem Soc 131:14208–14209
34. Yong KT, Roy I, Swihart MT, Prasad P (2009) Multifunctional nanoparticles as biocompatible targeted probes for human cancer diagnosis and therapy. J Mater Chem 19:4655–4672
35. Van Blaaderen A, Imhof A, Hage W, Vrij A (1992) Three-dimensional imaging of submicrometer colloidal particles in concentrated suspensions using confocal scanning laser microscopy. Langmuir 8:1514–1517
36. Verhaegh NA, Blaaderen AV (1994) Dispersions of rhodamine-labeled silica spheres: synthesis characterization, and fluorescence confocal scanning laser microscopy. Langmuir 10:1427–1438
37. Herz E, Burns A, Bonner D, Wiesner U (2009) Large stokes-shift fluorescent silica nanoparticles with enhanced emission over free dye for single excitation multiplexing. Macromol Rapid Commun 30:1907–1910
38. Wu C, Hong J, Guo X, Huang C, Lai J, Zheng J, Chen J, Mu X, Zhao Y (2008) Fluorescent core-shell silica nanoparticles as tunable precursors: towards encoding and multifunctional nano-probes. Chem Commun 6:750–752

39. Fedorenko SV, Bochkova OD, Mustafina AR, Burilov VA, Kadirov MK, Holin CV, Nizameev IR, Skripacheva VV, Menshikova AY, Antipin IS, Konovalov AI (2010) Dual visible and near-infrared luminescent silica nanoparticles. Synthesis and aggregation stability. J Phys Chem C 114:6350–6355
40. Wang L, Tan W (2006) Multicolor FRET silica nanoparticles by single wavelength excitation. Nano Lett 6:84–88
41. Chen X, Estévez MC, Zhu Z, Huang YF, Chen Y, Wang L, Tan W (2009) Using aptamer-conjugated fluorescence resonance energy transfer nanoparticles for multiplexed cancer cell monitoring. Anal Chem 81:7009–7014
42. Rampazzo E, Bonacchi S, Genovese D, Juris R, Montalti M, Paterlini V, Zaccheroni N, Dumas-Verdes C, Gilles C, Méallet-Renault R, Prodi L (2014) Pluronic-silica (PluS) nanoparticles doped with multiple dyes featuring complete energy transfer. J Phys Chem C 118:9261–9267

Application of Fluorescence in Life Sciences for Basic Research and Medical Diagnostics

Gerhard Hawa

Contents

Abstract This chapter summarizes the use of fluorescence for either diagnostic or basic research in modern life sciences. While the first section concentrates on microscopic applications for, e.g., monitoring of living cells, the second one focuses on analytical aspects, i.e., detection and quantitation of nucleic acids or proteins for research or diagnostic purposes. The state-of-the-art microscopic technologies (CLSM, TIRF, TPFM, STED) and analytical methods (RT-PCR, next-gen sequencing, multiplexing, microarrays) are discussed.

Keywords Apoptosis · Endocytosis · Gerontology · Microarrays · Multiplexing · Phenotype

G. Hawa (✉)
FIANOSTICS GmbH, Wiener Neustadt, Austria
e-mail: gerhard.hawa@fianostics.at

© Springer Nature Switzerland AG 2019   341
B. Pedras (ed.), *Fluorescence in Industry*, Springer Ser Fluoresc (2019) 18: 341–364,
https://doi.org/10.1007/4243_2018_5, Published online: 2 February 2019

1 Technologies Used for the Investigation of Cells

1.1 Fluorescence Microscopy (FM)

Fluorescence microscopy has become a very powerful and popular tool for modern cell biology, because it provides a window into the physiology of living cells at subcellular levels of resolution. This enables the study of diverse processes including protein location and associations, motility, and other phenomena such as ion transport and metabolism. A myriad of techniques have been developed over the last decade including the widespread use of fluorescent proteins [1], new ways for chemical synthesis of fluorophores with improved properties (i.e., stability, quantum yield, etc.) [2–4], the use of confocal microscopy or multiphoton microscopy, and the breaking of the diffraction limit for "super-resolution" which all together boosted application of FM.

In the most basic form (Fig. 1), FM involves exciting the fluorophore(s) in the sample of interest using a light source, a microscope, excitation and emission filters, and an objective.

One of the most significant advancements was the development of highly sensitive detectors like electron-multiplied, very low-noise cooled charge-coupled device detectors (EMCCD). They allow a very sensitive detection of low-light fluorescence because they are capable of integration of the emitted light over time. This is especially important, if the intensity of excitation light has to be reduced to avoid photobleaching.

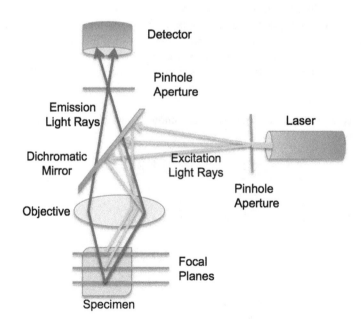

Fig. 1 Principle of fluorescence microscopy

In short the most used technologies and devices are:

- *Confocal Laser Scanning Microscopy (CLSM)*
  This technology, which greatly increased resolution compared to standard fluorescence microscopy, is based on passing a laser beam through a light source aperture which is then focused by an objective lens into a small area on the surface of the sample. By running the laser (scan) across the sample, the image is built up pixel-by-pixel by collecting the emitted photons from the fluorophores. The increase in resolution is mainly due to the exclusion of out-of-focus light, by the pinhole [5]. Only confocal light is collected [6, 7].
- *Total Internal Reflection Fluorescence (TIRF) Microscopy*
  If coherent light (i.e., a laser) is totally internal reflected at the interface of a transparent solid and a liquid, part of the electromagnetic field enters into the liquid. This field is called an evanescent wave, has the same frequency as the reflected light, and can be used for excitation of fluorescent molecules. This results in background suppression and further increase of resolution [8, 9].
- *Two-Photon Fluorescence Microscopy (TPFM)*
  This variant of FM uses near-infrared excitation light which can also excite fluorescent dyes. For each excitation, two photons of infrared light are absorbed. Using infrared light minimizes scattering in the tissue, and the background signal is strongly suppressed. Infrared light has an increased penetration depth and therefore can be a superior alternative to confocal microscopy for thicker specimens [10, 11].
- *Stimulated Emission Depletion (STED) Fluorescence Microscopy*
  Beside all the improvements in fluorescence microscopy, the problem of diffraction limitation remained, preventing the further improvement of resolution. This was overcome by the introduction of STED fluorescence microscopy. It uses intense laser light to switch off the fluorescence of neighboring molecules so that only fluorescence from the remaining excited dye molecules in the center of the excitation focus is then detected. By applying time gates to the collected data (gated STED or gSTED), an even further resolution enhancement is achievable.

While the first three methods did undoubtedly bring advances in resolution and clarity of the acquired images, the highest resolutions remained reserved for electron microscopy, which of course does not allow picturing living cells.

This was changed by the appearance of STED microscopy (Nobel Prize in Chemistry 2014) which was a major breakthrough for real live super-resolution imaging. It has been used to examine key biological processes that no other technique could have examined [12].

Many reviews have been written about all those methods, and readers are encouraged to consult the reference list for this subchapter, since it would be impossible to discuss this in detail here [13, 14].

344 G. Hawa

1.2 Flow Cytometry/Fluorescence-Activated Cell Sorting

Flow cytometry is a broadly used technique that utilizes laser-based technology to count, sort, and profile cells in a heterogeneous fluid mixture. Cells or other particles are suspended in a liquid stream and passed through a laser light beam one after the other. Their interaction with the light is measured by an electronic detection apparatus as light scatter and fluorescence intensity (Fig. 2).

While the scattered light yields information about particle size and form and thus is used for identification of what type of cell is detected, fluorescence signals are used to identify molecular components on the cell surface. This can be even done in a quantifying manner (e.g., how much of a certain cell surface protein is present), if a fluorescent label, or fluorochrome, is stoichiometrically bound to a cellular component. The fluorescence intensity will then represent the amount of that particular cell component.

Thus analytically speaking flow cytometry can be used to determine the phenotype and function of living cells.

This method can also be used to sort and separate different types of cells from each other. This is done by FACS (fluorescence-activated cell sorting), which further adds a degree of functionality. After, just like in conventional flow cytometry, forward-scatter, side-scatter, and fluorescent signal data are used to

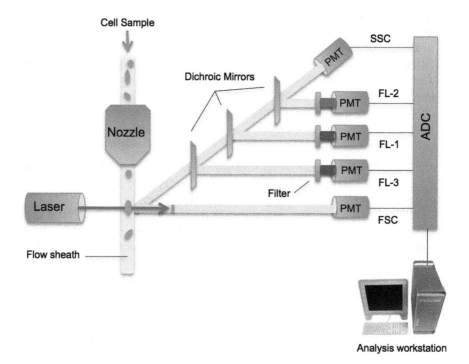

Fig. 2 Principle of flow cytometry

identify a certain cell type, and an electrical charge is applied on each cell so that cells will be sorted by charge (using electromagnets) into separate vessels when exiting the flow chamber.

Today "flow cytometry" and "FACS" are often used interchangeably, FACS being probably the most often used variant of this technique.

This physical sorting of cells out of a heterogeneous mixture is applied for a wide range of scientific fields. This includes basic research but also clinical applications.

Some of the most important applications are:

1.2.1 Clinical Usage

The main use in routine clinical laboratories is for the diagnosis, prognosis, and monitoring of disease. Except when reticulocyte analysis is concerned, all these applications involve the measurement of proteins in or on cells using immunofluorescence:

- *Analysis of leukemias and lymphomas*
  Differentiation of normal cells and their way to leukemic or lymphomic cells is tracked by the detection of specific surface markers (CDs, cluster of differentiation, e.g., CD7, CD15, CD45, etc.) [15–17]
- *Detection of minimal residual disease*
  Minimal residual disease (MRD) is disease beyond the limit of detection using conventional microscopy. Flow cytometric methods can detect far lower levels of disease, which can be important in the clinical management of leukemia [18, 19].
- *Stem cell enumeration for "stem cell rescue"*

  – Via CD34 detection of hemopoietic stem cells, they can be harvested from the peripheral blood as well as the marrow. After that, these cells can be used to repopulate a depleted bone marrow after, for example, high-dose chemotherapy [20].

- *T-cell crossmatch for donor/acceptor match in transplantation*

  – If the organ donor's lymphocytes are incubated with serum from the potential recipient of the graft, a mismatch can be easily identified by detecting bound immunoglobulins with a FITC-conjugated antihuman IgG antibody [21, 22].

- *Postoperative monitoring of organ transplantation*

  – There are a number of cell surface markers for early detection of rejection.

- *Detection of autoantibodies in autoimmune conditions*

  – Antihuman IgG antibodies are used to detect IgG (i.e., the autoantibodies) on the surface of the patient's cells. The method is very similar to those used for a T-cell crossmatch.

To the above list, other clinical applications could be added like detection and status assessment of HIV infection, feto-maternal hemorrhage, immunodeficiency diseases, reticulocyte analysis, platelet counting and function, and many others [23–27], which place flow cytometry as an integral part of modern clinical diagnostics.

1.2.2 Application in Basic Research

The use of flow cytometry in basic research has produced an endless number of papers, since nearly all topics of modern cell biology can be addressed via the expression of surface markers and morphological changes, and this is one of the major applications of FACS.

Starting with the probably most popular research areas like proliferation and apoptosis, flow cytometry was also used to study intracellular pH changes and drug uptake, to analyze and sort chromosomes, for following the binding and endocytosis of ligands, for tracking cells in vivo, and even to set up bead-based immunoassay (see Sect. 2.4).

Describing all that in detail would exceed the scope of this chapter by far, and the reader is asked to turn to the vast number of papers [28–30], dealing with basic research and FACS.

2 Molecular Analytics

2.1 Real-Time PCR

2.1.1 Principle

Although most readers may be familiar with the polymerase chain reaction (PCR), the corner stone of modern molecular biology, a few words to the basic mechanisms may be said here. PCR is a DNA amplification method that actually uses the same mechanism cells are using to replicate their genetic material. Its applications are manifold, starting from the simple enrichment of a rare piece of DNA up to finding traces of DNA on a crime scene.

The principle of PCR lies in the fact that the name-giving enzyme, the polymerase, is able to duplicate a piece of single-stranded DNA (ssDNA) provided that it has a double-stranded starting point (Fig. 3).

The first step in doing a PCR is always separating the strands at elevated temperature ("melting"). Adding small pieces of DNA, called primers, which are complementary to a certain part of the DNA that needs to be replicated, creates the double-stranded starting points for the polymerase ("annealing"). Raising the temperature again activates the polymerase, which completes replication by using single nucleotides also added to the reaction mix ("extension"). The sequence

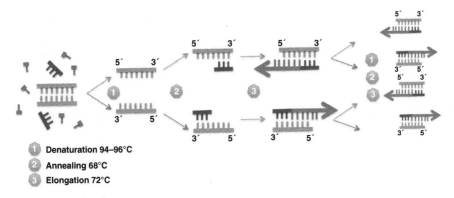

① Denaturation 94–96°C
② Annealing 68°C
③ Elongation 72°C

Fig. 3 Principle of PCR

from melting to annealing and extension is called a PCR cycle and can be repeated several times leading to an exponential increase of the number of double-stranded DNA (dsDNA) copies.

In the early days of PCR, the amount of DNA produced was quantified by tedious and labor-extensive gel electrophoresis.

Nowadays, the use of fluorescence allows direct monitoring of the DNA amplification progress as it happens (real-time PCR). There are many different techniques, but they all link the amplification of DNA to the generation of fluorescence, which can simply be monitored with a detector during each PCR cycle. In this way, as the number of copies increases, so does the fluorescence.

A variety of fluorescence molecules for DNA labeling in RT-PCR exist nowadays, but basically most of them belong either to the group of intercalating (e.g., SYBR Green, BEBO, YO-PRO-1, LC Green, SYTO-9) or hydrolysis probes (e.g., FAM, VIC, JOE, CY5, TaqMan probes).

While the first generate the signal by binding between the strands of a dsDNA, the second one starts to fluoresce when the polymerase hydrolyses the fluorescent molecule ("reporter") at one end of the primer probe, separating it from a suitable quencher molecule at the other end (Fig. 4).

Which of those options is better depends on experimental settings and goals, but both have in common that with increasing numbers of copies the fluorescence of the solution increases (Fig. 5).

2.1.2 Quantitative RT-PCR (qRT-PCR)

In addition to the advantage of online monitoring the course of the DNA amplification, RT-PCR also allows for exact quantitation of nucleic acids (DNA and RNA). In principle there are two types of quantitation strategies, absolute or relative. In absolute quantitation samples of known quantity are serially diluted and then amplified to generate a standard curve. Unknown samples are then quantified by comparison with this curve [31, 32].

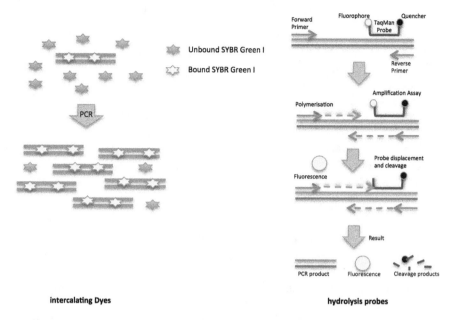

Fig. 4 Fluorescence detection of dsDNA by intercalating dyes and hydrolysis probes (reporter/quencher pairs)

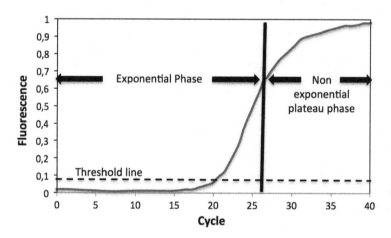

Fig. 5 Typical RT-PCR graph

Relative quantitation describes a real-time PCR experiment in which the DNA content in one sample is compared to the content in another sample. The results are expressed as fold change (increase or decrease).

So far we have only been discussing amplification of dsDNA. However, in basic science and probably generally the most often used application of RT-PCR is detection and quantitation of messenger RNA to compare gene activities under different cellular conditions.

In these cases another enzyme, called reverse transcriptase, is used in a first step to generate a DNA copy (cDNA) of the RNA strand. The second step then is amplification of the cDNA by a RT-PCR step as described above. Very often this method is referred to as "Two Step qRT-PCR."

A variety of parameters influence reproducibility and reliability of qRT-PCR like primer design, quality of enzyme preparations, and purity of the RNA/DNA preparations used, which cannot be discussed in detailed here since we want to focus on aspects of the use of fluorescence in RT-PCR.

In principle for RNA detection, the same fluorescent probes (i.e., intercalating or hydrolyzing) are used as described above. A special case is the multiplexing qRT-PCR [33, 34], where several different DNA sequences are detected simultaneously within the same sample (see also 2.4.3). In such experiments exclusively hydrolyzing probes of the reporter/quencher type must be used to distinguish between the different types of DNA sequences that are amplified. Of course, the reporter fluorophores in a multiplex reaction have to be spectrally distinct from each other so that the signal arising from each can be individually detected in a single channel. Similarly, the quencher molecules must be chosen carefully to avoid quenching at different wavelengths, which can complicate optical filtering. A remedy for this problem can be found in the use of the so-called dark quenchers which release energy in the form of heat rather than fluorescence, and therefore keep the overall background lower.

2.1.3 Application of RT-PCR

RT-PCR is probably the most used technology in basic genetic research and molecular biology. Just to give some examples, it has been used for infectious disease diagnostics [35, 36], genotyping [37–40], or polymorphism studies [41].

Limitations of the method are mainly methodological in nature, e.g., the influence of the cycler used [42] but not in terms of possible applications. Readers will find a vast field of publications covering nearly all thinkable applications.

2.2 Next-Generation Sequencing (NGS)

2.2.1 Introduction to NGS

The term next-generation sequencing (NGS), massively parallel sequencing, or deep sequencing describes a set of methods that, in contrast to the time-consuming traditional Sanger sequencing, allows rapid sequencing of genomes. In the Human Genome Project (HGP), it took over a decade to decipher the whole genetic material of our species. Nowadays, a person's whole genome can be obtained in days at a fraction of the costs associated with the HGP [43]. Hence, it is justified to say that NGS revolutionized genomic research [44].

There are several different NGS technologies using different approaches to sequencing. Some of the more prominent ones are explained in more detail below (Sect. 2.2.2).

What is similar in all of them is that they are capable of sequencing millions of small fragments of DNA in parallel, which is what differentiates them from the sequential and thus more time-consuming Sanger method. By computer-aided mapping of these fragments to the human reference genome, the genetic information is put together like a puzzle, and the individual information is gained. NGS can be used to analyze a whole genome or only parts of it, just depending on the research question under investigation.

From what has been said, it is clear that NGS has an enormous value in all types of genetic research, may it be in microbiology [45] (e.g., finding antibiotic resistance mutations) or in oncology (e.g., identifying oncogenes or patient-specific mutations for personalized treatments) [46, 47]. However, routine use in a clinical setting is still hindered by the high costs per patient ranging from one to several thousand US\$ and by the fact that broadly accepted international standards are still under discussion [48]. Nevertheless, NGS is used for a broad variety of clinical research and diagnostic purposes [49–54].

2.2.2 Widespread Systems in Industry

Illumina Sequencing

In principle, this method is just a variant of PCR using the incorporation of fluorescently labeled deoxyribonucleotide triphosphates (dNTPs) into the DNA template under investigation and taking snapshots at the point of incorporation during each cycle of DNA synthesis. The working steps include:

- *Library Preparation*—The DNA of interest is randomly fragmented and tagged with an adaptor sequence for cluster generation and amplified by PCR.
- *Cluster Generation*—The library is injected into a flow-through system, and the tagged sequences bind to a chip via small DNA sequences complementary to the library adapters. Each fragment is then amplified by solid phase PCR.
- *Sequencing*—All four nucleotides labeled with different fluorescent dyes are added. Because they act as elongation terminators (i.e., the DNA polymerase stops, after one nucleotide), only the next fitting one of the four bases is incorporated. After a washing step, to remove the other three nucleotides, the bound base is identified by its specific dye. Then the dye is cleaved, thus the terminator function removed and the cycle repeated.
- *Data Analysis*—Bioinformatic software is then used to, e.g., identify mutations by comparison with a reference genome sequence (Fig. 6).

As far as the aspect of fluorescence is concerned, the only thing to be considered here is that the spectra of the dyes used are different enough to be detected separately without any crosstalk.

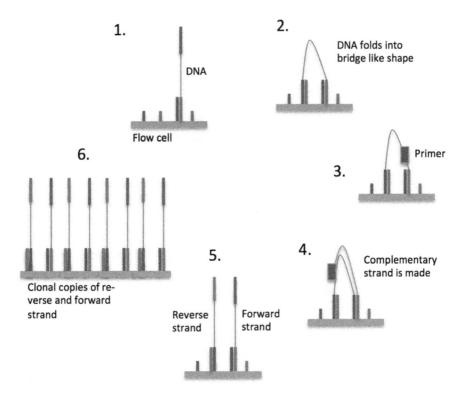

Fig. 6 Illumina next-gen sequencing

Roche 454 Sequencing

This system is mentioned here just for the sake of completeness, since it has luminescence detection but was discontinued by ROCHE in 2013. It is based on pyrosequencing as described by Petterson et al. already in 1993. Very briefly, the method consists of the release of pyrophosphate (PPi) when the right (i.e., complementary) nucleotide is incorporated ("sequencing by synthesis"). The PPi is then converted to ATP and used by the enzyme luciferase to generate a light peak.

Ion Torrent: Proton/PGM Sequencing

This is worth mentioning due to the widespread use of this system from Thermo Fisher Scientific. In principle it is a very sensitive semiconductor-based pH-meter, which detects the release of a hydrogen ion, when the right base is incorporated into a DNA strand. So the sequence information is generated in a digital manner with no light emission involved.

SOLiD Sequencing

It is another NGS method marketed by Thermo Fisher Scientific, which uses ligation of DNA fragments and di-base containing fluorescence-labeled probes (e.g., AT or GC) to obtain the sequence information.

As a first step, a library of DNA fragments is prepared from the sample. Then each fragment is coated to magnetic beads via a special DNA adaptor sequence. That way, sets of beads are created that carry only one DNA fragment. The DNA on the beads is then amplified by PCR and together with the beads transferred to a glass slide. Afterward, primers are added that bind to the adaptor sequence together with four fluorescently labeled di-base probes. From interrogation of two nucleotides at once, a very low degree of error arises compared to single nucleotide probe systems, because the likelihood of mis-hybridization is significantly lower.

2.3   Immunoassays

2.3.1   Principles of Immunoassays

Immunoassays (IAs) use the high specificity of antibodies, generated by animals in response to contact with of a foreign (i.e., not belonging to the specific organism) molecule (antigen). An antibody ideally binds only the antigen it was raised against. This makes them a perfect tool to detect such antigens in complex mixtures like body fluids. Immunoassays can be categorized by the mode of antigen binding (competitive/noncompetitive), the type of detection (direct/indirect), with or without solid support (homogenous/sorbent), or signal generation (radiometric/colorimetric/fluorescent, luminescent).

Since we focus here on the aspects of fluorescence applications (i.e., on a specific signal generation method), we will only discuss the most often used type called a sandwich immunoassay. For other variants of IAs, please turn to the vast amount of literature available [55–59].

A sandwich IA always consists of two antibodies (Fig. 7):

One is called the coating or capture antibody, and the other one is the detection antibody, which is responsible for signal generation. It can be done in homogenous solution or with a solid support, e.g., a microtiter plate.

If in the latter the signal is created with the help of an enzyme, then this resembles the most common form of an IA, the enzyme-linked immunosorbent assay (ELISA). The enzyme is used to generate the signal by converting its substrate into a colorimetric, luminescent, or fluorescent compound.

In the case of fluorescent IAs, the measurement signal can also result from a fluorescent molecule directly attached to the detection antibody, without the need for enzyme.

Specific advantages of fluorescence detection and the different types of fluorescence IAs are discussed below.

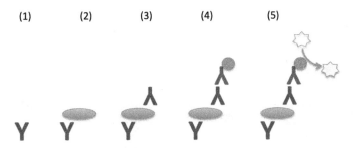

Fig. 7 Schematic principle of a sandwich immunoassay: (1) capture antibody coating, (2) addition of antigen, (3) "sandwich" with detection antibody, (4) reaction with enzyme-labeled secondary antibody, (5) substrate reaction

2.3.2 Advantages of Fluorescence Detection

In general, fluorescence detection on a molecular level is approximately a 1000 times more sensitive than colorimetric detection.

However, one has to distinguish between direct (i.e., detection with a fluorescent dye-labeled antibody) and indirect immunofluorescence (i.e., detection with an enzyme-labeled antibody using a chemifluorescent substrate).

When direct immunofluorescence is applied, the advantages are that there is no need for an additional reagent (i.e., the substrate), and fluorescent labels are more stable (at least when kept in the dark) than enzyme labels. It also allows multiplexing (i.e., the simultaneous detection of several analytes in one sample) by the use of antibodies labeled with dyes of different excitation and emission wavelength (see Sect. 2.4), something that would be impossible if an enzyme label was applied. It is also not possible to "over-incubate" or "overdevelop" such an assay, like it could happen during incubation with the enzyme substrate, leading to a loss of sensitivity because background becomes too high. Once equilibrium is reached, the amount of bound fluorescent-labeled antibody does not change anymore, and the signal stays constant. However, direct immunofluorescence suffers from the drawback that signal/analyte ratio normally is 1 or 2 at best.

Besides direct immunofluorescence also fluorescence resonance energy transfer (FRET) can be used as detection variant in such assays [60–62]. This even allows for homogeneous (i.e., not surface supported) setup of sandwich immunoassay, because the signal is only generated if donor and acceptor are in close proximity (Fig. 8).

Indirect immunofluorescence, as ELISAs generally do, uses the enzymatic reaction to amplify the signal, so that for each bound antibody 100–1000 molecules of substrate are converted, what may result in significantly increased sensitivity. In the end, which type of setup is more advantageous depends on the type of application. Point of care devices, where the addition of liquid substrates is not possible, will usually be realized with direct immunofluorescence (e.g., the

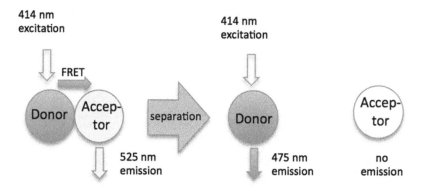

Fig. 8 Principle of fluorescence energy transfer

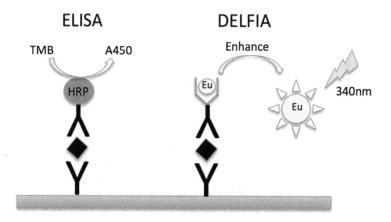

Fig. 9 Principle of DELFIA assays

Alere Triage® BNP Test for congestive heart failure). If the test environment and setup does allow the addition of reagents and performing washing steps and sequential incubation steps, indirect immunofluorescence may be a good choice (e.g., the DELFIA® immunoassays from PerkinElmer, where a time-resolved fluorescence (TRF) substrate is generated upon binding of the antibody to the analyte) (Fig. 9).

Time-resolved fluorescence [63–65] has the advantage that the emitted photons are not measured immediately, but after a certain time period, normally a few microseconds. This eliminates interference of scattered light from the excitation beam and background fluorescence and increases sensitivity. TRF need special "long glowing" fluorophores (e.g., europium chelates) to be performed. Of course, also fluorogenic substrates in combination with enzymes can be used in analogy to the chromogenic ones from classical ELISA to generate a signal.

2.3.3   Recent Advances

While the development of fluorescent dyes has a very long history, their importance has increased due to the recent advancement of new molecules which have been developed along with new biotechnological tools and devices. For example, laurdan, a naphthalene-based amphiphilic fluorescent dye having as characteristics the ability to penetrate membranes and a large Stokes shift, was developed to study membrane fluidity and dynamics, and its usage was made quite effective by the development of two-photon fluorescent microscopy, which enables the detection of signals with less background, less photodamage, and more depth discrimination.

A vast arsenal of different, very bright fluorescent dyes is available nowadays covering nearly all thinkable excitation an emission ranges and sold under various brand names (e.g., the Alexa Fluor from Thermo Fisher or DyLight from Abcam).

An interesting variant of fluorescent dyes are the so-called quantum dots. QDs can be defined as colloidal particles made of semiconductor materials with diameters ranging typically from 2 to 10 nm. The semiconductor particle (named "core") is usually coated by a layer of another semiconductor material (named "shell") which in general has a greater band gap than the core, rendering excellent optical properties. The QDs' core is responsible for the fundamental optical properties (i.e., light absorption and emission), and the shell is used to passivate the surface of the core with the goal to improve its optical properties and reduce chemical attack. After light absorption by the core material, an electron (e-) can be excited from the valence band onto the conduction band, leaving a hole (h+) in the valence band. When the electron returns to the valence band, fluorescence is emitted.

The shell physically separates the optically active core from its surrounding medium. As a consequence, the nanoparticles' optical properties become less sensitive to induced changes, for example, by the presence of oxygen or pH in the local environment. At the same time, the shell reduces the number of surface dangling bonds, which can act as trap states for electrons and minimize the QDs fluorescence efficiency.

Quantum dots are generally more stable and very often brighter than conventional fluorescent dyes.

Aside from the generation of various fluorescent dyes, there has also been some progress in making detection of fluorescence itself more sensitive. Significant improvements have been made either by the introduction of new highly efficient light sources (e.g., laser-LEDs) or detectors (e.g., CCD cameras), and work is still in progress.

Another interesting improvement of fluorescent detection lies in using metal (structure/particle)-enhanced fluorescence techniques or MEF [66–68] to lower detection limits. MEF has come more and more into focus in biomedical applications during the last decade. Briefly described, it is based on the fact that the excitation light interacts with the electrons in nanometer-sized metal particles, which result in very high, local electromagnetic fields (called localized surface plasmons, LSPs)

356   G. Hawa

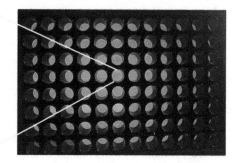

Fig. 10 Metal-enhanced fluorescence microtiter plate (MEF-MTP, to the right). The pop out shows the silver-coated nanostructure (to the left) at the bottom of the well that generates the MEF-effect (copyright FIANOSTICS GmbH 2017)

that dramatically increase the quantum yield of fluorescent dyes. Depending on the size, shape, and distance of the metal nanostructures, up to several hundredfold enhancements can be found. So far this technology mainly found academic use, because the required high reproducibility and scalability of metal nanostructure fabrication necessary for industrial application was not achieved. However, the first commercially available applications overcoming these problems have been recently developed [69].

By application of technologies originally developed for manufacturing of Blu-rays and DVDs (which are essentially nanostructured and metal coated plastics surfaces), MEF substrates with extremely high fluorescent enhancement and excellent reproducibility could be generated, which are easily scalable, i.e., producible in high numbers. This technology can be applied for any assay format (microplate, microarray, point of care, etc.), and readout can be performed with standard laboratory equipment, e.g., with a state-of-the-art fluorescent reader if a microplate format is used (Fig. 10).

Development of assays for this new technology platform is currently ongoing.

2.4 Multiplexing Assays

Fluorescence is an ideal detection method for second-generation multiplex assays, i.e., detection of several analytes in a single assay run within one sample.

The fact that fluorescent dyes of various Ex/Em wavelengths are available makes it possible to identify a specific molecule in a sample by identifying the fluorescent dye used for the detection of that molecule. In other words pairs of analyte/fluorophore are created, which can be individually detected, only limited by the number of different fluorescent dyes available or the resolution of the detector used.

2.4.1 Bead Based

The usage of beads has a long history in bioassay development. Agglutination tests or lateral flow devices have been based on polystyrol beads for decades now. Instead of a solid surface, e.g., a microplate well, the beads are used for binding the capture antibody (Fig. 11). Despite that, it is just a normal immunoassay as described above.

The main advantage is that beads have a much larger surface area than a solid surface, resulting in a very high antibody-binding capacity, which is beneficial in terms of the achievable detection limit and reaction kinetics.

Enzymes may be used as reporter molecules, but far more often, fluorescent dyes are applied. There are several commercial approaches that use beads for FIAs [70]. The probably best known is multiplexing with a combination of fluorescent/magnetic beads and fluorescent dyes developed by Luminex Corp., sold under the trade mark xMAP® [71].

xMAP technology applies two bead types—magnetic and polystyrene. The magnetic Magplex bead is a superparamagnetic 6.5 μm microsphere with a magnetic core and polystyrene surface. The polystyrene Microplex bead is a 5.6 μm polystyrene microsphere. Both bead types are internally dyed with precise proportions of red and infrared fluorophores. The differing proportions of the red and infrared fluorophores result in 100 unique spectral signature microspheres which are identified by the Luminex xMAP detection systems. The conjugation of a distinct monoclonal antibody to a distinct bead allows for analysis of multiple analytes in a single well.

The magnetic properties of the beads are not used for detection but simply to separate beads from excess reagents and sample during washing prior to readout, which is normally done on filter plates under low pressure conditions. For the assay readouts themselves, specially designed devices are needed that act like flow cytometers, in which the particles of the individual assays are transported to the readout within a flowing stream of assay buffer (Fig. 12).

The fluidics system of the reader aligns the beads into single file as they enter a stream of sheath fluid and then enter a flow cell. Once the beads are in single file within the flow cell, each bead is individually interrogated for bead color (analyte) and assay signal strength (PE fluorescence intensity).

Fig. 11 Solid surface (left) versus bead-based (right) IA

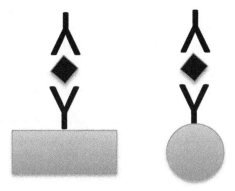

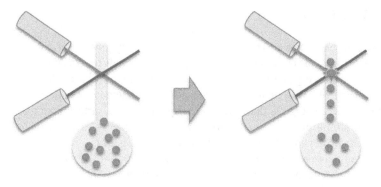

Fig. 12 Detection method of Luminex multiplexing assays

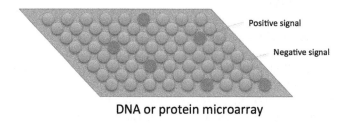

Positive signal

Negative signal

DNA or protein microarray

Fig. 13 Typical appearance of a microarray

A variety of assays have been developed for this system.

Other bead-based multiplexing assays are the FirePlex ones from Abcam or the BD™ cytometric bead assays (CBAs), both running on benchtop flow cytometers.

All of them have in common that a property of the bead itself is used to identify the type of assay and the fluorescent dye is used for quantification. Application of these platforms ranges over a broad variety of diagnostic applications [72–74].

2.4.2 Microarrays

A microarray is a sort of grid (e.g., on the bottom of a microplate or on a specially designed chip), where on each spot of the grid a different binding partner for various analytes of the sample is coated (Fig. 13).

By incubation of the microarray with a sample, the reaction with several molecules can be studied simultaneously. Since this is a method where localization of the signal identifies the analyte, direct fluorescence is the method of choice for detection.

A great variety of experimental setups and commercial applications exists nowadays. However it seems justifiable to separate microarray technology into two big groups, one that deals with detection of nucleic acids ("RNA/DNA chips") [75] and the other focused on measuring presence and concentration of proteins ("protein chips") [76]. Applications of microarray technology are manifold and include areas like endocrinology, microbiology, oncology, toxicogenomics, gerontology, neurology, and many other clinical areas [77–81].

Some important companies involved in providing microarray technology are Affymetrix, Illumina, Applied Microarrays, Clontech, R&D Systems, Sigma Life Sciences, and many more. A detailed description of this vast field would be out of the scope of this chapter.

2.4.3 Multiplex RT-PCR

Multiplex PCR is a widespread molecular biology technique for amplification of multiple targets in a single PCR experiment. In a multiplexing assay, more than one target sequence can be amplified by using multiple primer pairs in a reaction mixture. As an extension to the practical use of PCR (see Sect. 2.1), this technique has the potential to produce considerable savings in time and effort within the laboratory without compromising on the utility of the experiment [82–84].

Design of specific primer sets is essential for a successful multiplex reaction.

Multiplex RT-PCR is used for pathogen identification, high-throughput SNP genotyping, mutation analysis, gene deletion analysis, template quantitation linkage analysis, RNA detection, and forensic studies.

In terms of fluorescence use for detection, it is not different from singlex RT-PCR, so it is mentioned here only for the sake of completeness.

3 Summary

Technologies based on fluorescence detection have become an essential part of modern life sciences, and the field has seen a tremendous growth of new applications during the last two decades. A lot of those are pretty mature and widespread nowadays, and there may be only incremental improvements on procedures and experimental setup. However, as seen with metal-enhanced fluorescence (Sect.2.3.3), there is still room for new innovations that will complement our arsenal of fluorescence applications.

References

1. Shaner NC et al (2005) A guide to choosing fluorescent proteins. Nat Methods 2:905–909
2. Eisenstein M (2006) Helping cells to tell a colorful tale. Nat Methods 3:647–655
3. Coling D, Kachar B (1997) Theory and application of fluorescence microscopy. Curr Protoc Neurosci; Chapter 2:1. Unit 2
4. Suzuki T et al (2007) Recent advances in fluorescent labeling techniques for fluorescence microscopy. Acta Histochem Cytochem 40:131–137

5. Diaspro A et al (2006) Multi-photon excitation microscopy. Biomed Eng Online 5:36
6. Hibbs AR (2004) Confocal microscopy for biologists. Springer, New York
7. Pawley JB (ed) (2006) Handbook of biological confocal microscopy. Springer, New York
8. Fish KN (2009) Total Internal Reflection Fluorescence (TIRF) Microscopy. Current protocols in cytometry/editorial board;0 12:Unit12.18
9. Kudalkar EM, Davis TN, Asbury CL (2016) Single-molecule total internal reflection fluorescence microscopy. Cold Spring Harb Protoc 2016(5)
10. Svoboda K, Yasuda R (2006) Principles of two-photon excitation microscopy and its applications to neuroscience. Neuron 50:823–839
11. Wallace W et al (2001) A workingperson's guide to deconvolution in light microscopy. BioTechniques 31:1076–1080
12. Willig KI et al (2006) STED microscopy reveals that synaptotagmin remains clustered after synaptic vesicle exocytosis. Nature 440:935–939
13. Willig KI et al (2007) STED microscopy with continuous wave beams. Nat Methods 4:915–918
14. Alonso C (2013) An overview of stimulated emission depletion (STED) microscopy and applications. J Biomol Tech 24(Suppl):S4
15. Craig FE, Foon KA (2008) Flow cytometric immunophenotyping for hematologic neoplasms. Blood 111:3941–3967
16. Qadir M, Barcos M, Stewart CC, Sait SN, Ford LA, Baer MR (2006) Routine immunophenotyping in acute leukemia: role in lineage assignment and reassignment. Cytometry 70:329–334
17. Wood BL, Arroz M, Barnett D, DiGiuseppe J, Greig B, Kussick SJ, Oldaker T, Shenkin M, Stone E, Wallace P (2007) 2006 Bethesda international consensus recommendations on the immunophenotypic analysis of hematolymphoid neoplasia by flow cytometry: optimal reagents and reporting for the flow cytometric diagnosis of hematopoietic neoplasia. Cytometry B Clin Cytom 72(Suppl 1):S14–S22
18. Campana D, Coustan-Smith E (2004) Minimal residual disease studies by flow cytometry in acute leukemia. Acta Haematol 112:8–15
19. Kern W, Haferlach C, Haferlach T, Schnittger S (2008) Monitoring of minimal residual disease in acute myeloid leukemia. Cancer 112:4–16
20. Keeney M, Chin-Yee I, Weir K, Popma J, Nayar R, Sutherland DR (1998) Single platform flow cytometric absolute CD34+ cell counts based on the ISHAGE guidelines. Cytometry 34:61–70
21. Bell A, Shenton B, Garner G (1998) The flow cytometric crossmatch in solid organ transplantation. Proc RMS 33:219–220
22. Shanahan T (1997) Application of flow cytometry in transplantation medicine. Immunol Investig 26:91–101
23. Mandy F, Nicholson J, Autran B, Janossy G (2002) T-cell subset counting and the fight against AIDS: reflections over a 20-year struggle. Cytometry 50:39–45
24. Mandy F, Janossy G, Bergeron M, Pilon R, Faucher S (2008) Affordable CD4 T-cell enumeration for resource-limited regions: a status report for 2008. Cytometry 74(Suppl 1):S27–S39
25. Nance SJ, Nelson JM, Arndt PA, Lam HC, Garratty G (1989) Quantitation of fetal-maternal hemorrhage by flow cytometry. A simple and accurate method. Am J Clin Pathol 91:288–292
26. Davis BH, Bigelow NC (1994) Reticulocyte analysis and reticulocyte maturity index. In: Darzynkiewicz Z, Crissman HA (eds) Flow cytometry. Methods in cell biology, vol 42. Academic, San Diego, pp 263–274
27. Harrison P, Segal H, Briggs C, Murphy M, Machin S (2005) Impact of immunological platelet counting (by the platelet/RBC ratio) on haematological practice. Cytometry 67:1–5
28. Schröter C, Beck J, Krah S, Zielonka S, Doerner A, Rhiel L, Günther R, Toleikis L, Kolmar H, Hock B, Becker S (2018) Selection of antibodies with tailored properties by application of high-throughput multiparameter fluorescence-activated cell sorting of yeast-displayed immune libraries. Mol Biotechnol 60(10):727–735
29. Miura I, Takahashi N, Kobayashi Y, Saito K, Miura AB (2000 Oct) Molecular cytogenetics of stem cells: target cells of chromosome aberrations as revealed by the application of fluorescence in situ hybridization to fluorescence-activated cell sorting. Int J Hematol 72(3):310–317

30. Miura I, Kobayashi Y, Takahashi N, Saitoh K, Miura AB (2000) Involvement of natural killer cells in patients with myelodysplastic syndrome carrying monosomy 7 revealed by the application of fluorescence in situ hybridization to cells collected by means of fluorescence-activated cell sorting. Br J Haematol 110(4):876–879
31. Ramakers C, Ruijter JM, Deprez RH, Moorman AF (2003) Assumption-free analysis of quantitative real-time PCR data. Neurosci Lett 339:62–66
32. Gentle A, Anastaopoulos F, McBrien NA (2001) High-resolution semi-quantitative real-time PCR without the use of a standard curve. BioTechniques 31:502–508
33. Kim J, Jung S, Byoun MS, Yoo C, Sim SJ, Lim CS, Kim SW, Kim SK (2018) Multiplex real-time PCR using temperature sensitive primer-supplying hydrogel particles and its application for malaria species identification. PLoS One 13(1):e0190451
34. Li B, Liu H, Wang W (2017) Multiplex real-time PCR assay for detection of *Escherichia coli* O157:H7 and screening for non-O157 Shiga toxin-producing *E. coli*. BMC Microbiol 17:215
35. Rashed-Ul Islam SM, Jahan M, Tabassum S (2015) Evaluation of a rapid one-step real-time PCR method as a high-throughput screening for quantification of hepatitis B virus DNA in a resource-limited setting. Euroasian J Hepatogastroenterol 5(1):11–15
36. Mackay IM (2004) Real-time PCR in the microbiology laboratory. Clin Microbiol Infect 10:190–212
37. Mitchell LA, Phillips NA, Lafont A, Martin JA, Cutting R, Boeke JD (2015) pCRTag Analysis - a high throughput, real time PCR assay for Sc2.0 genotyping. J Vis Exp 99:52941
38. Wittwer CT, Reed GH, Gundry CN, Vandersteen JG, Pryor RJ (2003) High-resolution genotyping by amplicon melting analysis using LCGreen. Clin Chem 49:853–860
39. Guarnaccia M, Iemmolo R, Petralia S, Conoci S, Cavallaro S (2017) Miniaturized real-time PCR on a Q3 system for rapid KRAS genotyping. Sensors (Basel) 17(4):831
40. Nieto-Aponte L, Quer J, Ruiz-Ripa A, Tabernero D, Gonzalez C, Gregori J, Vila M, Asensio M, Garcia-Cehic D, Ruiz G, Chen Q, Ordeig L, Llorens M, Saez M, Esteban JI, Esteban R, Buti M, Pumarola T, Rodriguez-Frias F (2017) Assessment of a novel automatic real-time PCR assay on the Cobas 4800 analyzer as a screening platform for hepatitis C virus genotyping in clinical practice: comparison with massive sequencing. J Clin Microbiol 55(2):504–509
41. Deligezer U, Akisik E, Dalay N (2003) Genotyping of the MTHFR gene polymorphism, C677T in patients with leukemia by melting curve analysis. Mol Diagn 7:181–185
42. Zuna J, Muzikova K, Madzo J, Krejci O, Trka J (2002) Temperature nonhomogeneity in rapid airflow-based cycler significantly affects real-time PCR. BioTechniques 33:508–512
43. Bennett ST, Barnes C, Cox A et al (2005) Toward the 1000 dollars human genome. Pharmacogenomics 6:373
44. Goldfeder RL, Wall DP, Khoury MJ et al (2017) Human genome sequencing at the population scale: a primer on high-throughput DNA sequencing and analysis. Am J Epidemiol 186:1000
45. Deurenberg HR et al (2017) Application of next generation sequencing in clinical microbiology and infection prevention. J Biotechnol 243:16–24
46. Easton DF, Pharoah PD, Antoniou AC et al (2015) Gene-panel sequencing and the prediction of breast-cancer risk. N Engl J Med 372:2243
47. Rehm HL, Bale SJ, Bayrak-Toydemir P et al (2013) ACMG clinical laboratory standards for next-generation sequencing. Genet Med 15:733
48. Zehir A, Benayed R, Shah RH et al (2017) Mutational landscape of metastatic cancer revealed from prospective clinical sequencing of 10,000 patients. Nat Med 23:703
49. Rizzo JM, Buck MJ (2012) Key principles and clinical applications of "next-generation" DNA sequencing. Cancer Prev Res (Phila) 5:887
50. Taylor JC, Martin HC, Lise S et al (2015) Factors influencing success of clinical genome sequencing across a broad spectrum of disorders. Nat Genet 47:717
51. Biesecker LG, Green RC (2014) Diagnostic clinical genome and exome sequencing. N Engl J Med 370:2418
52. Lazaridis KN, Schahl KA, Cousin MA et al (2016) Outcome of whole exome sequencing for diagnostic odyssey cases of an individualized medicine clinic. The Mayo Clinic experience. Mayo Clin Proc 91:297

53. Vassy JL, Christensen KD, Schonman EF et al (2017) The impact of whole-genome sequencing on the primary care and outcomes of healthy adult patients: a pilot randomized trial. Ann Intern Med 167(3):159–169
54. Tarailo-Graovac M, Shyr C, Ross CJ et al (2016) Exome sequencing and the management of neurometabolic disorders. N Engl J Med 374:2246
55. Wild D (2005) The immunoassay handbook, 4th edn. Elsevier, London. ISBN: 9780080970370
56. Crowther JR (2009) Methods in molecular biology, the ELISA guidebook, 2nd edn. Humana Press, New York
57. Butler JE (1992) The behavior of antigens and antibodies immobilized on a solid phase. In: Van Regenmortel MHV (ed) Structure of antigens, vol 1. CRC Press, Boca Raton, pp 209–259
58. Lequin RM (2005) Enzyme immunoassay (EIA)/enzyme-linked immunosorbent assay (ELISA). Clin Chem 51(12):2415–2418
59. Engvall E, Perlmann P (1971) Enzyme-linked immunosorbent assay (ELISA) quantitative assay of immunoglobulin G. Immunochemistry 8(9):871–874
60. Kalarestaghi A, Bayat M, Hashemi SJ, Razavilar V (2015) Highly sensitive FRET-based fluorescence immunoassay for detecting of aflatoxin B1 using magnetic/silica core-shell as a signal intensifier. Iran J Biotechnol 13(3):25–31
61. Han J-H, Sudheendra L, Kennedy IM (2015) FRET-based homogeneous immunoassay on nanoparticle-based photonic crystal. Anal Bioanal Chem 407(18):5243–5247
62. Kattke MD, Gao EJ, Spsford KE, Stephenson LD, Kumar A (2011) FRET-based quantum dot immunoassay for rapid and sensitive detection of Aspergillus amstelodami. Sensors (Basel) 11(6):6396–6410
63. Farino ZJ, Morgenstern TJ, Vallaghe J et al (2016) Development of a rapid insulin assay by homogenous time-resolved fluorescence. PLoS One 11(2):e0148684
64. De Silva CR, Vagner J, Lynch R, Gillies RJ, Hruby VJ (2010) Optimization of time-resolved fluorescence assay for detection of Eu-DOTA-labeled ligand-receptor interactions. Anal Biochem 398(1):15–23
65. Lopez-Crapez E, Bazin H, Andre E, Noletti J, Grenier J, Mathis G (2001) A homogeneous europium cryptate-based assay for the diagnosis of mutations by time-resolved fluorescence resonance energy transfer. Nucleic Acids Res 29(14):e70
66. Gryczynski Z, Malicka J, Gryczynski I et al (2004) Metal-enhanced fluorescence: a novel approach to ultra-sensitive fluorescence sensing assay platforms. Proc SPIE Int Soc Opt Eng 5321(275):275–282
67. Aslan K, Geddes CD (2006) Microwave-accelerated and metal-enhanced fluorescence myoglobin detection on silvered surfaces: potential application to myocardial infarction diagnosis. Plasmonics 1(1):53–59
68. Aslan K (2010) Rapid whole blood bioassays using microwave-accelerated metal-enhanced fluorescence. Nano Biomed Eng 2(1):1–9
69. Hawa G et al (2018) Single step, direct fluorescence immunoassays based on metal enhanced fluorescence (MEF-FIA) applicable as micro plate-, array-, multiplexing- or point of care-format. Anal Biochem 549:39–44
70. Lim CT, Zhang Y (2007) Bead-based microfluidic immunoassays: the next generation. Biosens Bioelectron 22(7):1197–1204
71. Houser B (2012) Bio-Rad's Bio-Plex® suspension array system, xMAP technology overview. Arch Physiol Biochem 118(4):192–196
72. Khan IH, Mendoza S, Yee J et al (2006) Simultaneous detection of antibodies to six nonhuman-primate viruses by multiplex microbead immunoassay. Clin Vaccine Immunol 13(1):45–52
73. Rashtak S, Ettore MW, Homburger HA, Murray JA (2008) Combination testing for antibodies in the diagnosis of coeliac disease: comparison of multiplex immunoassay and ELISA methods. Aliment Pharmacol Ther 28(6):805–813
74. Lim K-H, Langley E, Gao F et al (2017) A clinically feasible multiplex proteomic immunoassay as a novel functional diagnostic for pancreatic ductal adenocarcinoma. Oncotarget 8(15):24250–24261

75. Rudi K, Rud I, Holck A (2003) A novel multiplex quantitative DNA array based PCR (MQDA-PCR) for quantification of transgenic maize in food and feed. Nucleic Acids Res 31(11):e62

76. Gannot G, Tangrea MA, Erickson HS et al (2007) Layered peptide array for multiplex immunohistochemistry. J Mol Diagn 9(3):297–304

77. Light M, Minor KH, DeWitt P, Jasper KH, Davies SJ (2012) Multiplex array proteomics detects increased MMP-8 in CSF after spinal cord injury. J Neuroinflammation 9:122

78. Bünger S, Haug U, Kelly M et al (2012) A novel multiplex-protein array for serum diagnostics of colon cancer: a case–control study. BMC Cancer 12:393

79. Welton JL, Brennan P, Gurney M et al (2016) Proteomics analysis of vesicles isolated from plasma and urine of prostate cancer patients using a multiplex, aptamer-based protein array. J Extracell Vesicles 5:10.3402

80. Purohit S, Li T, Guan W, Song X, Song J, Tian Y, Li L, Sharma A, Dun B, Mysona D, Ghamande S, Rungruang B, Cummings RD, Wang PG, She J-X (2018) Multiplex glycan bead array for high throughput and high content analyses of glycan binding proteins. Nat Commun 9:258

81. Shen H, Zhu B, Wang S et al (2015) Association of targeted multiplex PCR with resequencing microarray for the detection of multiple respiratory pathogens. Front Microbiol 6:532

82. Datukishvili N, Kutateladze T, Gabriadze I, Bitskinashvili K, Vishnepolsky B (2015) New multiplex PCR methods for rapid screening of genetically modified organisms in foods. Front Microbiol 6:757. https://doi.org/10.3389/fmicb.2015.00757

83. Olwagen CP, Adrian PV, Nunes MC et al (2017) Use of multiplex quantitative PCR to evaluate the impact of pneumococcal conjugate vaccine on nasopharyngeal pneumococcal colonization in African children. mSphere 2(6):e00404–e00417

84. Li B, Liu H, Wang W (2017) Multiplex real-time PCR assay for detection of *Escherichia coli* O157:H7 and screening for nony-O157 Shiga toxin-producing *E. coli*. BMC Microbiol 17:215

Photoluminescent Glasses and Their Applications

César A. T. Laia and Andreia Ruivo

Contents

Abstract Glass materials are very attractive for the development of eco-friendly, engineer safe, and fully recyclable smart materials. Photoluminescent glass applies these unique properties to photonics, lighting, and photovoltaics by applying light down-conversion from UV to visible or near-infrared light, suitable for devices, smart windows, and LEDs, among many other applications. Furthermore, enhanced optical properties can be achieved with enamel coatings or by deposition of phosphors, increasing the range of light harvesting of glass materials. This book chapter discusses current methods to synthesize photoluminescent glass and phosphors with a strong focus on the use of alternative raw materials and how they are introduced in such applications to achieve high photoluminescence performances (such as quantum efficiency, Stokes shifts, and brightness). Novel approaches such as quantum dots or photoluminescent zeolites promise new ways to develop luminescence,

C. A. T. Laia (✉)
LAQV@REQUIMTE, Chemistry Department, Faculty of Science and Technology,
Universidade NOVA de Lisboa, Caparica, Portugal

Research Unit VICARTE, Glass and Ceramics for the Arts, Faculty of Science and Technology,
Universidade NOVA de Lisboa, Caparica, Portugal
e-mail: catl@fct.unl.pt

A. Ruivo (✉)
Research Unit VICARTE, Glass and Ceramics for the Arts, Faculty of Science and Technology,
Universidade NOVA de Lisboa, Caparica, Portugal
e-mail: a.ruivo@campus.fct.unl.pt

© Springer Nature Switzerland AG 2019
B. Pedras (ed.), *Fluorescence in Industry*, Springer Ser Fluoresc (2019) 18: 365–388,
https://doi.org/10.1007/4243_2019_12, Published online: 12 March 2019

avoiding the use of critical raw materials such as lanthanides. Examples such as LEDs, light solar concentrators for photovoltaics, and art or design are given, showing the wide range of applications of optical smart glass.

Keywords Art · Light management applications · Luminescent glass · Phosphors · Photoluminescence

1   Luminescent Glass

1.1   Glass Structure

Glass is an attractive material, and innumerous glass pieces, from distinct historical periods, can be found all over the world. Besides glass unique characteristics such as transparency, translucency, brightness, and versatility, it is also a sustainable material that can be easily recycled.

Glass definition is a topic that has been discussed over time. The American Society for Testing and Materials (ASTM) defined glass as "an inorganic product of fusion which has cooled to a rigid condition without crystallizing" [1, 2]. However this definition was contested since many organic glasses are known, and glasses can also be prepared without being cooled from the liquid state. Another definition was stated by A. Paul, in 1990, according to which a glass is a state of matter that maintains the energy, volume, and atomic arrangement of a liquid but whose changes in energy and volume with temperature and pressure are similar in magnitude to those of a crystalline solid [3]. Further, J. E. Shelby defined glass as "an amorphous solid completely lacking in long range, periodic atomic structure, and exhibiting a region of glass transformation behavior." Any amorphous material (inorganic, organic, or metallic, formed by any technique) which exhibits glass transformation behavior is a glass [4]. More recently, A. K. Varshneya stated that the term noncrystalline solid includes two subclasses, *glass* and *amorphous solid*, and attributed the word *glass* to a solid having a liquid-like structure which continuously converts to a liquid upon heating [5].

The glass transformation range, or glass transition range, describes a thermodynamic region of the material where it occurs the phase transformation from a liquid to a glass structure and vice versa, presenting a strong volume change as it can be seen in the glass-forming volume-temperature diagram (Fig. 1). From this temperature range, it can be indicated the glass transition temperature (Tg) by intersecting the extrapolated liquid line with the glass state line.

Glass can be produced using diverse methods such as sol-gel process, chemical vapor deposition, and melt-quenching, the latter being the most common in industry. This method consists in fusion followed by cooling of inorganic raw materials at high temperatures. The raw materials are sources of several oxides, and these oxides can be divided in different groups [4]:

Fig. 1 Volume-temperature diagram for the formation of a crystal and a glass structure indicating the glass transition range, the glass transition temperature (Tg), and the crystal melting temperature (Tm). Adapted from [5]

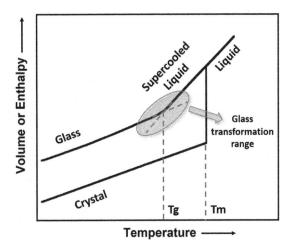

- *Glass former or network former*: The primary glass formers in commercial oxide glasses are SiO_2, B_2O_3, PbO, and P_2O_5, which by themselves can form single component glasses. The principal glass former component is silica, SiO_2. In silica glass silicon tends to form covalent bonds with four oxygens, forming a tetrahedron but without the long range periodic and symmetric organization observed in crystalline structures. Each oxygen placed in the corner of the tetrahedron is bonded to two silicon atoms.
- *Flux or network modifier*: Silica glass, for example, can only be synthesized at very high temperatures ($\approx 1700°C$), making it a very expensive process. To decrease the fusion temperature, network modifiers are added to the glass composition, e.g., Na_2O, K_2O, Li_2O, and PbO, which can act as both glass former and glass modifier. These oxides are responsible for the formation of non-bridging oxygens, which will interact through ionic bonds with the network modifier cations (1). These partial net ruptures which are necessary to lower the glass melting temperature can also decrease the glass stability, making the matrix more vulnerable to corrosion processes.

$$\equiv Si - O - Si \equiv +Na_2O \rightarrow \equiv Si - O^- Na^+ + Na^{+-}O - Si \equiv \qquad (1)$$

- *Stabilizer or intermediates*: They act with an intermediate character between the network former and the network modifier, e.g., CaO, Al_2O_3, and ZnO. These oxides cannot form alone a glass structure, but they can substitute partially the network formers.
- *Secondary components*: Have very specific roles, such as colorants (Fe, Co, Cu, Au), decolorants (CeO_2), opacifiers (F^-), and fining agents that promote bubble removal (Sb_2O_3, $NaNO_3$, NaCl, several fluorides, and sulfates).

Industrially the most common glass composition is the soda-lime silicate (a general structure can be found in Fig. 2) which is used in packaging, containers,

Fig. 2 Glass structure of a
soda-lime silicate glass

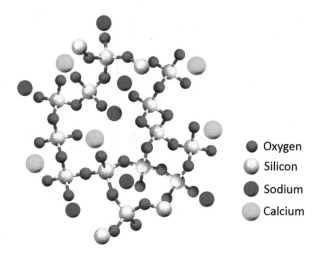

- Oxygen
- Silicon
- Sodium
- Calcium

and windows float glass. Nevertheless, borosilicate glasses are frequently used in Pyrex kitchen utensils, laboratory glassware, and optical applications, since they have better thermal shock properties and can withstand higher temperatures without deforming. Aluminosilicates are not so typically used, still due to their very high transformation temperatures, and softening points are useful in many applications such as halogen lamps, display glasses, and high-temperature thermometers.

1.2 Luminescence in Glass

The excitation energy obtained by a molecule when a photon is absorbed may be dissipated by different processes, such as fluorescence and phosphorescence [6, 7]. After the photon absorption, non-radiative transitions can occur. The non-radiative transition between two excited states of the same spin multiplicity is called internal conversion and of different spin multiplicity is intersystem crossing. Fluorescence is the emitted radiation originated from an excited state that has the same spin multiplicity, normally from S_1, to the ground state ($S_1 \rightarrow S_0$ relaxation). On the other hand, phosphorescence is originated from a de-excitation of an excited state with a different spin multiplicity from the ground state ($T_1 \rightarrow S_0$) [7]. A major difference between these two luminescence types is their characteristic decay times, since in fluorescence if the source of excitation is turned off, the emission decays very fast (10^{-9}–10^{-6} s), which is immediate for the human eye. In phosphorescent materials the emission decays much more slowly, existing numerous examples in which the emission can be observed by the naked eyes during several hours (many times called *afterglow*) [8, 9].

These photoluminescence processes can also occur in a glass matrix. Considering that glass manufacture is a very old practice, since, to the best of our knowledge, the most antique glass is dated from 7000 B.C. in Asia Minor [10]; luminescent glasses may be considered a very recent achievement. In history uranium-doped glasses are significantly known, due to their strong luminescence. One of the earliest reference concerning uranium glasses dates from 1817 where it is described that uranium oxide is responsible for a bright color in glass [11, 12]. Subsequently it continued to be used in a great variety of objects, such as lamps, jewels, plates, and beakers, which under UV light present a green luminescence. The intense green luminescence observed in these glasses is due to U^{6+} ions in the uranyl form $(UO_2)^{2+}$ [13].

Luminescent glasses are also commonly synthesized using lanthanide oxides. One of the first applications of rare earths in glasses is dated from 1880, when Otto Schott produced cerium oxide-doped glasses in order to study their effect in the glass mechanical and optical properties [14]. Nowadays several emissive centers can give rise to luminescent glasses; few examples are described afterward.

1.2.1 Luminescence Centers of Glass Materials

To achieve photoluminescent glass materials, the major limitation is the synthetic methods to produce glass. In industry glass is produced at temperatures above 1500°C, using an atmosphere that results from the combustion of fossil fuels which may induce either oxidative (i.e., with an excess of molecular oxygen) or reducing (i.e., with an excess of reducing substances such as carbon monoxide) atmosphere. At such drastic conditions, thermal and chemical degradation of organic compounds completely hampers their use as luminescent centers. Therefore only inorganic materials have the potential to dope glass to confer them photoluminescent properties.

We here now describe the solutions that may be used, including their limitations and advantages.

Lanthanides and Actinides (f–f Transitions)

Lanthanides and actinides are mostly elements of the f block of the periodic table. These elements are among those that were first described for the synthesis of luminescent glass. In both cases the electronic configuration is characterized by incomplete 4f (lanthanides) or 5f (actinides) electronic shells that lie deep in the atom structure. For that reason, the influence of the external environment is small, and forbidden f–f electronic transitions give rise to their photoluminescence [15, 16]. They are relatively stable from the electrochemical point of view, commonly in the +3 oxidation states with some notable exceptions. For the synthesis this is a very important advantage, because regardless of the environment, they end up in the same oxidation states with some exceptions.

Actinides may be embedded in the glass matrix, and there are indeed numerous descriptions of glass doped with uranium especially in historical glass crafts (see Fig. 3) [12]. In technology the application of glass science in the treatment and storage of nuclear waste materials is also an important topic [17]. However their radioactive properties are not suitable nowadays for most applications, and we will

Fig. 3 Uranium-doped glasses from a private collection under a UV light (ca. 370 nm). Photo taken by Filipa Lopes, 2007

not go deeply into their properties. Lanthanides do not have those limitations, and elements such as europium, terbium, samarium, dysprosium, or erbium are very popular in the industry [18]. They may be dispersed in the glass matrix, where *f–f* transitions take place, or as nanocrystals where charge-transfer processes are dominant. Using elements such as Er(III), near-infrared transitions are also observed [19], which leads to important light up-conversion applications; however, they are out of the scope of this chapter and are covered in other very recent reviews and book chapters [20–22].

The *f–f* electronic transitions are forbidden according to Laporte rule, since parity does not change between ground and excited states. For a given ion such as Eu(III), the $4f$ shell is not completely filled, and therefore the spin magnetic moment is not zero. The ground state of Eu(III) has a configuration $[Xe]4f^6$, with a $S = 3$ giving the electronic state $^7F_0$. Several *f–f* transitions can then take place in the UV-visible region of the light spectrum, but a change of the spin moment is unavoidable. For Eu(III), the transitions with higher oscillator strength are $^7F_0 \rightarrow {}^5L_6$ (circa 393 nm, 3.16 eV) and $^7F_0 \rightarrow {}^5D_2$ (circa 465 nm, 2.67 eV), and the excited state with lowest energy is $^5D_0$ from where photoluminescence occurs to "ground" electronic states ^7F_J ($J = 0, \ldots, 6$) [15]. At most the oscillator strength is about 1×10^{-6}, giving rise to very low extinction coefficients and long phosphorescence decays in the millisecond timescale. Since these are *f–f* transitions where ligands play a minor role ($4f$ orbitals are shielded by filled $5s$ and $5p$ electronic shells), they are also very narrow with full width at half maximum (FWHM) of a few nanometers, and extremely small Stokes shifts are observed (i.e., absorption and emission spectra of a given electronic transition such as $^7F_0 \rightarrow {}^5D_0$ and $^5D_0 \rightarrow {}^7F_0$ in the case of Eu(III) are completely overlapped) [15, 23].

These features implicate that very high concentrations should be used in order to ensure a meaningful brightness, up to 7% weight of, e.g., Eu_2O_3, and the same is true for other lanthanides [23]. Taking into account that lanthanides are critical elements where their availability is a concern and that at such high concentrations the impact on the glass structure may be undesirably high, this is a major drawback. To overcome this part, energy-transfer processes can increase greatly the lanthanides luminescence brightness, by using organic compounds such as β-diketonates or semiconductor metal oxides with large bandgap (above 3 eV). The use of organic compounds is, however, not achievable to confer luminescence to glass for the reasons stated above. With metal oxides, it is possible to achieve energy transfer to lanthanides which may be deposited by using coating technologies, but in such cases, charge-transfer processes are more suitable to achieve bright photo-luminescent materials.

Charge-transfer interactions—Oscillator strengths may be increased by more than 2 orders of magnitude by employing charge-transfer interactions. Although +3 oxidation states are the most stable ones, for some lanthanides such as europium, samarium, or dysprosium, +2 oxidation states can also be reached. In reducing conditions, glass doped with Eu(II) or Sm(II) can be obtained, and in those cases, the spectroscopy may be dramatically changed. Eu(II) has an $[Xe]4f^7$ configuration, and allowed $4f^7 \rightarrow 4f^65d^1$ transitions are possible in the visible region of the light spectra. Since $5d$ shells are no longer shielded, effects from the environment are strong, giving rise to broad absorption and emission spectra and *solvatochromic effects* (i.e., large shifts in the spectra are observed, as well as large Stoke shifts) [19]. The photoluminescence lifetimes are also shorter (microsecond timescale), decreasing the importance of non-radiative processes effectively and therefore increasing the photoluminescence quantum yields. For those reasons, this is applied in *phosphor* applications, in which a usually white pigment down-converts light from the UV region to the visible region with efficiencies approaching 100% (see below). Phosphors have still drawbacks, since they might suffer from thermal instabilities hampering their use to achieve bulk luminescence in glass, so they are deposited as coatings on glass surfaces.

Transition Metals

Here we refer to the elements of the d block, such as manganese or copper. Like lanthanides, one can also easily dope glass with such elements, and by doing that, we may synthesize glass with optical properties in the visible region [13]. Compared with lanthanides, these elements have a very rich electrochemistry where many different oxidation states can be achieved. In fact, a complete reduction of metals such as copper, silver, and gold is possible leading to the formation of metal nanoparticles dispersed in the glass matrix. This is a way to give color to glass, but their photoluminescence is extremely weak unless metal clusters are formed instead. In metal clusters, however, oxidation states are much more uncertain and in many cases under scrutiny. Oxidized states lead then to more standard transitions such as d–d and d–s transitions that will be described here. Some critical elements such as platinum and iridium have also photoluminescent properties; however, their applications in glass are scarce and therefore not covered here. Metal-to-ligand and ligand-to-metal charge-transfer transitions (MLCT and LMCT) also play a rather minor role in photoluminescent glass materials and therefore not described here.

d–d transitions—This situation is very similar to that described for *f–f* transitions. These are forbidden by Laporte rule, and they are also spin forbidden. The main difference comes from the fact that ligand crystal field plays a major importance here, leading to a splitting of *d* orbitals depending of the geometry and coordination around the metal center. A good example is manganese, where Mn(II) may be photoluminescent. For this oxidation state, we have a $[Ar]3d^5$ electronic configuration in the ground-state ^6S in the absence of a crystal field. In the presence of the crystal field, provided that the splitting has a low-energy difference, high-spin configuration is the ground state ($^6A_{1g}$ state), and the lowest-energy excited state is $^4T_{1g}$ state. Therefore a $^6A_{1g} \rightarrow {}^4T_{1g}$ is both Laporte and spin forbidden, and very high concentrations are required to obtain strong brightness (as it was discussed for the lanthanides). Further disadvantages are the rich electrochemistry of this type of metals (manganese can be in many oxidation states, including Mn(II), Mn(III), and Mn(IV)) and the amorphous nature of glass that may give rise to different coordination numbers of the metal with non-bridging oxygens within the glass structure (e.g., 4, 5, or 6) which has an obvious strong impact in the crystal field geometry. Still transition metals are much more affordable than lanthanides, and manganese is not listed as a critical raw material.

d–s transitions—These transitions are important when d^{10} electron configurations are obtained. This is the case of group 11 metals (copper, silver, and gold) in the oxidation state +1. Group 12 elements (zinc, cadmium, and mercury) in the oxidation state +2 also have the same electronic configuration. Generically, these are transitions of the type $nd^{10} \rightarrow nd^9(n + 1)s^1$ which are forbidden within Laporte rule but may be spin allowed. Since all electrons are paired, $S = 0$, we have singlet ground states. Therefore the spectroscopy is simplified and similar to what is generally observed with organic molecules. Intersystem crossing can occur in the excited state, generating long-lived triplet states with lifetimes in the microsecond timescale and significant photoluminescence.

These metals when combined with chalcogenides (oxygen, sulfur, selenium, and tellurium) develop semiconductor properties with variable bandgaps and luminescent properties. When their size is reduced to a few nanometers, the so-called quantum dots are formed with unique photophysics due to confinement effects in their electronic configurations normally increasing the bandgap by decreasing their size. Quantum dots, however, are usually made of group 12 elements (especially Cd(II)).

Group 11 elements are exemplified by the Cu(I) photoluminescence, which is very rich and used in organometallic and supramolecular chemistry for applications in, e.g., temperature-activated delayed fluorescence (TADF) or dye-sensitized solar cells (DSSC). In glass, cations such as Cu(I) can be dissolved within the glass matrix coordinated by four non-bridging oxygens. In glass Cu(I) solubility is very high because Cu(I) will have only a minor disruption of the atomic glass structure, and so heterogeneities are less probable. $^1A_{1g}$ is the ground state when Cu(I) is subjected to a crystal field, giving rise to *d–s* transitions to ^1E_g or $^1T_{2g}$ excited states with oscillator strength circa 0.002 (3 orders of magnitude greater f–f transitions). Solvatochromic effects are rather strong for Cu(I), which connected with their dissolution in amorphous glass give rise to extremely broad photoluminescence and very large Stokes shifts making them optimum candidates for light down-conversion applications in LEDs and luminescence solar concentrators. Ag(I) and Au(I) also display interesting properties, especially when aggregation as clusters is observed, including aggregation-induced emission (AIE) [24]. However clusters depart from pure d–s transitions due to splitting of the atomic orbital energies through copper-philic, argentophilic, and aurophilic interactions when metal-to-metal distances are below a threshold to allow electronic modifications. The mixing of d and s orbitals generates σ and σ^* molecular orbitals, giving rise to transitions of the type d_{σ^*}-s_σ, which may be the main responsible for their luminescence properties when clusters are formed.

Post-Transition Metals and Quantum Dots

Post-transition metals include elements from Group 13, 14, and 15 of the periodic table that display metallic properties such as gallium, indium, thallium, tin, lead, and bismuth. We may also include in this list elements from Group 12 (zinc, cadmium, and mercury), germanium, and antimony since they have some general chemical properties that are similar. As discussed in the Transition Metals section, Group 12 elements appear as d^{10} elements in their common oxidation state +2, but they make semiconductor materials with proper anions. From Group 13 GaN is a fine example of a semiconductor material with important optical properties, giving rise to the development of blue LEDs which are very important in the lighting industry. Sn(II), Pb(II), Sb(III), and Bi(III) have similar electronic configurations of the type ns^2 with s–p transitions allowed by parity and spin, and their oxides are photoluminescent. Very recently the combination of Pb(II) with halides became a hot topic in photovoltaics due to the discovery of perovskites solar cells. These perovskites have the general formula [cation][PbX$_3$] where X is an halide and the cation is an organic ammonium or Cs(I). In this case it is also possible to obtain quantum dots with bright photoluminescence. Combination of Pb(II) with chalcogenides such as sulfur is also described to give rise to photoluminescent quantum dots. The main challenge with quantum dots is that they are very difficult to synthesize in glass in order to give rise to bulk photoluminescence but may be used as coatings as well.

Synthesis of luminescent glass has been made essentially by using the desired luminescent centers in the batch composition; therefore all the glass bulk is luminescent. This can be a disadvantage when it is necessary to work glass after its production. Coatings application in the final phase of glass production, containing the abovementioned emissive centers, as surface coloring techniques can have several advantages, saving time, effort, and also synthesis and material costs. These coatings can be applied at lower temperatures, below or just above the glass transition temperature (Tg), avoiding glass deformation.

2 Application in Industry

2.1 Phosphors in LED Lighting: Solid-State Lighting

Luminescent glasses can be very useful in lighting industry as phosphors. Phosphors are usually described as solid materials that exhibit luminescence when exposed to radiation as ultraviolet, near-infrared radiation, or an electron beam [25, 26]. It consists of a host doped with small amounts of one or more activator ions that act as an emissive center whose excitation can occur directly or by energy transfer [27]. A phosphor should absorb the excitation energy and afterward emit light with a quantum efficiency as high as possible, approaching 100%. However, very high concentrations of these activator ions commonly give rise to luminescence suppression (quenching) [28]. Frequently rare earths are used as activator ions, but transition metal ions can also be used. The activator ions' emission can present a low interaction with the host, showing a very narrow band, as several trivalent lanthanide ions where the $4f$ shell is efficiently shielded by the closed $5s$ and $5p$ shells [15]. But activator ions can also suffer a strong influence from the host presenting a broad band, for example, Eu^{2+} and Ce^{3+}, where $4f$–$5d$ transitions are parity allowed, and since they are not protected like the f–f transitions, the band position of lower energy is strongly dependent on the ion local environment and the nature of their ligands [29].

Phosphors are used nowadays for innumerous applications such as cathode ray tubes, lighting, luminescent paints, laser dyes, plasma and field emission displays, neon tubing, etc. [25]. One of the most significant phosphor applications is on lighting, not only in fluorescent lamps but also in remote phosphor light-emitting diodes (LEDs). Since LEDs are currently replacing fluorescent lamps due to their several advantages such as higher light quality, high stability, and lower cost, they will be the ones discussed in this chapter. In white LEDs, phosphors can be used as a coating in order to convert near-UV or blue emission, for example, from InGaN LEDs to visible light. It is commonly used as a combination of the blue emitter with a yellow phosphor or of the UV emitter with (1) a mixture of different emitting phosphors (Fig. 2) or (2) with an intrinsic white phosphor [30, 31]. One example generally used for white LEDs consist of a blue LED chip covered with the yellow phosphor $Y_3Al_5O_{12}$:Ce^{3+} (YAG:Ce); however, this approach has a low color rendering index (CRI) and limited correlated color temperature (CCT) which can be improved with a red component [32]. The color rendering index defines the cooler fidelity where the color obtained with the test light source is compared with the one obtained by using a reference light source [33]. On the other hand, the correlated color temperature is the temperature of the Planckian radiator that produces a similar light to the one of the analyzed light source. White LEDs can also be produced by combining three different LED chips—red, green, and blue—giving rise to white light with high luminous efficiency and color rendering index (see Fig. 4). However these white LEDs have very complex drive circuits and a high cost [34]. The use of phosphors in LEDs, which are dispersed in an epoxy resin (or silicone) and protected

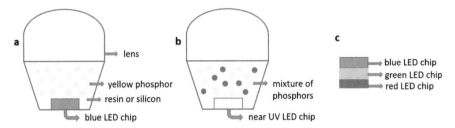

Fig. 4 Different LED structures that can produce white light. (**a**) Blue LED chip covered with the yellow phosphor, (**b**) near-UV LED chip covered with a mixture of yellow and blue phosphors, (**c**) combination of blue, green, and red LED chips. Adapted from [31]

by a lens, is a simpler system, less expensive which can also give rise to high-quality white light.

For these lighting purposes, phosphors need to satisfy different requests such as strong absorption of the LED light, a high quantum efficiency, brilliant color-rendering properties, a broad emission spectra in order to reproduce the spectral solar irradiance, high structural and chromatic stability to external factors, and small luminance saturation [34]. The equilibrium of light between the near UV or blue emission from the LED chips and the phosphors emission is crucial to obtain an adequate white light with proper color rendering index and color temperature [35].

As abovementioned rare-earth ions are frequently used in inorganic phosphors for the solid-state lighting field. These ions make possible to obtain a fine luminescence tunability and materials with a high optical gain [36]. Eu^{2+} ion, for example, can be found in several commercial phosphors for white LEDs since it can emit different luminescent colors in different hosts [37]. However, according to a 2017 European Commission report, rare earths are included on the list of critical raw materials which are materials with both economic importance and supply risk that have an associated risk of low substitution and low recycling rates [38]. The search of other elements to use as phosphors is extremely important since the worldwide availability of some rare earths is predicted to be insufficient in a short time. Pevitra et al. reported the synthesis of a new rare-earth-free yellow phosphor, $Ca_5Zn_{3.92}In_{0.08}(V_{0.99}Ta_{0.01}O_4)_6$, with quantum efficiencies up to 41% [39]. Other phosphors as Ga_2O_3-CdSe/CdS, $RbVO_3$, and $LiCa_3MV_3O_{12}$ (M = Zn and Mg), which give rise to white and blue luminescence, were also produced recently [40–42]. However some of these elements, such as indium and vanadium, are also pointed out as critical raw materials and also have some disadvantages as toxicity and cost. Other materials may overcome some of these potential disadvantages, using inexpensive and nontoxic raw materials while keeping the typical performances of commercial phosphors.

Phosphor coatings composed by organic compounds are also explored. These organic materials can be synthesized using cheap precursors, and they show high absorption efficiencies, broad emission band, the necessary brightness, and they can be tuned to obtain an acceptable color range by changing their functional groups. Several studies can be found about LED development by combining blue emission from InGaN with organic polymers giving rise to different colors [43, 44]. Li et al.

developed organic dyes which are excited by blue light using rhodamine B and fluorescein isothiocyanate in boron nitride forming a luminescent phosphor that can reach quantum efficiency's up to ca. 50% [45]. Yet frequently organic compounds are prone to suffer degradation, thus limiting their applications. Hybrid materials as metal-organic frameworks (MOFs) are also being explored, since the metal and the linker can be used to give rise to photoluminescence and together to obtain higher brightness and higher quantum yields [46, 47]. MOFs based on trivalent lanthanides are a very promising class of materials for addressing the challenges of luminescent centers. However, even in the case of MOFs, it is important to find rare-earth-free high-efficient phosphors. Gong et al. reported a blue-excitable yellow phosphor with internal quantum efficiency up to 90.7%, a very high value for rare-earth-free phosphors [48].

2.1.1 Luminescent Glasses as Phosphors

Potential alternative hosts to produce phosphors are glass materials, which are remarkably suited to enclose lanthanide ions but also to entrap other photoluminescent centers as shown in Section "Lanthanides and Actinides (f–f Transitions)". By using glasses it can be obtained a stable material with high quantum yields and bright luminescence and being an amorphous material, with short range order and a lack of order in a long range, has different sites available for the emissive activator, an advantage to achieve color tunability [49].

Glass can be used for LED phosphors using two different approaches:

1. Phosphor-in-glass (PiG): The organic resin or silicone where the phosphor is commonly placed has low thermal conductivity and low stability; therefore the use of a glass matrix to disperse the inorganic phosphors is being explored [34]. Glass materials to be applied in PiGs need to fill different features: (a) a low melting temperature, in order to make possible the mixture of the glass frit with the phosphors without occurring thermal degradation and chemical reactions between both, (b) has to be mechanically and chemically stable, (c) present a higher thermal conductivity, and (d) need to present a high refractive index, similar to the one of the phosphor, in order to decrease the scattering loss and increase the PiG transmittance [50, 51].

   Several authors developed innovative Ce:YAG PiG color converters for white LEDs, using different glass compositions as hosts such as a Sb_2O_3–B_2O_3–TeO_2–ZnO–Na_2O–La_2O_3–BaO glass or a more inexpensive glass as SiO_2–Al_2O_3–B_2O_3–ZnO–BaO, where the Ce:YAG phosphor was dispersed [52, 53]. This research showed that glass could be an alternative to the conventional resin, giving rise to white LEDs with very good features such as high humidity resistance, high thermal stability and good luminous efficacy, CCT, and CRI properties. Nevertheless the external quantum efficiency of the Ce:YAG phosphor decreases in these hosts, when compared to the powder [53]. In order to add the red component and therefore to optimize the CRI and CCT properties, other studies using a $Y_2Mg_2Al_2Si_2O_{12}:Ce^{3+}$ phosphor or adding a Mn^{2+} phosphor to the Ce:YAG dispersed in a glass host are also being made [54–56]. Taking in mind a more sustainable approach, by producing rare-earth-free phosphors, nanoporous materials such as zeolites can also be explored, since they can remarkably entrap photoluminescent nanoclusters and be dispersed in glass matrices. Silver zeolites, for example, have outstanding properties to be used in lighting, such as an emission spectra that covers all the visible range, very high external quantum efficiencies (EQE) that go up to 97%, large Stokes shifts, and high photostability [57]. Besides cations, luminescent anions can also be entrapped in zeolites with sodalite-cage-containing topologies, for example, sulfur and chloride anions [58]. Recently it was reported the development of photoluminescent sulfur-zeolite materials containing sulfide clusters as the emissive centers with a maximum intensity at around 650 nm when excited at 385 nm, synthesized using inexpensive raw materials, and showing significant emissive properties such as EQE up to 53%, large Stokes shifts, and high-temperature stability up to 450°C [59].

2. Luminescent glass as phosphor: Many luminescent glasses have already been produced having in mind their application as phosphors for lighting. The glass composition is doped with a small concentration of different emissive centers, commonly rare-earth ions, such as europium, samarium, terbium, and cerium in the oxidation state +3, forming an intrinsic luminescent glass [60, 61]. These studies are frequently focused not only on the production of white luminescence, by mixing different luminescent centers, but also on tuning the emission range by changing the excitation wavelength, since this tunable effect is easier to obtain in glasses and typically is not present in crystals. Zhang et al. already obtained a tunable emitting light by doping Eu^{2+} in a silicate glass (SiO_2–Li_2O–SrO–Al_2O_3–K_2O–P_2O_5), where the emission varies with the excitation wavelength [49]. More recently Tb^{3+} and Eu^{3+} ions were used to dope borosilicate glasses containing $CsPbBr_3$ quantum dots, which also presented an emission tunability, emitting light from green to red, but in this case by changing the europium and terbium ratio [62]. By combining these glasses with a blue LED chip, it is possible to obtain white luminescence with a good luminous efficiency, demonstrating that these materials are possible candidates to be used in white LEDs.

More sustainable raw materials can also be used to produce luminescent glasses for these applications. Several studies have been made by doping Sn^{2+}, Mn^{2+}, or Cu^+ in order to produce luminescent glasses with high Stokes shifts and EQEs up to 50%. A tunable broad emission was obtained by doping an oxyfluoride glass with a copper and manganese mixture reaching quantum yields of 56% and obtaining the white luminescence by changing the Cu/Mn ratio [63].

2.2   Smart Windows

Among the glass applications, windows appear as the crucial market for applications and are a key sector for sustainability in the energy sector of society. Heat exchange in windows is crucial not only for human comfort but also because large amounts of energy are consumed in heating and air conditioners. In recent years new products appeared with chromogenic windows that get darker under controlled ways in order to avoid solar light heating of buildings during summer while allowing it to go through when temperature is low. Thermochromic and electrochromic windows are leading technologies for this market and will be here described. Photovoltaic windows appear also as a leading content for such applications, and, more recently, luminescent solar concentrators (LSC) started to appear as a solution for light conversion technologies. We here describe briefly the current state of the art about electrochromic windows and LSC.

2.2.1   Electrochromic Windows

Electrochromism is a phenomenon in which the color of a given material or device may be changed by employing an external electric voltage. These color changes should be completely reversible. Electrochromism is achieved through *electrochromic devices* (ECD) which have an architecture that is similar to a battery. The standard configuration is achieved by employing a transparent electrode (e.g., a glass panel covered with a transparent conductive layer such as indium tin oxide, ITO) covered with a thin layer of an electrochromic material (that will change in different oxidation states), followed by an electrolyte layer and finally a counter-electrode (that can also be made with a glass/ITO substrate) [64, 65]. The most popular electrochromic materials are WO_3 [66], a metal-oxide material, Prussian blue, and polythiophene such as PEDOT [67].

Electrochromic windows are an intense research area nowadays, with a high number of patents per year. The laminated architecture of ECDs make them highly suitable for electrochromic windows, enabling a fine light management control of building interiors and saving energy in air conditioner in hot weather conditions [68].

2.2.2   Luminescent Solar Concentrators

The concept of luminescent solar concentrators (LSC) was first described by Garwin in 1960 [69]. In that seminal publication, it was described a luminescence medium panel that is covered with mirrors in three of the edges and a photovoltaic device in the remaining edge. Since this early description, progress moved slowly with key contributions by Weber and Lambe [70] and Batchelder et al. [71]. Key developments were needed, which were triggered by the appearance of dye-sensitized solar cells and the strong development of organic photovoltaics (OPV) by using semi-conductor polymers in the 1990s. This allowed the production of low-cost flexible devices that could be customized in a facile way. OPV devices are indeed very suitable for such applications, because they are produced in the industry using roll-to-roll printing technologies by companies such as InfinityPV. These OPV devices can be cut and assembled at a panel edge, and so the LSC can now be tested and optimized.

LSC requires a luminescent medium. Currently the preferential media are transparent plastic glasses, which are made from organic polymers [72, 73]. Inorganic glass examples are still not described, but several papers describe already materials with properties that can potentially be applied for LSC. For the luminescent centers, the following properties are preferred:

- Broad spectral absorption in the UV-visible region: This is an important property because it will ensure a larger light harvesting and increase the photoluminescence brightness of the medium.
- Low-medium coloration: This is in contradiction with the first property. If one desires efficiency, the resulting window will have a strong color which will decrease greatly natural illumination inside the buildings. If efficiency is not a factor, then UV and blue region (280–450 nm) light should be absorbed, but longer wavelengths should not, to ensure natural illumination.
- Large Stokes shifts: Absorption and emission should be completely decoupled in order to avoid inner-filter effects that will decrease the efficiency.
- High photoluminescence quantum yield: 100% fluorescence or phosphorescence quantum yields ensure maximum conversion of harvested light into luminescent light.
- Emission energy resonant with the PV-cell responsivity: A better performance will be achieved if the emission spectrum lies in the region where the PV IPCE is higher (normally close to the bandgap).
- Photochemical and thermal stability: For outdoors, this property is crucial to ensure a long lifetime of the LSC window.

Organic molecules can meet most of these criteria with the exception of the Stokes shifts (usually small) and photochemical stability (usually small as well). Quantum dots are however better because they have strong photochemical stability if they are properly encapsulated in the medium. Inorganic luminescent centers have strong advantages in these topics, but they usually do not have broad-absorption bands because electronic transitions are many times forbidden.

The ideal luminescent medium for LSC applications is therefore very difficult to obtain, regardless of the chosen approach. It is more sensible to develop a variety of solutions that would cover a range of situations found in real life. We now distinguish two set of solutions:

1. Organic LSC: These are constituted by an organic glass material (plastic) with fluorescent organic molecules or quantum dots. Several reviews describe this approach [73].

2. Inorganic LSC: Constituted by an inorganic glass (e.g., float glass or borosilicate glass) doped with luminescent centers.

Regardless of the solution used, the architecture is always similar. The light source will be mostly sunlight that is the result of black-body radiation with an effective temperature equal to 5772 K filtered by Earth's atmosphere. The ozone layer effectively blocks UV radiation with wavelengths below 280 nm. This light will be the excitation source of the luminescent medium. On the other hand, the LSC will be limited by the PV device used. In particular, the bandgap is crucial because lower light energies will not be converted to electricity if they are below the bandgap.

2.3   Art, Architecture, and Design

Materials with distinctive optical effects play an important role in the art field, not only to seduce the observers but also to illustrate the reality that surrounds us. The expression "smart materials," which are materials that can reversibly change their properties due to external stimuli, such as temperature, light, or application of an electric field, gained visibility around 1992, with the commercial emerging of "smart" snow skis [74, 75]. Quickly these materials started to be used not only in high-tech products but also in art, architecture, and design. Several artists have made artistic installations in museums and other public spaces using neon, fluorescent light tubes, and luminescent materials [76–78] showing that smart materials that change color when exposed to different lights can be very inspiring. Down-conversion materials as luminescent glasses are one example that can be explored, since luminescent color only appears under UV light. The light effects thus obtained improve the visual value of the artworks and seem to have extraordinary potential in architecture and design.

2.3.1   Luminescent Glass in Art

Over the centuries many artists have used glass as a material to produce unique art works, but only recently few artists start to explore the luminescence effect in glass. Susan Liebold, for example, is an artist that integrates in her flameworked pieces luminescent borosilicate glass that only becomes colored under UV light. Victor Engbers used uranium-doped glasses in his artistic installation *Greenhouse*, built for the Amsterdam Light Festival, and also in the exhibition "The Deep" (2015), in the ArtScience Museum in Singapore, the artist Lynette Wallworth used luminescent glasses in the *Hidden/Depths* multimedia installation. To the best of our knowledge, artists generally use glass doped with lanthanide and actinide oxides, due to their high luminescence intensity. Small concentrations of these oxides can be added to different glass batch compositions to be applied in artworks. Different lanthanide

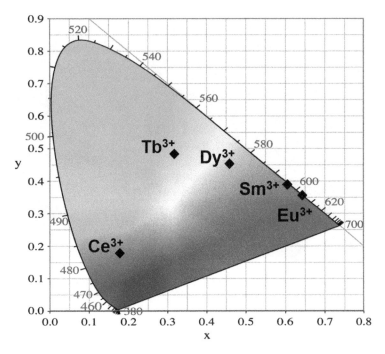

Fig. 5 CIE diagram with the color coordinates of soda-lime silicate glass samples doped with different lanthanide oxides: Eu_2O_3, Sm_2O_3, Dy_2O_3, Tb_4O_7, and CeO_2 ($\lambda_{exc} = 377$ nm)

oxides in a soda-lime silicate glass matrix give rise to different luminescent colors, orange-red (Eu_2O_3), green (Tb_4O_7), yellow (Dy_2O_3), orange (Sm_2O_3), blue (CeO_2), and violet (Tm_2O_3) (see Fig. 5), which are in agreement with what is reported for these elements in the trivalent state [13, 79, 80]. Europium in a borosilicate composition can produce two different colors, which correspond to two different oxidation states, Eu^{3+} (orange-red) and Eu^{2+} (blue), without using a reducing atmosphere [81]. In the international exhibition "Within Light/Inside Glass" (2015) in Venice and in Lisbon, different artists used the described luminescent glasses, illuminated by UV-LEDs, integrated on their artworks, Figs. 6 and 7 [77]. Teresa Almeida already explored the combination of these different luminescent colors and applied it to her artworks [82].

As described in Sect. 1.2, commonly it is cheaper and more practical to apply luminescence not in the glass bulk but as a coating. It is possible to find several fluorescent and phosphorescent non-vitreous paints to be applied in glass without being necessary to fire them, but artists several times aim to diffuse the paint into the glass substrate, and in this case, the use of an enamel in order to have the color incorporated into the glass is a common option. In spite the fact of the application of colored enamels in glass being of common use, the application of luminescent glass enamels is unusual. Stefano Rossi et al. developed a luminescent vitreous enamel where the photoluminescent pigment was based on a matrix constituted by Al_2O_3,

Fig. 6 *Questioning the answers.* Flame worked glass in the exhibition "Within Light/Inside Glass" in Venice. Robert Wiley 2015. Photo taken by Francesco Allegretto (2015)

SiO_2, SrO, MgO, CaO, and B_2O_3 doped with Eu_2O_3 and Dy_2O_3 [83]. Yet these enamels were fired around 850°C, and in glass art, it is very important to use enamels that can be fired below 565°C, allowing the enamel application in glass artworks without deforming their original shapes [84].

2.3.2 Luminescent Glass in Architecture and Design

Glass is already pointed out as a vital material for architecture as it can be used in windows, façade, or interior panels. Glass in architecture improves the quality of space, allowing to obtain illuminated areas, and by using contemporary types of glass may also contribute to energy saving as suggested in Sect. 2.2 when the subject of smart windows was discussed. One of the first large-scale glass buildings was the Crystal Palace, in Hyde Park, England, whose construction begun in 1850. More recently, contemporary glass adapted to a variety of architectural forms as it can be observed in different buildings such as the China Central Television (CCTV) in China, one project of the atelier OMA, and the Chanel store in Amsterdam entitled *A Crystal House*, of the atelier MVRDV, that was built with glass bricks.

In spite of the glass significance to architecture and the extended use of smart materials in this area in the so-called building envelopes which can react with the surrounding, as, for example, in convertible roofs or controlled room lighting [74], luminescent glasses were still not greatly explored. The company Veluna is one of the known companies that produces phosphorescent glass panels, bricks, and pebbles in which products were already applied as design elements in the Mt Sarakura Observatory in Kitakyushu, Japan. Filigree wallpaper by Juliet Quintero is another

Fig. 7 *Dendrogyra cylindricus.* Pâte de verre with luminescent glass in the exhibition "Within Light/Inside Glass" in Venice. Teresa Almeida 2015. Photo taken by Francesco Allegretto (2015)

example of the application of these materials on glass in a wallpaper design [74]. However, these photoluminescent materials are more commonly used in interiors, in walls (tiles and coatings), and also in more practical items as in security applications for exit corridors, outdoor walkways, and stairwells. For architecture and design issues, inorganic phosphorescent pigments, commonly constituted by earth alkaline metal aluminates activated with rare earths, zinc sulfide activated by copper, or silicates activated by magnesium or cerium, are extremely interesting since they can be applied in glass surfaces; however, the bulk glass does not have these intrinsic phosphorescent properties [85]. Besides the mentioned applications, luminescent glasses that glow under UV light are also used in other design products such as glasses, bottles, and glass labels as in the case of Dom Perignon champagne bottles.

3 Conclusion

Luminescent glasses are proven to be advantageous for the development of innovative eco-friendly luminescent materials with high durability, energy efficiency, and inexpensive production that will have wide consequences not only in technological applications but also for artistic uses. This chapter has specified several emitting species that can originate luminescent glasses, both in its matrix and as a coating. It focused not only on the use of rare-earth oxides but also of transition metals, post-transition metals, and quantum dots that can be a more sustainable solution. It has been discussed the application of these luminescent glasses in different areas: (1) LED lighting, by applying a known phosphor on a glass matrix or by using the luminescent glass as a phosphor; (2) smart windows, where luminescent glass can have a crucial role as a luminescent solar concentrator; and in (3) art, since it can be used in sculpture and painting, design, and architecture.

Luminescent glasses are starting to gain importance in the technological world and can be a solution for a more sustainable future as a fully recyclable material that can partially substitute plastics in different applications.

Acknowledgments The authors would like to thank the Associate Laboratory for Green Chemistry—LAQV and the research unit Glass and Ceramic for the Arts—Vicarte, which are financed by national funds from FCT/MCTES (UID/QUI/50006/2019 and UID/EAT/00729/2019). They would also like to thank the Portuguese FCT-MCTES for the financial support from PTDC/QEQ-QIN/3007/2014, and the EC is acknowledged for the INFUSION project grant N. 734834 under H2020-MSCA-RISE-2016 and DecoChrom project under Grant Agreement No. 760973.

References

1. Compilation of ASTM Standard Definitions (1990) 7th edn. ASTM, Philadelphia
2. Doremus RH (1994) Glass science, 2nd edn. Wiley, New York
3. Paul A (1990) Chemistry of glasses, 2nd edn. Chapman and Hall, New York
4. Shelby J (1997) Introduction to glass science and technology. The Royal Society of Chemistry, Cambridge
5. Varshneya A (2006) Fundamentals of inorganic glasses, 2nd edn. Society of Glass Technology, Sheffield
6. Turro N (1965) Molecular photochemistry. W.A. Benjamin, Massachusetts
7. Gilbert A, Baggott J (1995) Essentials of molecular photochemistry. Blackwell Science, Oxford
8. Qiu J, Makishima A (2003) Ultraviolet-radiation-induced structure and long-lasting phosphorescence in Sn^{2+}–Cu^{2+} co-doped silicate glass. Sci Technol Adv Mater 4:35–38. https://doi.org/10.1016/S1468-6996(03)00008-1
9. Sakai R, Katsumata T, Komuro S, Morikawa T (1999) Effect of composition on the phosphorescence from $BaAl_2O_4$: $Eu^{2+,}$ Dy^{3+} crystals. J Lumin 85:149–154. https://doi.org/10.1016/S0022-2313(99)00061-7
10. Kurkjian CR, Prindle WR (1998) Perspectives on the history of glass composition. J Am Ceram Soc 81:795–813. https://doi.org/10.1111/j.1151-2916.1998.tb02415.x
11. Lole F (1995) Uranium glass in 1817—a pre-Riedel record. J Glass Stud 37:139–140

12. Lopes F, Ruivo A, Muralha VSF, Lima A, Duarte P, Paiva I, Trindade R, Pires de Matos A (2008) Uranium glass in museum collections. J Cult Herit 9:e64–e68. https://doi.org/10.1016/j.culher.2008.08.009

13. Weyl WA (1951) Coloured glasses. Society of Glass Technology, Sheffield

14. Riker LW (1981) The use of rare earths in glass compositions. In: Gschneidner Jr KA (ed) Industrial applications of rare earth elements. American Chemical Society, Washington, pp 81–94. https://doi.org/10.1021/bk-1981-0164.ch004

15. Gorller-Walrand C, Binnemans K (1998) Spectral intensities of f–f transitions. In: Gschneidner Jr KA, Eyring L (eds) Handbook on the physics and chemistry of rare earths, vol 25. Elsevier Science B.V., Amsterdam, pp 101–264

16. Carnall WT (1979) The absorption and fluorescence spectra of rare earth ions in solution. In: Gschneidner Jr KA, Eyring L (eds) Handbook on the physics and chemistry of rare earths, vol 3. North-Holland Physics Publishing, Amsterdam

17. Donald IW, Metcalfe BL, Taylor RNJ (1997) The immobilization of high level radioactive wastes using ceramics and glasses. J Mater Sci 32:5851–5887. https://doi.org/10.1023/A:1018646507438

18. Charalampides G, Vatalis KI, Apostoplos B, Ploutarch-Nikolas B (2015) Rare earth elements: industrial applications and economic dependency of Europe. Procedia Econ Finan 24:126–135. https://doi.org/10.1016/S2212-5671(15)00630-9

19. Eliseeva SV, Bünzli JCG (2010) Lanthanide luminescence for functional materials and bio-sciences. Chem Soc Rev 39:189–227. https://doi.org/10.1039/b905604c

20. Zhang F (2015) Photon upconversion nanomaterials. Springer, Berlin

21. Ceroni P (2011) Energy up-conversion by low-power excitation: new applications of an old concept. Chem Eur J 17:9560–9564. https://doi.org/10.1002/chem.201101102

22. Liu X, Zhou J, Zhou S, Yue Y, Qiu J (2018) Transparent glass-ceramics functionalized by dispersed crystals. Prog Mater Sci 97:38–96. https://doi.org/10.1016/j.pmatsci.2018.02.006

23. Ruivo A, Muralha VSF, Águas H, Pires de Matos A, Laia CAT (2014) Time-resolved luminescence studies of Eu^{3+} in soda-lime silicate glasses. J Quant Spectrosc Radiat Transf 134:29–38. https://doi.org/10.1016/j.jqsrt.2013.10.010

24. Eichelbaum M, Rademann K, Hoell A, Tatchev DM, Weigel W, Stößer R, Pacchioni G (2008) Photoluminescence of atomic gold and silver particles in soda-lime silicate glasses. Nanotechnology 19:135701. https://doi.org/10.1088/0957-4484/19/13/135701

25. Yen WM, Shionoya S, Yamamoto H (2007) Phosphor handbook, 2nd edn. CRC Press, Boca Raton

26. Rong-Jun X, Hirosaki N, Li Y, Takeda T (2010) Rare-earth activated nitride phosphors: synthesis, luminescence and applications. Materials 3:3777–3793. https://doi.org/10.3390/ma3063777

27. Jüstel T, Nikol H, Ronda C (1998) New developments in the field of luminescent materials for lighting and displays. Angew Chem Int Ed 37:3084–3103. https://doi.org/10.1002/(SICI)1521-3773(19981204)37:22<3084::AID-ANIE3084>3.0.CO;2-W

28. Ronda CR (2008) Emission and excitation mechanisms of phosphors in luminescence: from theory to applications. In: Ronda CR (ed) Luminescence. Wiley-VCH, Weinheim. https://doi.org/10.1002/9783527621064.ch1

29. Lin H, Liang H, Zhang G, Tao Y (2012) A comparison of Ce^{3+} luminescence in $X_2Z(BO_3)_2$ (X=Ba, Sr; Z=Ca, Mg) with relevant composition and structure. J Rare Earths 30:1–5. https://doi.org/10.1016/S1002-0721(10)60627-8

30. McKittrick J, Shea-Rohwe LE (2014) Review: down conversion materials for solid-state lighting. J Am Ceram Soc 97:1327–1352. https://doi.org/10.1111/jace.12943

31. Pimputkar S, Speck J, Denbaars S, Nakamura S (2009) Prospects for LED lighting. Nat Photonics 3:180–182. https://doi.org/10.1038/nphoton.2009.32

32. Kim YH, Viswanath NSM, Unithrattil S, Kim HJ, Im WB (2018) Review-phosphor plates for high-power LED applications: challenges and opportunities toward perfect lighting. J Solid State Sci Technol 7:R3134–R3147. https://doi.org/10.1149/2.0181801jss

33. Khan TQ, Bodrogi P, Vinh QT, Winkler H (2014) LED lighting: technology and perception. Wiley, Weinheim. https://doi.org/10.1002/9783527670147
34. Wang L, Xie RJ, Suehiro T, Takeda T, Hirosak N (2018) Down-conversion nitride materials for solid state lighting: recent advances and perspectives. Chem Rev 118:1951–2009. https://doi.org/10.1021/acs.chemrev.7b00284
35. Chen L, Lin CC, Yeh CW, Liu RS (2010) Light converting inorganic phosphors for white light-emitting diodes. Materials 3:2172–2195. https://doi.org/10.3390/ma3032172
36. Chiriu D, Stagi L, Carbonaro CM, Ricci PC (2016) Strength and weakness of rare earths based phosphors: strategies to replace critical raw materials. Phys Status Solidi C 13:989–997. https://doi.org/10.1002/pssc.201600116
37. Zhang X, Wang J, Huang L, Pan F, Chen Y, Lei B, Peng M, Wu M (2015) Tunable luminescent properties and concentration-dependent, site-preferable distribution of Eu^{2+} ions in silicate glass for white LEDs applications. Appl Mater Interfaces 7:10044–10054. https://doi.org/10.1021/acsami.5b02550
38. Communication from the Commission to the European Parliament, the Council, the European Economic and Social Committee and the Committee of the regions on the 2017 list of critical raw materials for the EU, Brussels. https://eur-lex.europa.eu/legal-content/EN/TXT/?uri=COM:2017:0490:FIN
39. Pavitra E, Raju GSR, Park JY, Wang L, Moon BK, Yu JS (2015) Novel rare-earth-free yellow $Ca_5Zn_{3.92}In_{0.08}(V_{0.99}Ta_{0.01}O_4)_6$ phosphors for dazzling white light emitting diodes. Sci Rep 5:10296. https://doi.org/10.1038/srep10296
40. Stanish PC, Radovanovic PV (2016) Surface-enabled energy transfer in Ga_2O_3-CdSe/CdS nanocrystal composite films: tunable all-inorganic rare earth element-free white-emitting phosphor. J Phys Chem C 120:19566–19573. https://doi.org/10.1021/acs.jpcc.6b07035
41. Ishigaki T, Madhusudan P, Kamei S, Uematsu K, Toda K, Sato M (2018) Room-temperature solid state contact reaction synthesis of rare earth free $RbVO_3$ phosphor and their photoluminescence properties. J Solid State Sci Technol 7:R88–R93. https://doi.org/10.1149/2.0201806jss
42. Hasegawa T, Abe Y, Koizumi A, Ueda T, Toda K, Sato M (2018) Bluish-white luminescence in rare-earth-free vanadate garnet phosphors: structural characterization of $LiCa_3MV_3O_{12}$ (M = Zn and Mg). Inorg Chem 57:857–866. https://doi.org/10.1021/acs.inorgchem.7b02820
43. Hide F, Kozodoy P, DenBaars SP, Heeger AJ (1997) White light from InGaN/conjugated polymer hybrid light-emitting diodes. Appl Phys Lett 70:2664–2666. https://doi.org/10.1063/1.118989
44. Heliotis G, Gu E, Griffin C, Jeon CW, Stavrinou PN, Dawson MD, Bradley DDC (2006) Wavelength-tunable and white-light emission from polymer-converted micropixellated InGaN ultraviolet light-emitting diodes. J Opt A Pure Appl Opt 8:S445–S449. https://doi.org/10.1088/1464-4258/8/7/S20
45. Li J, Lin J, Huang Y, Xu X, Liu Z, Xue Y, Ding X, Luo H, Jin P, Zhang J, Zou J, Tang C (2015) Organic fluorescent dyes supported on activated boron nitride: a promising blue light excited phosphors for high-performance white light-emitting diodes. Sci Rep 5:8492. https://doi.org/10.1038/srep08492
46. Rocha J, Carlos LD, Paz FAA, Ananias D (2011) Luminescent multifunctional lanthanides-based metal-organic frameworks. Chem Soc Rev 40:926–940. https://doi.org/10.1039/c0cs00130a
47. Carlos LD, Ferreira RAS, Bermudez VZ, Ribeiro SJL (2009) Lanthanide-containing light-emitting organic–inorganic hybrids: a bet on the future. Adv Mater 21:21509–21534. https://doi.org/10.1002/adma.200801635
48. Gong Q, Hu Z, Deibert BJ, Emge TJ, Teat SJ, Banerjee D, Mussman B, Rudd ND, Li J (2014) Solution processable MOF yellow phosphor with exceptionally high quantum efficiency. J Am Chem Soc 136:16724–16727. https://doi.org/10.1021/ja509446h
49. Zhang X, Wang J, Huang L, Pan F, Chen Y, Lei B, Peng M, Wu M (2015) Tunable luminescent properties and concentration-dependent, site-preferable distribution of Eu^{2+} ions in silicate glass

for white LEDs applications. ACS Appl Mater Interfaces 7:10044–10054. https://doi.org/10.
1021/acsami.5b02550

50. Peng Y, Mou Y, Wang H, Zhuo Y, Lid H, Chen M, Luo X (2018) Stable and efficient
all-inorganic color converter based on phosphor in tellurite glass for next-generation laser-
excited white lighting. J Eur Ceram Soc 38:5525–5532. https://doi.org/10.1016/j.jeurceramsoc.
2018.08.014

51. Xu X, Li H, Zhuo Y, Li R, Tian P, Xiong D, Chen M (2018) High refractive index coating of
phosphor-in-glass for enhanced light extraction efficiency of white LEDs. J Mater Sci
53:1335–1345. https://doi.org/10.1007/s1085

52. Zhang R, Lin H, Yu Y, Chen D, Xu J, Wang Y (2014) A new-generation color converter for
high-power white LED: transparent Ce^{3+}:YAG phosphor-in-glass. Laser Photonics Rev
8:158–164. https://doi.org/10.1002/lpor.201300140

53. Zhang X, Yu J, Wang J, Lei B, Liu Y, Cho Y, Xie RJ, Zhang HW, Li Y, Tian Z, Li Y, Su Q
(2017) All-inorganic light convertor based on phosphor-in-glass engineering for next-
generation modular high-brightness white LEDs/LDs. ACS Photonics 4:986–995. https://doi.
org/10.1021/acsphotonics.7b00049

54. Chen Z, Wang B, Li X, Huang D, Sun H, Zeng Q (2018) Chromaticity-tunable and thermal
stable phosphor-in-glass inorganic color converter for high power warm w-LEDs. Materials
11:1792. https://doi.org/10.3390/ma11101792

55. Wang B, Lin H, Xu J, Chen H, Wang Y (2014) $CaMg_2Al_{16}O_{27}$:Mn^{4+}-based red phosphor: a
potential color converter for high-powered warm W-LED. ACS Appl Mater Interfaces
6:22905–22913. https://doi.org/10.1021/am507316b

56. Chen H, Lin H, Xu J, Wang B, Lin Z, Zhou J, Wang Y (2015) Chromaticity-tunable phosphor-
in-glass for long-lifetime high-power warm w-LEDs. J Mater Chem C 3:8080–8089. https://doi.
org/10.1039/C5TC01057H

57. Coutiño-Gonzalez E, Baekelant W, Steele JA, Kim CW, Roeffaers MBJ, Hofkens J (2017)
Silver clusters in zeolites: from self-assembly to ground-breaking luminescent properties. Acc
Chem Res 50:2353–2361. https://doi.org/10.1021/acs.accounts.7b00295

58. Kirk RD (1955) The luminescence and tenebrescence of natural and synthetic sodalite. Am
Mineral 40:22–31

59. Ruivo A, Coutino-Gonzalez E, Santos MM, Baekelant W, Fron E, Roeffaers MBJ, Pina F,
Hofkens J, Laia CAT (2018) Highly photoluminescent sulfide clusters confined in zeolites.
J Phys Chem C 122:14761–14770. https://doi.org/10.1021/acs.jpcc.8b01247

60. Zhu C, Yang Y, Liang X, Yuan S, Chen G (2007) Rare earth ions doped full-color luminescence
glasses for white LED. J Lumin 126:707–710. https://doi.org/10.1016/j.jlumin.2006.10.028

61. Silva Jr CM, Bueno LA, Gouveia-Neto AS (2015) Er^{3+}/Sm^{3+}- and Tb^{3+}/Sm^{3+}-doped glass
phosphors for application in warm white light-emitting diode. J Non-Cryst Solids 410:151–154.
https://doi.org/10.1016/j.jnoncrysol.2014.08.054

62. Cheng Y, Shen C, Shen L, Xiang W, Liang X (2018) Tb^{3+}, Eu^{3+} co-doped $CsPbBr_3$ QDs glass
with highly stable and luminous adjustable for white LEDs. ACS Appl Mater Interfaces
10:21434–21444. https://doi.org/10.1021/acsami.8b05003

63. Xu DK, Shi YF, Peng XS, Wei RF, Hu FF, Guo H (2018) Tunable broad photoluminescence in
Cu^+/Mn^{2+} co-doped oxyfluoride glasses sintered in air atmosphere. J Lumin 202:186–119.
https://doi.org/10.1016/j.jlumin.2018.05.050

64. Monk P, Mortimer R, Rosseinsky D (2007) Electrochromism and electrochromic devices.
Cambridge University Press, Cambridge

65. Aliprandi A, Moreira T, Anichini C, Stoeckel MA, Eredia M, Sassi U, Bruna M, Pinheiro C,
Laia CAT, Bonacchi S, Samorì P (2017) Hybrid copper-nanowire–reduced-graphene-oxide
coatings: a "green solution" toward highly transparent, highly conductive, and flexible electrodes for (opto)electronics. Adv Mater 29:1703225. https://doi.org/10.1002/adma.201703225

66. Costa C, Pinheiro C, Henriques I, Laia CAT (2012) Inkjet printing of sol–gel synthesized
hydrated tungsten oxide nanoparticles for flexible electrochromic devices. ACS Appl Mater
Interfaces 4(3):1330–1340. https://doi.org/10.1021/am201606m

67. Kraft A (2019) Electrochromism: a fascinating branch of electrochemistry. ChemTexts 5:1. https://doi.org/10.1007/s40828-018-0076-x

68. Xia X, Ku Z, Zhou D, Zhong Y, Zhang Y, Wang Y, Huang MJ, Tu J, Fan HJ (2016) Perovskite solar cell powered electrochromic batteries for smart windows. Mater Horiz 3:588–595. https://doi.org/10.1039/C6MH00159A

69. Garwin RL (1960) The collection of light from scintillation counters. Rev Sci Instrum 31:1010–1011. https://doi.org/10.1063/1.1717105

70. Weber WH, Lambe J (1976) Luminescent greenhouse collector for solar radiation. Appl Opt 15:2299–2300. https://doi.org/10.1364/AO.15.002299

71. Batchelder JS, Zewail AH, Cole T (1981) Luminescent solar concentrators. 2: experimental and theoretical analysis of their possible efficiencies. Appl Opt 20:3733–3754. https://doi.org/10.1364/AO.20.003733

72. Correia SFH, Bermudez VZ, Ribeiro SJL, André PS, Ferreira RAS, Carlos LD (2014) Luminescent solar concentrators: challenges for lanthanide-based organic–inorganic hybrid materials. J Mater Chem A 2:5580. https://doi.org/10.1039/c3ta14964a

73. Debije MG, Verbunt PPC (2012) Thirty years of luminescent solar concentrator research: solar energy for the built environment. Adv Energy Mater 2:12–35. https://doi.org/10.1002/aenm.201100554

74. Ritter A (2009) Smart materials in architecture, interior architecture and design. Birkhäuser, Basel

75. Addington DM, Schodek DL (2005) Smart materials and new technologies for the architecture and design professions. Architectural Press, Oxford

76. Shiess C (1994) The light artist anthology-neon and related media. ST Publications, Cincinnati

77. Exhibition Catalogue (2015) Within light/inside glass, an intersection between art and science. Fundação Millennium Gallery, Lisbon

78. Reis PC (1999) In: Sternberg M, Tighe H (eds) Pedro Cabrita Reis: on light and space, catalogue of exhibition at Museum Moderner Kunst Wien and Museu de Arte Contemporânea de Serralves. Charta, Milan

79. Almeida T, Ruivo A, Pires de Matos A, Oliveira R, Antunes A (2008) Luminescent glasses in art. J Cult Herit 9:e138–e142. https://doi.org/10.1016/j.culher.2008.06.002

80. Gorller-Walrand C, Binnemans K (1998) Handbook on the physics and chemistry of rare earths, vol 25. Elsevier Science B.V., Amsterdam, pp 101–264

81. Ruivo A, Almeida T, Quintas F, Wiley R, Troeira M, Paulino N, Laia CAT, Queiroz CA, Pires de Matos A (2013) Colours of luminescent glasses for artworks. In: MacDonald L, Westland S, Wuerger S (eds) Proceedings of the 12th international AIC colour congress, pp 885–888

82. Almeida T (2013) Art/science: a case study of luminescent vitreous materials. In: Proceedings of the VI world congress on communication and arts, Geelong, 4–7 Apr 2012. https://doi.org/10.14684/WCCA.6.2013.12-16

83. Rossi S, Quaranta A, Tavella L, Deflorian F, Compagnoni AM (2015) Innovative luminescent vitreous enameled coatings. In: Tiwari A, Rawlins J, Hihara LH (eds) Intelligent coatings for corrosion control. Butterworth-Heinemann, Oxford. https://doi.org/10.1016/C2012-0-06936-0

84. Stone G (2000) Firing schedules for glass: the kiln companion, 1st edn. Igneous Glassworks, Melbourne

85. Bamfield P, Hutchings MG (2010) Chromic phenomena—technological applications of colour chemistry, 1st edn. RSC Publishing, Cambridge

Luminescence-Based Sensors
for Aeronautical Applications

Bruno Pedras, Guillermo Orellana, and Mário Nuno Berberan-Santos

Contents

Abstract Aeronautical industry deals with extremely complex and sensitive issues, the most critical of which being arguably aircraft safety during flight. For a variety of reasons and through different pathways, the integrity of an airplane might be compromised in such a way that passengers and crew might be at risk. In the case of military aircraft, this risk is largely aggravated in war zones.

Different aspects are addressed when designing an aircraft, and, in each of those areas, maintenance will eventually be required. Therefore, to keep an effective and affordable maintenance, real-time information for each of the different components of the aircraft must be provided. Sensors designed specifically for each parameter of interest are to be included in the primary design of the airplane. Optochemical sensing plays an important role in a vast number of fields, in most cases overcoming the limitations of other types of sensors. Luminescence-based sensors constitute a specific type of optochemical sensors, displaying a bunch of characteristics that render them unrivalled in terms of performance, sensitivity, reversibility, ease of

B. Pedras (✉) and M. N. Berberan-Santos
CQFM-IN and iBB-Institute for Bioengineering and Biosciences, Instituto Superior Técnico, Universidade de Lisboa, Lisboa, Portugal
e-mail: bruno.pedras@tecnico.ulisboa.pt

G. Orellana
Chemical Optosensors and Applied Photochemistry Group (GSOLFA), Department of Organic Chemistry, Faculty of Chemistry, Complutense University of Madrid, Madrid, Spain

© Springer Nature Switzerland AG 2019   389
B. Pedras (ed.), *Fluorescence in Industry*, Springer Ser Fluoresc (2019) 18: 389–412,
https://doi.org/10.1007/4243_2019_11, Published online: 6 March 2019

miniaturization, and low cost. This chapter addresses the most relevant advances in luminescence-based sensors for molecular oxygen (both in fuel tanks and as pressure-sensitive paints), corrosion and wear, hydraulic fluid monitoring, and other parameters relevant to the aeronautical industry.

Keywords Aircraft · Corrosion · Fluorescence · Fuel tank · Hydraulic fluid · Inerting · Luminescence · Oxygen · Pressure-sensitive paint · Sensors

1 Introduction

Aircraft design aims at optimal flight performance for given input mission constraints, taking into account maintenance and operational costs, and passenger and crew safety. For these reasons, real-time information on aircraft systems and components is needed, not only for decision-making but also for automated feedback loops. Sensors designed specifically for all parameters of interest are therefore used. In several cases, luminescence-based sensors are the best choice.

2 Luminescence-Based Sensors in Aeronautic Industry

Luminescence-based sensors are a specific type of optochemical sensors, combining characteristics that render them unique, specifically owing to their light weight, reliability, sensitivity, low cost, and non-electrical nature. Some are routinely used; others are still in the test phase.

2.1 Oxygen Sensors

2.1.1 Fuel Tank Inerting Systems

One of the major targets in aircraft safety is suppression of the flammability that may result from ignition of the mixture of fuel vapors and molecular oxygen in aircraft fuel tanks. Given the possible presence of ignition sources (e.g., lightning, static electricity), there is a need to reduce the oxygen concentration to safe levels, i.e., significantly below the atmospheric composition (21% at all relevant altitudes) (Fig. 1). This issue becomes even more critical in the case of military aircraft, where a spontaneous explosion of fuel vapors in the ullage may occur if enemy fire penetrates the fuel tank. In commercial aircrafts, the inclusion of tank inerting systems was hampered by both the cost of equipment and its operation.

However, the need to implement this type of systems in commercial aircrafts was fostered by the crash of TWA800 flight in July 1996, which was attributed to the

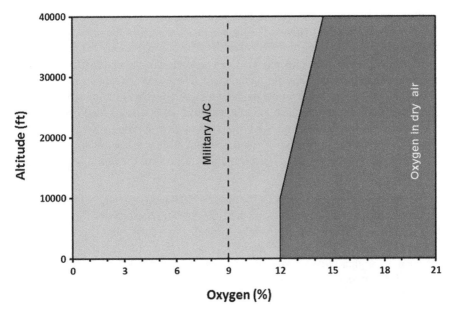

Fig. 1 Schematic representation of the safe oxygen levels in fuel tank inerting, according to [1]. The tank is considered inert when the bulk average oxygen concentration within each compartment of the tank is ≤12% from sea level up to 10,000 feet altitude, then linearly increasing from 12% at 10,000 feet to 14.5% at 40,000 feet altitude and extrapolated linearly above that altitude

explosion of the center fuel tank. Furthermore, since the wings contain composite materials, it is necessary to reduce the oxygen concentration in the wing tanks as well.

Up to the 1960s, polyurethane reticulated foam placed inside the fuel tanks was the only available technology to prevent explosions in military aircraft. This approach presented pitfalls such as the reduction of fuel capacity and the added weight and was only operational for about 5 years. Early inerting techniques included gaseous and liquid nitrogen systems [2], which were later replaced by an alternative approach based on Halon 1301 (bromotrifluoromethane) in the early 1970s [3]. Due to its potential for damaging the ozone layer, and following recommendations from the Environmental Protection Agency, Halon inerting systems were discontinued and replaced by environment-friendly alternatives. Next-generation systems used nitrogen-enriched air (NEA) for the oxygen content reduction. According to their operating protocol, two types of NEA-based inerting system are commonly used:

– *Ground-based inerting*, which is performed while the aircraft is parked and addresses the oxygen dissolved in the fuel while it is supplied to the aircraft (in the refueling process), which will eventually be released into the ullage upon take-off. This type of inerting comprises "scrubbing" the fuel (by means of a device that sparges NEA through the fuel, displacing the dissolved oxygen to the ullage) and washing the ullage with NEA. Usually, for short-duration flights, ground-based inerting is enough to guarantee a flammability-safe oxygen level [4, 5].

- *Onboard inerting* uses onboard inert gas generating systems (OBIGGS), which continuously produce NEA during the flight. These systems rely on air separation modules containing hollow fiber membranes that selectively remove oxygen from engine bleed air, which is supplied after being cooled via a heat exchanger to temperatures lower than 93.3 °C. Prototypes of these inerting systems have been effectively tested on Boeing 747 and 737 and Airbus A320 aircraft [6] and are used in all recent models. Figure 2 depicts a generic scheme of an OBIGGS system.

For the early versions of OBIGGS, the key challenge was whether it would be able to provide enough NEA in cases where the aircraft is descending. In this specific case, exterior air enters the fuel tanks through the vent system to suppress the pressure difference between the outside of the aircraft and the ullage, causing a considerably higher pressure than the one that can be provided by the engine bleed air for NEA production (while descending, the aircraft power settings are reduced). Moreover, at this point, the fuel tanks may not be full, enhancing the ullage free

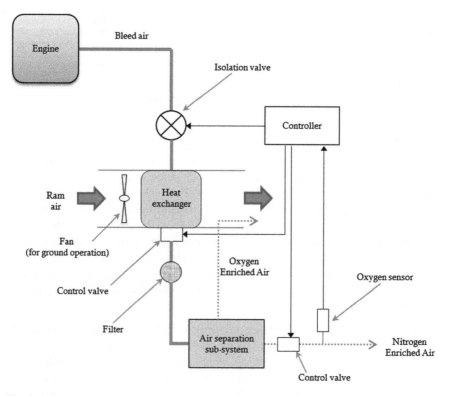

Fig. 2 Schematic representation of a generic OBIGGS system. Adapted from [7]

volume, a situation that considerably worsens the problem. The solution was found in later versions by incorporating pressurized NEA onboard that could be used if needed.

The currently proposed FAA regulation to decrease the flammability risk is to reduce the oxygen levels down to 12%, by in-tank dilution with inert gas [8]. If the oxygen levels are not directly monitored, the production of NEA is continuous and excessive. This continuous operation of the OBIGGS becomes expensive from the point of view of fuel and filters economy, and, for this reason, it is of utmost importance to know the exact oxygen concentration at any given time in the fuel tanks in order to optimize the OBBIGS performance.

2.1.2 Oxygen Measurement in Aircraft Fuel Tanks

There are several methods for measuring the oxygen concentration; not all of them are compatible with aircraft fuel tanks. For instance, electrochemical cells exhibit the disadvantage of wiring and can possibly constitute a source of ignition. Besides, the internal electrolytes are not compatible with negative temperatures. In this regard, optical sensors and, particularly, luminescence-based optosensors overcome the pitfalls of electrochemical sensors, not only in terms of safety (not having electrical components at the sensitive terminal) but also in terms of sensitivity to very low oxygen concentrations. Furthermore, they are resistant to biofouling and can operate reversibly in a wide range of temperatures.

The principle that underlies molecular oxygen optosensing is the collisional quenching by this analyte of a photoexcited luminescent indicator dye embedded into a thin O_2-permeable polymer layer, the latter playing an essential role in the overall performance of the sensor [9–11]. The longer the electronic excited state lifetime of the indicator dye, the higher will be its sensitivity to O_2. The unquenched lifetime of an ideal oxygen-sensing dye is expected to be long (hundreds of nanoseconds to ca. 1 millisecond), the reason for which phosphorescent dyes (e.g., Ru(II) polypyridyl complexes or Pt(II) and Pd(II) metalloporphyrins) are indicators commonly employed for this type of sensors, conveniently immobilized in suitable oxygen-permeable materials [12]. In addition to good permeability to O_2 (to allow short response times), the polymer support must be transparent to both the excitation and emission light. A wide range of polymer supports are available for this purpose, spanning from organic [13–19] (e.g., polystyrene, PVC, silicone, methacrylates) to inorganic [20–24] (silica sol-gels), to hybrid (e.g., organically modified ceramics or "ormocers") materials. The indicator dye may also be adsorbed onto silica or aluminosilicate nano-/microparticles to increase its compatibility with the polymer thin film and to increase the dye sensitivity to O_2 [19]. However, the polymer matrix should be chemically inert with respect to the jet fuel and belong to a class of materials that is allowed for aircraft fuel tank use. Fluoropolymers [25] and their derivatives are examples of such materials (their films exhibiting low polarity and strong hydrophobicity, allowing a rapid diffusion of oxygen, high stability toward temperature and oxidation, and a good resistance to jet fuel).

Two approaches are usually followed for the measurement of oxygen concentration with luminescent indicator dyes. The first one involves measuring the luminescence *intensity*, and it is susceptible to uncontrolled variations due to a broad variety of reasons, such as dye leaching and/or photobleaching, intensity of the light source or detector drifts, and loss of transmission of the optical components, among others. The second approach involves measuring luminescence *lifetime* (τ equal to the reciprocal of the first-order excited state decay process), an intrinsic property of each luminophore that is largely independent of dye concentration and hence not affected by the mentioned variations. Lifetime-based measurements are thus preferable to intensity-based ones.

The dynamic (or "collisional") quenching of the indicator electronic excited state by molecular oxygen is, assuming a homogeneous microenvironment, described by the Stern-Volmer Eq. (1) [26]:

$$\frac{I_0}{I} = \frac{\tau_0}{\tau} = 1 + K_{SV}[O_2] = 1 + k_q\tau_0[O_2] \tag{1}$$

where I and τ represent the luminescence intensity and lifetime of the indicator dye, respectively (the subscript "0" denotes the absence of O_2), and K_{SV} is the so-called Stern-Volmer constant (which is equivalent to the product of the diffusion-controlled bimolecular quenching constant, k_q, and the indicator dye emission lifetime in the absence of oxygen, τ_0). A plot of τ_0/τ versus $[O_2]$ under strict homogeneous conditions yields a linear dependency with a slope equal to K_{SV} and an intercept of unity.

Nevertheless, direct *pump-and-probe* luminescence lifetime measurements frequently involve complex and expensive instrumentation (e.g., a laser light source and a fast digitizing system), preventing a widespread implementation of this technique. In this regard, luminescence *phase-shift* measurements are more affordable and adequate to the purpose [27]. With this technique, the excitation source (usually a light-emitting diode or "LED") is modulated at an optimal frequency ($f \sim 1/\tau$), so that the luminescence becomes modulated at the same value but phase-shifted relative to the former due to the non-instantaneous deactivation of the excited state. This phase-shift value, ϕ, which is related to the excited state lifetime according to Eq. (2), provides a measurement of the analyte of interest, after proper calibration [28–30].

$$\tan\phi = 2\pi f\tau \tag{2}$$

Combining Eqs. (1) and (2), oxygen concentration can be determined from the phase shift in quasi-real time.

In some cases, departure from linearity is observed in the Stern-Volmer plots, and the kinetics is no longer adequately described by the simplest form of the Stern-Volmer equation. In this case, a modified version of the SV equation is used that reflects the different accessibilities of the quencher resulting from the different microenvironments of the luminophore molecules. The most widespread model

[31–33] is a simplified one that considers the existence of two independent quenching sites, each with its own Stern-Volmer constant, that assumes purely dynamic quenching, as represented in Eq. (3):

$$\frac{I_0}{I} = \frac{\tau_0}{\tau} = \left(\frac{f_{01}}{1 + K_{SV1}[Q]} + \frac{f_{02}}{1 + K_{SV2}[Q]}\right)^{-1} \tag{3}$$

where f_{01} is the fraction of sites with Stern-Volmer constant K_{SV1} and $f_{02} = 1 - f_{01}$ is the fraction of sites with Stern-Volmer constant K_{SV2}.

Two physical parameters must be taken into account when measuring oxygen. The first one is *temperature*, which strongly affects the luminescence decay times. In most cases, mathematical algorithms may be used to perform temperature corrections within the sensor, or, alternatively, a temperature-sensitive (but oxygen insensitive) sensing layer can be added to the membrane (the resulting sensor is then called a dual sensor, since it will measure both parameters). The other variable that needs to be considered is the total *pressure*, p, which varies with altitude, according to Eq. (4) [34]:

$$p(z) = p(0)\left(1 - \frac{\beta z}{T_0}\right)^{mg/k\beta} \tag{4}$$

where $T_0 = 288$ K and $\beta = 6.5$ K km^{-1}. Indeed, regulations are formulated in terms of the oxygen volume fraction (or percentage), but the sensor measures oxygen concentration. Assuming an ideal gas behavior, the O_2 mole fraction $F(O_2)$ can be expressed as in Eq. (5):

$$F(O_2) = \frac{p(O_2)}{p} = [O_2]RT/p \tag{5}$$

Combining Eqs. (4) and (5), it is possible to calculate the partial pressures of oxygen in dry air for different representative altitudes (Table 1).

The oxygen sensor may consist of an oxygen-sensitive layer placed inside the aircraft fuel tank and interrogated through a bundle of optical fibers located in an adjacent compartment (preventing the contact of the instrumentation with jet fuel). The optical system may be formed by a LED acting as low-voltage excitation source, whose light is guided by the optical fiber to the sensing polymer disk. The resulting emitted radiation is collected by another (or the same) optical fiber and directed to a photodiode detector. The signal is then processed and converted to oxygen concentration in real time. From the measured temperature and total pressure, the oxygen fraction is finally obtained. A generic scheme is depicted in Fig. 3.

A considerable number of systems have been developed and patented throughout the years, both by academic and industrial researchers alike. Susko [35] describes an inerting system for aircraft fuel tank containing an integrated oxygen sensor, which requires constant temperature. This condition is attained by means of a receptacle

Table 1 Partial pressure of oxygen in dry air for different representative altitudes

Altitude (ft/m)	Atmospheric pressure (mmHg/Pa)	Ambient O$_2$ 20.95% (mmHg)	1% O$_2$ (mmHg)
0/0	760/101,325	159	7.6
5,000/1,524	632/84,291	132	6.3
10,000/3,048	522/69,654	109	5.2
12,000/3,658	483/64,410	101	4.8
15,000/4,572	429/57,147	90	4.3
20,000/6,096	349/46,524	73	3.5
25,000/7,620	282/37,560	59	2.8
30,000/9,144	225/30,049	47	2.3
35,000/10,668	179/23,803	37	1.8
40,000/12,192	140/18,653	29	1.4

Fig. 3 Schematic depiction of a generic oxygen optode for aircraft fuel tank operation

that coats the probe, significantly hampering the desired miniaturization. The sensor can operate at different temperatures, although one at a time, i.e., a thermostat is programmed for a desired temperature, and the oxygen measurement is performed under those conditions. The effect of sudden variations in a wide range of temperatures (as is the case under in-flight conditions) is not described in this patent. Moreover, no information concerning the sensor resistance to jet fuel or the possible photobleaching of the luminescent dye is provided, as there are no data regarding its composition (commercially available sensors, such as the Ocean Optics FOXY Fiber Optic Oxygen Sensor, are suggested).

The sensor described by Martin and Engebretson and assigned to Dakota Technologies Inc. [25] is based on perfluorinated indicators and matrices that exhibit mechanical and chemical durability, among other advantages. As a pitfall, the manufacturing of this system is limited by the requirement of perfluorinated solvents. This disadvantage is partly solved in the systems developed by Lam et al. [36, 37], assigned to Airbus Operations Ltd, where the perfluorinated indicators are covalently attached to functionalized silica matrices. Thibaud [38] describes an inerting system with an oxygen sensor placed in the air separation modules of NEA production. This sensor is generically described as being "metal oxide-based." The optical sensors described by Kazemi and coworkers [39, 40] are based on Ru(II) complexes deposited on a glass support; the mechanical resistance under in-flight conditions is not specified. No reference is made to the sensor photostability or to its resistance to jet fuel. The range of investigated oxygen compositions is between 0 and 10%, and the temperature interval from -18 to 70 °C. The measured property is luminescence intensity that, as mentioned above, is prone to fluctuations, unlike the luminescence lifetime. Some of these issues are tackled in the sensors developed by Mendoza and coworkers [41, 42]. However, the polymer matrices composition and the indicator dyes are not disclosed. The paper mentions, although without detail, resistance to jet fuel and response to O_2 levels above the atmospheric one (21%). An example of a sensor based on organically modified silica ("ormosil") containing an indicator dye has been introduced by Goswami and coworkers at SPIE Conferences [43, 44], reporting on a collaborative work between the Boeing Co. and InnoSense LLC. More recently, Pedras and coworkers developed a sensor consisting of a polyurethane-based matrix doped with erythrosine B as luminescent indicator [45]. The sensor displays complete resistance to jet fuel, no dye leaching, and excellent photostability. The tested temperature range was between -60 and 60 °C, and the oxygen composition from 0 to 21%. To decrease the complexity of the sensors, two methods have been described that do not require a sensing membrane for oxygen detection in the fuel tank. Gord et al. [46] have disclosed a method whereby a luminescent indicator is directly dissolved in the fuel. This method was tested in fuel samples, but its application in a real fuel tank is questionable due to the amount of indicator that would be needed, together with the probable fuel contamination with this type of substance. More recently, Martelo and Berberan-Santos developed a method that relies on the intrinsic fluorescence of jet fuel upon UV irradiation, which is related to the dissolved oxygen concentration [47, 48].

Alternatively to luminescence-based sensors, but still worth mentioning for the sake of comparison in the field of oxygen sensors for aircraft fuel tanks, the systems developed by Chen and Silver (assigned to Southwest Sciences, Inc.) [49], McCaul and Winsemius (assigned to Oxigraf, Inc.) [50], Hedges et al. (assigned to The Boeing Company) [51], and Chabanis et al. [52] use as excitation source a laser diode whose output wavelength corresponds to an oxygen absorption band. The detection is based on radiation absorption, which is less sensitive than luminescence detection and prone to interferences.

2.1.3   Pressure-Sensitive Paints

Another application of oxygen sensing in aeronautics are the so-called pressure-sensitive paints (PSPs), also used in the automotive industry. In this application, the aircraft (or a model thereof) external surface is covered (sprayed or painted) with a polymer thin film containing a luminescent indicator dye and then subjected to an air flow while being simultaneously illuminated, so that the dye molecules are excited. The oxygen flow through the sensing membrane quenches the excited molecules according to the partial oxygen pressure at each point of the surface. The measurements of surface pressure obtained with this technique overcome the limitations of conventional methods based on pressure taps or transducers installed only at discrete points of the model. In this way, using a charge-coupled device (CCD) camera, a complete spatial distribution of the surface pressure can be obtained, only limited by the resolution of the imaging device. Figure 4 represents a model mounted in the test section of a wind tunnel and an image of the surface pressure distribution obtained on a PSP-coated model.

In 1980, Peterson and Fitzgerald [54] performed the first reported airflow studies over thin-layer chromatography plates previously died with Fluorescent Yellow, based on an observation described by Kautsky and Hirsch [55] in 1935 that stated the reduction of luminescence intensity of silica-adsorbed organic dyes upon exposure to oxygen. Peterson's experiment gave rise to the conception of PSPs, which were first tested at the Central Aerohydrodynamic Institute, in Moscow, in 1981 [56]. Ever since, this technique has been applied and developed worldwide, in aeronautical research facilities such as the National Aeronautics and Space Administration (NASA) [57], the French Aerospace Lab (ONERA) [58], the Japan Aerospace Exploration Agency (JAXA) [59], and the German Aerospace Center (DLR) [53].

The fundamental measuring principle underlying pressure-sensitive paints, like in the case of fuel tank oxygen sensors described in the previous section, is the quenching of luminescence by molecular oxygen. Therefore, the sensitivity of the PSP toward oxygen (and hence its performance) is defined by the Stern-Volmer constant, K_{SV}, as it has been mentioned above.

The polymer in which the luminophore is dispersed plays a key role in PSPs, since it influences its sensitivity to O_2. An ideal polymer for this application should display a high solubility (S) and diffusion (D) coefficients for oxygen, which multiplied define the permeability coefficient P ($P = SD$). The permeability coefficient (which serves to quantify the diffusion of oxygen through the polymer) can be expressed as:

$$P = \frac{(\text{thickness of polymer film})(\text{volume of oxygen})}{(\text{area})(\text{time})(\text{pressure drop across film})} \tag{6}$$

where the volume of oxygen is determined at standard temperature and pressure conditions. Currently, a wide variety of materials can be used in the formulation of

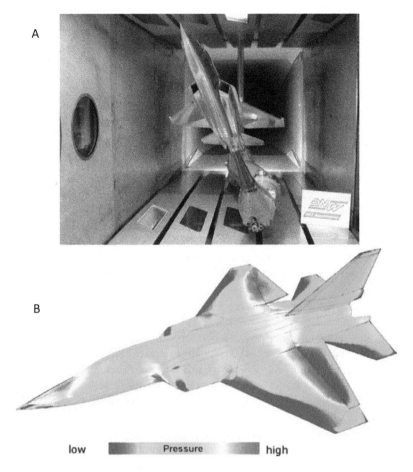

Fig. 4 (**a**) Model in the test section of a wind tunnel. (**b**) Pressure distribution on a PSP-coated model. Reprinted from [53] with kind permission. Copyright (2005) Springer Science + Business Media

PSPs [60, 61], having in mind most characteristics that are common to the sensors described in the previous section. For a complete description of the technical aspects of PSPs, see the book by Liu and Sullivan [62].

Unavoidable cross-sensitivity toward temperature is also an issue to be taken into account (as for the fuel tank sensors). Throughout the coated surface, not all the sites display the same temperature, which will influence the luminescence intensity and lifetime. This issue can be solved by measuring simultaneously temperature and oxygen, using a dual luminophore pressure-sensitive paint (DL-PSP) [63, 64]. Moreover, the measurement of temperature distributions on surfaces is also an active topic, closely related to PSPs. By dispersing luminescent temperature indicators in polymer binders, the so-called temperature-sensitive paints (TSPs) are produced. In the latter case, the polymer supports have to exhibit very low oxygen permeability coefficients, in order to avoid oxygen quenching of the indicator dye.

As imaging technology progresses, intensity-based PSPs are being replaced by time-resolved lifetime imaging, since the latter is less prone to error [65].

2.2 Sensors for Corrosion and Wear Detection

Corrosion is a significant issue in the aircraft industry. It can be responsible for an increase in the repair and maintenance budgets, reduce the aircraft duration of service, and ultimately precede fatigue cracking, leading to structural failures that would jeopardize its operational safety. Aging of the aircraft structures considerably enhances the probability of corrosion-induced damage, reason for which an early detection [66] and continuous monitoring by manufacturers and maintenance providers [67, 68] is essential to prevent aircraft degradation, with all its inherent consequences. The "Aloha Incident" that took place on April 28, 1988, raised the alarm on the insufficiency and inadequacy of inspection procedures of some airlines. In this event, a Boeing 737-200 experienced a major structural failure that led to tearing of a major portion of the upper fuselage in full flight at 24,000 feet. The pilot managed to land the plane, but one person died. The official accident report emphasizes "multiple site fatigue cracking of the fuselage lap joints" [69].

The ability to monitor corrosion of both civil and military aircraft has been the driving force for the development of numerous corrosion sensors using different monitoring techniques [70, 71]. Due to the scope of this chapter, we will exclusively focus on luminescence-based optical sensors, either applied in smart coatings or by other procedures such as functionalized optical fibers. Different types of corrosion may occur [72, 73], which lead to a variety of strategies depending on the location and measured analyte. All of those have in common the fact that corrosion is an electrochemical process, for which there are at least three types of luminescent indicator dyes that can be applied: pH indicators, which account for acidity/alkalinity changes, ion indicators for detection of metal ions, and redox indicators for electron transfer. Metal alloys used in aircraft structures are thermodynamically unstable with respect to their oxides, sulfides, and some ionic forms in solution. Hence, in the presence of oxygen and water, even subject to corrosion protection, they slowly degrade into corrosion products such as hydrated metal oxides, a process aggravated by mechanical and thermal stress.

Many aircraft components are derived from aluminum-based alloys. For this reason, one of the products of corrosion will be aluminum cations, and typical indicators are organic molecules that can efficiently chelate these ions forming a complex that can either result in chelation-enhanced fluorescence (CHEF) or chelation-enhanced quenching (CHEQ). In the case of corrosion of steel components, a typical indicator should change its fluorescence properties upon complexation with ferrous and/or ferric ions. In terms of location, corrosion can occur anywhere in the metal parts of the aircraft. Areas that are more susceptible to corrosion, such as lap joints and under sealant beads, are sometimes more difficult to access. For these more challenging areas, optical fiber sensors can be used, which are readily embedded in the usually inaccessible areas and allow distributed sensing. Alternatively, the luminescent indicator dye can be incorporated into polymer coating formulations reactive to the underlying corrosion events. According to the aimed analyte, location, or distribution in each specific case, different combinations of the previously mentioned features have been developed throughout the years, giving rise to a large variety of luminescence-based corrosion sensors. Some representative examples are reviewed below.

McAdam et al. [74] have developed a fiber-optic sensor for aluminum detection consisting of a polymer (PMMA and polyurethane were both tested, the latter giving the best results) doped with 8-hydroxyquinoline which, in the presence of Al^{3+} ions, forms a strongly fluorescent octahedral complex. This doped polymer is dip-coated in the distal end of the optical fiber, creating a local (point) sensor. A later version [75] tackled this limitation by using the concept of distributed detection of fluorescence, i.e., coating the fiber along its length, in order to enhance the interaction between the light and the fluorophore. This concept had been previously developed by Sinchenko et al., initially using intensity-based measurements [76], and, in later stages of development, applying it to time-resolved fluorescence sensing [77–79]. A similar principle, but based on phase-resolved fluorescence, had been previously tested by Chin et al. [80] by embedding an optical fiber with near-infrared dyes to detect aluminum and iron cations that act as quenchers of the luminescent dyes according to linear Stern-Volmer plots. Panova et al. [81] monitored three different types of corrosion (galvanic, crevice, and pitting) by using an optical fiber with an immobilized pH-sensitive fluorescent dye (in this case a fluorescein derivative) which, coupled to a CCD camera, allowed the collection of fluorescence images as a function of time. Galvanic corrosion was measured using a copper/aluminum galvanic pair, while crevice and pitting employed a stainless steel surface. The changes in fluorescence resulted from pH increases at cathodic sites and pH decreases at anodic sites, allowing the monitoring of local chemical concentrations and simultaneous observation of corrosion sites in real time. Another example of optical fiber aluminum detection has been reported by Szunerits and Walt [82] and is based on the pH-sensitive seminaphtho-fluorescein 5-(and-6-)carboxysuccinimidyl ester (SNAFL-SE) fluorescent dye and the aluminum-chelating morin dye.

Fluorescent sensor coatings for detection of corrosion in aluminum alloys Al 2024 and Al 2024T3 were developed by Li et al. [83] and Sibi and Zong [84], respectively. More recently, an epoxy coating doped with 8-hydroxyquinoline for ferric ion detection in steel corrosion has been described by Roshan et al. [85], where the dye exhibits a CHEF phenomenon due to complexation with ferric ions. The same effect is observed in the study performed by Bryant and Greenfield [86] but using 8-hydroxyquinoline-5-sulfonic acid as a probe for aluminum corrosion by CHEF and 9-anthryl-5-(2-nitrobenzoic acid)disulfide for iron through quenching via cleavage of the disulfide bond. The work developed by Zhang and Frankel [87] presents an acrylic-based coating system, modified with colorimetric and fluorescent pH indicators, to detect the pH increase due to the cathodic reaction that goes along with the anodic corrosion reaction of the studied Al alloy. Augustyniak et al. developed "turn-on" sensors in smart coatings for detection of steel [88] and aluminum [89] corrosion, which are further referenced by the author [90]. Epoxy-polyamide coatings with the fluorescent indicators 7-amino-4-methylcoumarin (for an aluminum-coated alloy) and 7-diethylamino-4-methylcoumarin (for coated steel) were the subject of an extensive work by Liu [91], in which the indicator dyes displayed opposite fluorescence behaviors: while the aluminum indicator exhibits fluorescence quenching due to the low pH values at the anodic corrosion sites [92], the ferrous/ferric ions indicator undergoes fluorescence enhancement due to an increase of pH [93]. A corrosion monitoring system for aerospace primers in which CdSe/ZnS quantum dots are embedded in these coatings and interrogated has been introduced by Trinchi et al. [94]. Liu et al. developed a fluorescent pH probe for iron corrosion detection in which 5,6-carboxyfluorescein undergoes quenching inside corrosion pits due to the lowering of pH produced by release of hydrons from iron corrosion [95]. The Corrosion Technology Laboratory at NASA Kennedy Space Centers has developed a smart coating technology based on micro-capsules that contain corrosion indicators and inhibitors whose release is triggered upon crossing the corrosion onset [96–101]. Coumarin, 7-hydroxycoumarin, and Rhodamine B are among the selected fluorescent indicators.

Tribological coatings, used to reduce wear and friction in turbine fans and jet engines, can be functionalized to become "smart" coatings and provide indication of the remaining wear life in aircraft parts. Muratore et al. [102] have described a system where erbium- and samarium-doped yttria-stabilized zirconia (YSZ) sensor layers were embedded in a solid lubricant coating (molybdenum disulfide), at different depths. The sensor layers were excited by green laser light, and their different luminescence spectra provided a measure of wear depth.

2.3 Sensors for Hydraulic Fluid Monitoring

Hydraulic systems constitute a crucial element of an aircraft, since they are present in almost all the moving components, being responsible for the force transmission between different points through a fluid that operates the otherwise rigid or immovable units (e.g., doors, flight controls, brakes, flaps, landing gear). A current commercial airplane operating various hydraulic systems may contain hundreds of liters of hydraulic fluid. The physicochemical condition of the latter must be monitored by sampling it on a regular basis to determine suspended particles, conductivity, moisture, and acidity, among other indicator parameters. The performance of the entire aircraft hydraulic system is affected by the condition of the hydraulic fluid, and, if degradation goes undetected, it may cause damages with serious consequences.

The primitive hydraulic fluids were petroleum- or vegetable-derived oils. Accidents with those fluids, such as fires or explosions, led manufacturers to search for fire-resistant alternatives. Phosphate ester-based hydraulic fluids were first manufactured in the late 1940s and found widespread use in the late 1950s, currently being present in the majority of commercial aircraft. Although phosphate ester-based fluids are fire-resistant, they possess the drawback of being hygroscopic, a feature that can severely compromise its performance.

A hydraulic fluid can be rendered unsuitable either by contamination (e.g., particle contamination) or degradation. In particular, phosphate esters are produced by reaction between alcohols and phosphoric acid. Their degradation by hydrolysis (based on the reverse reaction between water and the phosphate ester) originates free alcohols and the corrosive phosphoric acid and hydrogen phosphates. In the specific case of Skydrol® fluid, one of the side groups is a phenyl group, which therefore forms free phenol when hydrolyzed (Fig. 5).

The phenol molecule exhibits fluorescence between 270 and 350 nm. This feature can be used to monitor the hydraulic fluid degradation as a function of phenol emission [103]. The same property allows the indirect determination of the fluid total acid number (TAN), since phosphoric acid is one of the by-products of the hydrolysis reaction [104]. The fluorescence intensity increases with increasing acid contamination, since more phenol molecules are formed as by-products.

Although other fluorescence-based sensing methods for hydraulic fluid or oil degradation can be found [105–108], this technique still needs to be further implemented in order to compete with the infrared sensors for aviation hydraulic fluid monitoring already at the prototype stage [109, 110]. Research and development to this stage was also carried out by Orellana and coworkers, in collaboration with Airbus SAS and CESA (now Héroux-Devtek Spain), with the successful development and testing of luminescent sensors for *onboard* water/acidity/O_2 monitoring in the hydraulic fluid of commercial aircraft [111] (Fig. 6). The innovative sensors capitalize on luminescent Ru(II) polypyridyl complexes whose molecular structure was tailored to provide maximum sensitivity and selectivity to each of the analytes. Emission lifetime measurements by phase-sensitive detection ensure the necessary long-term signal stability. In addition to functional tests at different temperatures, the manufactured prototypes, including sensitive layers, optics, and electronics, have passed already endurance, fatigue, and vibration tests. Application of the luminescent sensors to development of small ground service mobile laboratory units is also currently under way.

2.4 Other Luminescence-Based Sensors

Cabin air quality monitoring and control is essential in passenger aircrafts. A high quality is ensured by continuously replacing a fraction of the circulating air with fresh exterior air. In this process at cruise altitude, cold air from the outside is compressed inside the engines, adiabatically heated, and then cooled again, and *ozone* must be catalytically converted to molecular oxygen. If leaks in the bleed air system or saturation of the catalytic converter occur, air quality becomes compromised. Likewise, fuel leakage can be a critical issue, for which the detection of *hydrocarbons*, H_2, and NO_2 (in the case of rocket fuel, the decomposition of hydrazine gives rise to nitrogen dioxide) in trace amounts becomes important. All

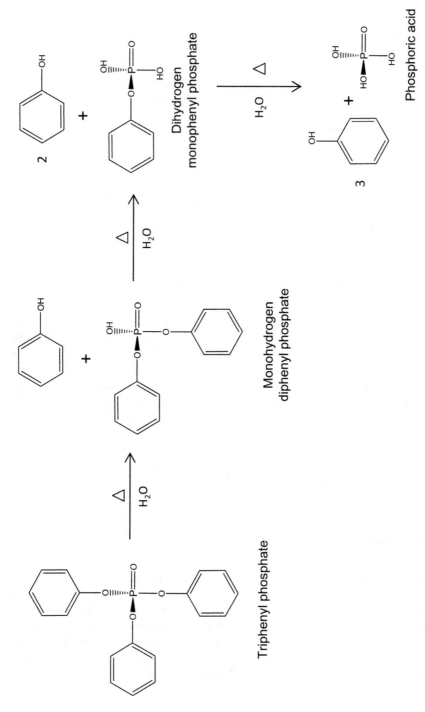

Fig. 5 Hydrolysis of a triphenyl phosphate ester-based hydraulic fluid, with the subsequent formation of phenol and phosphoric acid

Fig. 6 Luminescent sensors manifold for onboard hydraulic fluid simultaneous monitoring of water/acidity/ O_2 levels developed at Complutense University of Madrid under the "SuperSkySense" project [111]

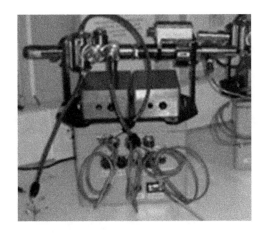

the aforementioned parameters combined are part of several works on multi-parameter optochemical transducers based on the photoluminescence of GaN/InGaN nanowires and quantum dots [103, 112–117].

One of the primary causes of acid rain is *sulfur* present in combustion fuels. Fluorescence-based sensing of sulfur in motor fuels and oils is determined by the American Society for Testing and Materials (ASTM) [118]. According to this method, a sample is burned to oxidize sulfur to sulfur dioxide, after which the combustion gases are irradiated with ultraviolet light, and the fluorescence of sulfur dioxide is measured. Another potential hazard, especially in spacecraft, is *formaldehyde*. The development of sensors based on fluorescent dyes that bind reversibly to formaldehyde allows a proper determination of this analyte in concentrations down to 0.5 ppm [119].

3   Conclusions/Perspectives/Outlook

Luminescence has been shown to be very useful in aeronautics, the applications of which span from aircraft design (temperature- and pressure-sensitive paints for wind tunnel prototype testing) to flight safety assurance (O_2 sensing in inerted fuel tanks, hydraulic fluid degradation, corrosion and cabin air monitoring). Despite the significant progress in this field experienced along the last 15 years, further development is required to bring the scientific and technical advances to the market. The stringent requirements imposed by the aviation industry in terms of materials, durability, electromagnetic immunity, vibration resistance, weight, pressure, and temperature conditions, among others, are certainly a handicap to go across the deep valley that separates research from market. Nevertheless, the future looks promising for luminescent sensors in aeronautical applications due to the extraordinary sensitivity (down to a single photon), selectivity (excitation and emission wavelengths can be tuned to avoid interferences), and versatility (the emission intensity, lifetime, and polarization may all be measured individually or combined) of this optical technique.

Acknowledgments B.P. acknowledges FCT-Portugal and IST-ID for the research contract IST-ID/104/2018. G. O. acknowledges the European Commission (STREP-FP6-2005-AERO-1 "SUPERSKYSENSE" project), the Spanish Ministry for Economy and Competitiveness (CDTI CENIT-E2010 "PROSAVE" project), CESA, and Aerlyper companies for their generous funding of UCM research on luminescent sensors for aeronautical applications. M.N.B.S. and B.P. acknowledge FCT-Portugal for FAPESP/20107/2014.

References

1. Federal Regulation 73 (140) (2008) 42444; EASA CS-25, Amendment 6, 2009
2. SAE International (2008) Aerospace Information Report AIR 1903 Aircraft Inerting Systems. https://doi.org/10.4271/AIR1903
3. Klein JK (1981) The F-16 Halon Tank Inerting System. Aircraft Systems and Technology Conference, AIAA-81-1638. https://doi.org/10.2514/6.1981-1638
4. Burns M, Cavage WM (2001) Ground and flight testing of a Boeing 737 center wing fuel tank inerted with nitrogen-enriched air. Federal Aviation Administration Report No.: DOT/FAA/AR-01/63
5. Cavage WM (2002) Ground-based inerting of commercial transport aircraft fuel tanks. In: RTO-MP-103, RTO AVT specialists' meeting on fire safety and survivability, Aalborg, Denmark, 23–26 Sept 2002
6. Burns M, Cavage WM, Hill R, Morrison R (2004) Flight-testing of the FAA onboard inert gas generation system on an Airbus A320, Federal Aviation Administration Report No.: DOT/FAA/AR-03/58
7. Langton R, Clark C, Hewitt M, Richards L (2009) Aircraft fuel systems. In: Moir I, Seabridge A, Langton R (eds) Aerospace series, Wiley, Chichester, ISBN 978-0-470-05708-7
8. Summer SM (2003) Limiting oxygen concentration required to inert jet fuel vapors existing at reduced fuel tank pressures, Federal Aviation Administration Technical Note No.: FAA/AR-TN02/79
9. Quaranta M, Borisov SM, Klimant I (2012) Indicators for optical oxygen sensors. Bioanal Rev 4:115–157. https://doi.org/10.1007/s12566-012-0032-y
10. Orellana G, Moreno-Bondi MC, García-Fresnadillo D, Marazuela MD (2005) The interplay of indicator, support and analyte in optical sensor layers. In: Orellana G, Moreno-Bondi MC (eds) Frontiers in chemical sensors: novel principles and techniques, Springer, Berlin, pp 189–227
11. Amao Y (2003) Probes and polymers for optical sensing of oxygen. Mikrochim Acta 143:1–12
12. Wang X, Chen H, Zhao Y, Chen X, Wang X (2010) Optical oxygen sensors move towards colorimetric determination. Trends Anal Chem 29:319–338. https://doi.org/10.1016/j.trac.2010.01.004
13. Garcia-Fresnadillo D, Marazuela MD, Moreno-Bondi MC, Orellana G (1999) Luminescent Nafion membranes dyed with ruthenium(II) complexes as sensing materials for dissolved oxygen. Langmuir 15:6451–6459
14. Hartmann P, Leiner MJP (1995) Luminescence quenching behavior of an oxygen sensor based on a Ru(II) complex dissolved in polystyrene. Anal Chem 67:88–93

15. Mills A, Thomas M (1997) Fluorescence-based thin plastic film ion-pair sensors for oxygen. Analyst 122:63–68
16. Di Marco G, Lanza M, Campagna S (1995) Luminescent Ru(II)-polypyridine complexes in poly-2-hydroxyethylmethacrylate matrices as oxygen sensors. Adv Mater 7:468–471
17. Ishiji T, Kudo K, Kaneko M (1994) Microenvironmental studies of an Ru(bpy)$^{2+}$ luminescent probe incorporated into Nafion film and its application to an oxygen sensor. Sens Act B 22:205–210
18. Li X, Ruan F, Wong Y (1993) Optical characteristics of a ruthenium(II) complex immobilized in a silicone rubber film for oxygen measurement. Analyst 118:289–292
19. Orellana G, López-Gejo J, Pedras B (2014) Silicone films for fiber-optic chemical sensing. In: Tiwari A, Soucek MD (eds) Concise encyclopedia of high performance silicones, Wiley, Hoboken, pp 339–353. https://doi.org/10.1002/9781118938478.ch22. ISBN: 9781118938478
20. Meier B, Werner T, Klimant I, Wolfbeis OS (1995) Novel oxygen sensor material based on a ruthenium bipyridyl complex encapsulated in zeolite Y: dramatic differences in the efficiency of luminescence quenching by oxygen on going from surface-adsorbed to zeolite-encapsulated fluorophores. Sens Act B 29:240–245
21. Matsui K, Sasaki K, Takahashi N (1991) Luminescence of Tris(2,2′-bipyridine)ruthenium (II) in sol-gel glasses. Langmuir 7:2866–2868
22. Carraway ER, Demas JN, DeGraff BA (1991) Photophysics and oxygen quenching of transition-metal complexes on fumed silica. Langmuir 7:2991–2998
23. Chan C, Chan M, Zhang M, Lo W, Wong K (1999) The performance of oxygen sensing films with ruthenium-adsorbed fumed silica dispersed in silicone rubber. Analyst 124:691–694
24. Diaz-Garcia ME, Pereiro-Garcia R, Velasco-Garcia N (1995) Optical oxygen sensing materials based on the room-temperature phosphorescence intensity quenching of immobilized Erythrosin B. Analyst 120:457–461
25. Martin TL, Engebretson DS (2006) Sensors for measuring analytes. US Patent Appl Pub 0171845 A1, 3 Aug 2006
26. Valeur B, Berberan-Santos MN (2012) Molecular fluorescence – principles and applications, 2nd edn. Wiley-VCH, Weinheim
27. Panahi A (2009) Fiber optic oxygen sensor using fluorescence quenching for aircraft inerting fuel tank applications. Proc SPIE 7314:73140D
28. McDonagh C, Kolle C, McEvoy AK, Dowling DL, Cafolla AA, Cullen SJ, MacCraith BD (2001) Phase fluorometric dissolved oxygen sensor. Sens Act B 74:124–130
29. Trettnak W, Kolle C, Reininger F, Dolezal C, O'Leary P (1996) Miniaturized luminescence lifetime-based oxygen sensor instrumentation utilizing a phase modulation technique. Sens Act B 36:506–512
30. Chu C, Lin K, Tang Y (2016) A new optical sensor for sensing oxygen based on phase shift detection. Sens Act B 223:606–612
31. Carraway ER, Demas JN, DeGraff BA, Bacon JR (1991) Photophysics and photochemistry of oxygen sensors based on luminescent transition-metal complexes. Anal Chem 63:337–342
32. Mills A (1998) Optical sensors for oxygen: a log-gaussian multisite-quenching model. Sens Act B 51:69–76
33. Hartmann P, Leiner MJP, Lippitsch ME (1995) Response characteristics of luminescent oxygen sensors. Sens Act B 29:251–257
34. Berberan-Santos MN, Bodunov EN, Pogliani L (1997) On the barometric formula. Am J Phys 65(5):404–412
35. Susko K (2003) On-board fuel inerting system. US Patent 6,634,598 B2, 21 Oct 2003
36. Lam JK-W, Osborne D, Ratcliffe NM (2010) Monitor and a method for measuring oxygen concentration. US Patent Appl Pub 0018119 A1, 28 Jan 2010
37. Lam JK-W, Ratcliffe NM, Smith S (2011) Method and apparatus for monitoring gas concentration in a fluid. US Patent 8,081,313 B2, 20 Dec 2011
38. Thibaud C (2017) Oxygen sensing for fuel tank inerting system. US Patent Appl Pub 0014752 A1, 19 Jan 2017

39. Kazemi AA, Mendoza E, Goswami K, Kempen L (2013) Fiber optic oxygen sensor detection for harsh environments of aerospace applications. Proc SPIE 8720:872002
40. Kazemi AA, Goswami K, Mendoza EA, Kempen LU (2007) Fiber optic oxygen sensor leak detection system for space applications. Proc SPIE 6758:67580C
41. Mendoza EA, Kempen C, Sun S, Esterkin Y (2014) Highly distributed multi-point, temperature and pressure compensated, fiber optic oxygen sensors (FOxSense™) for aircraft fuel tank environment and safety monitoring. Proc SPIE 9202:92021M
42. Mendoza EA, Esterkin Y, Kempen C, Sun S (2011) Advances towards the qualification of an aircraft fuel tank inert environment fiber optic oxygen sensor system. Proc SPIE 8026:802604
43. Goswami K, Sampathkumaran U, Alam M, Tseng D, Majumdar AK, Kazemi AA (2006) Ormosil coating-based oxygen sensor for aircraft ullage. Proc SPIE 6379:637909
44. Goswami K, Sampathkumaran U, Alam M, Tseng D, Majumdar AK, Kazemi AA (2007) Nanomaterial-based robust oxygen sensor. Proc SPIE 6758:67580F
45. Pedras B, Berberan-Santos MN, Baleizão C, Farinha JPS (2018) Sensores de oxigénio luminescentes não-metálicos para tanques de combustível de aeronaves e o seu método de funcionamento. Patente PT 110889
46. Gord JR, Buckner SW, Weaver WL, Grinstead KD Jr (1999) Optical method for quantitating dissolved oxygen in fuel. US Patent 5,919,710, 6 July 1999
47. Martelo L, Berberan-Santos MN (2018) Método ótico para a medição da concentração de oxigénio em sistema de combustível. Patente PT 109877
48. Martelo L, Berberan-Santos MN (2018) Device and method for measuring the spatial distribution of the concentration of compounds and their mixtures in a fluid and/or for determining the fluid level. Patent WO/2018/138649, 2 Aug 2018
49. Chen S-J, Silver JA (2008) Oxygen sensor for aircraft fuel inerting systems. US Patent 7,352,464 B2, 1 Apr 2008
50. McCaul BW, Winsemius TM (2014) Oxygen sensor for tank safety. US Patent 8,667,977 B1, 11 Mar 2014
51. Hedges DE, Holland MJ, Rhodes SS, Klemisch JD, Pavia JC, Henry TM, Anderson CJ, McCaul BW, Winsemius TM, Thorson EK (2015) Oxygen analysis system and method for measuring, monitoring and recording oxygen concentration in aircraft fuel tanks. US Patent Appl Pub 0219554 A1, 6 Aug 2015
52. Chabanis G, Fleischer M, Mangon P, Meixner H, Strzoda R (2004) Device and method for monitoring the oxygen concentration in an aircraft fuel tank. PCT WO 2004/113169 A1, 29 Dec 2004
53. Klein C, Engler RH, Henne U, Sachs WE (2005) Exp Fluids 39:475–483
54. Peterson JI, Fitzgerald RV (1980) Rev Sci Instrum 51:670–671
55. Kautsky H, Hirsch A (1935) Z Anorg Allg Chem 222:126–134
56. Pervushin GE, Nevsky LB (1981) Composition for indicating coating (in Russian). Patent of USSR-SU 1065452
57. Baron AE, Danielson JDS, Gouterman M, Wan JR, Callis JB, McLachlan B (1993) Rev Sci Instrum 64:3394–3402
58. Merienne M-C, Le Sant Y, Ancelle J, Soulevant D (2004) Meas Sci Technol 15:2349–2360
59. Nakakita K, Kurita M, Mitsuo K, Watanabe S (2006) Meas Sci Technol 17:359–366
60. Stich MIJ, Wolfbeis OS (2008) Fluorescence sensing and imaging using pressure-sensitive paints and temperature-sensitive paints. In: Resch-Genger U (ed) Standardization and quality assurance in fluorescence measurements I. Springer series on fluorescence, vol 5. Springer, Berlin, pp 429–461
61. Gregory JW, Asai K, Kameda M, Liu T, Sullivan JP (2008) A review of pressure-sensitive paint for high-speed and unsteady aerodynamics. Proc IMechE Part G J Aero Eng 222:249–290
62. Liu T, Sullivan JP (2005) Pressure and temperature sensitive paints. Springer, Berlin. https://doi.org/10.1007/b137841. ISBN 978-3-540-26644-0

63. Koese ME, Carrol BF, Schanze KS (2005) Preparation and spectroscopic properties of multiluminophore luminescent oxygen and temperature sensing films. Langmuir 21:9121–9129
64. Stich MIJ, Nagl S, Wolfbeis OS, Henne U, Schaeferling M (2008) A dual luminescent sensor material for simultaneous imaging of pressure and temperature on surfaces. Adv Funct Mater 18:1399–1406
65. Mitsuo K, Asai K, Takahashi A, Mizushima H (2006) Advanced lifetime PSP imaging system for pressure and temperature field measurement. Meas Sci Technol 17:1282–1291
66. Bartelds G, Heida JH, McFeat J, Boller C (2004) In: Staszewski WJ, Boller C, Tomlinson GR (eds) Health monitoring of aerospace structures: smart sensor technologies and signal processing. Wiley, New York, pp 1–28
67. Hall J (1993) Corrosion prevention and control programs for Boeing airplanes. Technical Paper 931259, SAE International. https://doi.org/10.4271/931259
68. Trego A, Price D, Hedley M, Corrigan P, Cole I, Muster T (2007) Development of a system for corrosion diagnostics and prognostic. Corrosion Rev 25(1–2):161–177
69. National Transportation Safety Board (1989) Aircraft Accident Report, Aloha Airlines Flight 243, Boeing 737-200, N73711, Near Maui, Hawaii, 28 Apr 1988. http://libraryonline.erau.edu/online-full-text/ntsb/aircraft-accident-reports/AAR89-03.pdf
70. Harris SJ, Mishon M, Hebbron M (2006) Corrosion sensors to reduce aircraft maintenance. In: RTO-MP-AVT-144 Workshop on Enhanced aircraft platform availability through advanced maintenance concepts and technologies. Vilnius, Lithuania, 3–5 Oct 2006
71. Rinaldi G (2009) A literature review of corrosion sensing. methods. Defence R&D Canada – Atlantic, Technical Memorandum. DRDC Atlantic TM 2009-082
72. Ford T (1999) Corrosion detection and control. Aircraft Eng Aero Tech 71:249–254
73. Zarras P, Stenger-Smith JD (2014) Corrosion processes and strategies for prevention: an introduction. In: Makhlouf ASH (ed) Handbook of smart coatings for materials protection. Woodhead Publishing Series in Metal and Surface Engineering 64, Chapter 1, pp 3–28
74. McAdam G, Newman PJ, McKenzie I, Davis C, Hinton BRW (2005) Fiber optic sensors for detection of corrosion within aircraft. Struct Health Monit 4:47–56
75. Kostecki R, Ebendorff-Heidepriem H, Davis C, McAdam G, Wang T, Monro TM (2016) Fiber optic approach for detecting corrosion. Proc SPIE 9803
76. Sinchenko E, McAdam G, Davis C, McDonald S, McKenzie I, Newman PJ, Stoddart PR (2006) Optical fibre techniques for distributed corrosion sensing. In: ACOFT/AOS proceedings, Melbourne, Australia, 10–13 July 2006, pp 84–86. https://doi.org/10.1109/ACOFT.2006.4519220
77. Sinchenko EI, Gibbs WEK, Stoddart PR (2008) Fluorescence-based distributed chemical sensing for structural health monitoring. Proc SPIE 7268
78. Sinchenko E, Gibbs WEK, Davis CE, Stoddart PR (2010) Characterization of time-resolved fluorescence response measurements for distributed optical-fiber sensing. Appl Optics 49:6385–6390
79. Sinchenko E (2013) Fiber optic distributed corrosion sensor. PhD Dissertation, Centre for Atom Optics and Ultrafast Spectroscopy, Faculty of Engineering and Industrial Science, Swinburne University of Technology, Melbourne, Australia
80. Chin R, Hallidy W, Cruce C, Salazar N, Jamison K, Patonay G, Strekowski L, Gorecki T, Maklad M (1996) NIR-PRFS fiber optic corrosion sensor. Proc SPIE 2682:275–287
81. Panova AA, Pantano P, Walt DR (1997) In situ fluorescence imaging of localized corrosion with a pH-sensitive imaging fiber. Anal Chem 69:1635–1641
82. Szunerits S, Walt DR (2002) Aluminum surface corrosion and the mechanism of inhibitors using pH and metal ion selective imaging fiber bundles. Anal Chem 74(4):886–894
83. Li S-M, Zhang H-R, Liu J-H (2006) Preparation and performance of fluorescent sensing coating for monitoring corrosion of Al alloy 2024. Trans Nonferrous Met Soc China 16: s159–s164
84. Sibi MP, Zong Z (2003) Determination of corrosion on aluminum alloy under protective coatings using fluorescent probes. Prog Org Coat 47:8–15

85. Roshan S, Dariani AAS, Mokhtari J (2018) Monitoring underlying epoxy-coated St-37 corrosion via 8-hydroxyquinoline as a fluorescent indicator. Appl Surf Sci 440:880–888

86. Bryant DE, Greenfield D (2006) The use of fluorescent probes for the detection of under-film corrosion. Prog Org Coat 57:416–420

87. Zhang J, Frankel GS (1999) Corrosion-sensing behavior of an acrylic-based coating system. Corrosion 55(10):957–967

88. Augustyniak A, Tsavalas J, Ming W (2009) Early detection of steel corrosion via "turn-on" fluorescence in smart epoxy coatings. ACS Appl Mater Interfaces 1(11):2618–2623

89. Augustyniak A, Ming W (2011) Early detection of aluminum corrosion via "turn-on" fluorescence in smart coatings. Prog Org Coat 71:406–412

90. Augustyniak A (2014) Smart epoxy coatings for early detection of corrosion in steel and aluminum. In: Makhlouf ASH (ed) Handbook of smart coatings for materials protection. Woodhead Publishing Series in Metal and Surface Engineering 64, Chapter 21, pp 560–585

91. Liu G (2010) Fluorescent coatings for corrosion detection in steel and aluminum alloys. PhD Dissertation, University of Texas at Austin

92. Liu G, Wheat HG (2009) Use of a fluorescent indicator in monitoring underlying corrosion on coated aluminum 2024-T4. J Electrochem Soc 156(4):C160–C166

93. Liu G, Wheat HG (2010) Coatings for early corrosion detection. ECS Trans 28(24):239–247

94. Trinchi A, Muster TH, Hardin S, Gomez D, Cole I, Corrigan P, Bradbury A, Nguyen T-L, Safai M, Georgeson G, Followell D (2012) Distributed quantum dot sensors for monitoring the integrity of protective aerospace coatings. In: 2012 IEEE Aerospace conference, 3–10 Mar 2012, pp 1–9. https://doi.org/10.1109/AERO.2012.6187252

95. Liu X, Spikes H, Wong JSS (2014) In situ pH responsive fluorescent probing of localized iron corrosion. Corr Sci 87:118–126

96. Li W, Buhrow JW, Jolley ST, Calle LM (2014) pH-sensitive microparticles with matrix-dispersed active agent. US Patent 8,859,288 B2. 14 Oct 2014

97. Calle LM, Li WN, Buhrow JW, Perusich SA, Jolley ST, Gibson TL, Williams MK (2015) Elongated microcapsules and their formation. US Patent 9,108,178 B2. 18 Aug 2015

98. Calle LM, Li W, Buhrow JW, Jolley ST (2016) Hydrophilic-core microcapsules and their formation. US Patent 9,227,221 B2. 5 Jan 2016

99. Calle LM, Li W, Buhrow JW, Jolley ST (2016) Hydrophobic-core microcapsules and their formation. US Patent 9,233,394 B2. 12 Jan 2016

100. Calle LM, Li W (2010) Coatings and methods for corrosion detection and/or reduction. US Patent 7,790,225 B1. 7 Sept 2010

101. Calle LM, Li W (2014) Microencapsulated indicators and inhibitors for corrosion detection and control. In: Makhlouf ASH (ed) Handbook of smart coatings for materials protection. Woodhead Publishing Series in Metal and Surface Engineering 64, Chapter 15, pp 370–422

102. Muratore C, Clarke DR, Jones JG, Voevodin AA (2008) Smart tribological coatings with wear sensing capability. Wear 265:913–920

103. Paul S (2014) Optochemical sensor systems for aerospace applications. PhD Dissertation, Justus-Liebig-Universität Giessen, pp 37–40

104. Muller G, Sumit P, Helwig A (2011) Sensor and method for online monitoring of the acid value of a hydraulic fluid in an hydraulics system in an aircraft. Patent EP2378277A2, 14 Apr 2011

105. Markova LV, Myshkin NK, Ossia CV, Kong H (2007) Fluorescence sensor for characterization of hydraulic oil degradation. Tribol Ind 29(1&2):33–36

106. Kong H, Yoon ES, Han HG, Markova L, Semenyuk M, Makarenko V (2008) Method and device for monitoring oil oxidation in real time by measuring fluorescence. US Patent 7,391,035 B2. 24 Jun 2008

107. Kong H, Han H-G, Yoon E-S, Lyubov M, Myshikin N, Semenyuk M (2005) Apparatus for measuring oil oxidation using fluorescent light reflected from oil. US Patent Appl Pub 2005/0088646 A1, 28 Apr 2005

108. Lewis DE, Utecht RE, Judy MM, Matthews JL (1995) Fluorescent method for monitoring oil degradation. US Patent 5,472,878. 5 Dec 2005

109. Paul S, Legner W, Krenkow A, Müller G, Lemettais T, Pradat F, Hertens D (2010) Chemical contamination sensor for phosphate ester hydraulic fluids. Int J Aerospace Eng 2010, Article ID 156281. Hindawi Pub Corp. https://doi.org/10.1155/2010/156281

110. Bley T, Steffensky J, Mannebach H, Helwig A, Müller G (2016) Degradation monitoring of aviation hydraulic fluids using non-dispersive infrared sensor systems. Sens Act B Chem 224:539–546

111. https://cordis.europa.eu/docs/publications/1228/122807231-6_en.pdf. Accessed 20 Nov 2018

112. Maier K, Helwig A, Müller G, Becker P, Hille P, Schörmann J, Teubert J, Eickhoff M (2014) Detection of oxidizing gases using an optochemical sensor system based on GaN/InGaN nanowires. Sens Act B Chem 197:87–94

113. Maier K, Helwig A, Müller G, Hille P, Teubert J, Eickhoff M (2017) Competitive adsorption of air constituents as observed on InGaN/GaN nano-optical probes. Sens Act B Chem 250:91–99

114. Paul S, Maier K, Das A, Furtmayr F, Helwig A, Teubert J, Monroy E, Müller G, Eickhoff M (2013) III-nitride nanostructures for optical gas detection and pH sensing. Proc SPIE 8725

115. Paul S, Helwig A, Müller G, Furtmayr F, Teubert J, Eickhoff M (2012) Opto-chemical sensor system for the detection of H_2 and hydrocarbons based on InGaN/GaN nanowires. Sens Act B Chem 173:120–126

116. Becker P, Eickhoff M, Helwig A, Müller G, Paul S, Teubert J (2014) Invention relating to gas sensors. Patent EP2790009A1. 15 Oct 2014

117. Eickhoff M, Helwig A, Müller G, Teubert J, Paul S, Wallys J (2015) Optical pH value sensor. Patent EP2917726A1. 16 Sept 2015

118. ASTM D5453-00 (2000) Standard test method for determination of total sulfur in light hydrocarbons, motor fuels and oils by ultraviolet fluorescence, ASTM International, West Conshohocken, PA, www.astm.org. https://doi.org/10.1520/D5453-00

119. Patty KD, Gregory DA (2008) Optical detection of formaldehyde. Proc SPIE 6958. https://doi.org/10.1117/12.784409

Index

CPSIA information can be obtained
at www.ICGtesting.com
Printed in the USA
LVHW082319190619
621807LV00003B/43/P

9 783030 200329